VoIP VOICE AND FAX
SIGNAL PROCESSING

VoIP VOICE AND FAX SIGNAL PROCESSING

Sivannarayana Nagireddi, PhD

A JOHN WILEY & SONS, INC., PUBLICATION

For general information on our other products and services or for technical support, please contact our Customer Care Department within the United States at (800) 762-2974, outside the United States at (317) 572-3993 or fax (317) 572-4002.

Wiley also publishes its books in a variety of electronic formats. Some content that appears in print may not be available in electronic formats. For more information about Wiley products, visit our web site at www.wiley.com.

Library of Congress Cataloging-in-Publication Data:

Nagireddi, Sivannarayana.
    VoIP voice and fax signal processing / Sivannarayana Nagireddi.
      p. cm.
    Includes bibliographical references and index.
    ISBN 978-0-470-22736-7 (cloth)
  1. Internet telephony. 2. Facsimile transmission. 3. Signal processing—Digital techniques. I. Title.
    TK5105.8865.S587 2008
    621.385—dc22

                                    2008007582

Printed in the United States of America

This book is dedicated to
· *VoIP and **Signal Processing Contributors***
· my ***Teachers***

CONTENTS

6   ECHO Cancellation   **113**

ACKNOWLEDGMENTS

I incorporated points that came from several VoIP and signal processing contributing members, as well as from interactions with customers, service providers, third-party developers, interoperability events, publications, standards, recommendations, and conference contributions. I enjoyed the interactions with several contributors from all across the world, and I am grateful for their several decades of contributions, hard work, and foresight in advancing VoIP and signal processing.

I sincerely thank Prof. V. John Mathews, Prof. D. C. Reddy, and Dr. V. V. Krishna for their close technical and personal guidance while going through various stages of compiling this publication.

Several members devoted time in reviewing the material. I thank Dhruva Kumar N and Vasuki MP (Encore Software, India) for reviewing fax chapters and sharing several technical views; Simon Brewer (Analog Devices, Inc.) and his team members for sharing several technical views and knowledge. I would like to thank my colleagues Darren Hutchinson, Chris Moore, Sreenivasulu Kesineni, James Xu, and A.V. Ramana for reviewing some of the chapters.

At Ikanos Communications, Inc., several members provided encouragement for this effort. I thank Sam Heidari, Sanjeev Challa, Ravi Selvaraj, Dean Westman, Michael Ricci, Fred Koehler, Sandeep Harpalani, Ravindra Bhilave, Margo Westfall, Noah Mesel, and my software team members.

Special thanks to the following team members: Venkateshwarlu Vangala, Vijay S. Kalakotla, Hemavathi Lakkalapudi, J. Radha Krishna Simha and S.Venkateswara Rao for compiling some of the sections, several deep technical discussions, and technical review of chapters. I would like to recognize the persistent efforts of Hemavathi Lakkalapudi that helped me in concluding several chapters in a timely manner, validating several illustrations, and tables, and a lot of editing and review work; my appreciation also goes to J. Radha Krishna Simha for verifying some of the algorithms and formulating the results.

I am indebted to my wife Vijaya for her persistent encouragement, accommodating my tight schedules and taking care of several responsibilities to make this publication happen, and to my daughter Spandana and son Vamsi Krishna for their continued encouragement.

I would like to thank my friends, especially to Sushil Gote, for reviewing several chapters. I also thank several agencies in granting permissions to use their technical material, as well as the John Wiley editorial staff for their friendly support in completing this publication.

SIVANNARAYANA NAGIREDDI

ABOUT THE AUTHOR

Sivannarayana Nagireddi, PhD, is currently working as the architect of voice over IP solutions at Ikanos Communications, Inc., and leads DSP and VoIP team. Dr. Sivannarayana and his team developed VoIP solutions including signal processing algorithms for voice and fax enabled residential gateway processors, which have been deployed by telecommunications providers.

Sivannarayana has been working on digital signal processing and systems for the last 22 years. His contributions in voice and VoIP started in 1999 with Encore Software, India. In early 2000, he built a DSP team for voice applications for Chiplogic India, and later on by mid-2000, he started managing VoIP solutions for Chiplogic USA. During the merger of Chiplogic with Analog Devices, Inc., he continued his VoIP solutions effort for Analog Devices, Inc. After working for 5 years at Analog Devices, Inc., he moved to Ikanos Communications, Inc., at the time of the acquisition of the network processor and ADSL ASIC product lines from Analog Devices, Inc.

Prior to contributions into voice and VoIP applications, for about 13 years from 1986 to 1999, he was working on signal processing algorithms and building systems for communication, radars, image processing, and medical applications.

Sivannarayana graduated with a degree in engineering from the Institute of Electronics and Telecommunications Engineering (IETE), New Delhi, India, in 1985. He received a Masters degree in electronics and communications engineering (ECE) from Osmania University, India. He was then awarded the PhD from the ECE Department, Osmania University, with a focus on wavelet signal processing applications.

His favorite topics are time-frequency analysis and communication signal processing, as well as building complete systems and supporting them for successful use. He is a member of the IEEE, a Fellow of IETE-India, and a reviewer for *Medical Engineering & Physics Journal* (Elsevier-UK).

PREFACE

Voice over IP (VoIP) gained popularity through actual deployments and by making use of VoIP-based telephone and fax calls with global roaming and connectivity via the Internet. Several decades of effort have gone into VoIP, and these efforts are benefitting real applications. Several valuable books have been published by experts in the field. While I was building the team, and training them, and conducting several design and support phases, I felt like a consolidated view and material on VoIP voice and fax signal processing was missing. Several contributions in the form of white papers, application notes, data sheets, standards, several books at the system level, and specialized books on signaling, speech compression, echo cancellation, and voice quality exist. Fax processing is available in books mainly for a public switched telephone network (PSTN), several white papers on fax over IP (FoIP), and a lot of ITU recommendations.

In this book, I am trying to bring out a consolidated view and basic approach with interpretation on popularly used techniques mapped to VoIP voice and fax signal processing. As a summary, this book broadly covers topics such as PSTN and VoIP overview, VoIP infrastructure, voice interfaces, voice signal processing modules and practical aspects, wideband voice, packetization, voice bit rate on multiple network interfaces, testing at module level and as a total VoIP system, fax on PSTN, FoIP processing, FoIP anomalies, testing, FoIP bit rates, miscellaneous topics that include country-specific deviations, bandwidth issues, voice quality improvements, processors and OS, and FAQs on VoIP and FoIP.

This book is organized into 22 chapters. In Chapter 1, PSTN interfaces, transmission requirements, as well as power and quantization levels are presented to create continuity for the subsequent chapters. In Chapter 2, connectivity between PSTN and VoIP, VoIP infrastructure and their architectures, pictures and interfaces of some of the practically deployed boxes, and their functions are presented. Software at block level for voice and fax, acoustic and network interfaces, VoIP signaling, and end-to-end VoIP call flow are also given in this chapter. Even though the first two chapters are introductory, several concepts required for subsequent chapters are systematically presented.

In Chapter 3, the popular voice compression codecs considered for VoIP deployment and their voice quality considerations one presented. Chapter 4 is on VAD/CNG for saving Internet bandwidth. Various inter-operation issues and testing is also given in this chapter. Chapter 5 is on packet loss concealment that improves voice quality in packet loss conditions. These three chapters are presented in a row to deal with voice compression and its extensions. Required overview on software, testing, complexity, quality, and their dependencies are also presented in these three chapters.

Echo cancellation is a big topic with several books exclusively written on that topic. I covered in Chapter 6 concepts mapped to telephones, telephone interfaces, VoIP CPE echo generation, rejection, and testing. DTMF is more of a time-frequency analysis problem with time sensitivity for generation, detection and rejection operations. In Chapter 7, a consolidated view of DTMF with illustrations and mathematical derivations for tones generation, detection, and rejection is given. Required emphasis on testing and country-specific deviations are also given in Chapter 7. As an extension on DTMF, Chapter 8 presents about different caller ID features that have close relations with basic tones, DTMF, phone and interfaces, various timing formats, caller ID and call progress tones detection, and working principles. Chapter 9 is on wideband voice with an example created using a VoIP adapter that addresses both narrow and wideband combinations. Wideband voice provides higher quality and is expected to be widely available in terminals such as IP phones, WiFi phones, and multimedia terminals.

Chapter 10 is on RTP, RTCP, packetization, packet impediments, and jitter buffers. On jitter buffers, several details are provided with illustrations, mathematical formulations, algorithms, various modes of operations, and helpful recommendations included. The VoIP bit rates from various codecs, network interfaces, and recommendations from practical deployments are given in Chapter 11. The network bit rate is usually given up to VoIP headers. In this book, interface headers, exact calculations, and tables with codec, packetization, and network interfaces are presented. Some clock options and interpretation of clock influences with simple calculations are given in Chapter 12. VoIP quality is influenced by the clock oscillator frequency and its stability. In Chapter 13, a high-level description of the VoIP voice tests and some of the instruments used for testing are presented.

Chapters 14–16 are dedicated to fax signal processing. In Chapter 14, a fax operation on PSTN, an end-to-end fax call, fax call phases, different fax call set-up tones, modulations, and demodulation schemes are presented that provide the background for FoIP. Chapter 15 is mainly on FoIP and gives an introduction to modem over IP at a high-level. The end-to-end VoIP fax call is given with SIP signaling in several diagrams for easy understanding of FoIP. The conditions for successful fax and modem calls and interoperability issues in FoIP are highlighted along with testing. A real-time VoIP fax is sent as a G.711 voice call or T.38 fax relay. In the literature, FoIP detailed bandwidth calculations are not listed. G.711 takes a lot of bit rate, whereas T.38 takes a small fraction of it. In Chapter 16, detailed headers and bandwidth calculations on Ethernet and DSL interfaces for various fax modulation rates and redundancy levels are given.

Similar to PSTN, VoIP has several dependencies for multiple country deployments that are discussed in Chapter 17. Each country and region has several deviations in its central office configurations, such as transmission lines, telephone impedances, tones, and acoustics. Chapter 18 is on IPQoS issues related to the bandlimited network, delay, and jitter for voice packets. Interpretation of the bandlimited nature, bandwidth, delay calculations, and recommendations for various packet sizes as a trade-off among packet sizes, delays, and fragmentation are given in this Chapter 18. The goal here is to improve the voice quality. Architectural, hardware processors, processing, and operating system considerations for VoIP are given in Chapter 19. Chapter 20 discusses consolidation of voice quality evaluation as well as various quality assessments through subjective, PESQ, and E-model. A list of major contributors of quality degradation and improvement options are included in this chapter.

Several questions and answers on voice and VoIP are provided in Chapter 21. About 100 questions and answers are given that systematically cover the topics listed in this book and are supplemented with several points that could not be directly addressed in continuity. Similarly, a fax FAQ section is given in Chapter 22. My expectation is that a sequential reading of these fax FAQs will give a quick overview of the fax processing flow in PSTN and FoIP.

The algorithms and mathematics are made fairly simple like arithmetic, and they are supplemented with several illustrations, direct results in tables, and summaries or recommendations on various aspects. Several FAQs in Chapters 21 and 22 will help for easy reading of the book. I tried to make this book simple to understand by many readers across several roles. I hope this book will help in understanding voice and fax signal processing for many new engineers, new contributors of VoIP, and students at the graduate and postgraduate level, as well as for managers, business, sales, and marketing teams, customers, and service providers.

In conclusion, several books are forthcoming that are going to address voice quality in general and wideband voice in particular. The contributions on wideband voice and signal processing techniques that are expected will create more natural conversation with a higher mean opinion score.

GLOSSARY

3GPP Third-generation partnership project

A Advantage factor (in R-factor)

AAL5 ATM adaptation layer 5

ABNF augmented Backus–Naur form

AC alternating current

ACELP algebraic code excited linear prediction

ACK acknowledgment

ACR absolute category rating

ADC analog-to-digital converter

ADPCM adaptive differential pulse code modulation

ADSL asymmetric DSL

ADSL2 asymmetric DSL 2

AFE analog front end

AGC automatic gain control

AJB adaptive jitter buffer

A-law logarithmic 64-kbps compression, which is the same as G.711 PCMU

ALC automatic level control

ALG application level gateway

ALU arithmetic logic unit (ALU)

AM amplitude modulation

AMR adaptive multi rate

AMR-HR AMR half rate

AMR-FR AMR full rate

AMR-NB adaptive multirate narrowband

AMR-WB adaptive multirate wideband

ANS answer tone, which is the same as CED

/ANS ANS with phase modulation

ANSam ANS tone with amplitude modulation

/ANSam   ANS tone with amplitude and phase modulation
ANSI   American National Standards Institute
APP   application-specific function
ARQ   automatic repeat request
ASN   abstract syntax notation
ASN.1   Abstract syntax notation.1
ATM   asynchronous transfer mode
ATT   American Telephone and Telegraph

BCG   bulk call generator
B-Channel   Bearer Channel
BNLMS   block normalized least mean square
BORSHT   battery, overvoltage protection, ringing, supervision, hybrid, and test functions (in the telephone interface)
BPF   band-pass filter
BPI   baseline privacy interface
BPSK   binary phase-shift keying
BRI   basic rate interface
BT   British Telecom
BurstR   burst ratio
BW   bandwidth
Byte or byte   8-bits of data

CA   call agent
CAR   receiving terminal activation signal (Japan-caller ID)
CAS   CPE alerting signal
CAS   channel-associated signaling
CC   CSRC count
CCA   Cable Communications Association
CCITT   Committee Consultative International Telegraph and Telephone
CCR   comparison category rating
CED   called terminal identification tone
CELP   code excited linear prediction
CFR   confirmation to receive
CID   caller identity delivery or caller ID
CIDCW   calling identity delivery on call waiting or caller ID on call waiting
CI   call indication
CJ   CM terminator
CLASS   custom local area signaling services

CLI   caller line identification

CLIP   caller line identity presentation

CLIR   caller line identification restriction

CLR   circuit loudness rating

CM   call menu

CM   cable modem

CMOS   comparison mean opinion score

CMTS   cable modem terminal system

CND   calling number display (on CPE)

CND   calling number delivery (on CO)

CN   comfort noise

CNG   calling tone in fax call

CNG   comfort noise generation

CO   central office

codec   voice coder (compression) and decoder (decompression) (in this book)

CODEC   COder (hardware ADC) and DECoder (hardware DAC) or SLAC (in this book)

Coef   coefficient

Compander   compressor and expander

Cos(...)   cosine function

CP   call progress

CPE   customer premises equipment

CPI   common part indicator

CPTD   call progress tone detection

CPTG   call progress tone generation

CPU   central processing unit

CRC   cyclic redundancy check

CRLF   carriage return line feed

CRP   command repeat

CS-ACELP   conjugate-structure algebraic-code-excited linear-prediction

CSI   called subscriber identification

CRLF   carriage return line feed

CSeq   command sequence

CSRC   contributing sources

CT   call tone

CTC   continue to correct

CTR   continue to correct response

DA   destination address

DAA   digital access arrangement

DAC   digital-to-analog converter

dB   deciBel

dBm   decibel power with 1 milliWatt reference power

dBm0   dBm of the signal that would be measured at the relevant 0-dBr level reference point

dBov   dB relative to the overload point of the digital system

dBr   power with zero-level point (used to refer to relative power level)

dBrnc   noise power with 1 picoWatt reference and c-message filter weighting

dBp   noise power with psophometric weighting

dBSPL   The sound pressure with $20\,\mu$Pa (microPascal) as reference

dBV   RMS voltage in dB with 1-V RMS as reference

D-Channel   Data channel

DC   direct current

DCE   data communications equipment

DCME   digital circuit multiplication equipment

DCT   discrete cosine transform

DCN   disconnect

DCR   degradation category rating

DCS   digital command signal

DDR   double data rate (memory)

DECT   digital enhanced cordless telecommunications

DESA   discrete energy separation algorithm

DFT   discrete Fourier transforms

DIS   digital identification signal

DLC   digital loop carrier

DM   data memory (in processors)

DMA   direct memory access

DMIPS   Dhrystone MIPS

DMOS   degradation mean opinion score

DOCSIS   data over cable service interface specifications

dpi   dots per inch

DS   digital signaling

DS3   digital Service, Level 3

DSL   digital subscriber line

DSLA   digital speech level analyzer

DSLAM DSL access multiplexer (central office equipment for DSL service)

DSP digital signal processor

DT double talk

DTC digital transmit command

DTD double-talk detector

DT-AS dual-tone alerting signal

DTE data terminal equipment

DTMF dual-tone multifrequency

DTX discontinuous transmission

E1 E-carrier digital signaling

E-model Electrical-model

EBI even bits inversion

EBIU extended bus interface unit

EC echo canceller

ECM error correction mode

EN enterprise networks

EOL end of line

EOM end of message

EOP end of procedure

EOR end of retransmission

ERL echo return loss

ERLE echo return loss enhancement

ERR end of retransmission response

ETSI European Telecommunications Standards Institute

EV embedded variable

Fax facsimile (Facsimile meaning "a copy")

FaxLab fax testing instrument from Qualitylogic

FCD facsimile-coded data

FCF facsimile control field

FCS frame check sequence

FDM file diagnostic message

FEC forward error correction

FFT fast Fourier transform

FGPS physical layer overhead F—FEC, G—Guard Time, P—Preamble, S—Stuffing bytes

FIF facsimile information field

FIR   finite impulse response
FJB   fixed jitter buffer
FM   frequency modulation
FMC   fixed mobile convergence
FoIP   fax over IP
FOM   figure of merit
FSK   frequency-shift keying
FT   French Telecom
FTT   fail to train
FXO   foreign exchange office
FXS   foreign exchange subscriber or station

G1   Group-1 facsimile
G3   Group-2 facsimile
G3   Group-3 facsimile
G3C   Group 3C facsimile
G3FE   Group-3 facsimile equipment
G4   Group-4 facsimile
G711WB   wideband embedded extension for G.711 PCM
GDMF   Generic data message format
GIPS   Global IP sound
GoB   Good or better
GPS   Global positioning system
GR   General requirements
GSM   Global system for mobile communications
GUI   Graphic user interface
GW   Gateway

H registers   echo canceller filter memory
HCS   header check sum
HDLC   high-level data link control
HEC   header error control
HG   home gateway (CPE)
HPF   High-pass filter
HTTP   Hypertext transfer protocol
Hz   Hertz, frequency in cycles per second

IAD   integrated access device
IAF   Internet-aware fax device

ID   identity delivery

IDE   integrated development environment

IDMA   internal direct memory access

IEEE   Institute of Electrical and Electronic Engineers, Inc.

IETF   Internet Engineering Task Force

IFP   internet facsimile protocol

IFT   internet facsimile transfer

IIR   infinite impulse response

iLBC   internet low-bit-rate codec

IMS   IP multimedia system

IP   Internet Protocol

IPC   interprocessor communication

iPCM   internet PCM

IPoA   IP over ATM

IPSec   IP security

IPQoS   IP quality of service

IPv4   IP version 4

IPv6   IP version 6

IRS   intermediate reference system

iSAC   internet speech audio codec

ISDN   integrated service digital network

ISI   inter-symbol interference

ISO   International Standards Organization

ISP   Internet service provider

ITU   International Telecommunications Union

IVR   interactive voice response

J1   J carrier digital signaling

JB   jitter buffer

JBIG   joint bilevel image experts group

JM   joint menu signal

JPEG   joint photographic experts group

JTAG   joint test action group

kbps   kilo (1000) bits per second

kHz   kilo-Hz or kilo Hertz

L16   linear 16 bit (used in Audio)

LAN   local area network

LAPD   Link Access Protocol—Channel D
LCD   liquid crystal display
LD-CELP   low-delay code excited linear prediction
LEC   line echo cancellers
LMS   least mean squares
LP   linear prediction
LPC   linear prediction coefficients
LPF   low-pass filter
LQ   listening quality
LR   loudness rating
Lret   returned echo level
LS   least significant
LSB   least significant byte
LSF   line spectral frequencies
LSP   line spectrum pairs
LSTR   listener side tone rating

mA   milliAmpere
MAC   media access control
MAC   multiplier and accumulator (in processors)
MAC OH   MAC layer overhead
MAN   metropolitan area networks
Mbps   mega bits per second
MCF   message confirmation
MCPS   million cycles per second
MCU   multipoint control units
MDCT   modified discrete cosine transform
MDMF   multiple data message format
Mega   one million
MEGACO   media gateway and a media gateway controller
MF   multifrequency
MFPB   multifrequency push button
MG   media gateway
MGC   media gateway controller
MGCP   Media Gateway Control Protocol
MH   modified Huffman
MHz   mega (one million) Hz
MI   multiple instance
MII   media independent interface

milli $1/1000^{th}$ or 10^{-3}

MIME multipurpose Internet mail extensions

MIPS million instructions per second

MIPS machine without interlocked pipeline stages (processor)

MMR modified modified read

MoIP modem over IP

MOS mean opinion score

MOS-CQ MOS-conversational quality

MOS-LQ MOS-listening quality

MP-MLQ multipulse maximum likelihood quantization

MPS multipage signal

MR modified read

ms millisecond ($1/1000^{th}$ of second)

MS most significant

MSB most significant byte

MSLT minimum scan length time

MSN Microsoft network

mV milliVolt (10^{-3} Volts)

mW milliWatt (10^{-3} Watts)

NAT Network address translation

NB narrowband

NGDLC next-generation DLC

NLP nonlinear processing

ns nanoseconds (10^{-9} seconds)

NSC nonstandard facilities command

NSF nonstandard facilities

NSS nonstandard setup

NTP network timing protocol

NTR network timing reference

NTT Nippon Telegraph and Telephone

nW nanoWatt (10^{-9} Watts)

OLR overall loudness rating

OS operating system

OSI open switching interval

OSI open system interconnection

PAMS perceptual analysis measurement system

Params. Parameters

PAR   peak-to-average ratio

PBX   private branch exchange

PC   personal computer

PCI   peripheral component interconnect

PCM   pulse code modulation

PCMA   PCM A-law (G.711 A-law)

PCMU   PCM μ-law (G.711 μ-law)

PCM4   PCM channel measuring test set

PDU   protocol data unit

PESQ   perceptual evaluation of speech quality

PHS   Payload header suppression

PID   procedure interrupt disconnect

PIN   permanent identification number

PLC   packet loss concealment

PLL   phase locked loop

PM   phase modulation

PM   program memory (in processors)

PON   passive optical network

POTS   plain old telephone service

PoW   poor or worse

PPM   parts per million

PPPoA   point-to-point protocol over ATM

PPPoE   point-to-point protocol over Ethernet

PPR   partial page request

PPS   partial page signal

PPS-EOM   partial page signal—End of message

PPS-EOP   partial page signal—End of page

PPS-MPS   partial page signal—multipage signal

PPS-NULL   partial page signal NULL

PRI   primary rate interface

PRI-MPS   procedure interrupt—multipage signal

ps   picoseconds (10^{-12} seconds)

PSK   phase-shift keying

PSQM   perceptual speech quality measure

PSTN   public switched telephone network

PT   payload type

pW   picoWatt (10^{-12} Watts)

PWD   password

QAM   quadrature amplitude modulation

Qdu   quantization distortion unit

QMF   quadrature mirror filter

QoS   quality of service

QPSK   quadrature phase-shift keying

R-factor   Rating factor

RAM   remote access multiplex (in DSLAM)

RAS   remote access server (in modem)

RCP   return to control for partial page

Rec.   recommendation

RED   redundancy

REN   ringer equivalence number

RFC   request for comments

RG   residential gateway

RISC   reduced instruction set computer

RI-TCM   rotationally invariant TCM

RJ-11   registered jack-11 (telephone connector)

RJ-45   registered jack-45 for Ethernet and T1/E1 connection

RLR   receive loudness rating

RLS   recursive least squares

RMS   root mean square

RNR   receive not ready

ROH   receiver Off-Hook

RP-AS   ringing pulse-alerting signal

RR   receive ready

RS-232   recommended standard-232 (serial port)

RSTR   reset button on the system

RTC   return to control

RTCP   RTP Control Protocol

RTCP-XR   RTCP-Extended Report

RTN   Retrain negative

RTP   Retrain positive

RTP   Real-Time Transport Protocol

Rx   receive

s   second(s)

SA   source address

SAR   segmenting and reassembly

SAS   subscriber alerting signal

SB-ADPCM   sub-band-adaptive differential pulse code modulation

SDES   source description

SDIO   secured digital input output

SDMF   single data message format

SDP   Session Description Protocol

SDRAM   synchronous dynamic random access memory

Sec/sec/s   time in seconds

SEP   selective polling

SG3   supergroup-3

SG-12   ITU study group-12

Sgn   sign calculation

SID   silence insertion description

Sin(…)   sine wave function

SIP   Session Initiation Protocol

SLAC   subscriber line access circuit

SLIC   subscriber line interface circuit

SLR   sending loudness rating

SME   short messaging entity (in SMS)

SMS   short message service

SMTP   simple mail transfer protocol

SN   sequence number

SNMP   Simple network management protocol

SNR   signal-to-noise ratio

SPCS   stored program control system

SPI   serial peripheral interface

SPL   sound pressure level

SQTE   speech quality test events

SR   sender report

SRAM   synchronous random access memory

SRL   singing return loss

SRL-Hi   SRL high frequency

SRL-Lo   SRL low frequency

SS7   signaling system 7

SSRC   synchronization source

STD   signal to total distortion

STFT   short-time Fourier transforms

STL   software tool library

STMR side tone masking rating
STUN simple traversal of UDP through NAT
SUB subaddress

T type of payload
T1 T-carrier digital signaling
TAS TE alerting signal
TBR technical basis for regulation
TCF training check field
TCLw weighted terminal coupling loss
TCM Trellis-coded modulations
TCP Transmission Control Protocol
TDAC time-domain alias cancellation
TD-BWE time domain-bandwidth extension
TDM time division multiplex
TE terminal equipment
TELR talker echo loudness rating
TK Teager–Kaiser
TIA Telecommunications Industry Association
TLS transport layer security
TPKT transport protocol data unit packet
TR technical reference
TR-57 technical reference-57 (Telcordia/Bellcore document for DLCs)
TSA time slot allocation
TSI transmitting subscriber identification
TTA Telecommunication Technology Association
TTC Telecommunication Technology Committee
Tx transmit

UA user agent
UAC user agent client
UAS user agent server
UDP User Datagram Protocol
UDPTL UDP transport layer
UMTS universal mobile telecommunication system
URI uniform resource identification
URL uniform resource locator
USB universal serial bus

UTC coordinated universal time, previously known as Greenwich mean time (GMT)

UU user to user

V Volts–unit of voltage

VAD voice activity detection

VCO voltage controlled oscillator

VDSL very high-speed DSL

VLAN virtual LAN

VLSI very large-scale integration

Vocoder voice coder

VoATM voice over ATM

VoIP voice over IP

VQ vector quantization

WAN wide area network

WB wideband

WEPL weighted echo path loss

WiFi wireless fidelity (IEEE 802.11 series)

WiMax worldwide interoperability for microwave access

WLAN wireless LAN

XDSL Any DSL

XNOR inverse of the exclusive OR

XOR exclusive OR

μ-law logarithmic 64-kbps compression, which is the same as G.711 PCMU

μF micro Farad

μs micro seconds (10^{-6} seconds)

μW microWatt (one/million of Watt)

Ω ohms (impedance units)

1

PSTN BASIC INFRASTRUCTURE, INTERFACES, AND SIGNALS

Telephones, fax, and dial-up modems are popularly used with a two-wire TIP-RING foreign exchange subscriber (FXS) interface that supplies a battery. The interface inside the phone or fax machine is a foreign exchange office (FXO). In some countries, a four-wire integrated services digital network (ISDN) is used for telephone services. A T1/E1 family of interfaces is used mainly for higher channel communication. In an office environment, a user may get telephone service through a public switched telephone network (PSTN) central office (CO) or a private branch exchange (PBX) system resident close to the office phones. PBX systems may use multiple FXS or digital phone interfaces for connecting to the user and FXO, ISDN, or T1/E1 family of interfaces to communicate with the nearest PSTN CO or digital loop carrier (DLC). A DLC resides close to the subscribers and extends the reach of central offices. For inter-regional services, the local CO will route the calls to the destination CO. The destination CO then terminates the call directly or through the local DLC. Several handbooks and documents are available on this subject [Freeman (1996), Bellamy (1991), ITU-Handbook (1992)]. The combinations and possibilities in service vary with each service provider and country. VoIP service and user interfaces are closely related to historical PSTN services. In this chapter, an overview of the PSTN telephone infrastructure and some of interfaces and voice signal characteristics is provided to create continuity and to map the PSTN functionality with VoIP infrastructure.

VoIP Voice and Fax Signal Processing, by Sivannarayana Nagireddi
Copyright © 2008 by John Wiley & Sons, Inc.

1.1   PSTN CO AND DLC

In this section, an overview of central offices and DLCs is presented. An analog CO provides services to the user, and in recent times, analog central offices were replaced with a combination of digital CO and DLC.

1.1.1   Analog CO

Analog telephony requires a two-wire analog TIP-RING interface. Several years back, the PSTN CO was directly providing several pairs of analog lines to the closest junction box, and from there, individual TIP-RING wire pairs were being distributed to the subscriber [URL (IEC-DLC)]. When a subscriber is far (5000 to 15,000 feet) from the CO, long-distance analog lines can create distortions and signal attenuation. To counter this problem, bigger diameter wires and compensating loading coils were used with analog CO. Sometimes the line voltage is increased at the CO to cater for voltage drop over long lines. For a coverage area of a few miles in diameter, this approach may still be used. Overall, using long lines from the CO is a costly effort, deployment may not scale up, and voice quality degrades. Additionally, fax and modem calls may operate at a lower speed on long analog lines. For a growing customer base, an analog CO may not scale up properly. A VoIP adapter providing telephone service to the end user closely resembles the analog CO by providing telephone interfaces like PSTN DLC or CO and by allowing voice calls through an Internet connection.

1.1.2   Digital CO and DLC

In recent years, PSTN systems migrated to ITU-T-G.711 A-law/μ-law compression [ITU-T-G.711 (1988)]-based synchronous digital communication. Long-distance analog line pairs were replaced with digital distribution boxes. A PSTN CO will connect a few digital lines (T1/E1 family interfaces) to the intermediate DLC box positioned close to the subscriber, and the DLC will then distribute analog TIP-RING lines for the last mile of telephone service. Digital CO along with DLC is shown in Fig. 1.1. With the introduction of the CO and DLC combination, the signal quality is greatly improved, even for the users located far from the CO. Each DLC will take care of several hundred users, and if necessary, the DLC is capable of sending the required high enough voltage to overcome any remaining distance problems. The other side of the DLC is connected to the CO. DLCs may have a basic T1/E1 family of interfaces at the first level. Advanced services, including the support of several offices, PBX systems, multimedia, and Internet capabilities are supported through next-generation DLC (NGDLC) as given in [URL (IEC-DLC)]. NGDLC will need wider bandwidth for communication between the DLC and the CO with fiber being the most popular interface for deployment. In some locations, fiber may not available, leading to a choice of other interfaces,

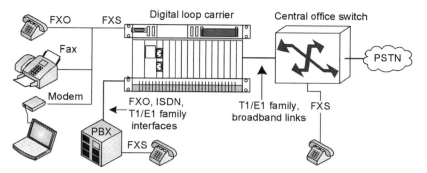

Figure 1.1. PSTN digital office and interfaces.

including coaxial cable, digital subscriber line (DSL), or a combination of these that are used for higher end requirements with NGDLC. NGDLC are deployed for multiservices of voice, video and data as well as various Internet services with the right interfaces.

1.2   PSTN USER INTERFACES

PSTN end users will get services through FXS, ISDN, and the T1/E1 family of interfaces [URL (TIA-496B), URL (T1/E1), URL (ISDN)]. These interfaces are also used in the migration of VoIP voice and fax solutions. As shown in Fig. 1.1, the telephone, fax, and dial-up modem are connected on an FXS interface. The FXO interface is part of a telephone, fax, and modem. Some COs provide an ISDN interface for residential applications. The T1/E1 families of interfaces are mainly used with PBX and enterprise services.

1.2.1   FXS and FXO Analog Interfaces

A PSTN wall socket is the FXS interface given to the subscriber for connecting a telephone. FXS is the two-wire TIP-RING interface provided by the PSTN CO or DLC. This interface is used for connecting telephones, fax machines, and dial-up modems. FXS supplies battery voltage, high-voltage ring, and sufficient current to drive three to five parallel phones. A subscriber line interface circuit (SLIC) and a subscriber line access circuit (SLAC) are the main components of the FXS interface. SLIC consists of a two-to-four-wire hybrid and of high-voltage electronics. SLAC is the interface between a SLIC analog signal and processor digital interface.

The FXO receives battery voltages from the PSTN FXS interface. Sometimes the FXO interface is known as a passive interface, which means the FXO will not generate a high-voltage battery on analog TIP-RING interfaces. The FXO interface is available on the TIP-RING connections from a phone, fax machine, or modem. Subscribers can connect this FXO interface to the FXS

Table 1.1. FXS and FXO Basic Differences

Attribute	FXS	FXO
Main functional category	This works like a central office	This works like an electronic phone
Location in the system	PSTN CO/DLC, PBX, or VoIP adapter gives this interface	Telephones, fax, and dial-up modem are having FXO interface
Where is it connected?	Connected to telephones, faxes, or modems	Central office, PBX, VoIP adapter
Active/passive port	Active port, gives battery voltages of the order of −24 to −72 V, usually of −48 V	Passive, can receive voltages
Drive capability	Can drive 3 to 5 telephones, or fax machine/modem	Passive interface, present inside phone, fax, or modem
Popular name of the device used on the interface	SLIC is the front end device on DLC or VoIP adapter that drives phones	Passive circuit and DAA is inside the phone, fax, or modem

interface provided by the central office, PBX, or VoIP adapter. PBX systems will use FXS on one side to connect to multiple phones and may use multiple FXO interfaces to connect to PSTN CO or DLC. The other popular name used for this FXO interface is the digital access arrangement (DAA). The DAA name is mainly associated with dial-up modems. In simplest form, multiple FXOs can be connected on one FXS interface. FXS is the main active battery source to supply current to multiple FXO interfaces. FXS and FXO are simple interfaces. Some of the functional differences of FXS and FXO are listed in the Table 1.1.

1.2.2   SLAC, CODEC and codec–Clarifications on Naming Conventions

As explained in Section 1.2.1, telephone interfaces use SLIC and SLAC. SLIC converts a signal from two-wire TIP-RING interface to a four-wire interface and gives it to SLAC for sampling. Currently, the name hardware CODEC is more popular than SLAC. SLAC/CODEC has several functions as explained in Chapter 17. CODEC in a simple configuration consists of a COder [hardware with an analog-to-digit converter (ADC)], and DECoder [hardware with digital-to-analog converter (DAC)] that samples the analog signal at 8- or 16-kHz. FXO interfaces use DAA that also makes use of ADC and DAC for sampling. The name SLIC-SLAC use UPPER-case alphabets. The usage of SLAC equivalent part "CODEC" with SLIC-SLAC and SLIC-CODEC is more common.

The name "codec" with lower-case alphabets is popular for compression. The codec has "encoder for compression" and "decoder for decompression" of digital samples. The names voice codec, speech codec, compression codec, and low-bit-rate codec or simply with lower case "codec" denotes samples compression through computation. Through out this book, the same naming conventions are used.

No hard rules exist for this naming convention. In the literature, lower case name "codec" or "Codec" is also considered for hardware sampling operations, whereas upper-case "CODEC" name is not used to represent compression. Refer the context–sampling hardware (CODEC) or samples compression (codec) for correct interpretation.

1.2.3 TIP-RING, Off-Hook, On-Hook, and POTS Clarifications

The terms TIP, RING, off-hook, and on-hook are used frequently in PSTN and VoIP services. A summary of these keywords is given here. TIP and RING are the pair of wires used for telephone connection. These names originated in reference to the phone plug used to make connections in manual switchboards. In manual switchboards, one side of the line makes contact with the TIP of the plug and the other with the RING contact immediately behind the TIP of the plug [URL (POTS)]. Sleeve is usually of a common point or ground, and this is not used in the current two-wire TIP-RING-based system. Currently, all the telephone exchanges are electronic with digital switching. Even in operator-assisted PBX systems, an operator is using telephone keys or soft keys for making the connection. Normal telephones are not now using TIP, RING, or sleeve connectors. Some instruments for telephone measurements [URL (Sage935)] are using TIP-RING sleeve connectors. The names "TIP, RING" continued for the two wires used for connecting telephones and central offices. A picture to represent this TIP-RING connector is available at reference [URL (POTS)].

The terms "off-hook" and "on-hook" were used in the early days of telephone usage. They refer to the handset position with respect to the cradle of the telephone. In the early telephones, the phone handset was lifted from the hook and was called off-hook. Keeping the back of phone on the hook was called on-hook. On-hook disconnects the phone connection. Off-hook is for getting the dial tone or for continuing the voice call. This hook switch can be clearly seen on most public phones mounted vertically. On public phones, the handset is physically hanging on the hook switch. Lifting the handset from the hook switch is off the hook, and putting the handset back on the hook is the on the hook. A hook switch is also available in handset-based modern phones. The hook may be a little button that pops up on lifting the handset and goes down when putting back the handset.

POTS is a plain old telephone service that was using TIP-RING for main signals, a rotary wheel for pulse dialing, and a bell for ringing. The widely used analog phones of today are also using two-wire TIP-RING with few differences in dial pad, ringer, and connectors. The functions of TIP-RING, dialing, ring, and so on, have remained same. Hence, the current analog phones are also referred as POTS phones. The names "analog phone", "normal phone", "house phone", "POTS phone", "two-wire phone", or simply "phone" are used interchangeably. In this book, the names "ring" or "Ring" are used for phone ringing alert, and name "RING" is used for telephone physical interface wire.

1.2.4 ISDN Interface

ISDN [URL (ISDN)] gives the highest possible PSTN quality than analog telephony. ISDN is used as a reference for comparing voice call quality of narrowband (300 to 3400 Hz) voice. It is also used for high-quality voice and fax at the highest speed of 33,600 bits per second (bps) and for dial-up modem at higher data rates of 56 kilo bits per second (kbps) from one of the two bearer (B)-channels. ISDN is a four-wire interface with a separate send and receive pair of wires. The physical interface of ISDN is of analog, but the transmission content is a bit stream of information. This interface in a system allows simultaneous voice conversations and data communication. With ISDN, voice and data are carried by two B-channels occupying a bandwidth of 64 kbps per channel. Some PSTN switches may limit B-channels to 56 kbps. A data (D) channel handles signaling and data at 16 kbps. There are two types of ISDN services: basic rate interface (BRI) and primary rate interface (PRI). BRI consists of 2B + D, two 64 kbps B-channels and one 16 kbps D-channel for a total of 144 kbps. The basic service is intended to meet the needs of individual users. For enterprise use, PRI is used, and this goes by the name H-series with a suffix based on the channel density. H-channels provide a way to aggregate several B-channels. They are classified as H_0 = 384 kbps (six B-channels), H_{10} = 1472 kbps (23 B-channels), H_{11} = 1536 kbps (24 B-channels), and H_{12} = 1920 kbps (30 useful B-channels). H_{12} is also known as international E1.

1.2.5 T1/E1 Family Digital Interface

T1/E1 is a four-wire interface [URL (T1/E1)]. It uses two wires for send and two wires for receive. T1 is used in North America, and E1 is used in Europe with a T1 equivalent interface named J1 used in Japan. In most of the systems, the same infrastructure and interfaces will work for both T1 and E1 without changing hardware electronics. These interfaces are used for enterprise solutions for voice and data communications. The physical connector interface will look similar to the Ethernet interface. Sometimes two coaxial cables are also used for this interface. PSTN DLC/NGDLC or CO will be supplying this interface to the user. Offices also use the T1/E1 interface for Internet service. In general, T1/E1 is used for voice, data, or a combination of voice and data.

At PSTN CO or DLC, the voice signal coming over the telephone line is sampled at 8 kHz, digitized, and compressed into eight bits producing 64 kbps per line or channel. This per-channel rate of 64 kbps is defined on digital signal 0 (DS0). In T1, 24 DS0 channels are multiplexed into a single digital stream.

The resulting data stream is of $24 \times 8 \times 8\,kHz = 1536\,kbps$ making each frame data 192 bits ($1536\,kbps/8000$). In T1, one-bit synchronization is used for every frame at $8\,kHz$ making each frame data 193 bits and total bit rate $1.544\,Mbps$. Of 24 channels of T1, one channel is usually allocated for telephone signaling. This process is represented as 23B + D channels, meaning one D-channel is used for data or signaling. In E1, 32 digitized channels are multiplexed into a single digital data stream, resulting in 32×8 bits $\times 8\,kHz = 2048\,kbps$. The resulting TDM frame size is 256 bits and frames are at $8\,kHz$.

1.3 DATA SERVICES ON TELEPHONE LINES

Telephone, fax, and modem use voice band frequencies (300 to 3400 Hz) on the telephone wires. Fax and modem are data terminals. Fax is an independent device that sends data modulated into voice band. A dial-up modem also modulates data into voice band. A dial-up modem is used for Internet connectivity and data transfer. Dial-up modems provide data connectivity up to 56 kbps. Some more details on dial-up modems are given in Chapter 15. Telephones, fax machines, and dial-up modems use telephone lines one at a time, but they can work simultaneously with a broadband service like xDSL.

For VoIP, Internet connectivity is essential. A dial-up modem is one of the basic and most popular options for Internet connectivity. It can provide data connectivity of up to 56 kbps, which is sufficient only with a higher compression VoIP voice call. DSL can provide much higher bandwidth compared with the dial-up modems, and hence, it is used in several deployments for Internet connectivity. In theory, both dial-up and DSL can be used together, but when DSL service is used, a dial-up modem may not be required.

1.3.1 DSL Basics

DSL service is provided through the existing analog telephone two-wire interface or on four-wire ISDN lines. Voice, fax, and modem signals occupy a 0- to 4-kHz band. Telephone wires are capable of carrying wider bandwidth signals to a distance of several thousand feet. DSL technology is based on discrete multi-tone (DMT) modulation using orthogonal frequency division multiplexing (OFDM). It operates on an array of N relatively lower rate transceivers in parallel to achieve an overall high rate on a single line. These N information streams are kept separated from one another by sending them on N different subcarriers called sub-bands or subchannels. Thus, DMT is a very flexible modulation scheme in which each subchannel can carry a different number of bits depending on the channel conditions. The subchannel that supports a higher signal-to-noise ratio (SNR) is loaded with a larger number of bits [Goralski (1998)].

The asymmetric digital subscriber line (ADSL) given in recommendation ITU-T-G.992.1 (1999) specifies DMT modulation with a total of 256 subchannels centered at frequencies $(m)(f_s)$, where m is a subchannel from 0 to 255 and f_s is 4312.5 Hz. Thus, it occupies a bandwidth of (256) (4312.5) = 1.104 MHz on the telephone line. Not all of these subchannels can be used in practice, as voice band splitting filters need to be employed to separate the DSL band from that of the voice band. The recommendation specified for either frequency separation of the downstream and upstream channels or separation by echo cancellation. The maximum number of bits allowed on each subchannel is 15, thereby limiting the theoretical maximum downstream rate that can be achieved to (15 bits per subchannel) (255 subchannels) (4000 symbol rate) = 15.3 Mbps. The G.992.1 recommendation specifies the downstream rate of 8 Mbps and the upstream rate of 1 Mbps.

The asymmetric DSL2 plus (ADSL2+) family [ITU-T-G.992.3 (2005), G.992.5 (2005)] occupies analog signal bandwidth from 25 kHz to 2.2 MHz to support downstream data rates up to 24 Mbps. Very high-speed DSL (VDSL) [ITU-T-G.993.1 (2004), G.993.2 (2006)] will occupy analog bandwidth up to 17 MHz and some versions up to 30 MHz to support data rates of 50 to 100 Mbps and more. In summary, voice services are continued at low frequency while using the same telephone line for various flavors of DSL service.

The DSL access multiplexer (DSLAM) is the CO for DSL services. DSL services do not need a high-voltage battery for operation. Hence, isolation in combining and splitting is essential between high voltage for normal telephone operation and low voltage for wideband DSL signals. DSL has to be combined on analog signal lines only. Different combining techniques are used to provide DSL service. Some popular techniques of combining DSL are direct analog combining, remote DSLAM line based, ADSL lines cards, and remote access multiplex (RAM) [URL (IEC-ADSL)]. Overview on DSL combining methods are given below. It is expected that these techniques and technologies will keep growing and changing with time.

Direct Analog Combining. This technique is applicable for the analog COs that distribute direct analog lines. DSL signals are directly combined on the analog lines with isolation splitter-combiners. This type of operation is possible if the coverage distance is less and the bandwidth requirements are of a few Mbps. Most central offices extend services through DLCs; hence, this type of DSL support is not popular.

Remote DSLAM-Based Combining. In this scheme, a complete DSLAM is colocated along with DLCs or is positioned close to the DLC based on the available installation options. DSLAMs are usually maintained in a controlled environment and are managed remotely. The service can be used with any PSTN lines and several flavors of DSL. The main disadvantage of this method is the cost involved to deploy and maintain. It will also require long wiring if DLC and DSLAM are not located in the same racks.

ADSL Line Card Solutions. DLC will have multiple slots. DSLAM in line cards are arranged in either a separate rack similar to DLC or some of the

DLC box slots be reserved for the DSL line cards. Keeping the DSL cards in DLC may minimize the wiring requirements. Management of DSL and DSLAM line cards will be under the same framework. Many deployments use this approach. DSL flavors keep changing with time. Upgrading the line cards in the field is a major concern of this deployment, and multiple user cards may go through different interoperability issues.

RAM. RAM performs similar functions of providing DSL service. RAM integrates closely with the DLCs. It is a separate small box. The RAM boxes are also called pizza boxes based on their shape. Pizza boxes are stacked and positioned along with the DLCs. RAM-based boxes have the advantage of both remote DSLAM and ADSL line card solutions. These boxes are easy to upgrade, change with another version of box, and scale in density.

1.4   POWER LEVELS AND DIGITAL QUANTIZATION FOR G.711 μ/A-LAW

Telephone line electrical signal power levels are presented in dBm units. Power level expressed in dBm is the power in decibel (dB) with reference to one milliwatt (mW). Power 0 dBm represents one milliwatt of power. One mW is a strong speech level for listening. Usual telephone conversion levels are close to −16 dBm (25 μW; μW is microwatt). ITU-T-G.711 and ITU-T-G.168 (2004) recommendations listed speech power levels and corresponding digital quantization mapping for 8-bit A-law and μ-law. A-law and μ-law (pronounced as Mu) are logarithmic compression methods explained in Chapter 3 under G.711 compression. Telephone interfaces have SLIC and SLAC for converting TIP-RING signals to digital samples. SLIC converts a signal from two-wire TIP-RING interface to a four-wire interface to SLAC. In recent literature, the hardware CODEC name is used instead of SLAC. CODEC consists of a COder, DECoder hardware with an analog-to-digital converter (ADC) and a digital-to-analog converter (DAC) that samples the signals at 8 or 16 kHz. Sampled signals are given to the processor for additional processing. In this chapter, signal levels and the mapping to 16-bit numbers in A-law and μ-law, are given with reference to dB and dBm scale.

1.4.1   μ-Law Power Levels and Quantization

The maximum μ (read as Mu)-law undistorted power level for the sine wave is 3.17 dBm. Power of 3.17 dBm is 2.0749 mW in 600 Ohms (Ω) of telephone impedance. Impedance of 600 Ω is applicable to North America and a few other countries like Japan and Korea. A sine wave of 2.0749 mW in 600 Ω develops 1578 milliVolts (mV) peak or a 3.156 Volts peak-to-peak sine wave amplitude. Voltage of 1578 mV has to be quantized as 8159 amplitude [ITU-

T-G.711 (1988), ITU-T-G.168 (2004)] in 16-bit binary numbering format. Assuming linear mapping for calculation purposes, the quantization level is $1578\,mV/8159 = 0.1934\,mV$. Once voice samples are available as per the above quantization of one level per $0.1934\,mV$, the following formula given in ITU-T-G.168 is used for estimating the voice power over a block of samples:

$$S(k) = 3.17 + 20\log\left[\frac{\sqrt{\dfrac{2}{n}\sum_{i=k}^{k-n+1} e_i^2}}{8159}\right] \quad \text{μ-law encoding} \qquad (1.1)$$

$S(k)$ = signal level in dBm
e_i = linear equivalent of the pulse codec modulation (PCM) encoded signal at sample index i
k = discrete time index or starting index for block of samples
n = block size of samples or number of samples over which root mean square (RMS) measurement is made.

In general, N will be used in signal processing literature for representing blocks of samples and n will be used for an index within the block of samples. In the above equation, to maintain consistency with the ITU recommendations, n is used for denoting block size and i is used for the index.

A sine wave with a peak amplitude of 8159 results in 3.17 dBm without any dependency on frequency. μ-law coding distorts the sine wave if it exceeds 3.17 dBm. Square wave power is 3 dB more than sine wave power, and an amplitude of 8159 of square wave results in 6.17 dBm. Alternatively, a constant value of 8159 (constant voltage) is 3 dB more power than a sine wave of 8159. Sometimes use of a constant value and calibrating with reference to a constant of 8159 as 6.17 dBm can help for calibration of the algorithms and power estimation blocks. In practice, lower amplitude is used for such calibrations in the range of −13 to −10 dBm instead of 6.17 dBm. Coding with μ-law supports the SNR up to 37 to 41 dB (for a 0- to −30-dBm signal power level) for the useful speech signal levels.

1.4.2   A-Law Power Levels and Quantization

An A-law maximum undistorted power level for the sine wave is 3.14 dBm. Power of 3.14 dBm is 2.06 mW in 600 Ω of telephone impedance. A sine wave of 2.06 mW in 600 Ω will develop 1572 mV peak or a 3.144 Volts peak-to-peak sine wave amplitude. Voltage of 1572 mV has to be quantized to 4096 amplitude levels. Quantization level is $1572\,mV/4096 = 0.3838\,mV$. Once voice samples are available as per the above quantization of one level per $0.3838\,mV$, the following formula given in ITU-T G.168 (2004) is used for estimating the voice power over a block of samples:

$$S(k) = 3.14 + 20 \log \left[\frac{\sqrt{\frac{2}{n} \sum_{i=k}^{k-n+1} e_i^2}}{4096} \right] \quad \text{A-law encoding} \quad (1.2)$$

The symbols used in this equation are the same as those used in the μ-law power representation in Eq. (1.1). For a sine wave of peak 4096, this calculation results in 3.14 dBm with out any dependency on frequency. A-law starts distortion to sine wave if it exceeds 3.14 dBm. An amplitude of 4096 for the square wave results in 6.14 dBm. As an example, substituting one complete cycle of a 2-kHz sine wave at 4096 amplitude (four samples for one complete cycle are 0, 4096, 0, −4096) in A-law power calculation gives 3.14 + 0 dBm = 3.14 dBm. Substituting four constant values of 4096 in A-law power calculation will give 3.14 + 3.0 dBm = 6.14 dBm. A-law supports SNR up to 37 to 41 dB (this varies with signal power level) for the useful speech signal levels.

1.5 SIGNIFICANCE OF POWER LEVELS ON LISTENING

Table 1.2 is given for different power levels starting from +3.17 dBm and down to −66 dBm in a few suitable steps. In relation to Table 1.2, summary points are given here [ITU-Handbook (1992)]. These levels are mainly mapped to the telephone interfaces and voice chain processing built for VoIP. In VoIP, power levels play a role in operation of some voice processing operations such as voice activity detection (VAD)/comfort noise generation (CNG), automatic gain control (AGC), echo level, and ideal channel noise. PSTN systems do not have any power-sensitive module operations such as VAD/CNG. In voltage scale, a 0.775-Volt RMS sine wave is the same as 0 dBm, terminated in 600-Ω impedance. The usual speech conversation level is −16 dBm in power. Some members may speak at a low amplitude below −24 dBm, which can also happen because of positioning of acoustic interfaces during conversation. This problem demands more effort by the listener. Automatic gain control (AGC), if employed, starts working from this power level. As per the ITU-T-G.169 (1999) recommendation, a maximum gain of 15 dB is used on low-level speech signals. A peak-to-average (PAR) level of speech is 15 to 18 dB [ITU-Handbook (1992)] for 0.1% of the active speech time. An average speech level of −12 dBm with 15 dB peak to average can preserve the speech as undistorted.

In practice, speech signals are gain controlled up to speech levels of −42 to −40 dBm. Below −42 dBm, it is difficult to continue the communication with required perception of speech. Lower power levels below −42 dBm are mapped to the region of VAD/CNG zones in VoIP. In PSTN, this type of silence region and power distinction is not present. VAD/CNG algorithms given in Chapter 4 track up to −60 dBm of power. Note that VAD/CNG algorithms use units of

Table 1.2. Summary on Useful Power Levels and Corresponding A- and μ-Law Mapping Values

Power in dBm	Power	Sine wave peak voltage in 600 Ω (milliVolts)	A-law peak 16-bit number	μ-law peak 16-bit number	Remarks on power levels
3.17	2.0749 mW	1578	Distortion starts	8159	μ-law undistorted sine wave
3.14	2.0606 mW	1572	4096	8131	A-law undistorted sine wave
0	1.0000 mW	1095	2853	5664	Loud speech
−3	0.5012 mW	776	2020	4010	
−6	0.2512 mW	549	1430	2839	
−9	0.1259 mW	389	1012	2010	
−12	63.1 μW	275	717	1423	Usual conversation level
−15	31.6 μW	195	507	1007	is −16 dBm
−18	15.8 μW	138	359	713	
−20	10.0 μW	110	285	566	
−21	7.9 μW	97.6	254	505	
−24	4.0 μW	69.1	180	357	Low-level speech, AGC action starts from −24 dBm
−27	2.0 μW	48.9	127	253	
−30	1.0 μW	34.6	90	179	
−33	0.5 μW	24.5	64	127	
−36	0.25 μW	17.4	45	90	
−39	0.125 μW	12.3	32	64	
−42	62.5 nW	8.7	23	45	AGC action ends, VAD action may start
−50	10.0 nW	3.5	9	18	Instruments maintain reasonable accuracy up to −52 dBm
−60	1.0 nW	1.1	3	6	Long delayed echoes are perceived as audible
−66	0.25 nW	0.5	1	3	Close to ideal channel noise of −68 dBm (−68 dBm = 20 dBrnc)

Abbreviations: mW, milli Watt; μW, micro Watt; nW, nano Watt

dBov, and the suffix ov stands for overload. It is not related to units of voltage, and it is related to 6.17 dBm = 0 dBov in μ-law compression. More details on these aspects are given in Chapter 4.

When round-trip delay is more, echoes are clearly perceivable. Echo levels given in Chapter 6 are suppressed to lower than −65 dBm. At this lower power level of −65 dBm, one or two bits in a 16-bit numbering system will appear as listed in Table 1.2. An echo power level of even −60 dBm is considered disturbing in VoIP and inter-regional PSTN calls. When no voice is speaking into the telephone or the phone is in mute, some noise overrides in the telephone system because of front-end coupling and power supplies. This noise is called ideal channel noise. These noise levels are maintained to below −68 dBm and preferably to −80 dBm. Noise power is expressed in picoWatts (1 picoWatt = 10^{-12} Watt) on a dB scale as dBrnC; suffix "C" denotes the C-message filter. A noise level of 90 dBrn is equivalent to 1 mW or 0 dBm [Bellamy (1991)].

1.6   TR-57, IEEE-743, AND TIA STANDARDS OVERVIEW

In PSTN, communication from the originating CO or DLC to the destination DLC is synchronous digital communication. This process has been well established over the last several decades. PSTN transmission follows some common standards such as ITU and IEEE and country-specific standards like TR-57 and TIA standards [TR-NWT-000057 (1993), TIA/EIA-470C (2003), IEEE-STD-743 (1995)]. PSTN manufacturers use TR-57 in North America to specify the signaling and transmission requirements for DLC. The FXS interface on VoIP adapters works similar to DLC or analog CO providing the TIP-RING interface. Assuming other favorable conditions for VoIP voice quality, as explained in Chapter 20, ensuring TR-57 transmission characteristics helps in improving voice quality. TR-57 is used in North America, and the current standard is GR-57 [GR-57-CORE (2001)]. Several other standards are followed in different countries. For example, Japan makes use of the NTT PSTN specification [URL (NTT-E)], and France uses the France Telecom STI series [FT ITS-1 (2007)] of specifications. Most of these country specifications are closely related to each other with only a few deviations like end-to-end losses, transmission filter characteristics, impedance, and so on. TR-57 transmission specifications can also be followed in achieving voice quality in VoIP systems. In VoIP, some signaling events as call progress tones used in establishing end-to-end calls may go through longer delays than PSTN specification. These deviations will tend to develop during the call establishment phase. Meeting transmission characteristics [TIA/EIA-912 (2002)] can take care of established VoIP call voice quality.

1.6.1   TR-57 Transmission Tests

TR-57 or GR-57 transmission summary points and high-level interpretation are given in this section. [Courtesy: Text and parameter values with TR-57 reference are from "TR-NWT-000057 (1993), Functional criteria for digital loop carrier systems—Chapter 6"; used with the permission from Telcordia, NJ, www.telcordia.com]. It is suggested to refer to TR-57 [TR-NWT-000057 (1993)] or the recent document GR-57 [GR-57-CORE (2001)] for the complete requirements and detailed specifications referred to in this book. Integrating these test specifications into PSTN and VoIP helps in maintaining the quality of voice and fax calls. Refer to instrument manuals [URL (Sage-options)] for details on measurements and interpretation.

Return Loss. Return loss happens because of an impedance mismatch of the terminations and the source. It is expressed as the ratio of the outgoing signal to the reflected signal. It is given as echo return loss (ERL) and singing return loss (SRL) with SRL at both high (Hi) and low (Lo) frequencies. ERL is the frequency-weighted average of the return losses over the 3-dB bandwidth points of 560 to 1965 Hz. Higher ERL is advantageous and gives lower echo. ERL in dB has to be greater than 18 dB, and this level avoids echo issues in the intraregional calls. In numerical terms, 18-dB ERL means 1/64th of the power or 1/8th of the voltage returning as echo. It is possible to achieve ERL up to the extent of 24 dB in practical systems with short-loop applications with VoIP gateways. To clarify on a relative scale, an ERL of 24 dB is better than an ERL of 18 dB by 6 dB. SRL is created by a resonance effect at selected frequencies. At some frequencies, the round trip of the echo is created with multiples of 360° of phase shift, which makes oscillations sustain for a longer duration. This phenomenon is called "line is on singing." SRL-Lo is the singing return loss for 3-dB bandwidth low frequencies in the range of 260 to 500 Hz. SRL-Hi is for the high-frequency singing return loss measured in the band of 2200 to 3400 Hz. Details on the band-pass characteristics are given in IEEE STD-743 [(1995), URL (Midcom)]. The SRL has to be greater than 10 dB. In general, SRL-Hi will be higher than SRL-Lo. Lower dB values of SRL-Hi, SRL-Lo creates annoyance to both talker and listener of the voice call.

Longitudinal Balance. Balance is the match between TIP and RING lines. It provides the discrimination for the common mode signal on the TIP and RING interface or indicates how well the TIP-RING is matching. It is evaluated as the ratio of longitudinal to metallic balance. Longitudinal is a common mode signal, and metallic is a differential signal. In the measurements, a common or ground point is required in addition to TIP-RING lines, as given in reference [URL (Sage-Balance)]. The preferred balance is greater than 65 dB at 1 kHz, but the acceptable balance is 63 dB. In good implementations, a balance of 68 dB is achieved.

Total Loss. It identifies that the transmitter and receiver are having a required loss. The end-to-end losses have to be within the acceptable limits, including the network cables loss. PSTN systems achieve their highest end-to-end loss of 8 dB in off-hook, and these losses vary with country-specific requirements. These losses also influence the loudness rating (given in Chapters 17 and 20) that has a close relationship with voice quality.

Off-Hook Frequency Response. Frequency response has to be flat to meet set tolerances for most of the voice band. This test also ensures that frequency response roll off did not happen in the 400- to 2800-Hz band. Frequency response is checked at 0 dBm, and the flatness goals are of ±0.5 dB. The reference for comparison of flatness is at 1004 Hz. With reference to a 1004-Hz frequency response, deviations from 400 to 2800 Hz have to be within ±0.5 dB as optional and −0.5 to +1.0 as required. In actual practice, the frequency response is taken from 100 Hz to 3900 Hz in suitable frequency steps. The values between 400 and 2800 Hz are used for comparing the frequency response.

60-Hz Signal Loss. Signal loss at 60 Hz is not desirable in speech transmission, but this "mains hum" can unfortunately be introduced from the AC mains power supply (110 V, 60 Hz). This test is to ensure that a 60-Hz contribution is at least 20 dB lower than a frequency response at 1004 Hz. This 60-Hz contribution has to be treated before sampling. If end-to-end loss at 1004 Hz is "X" dB, loss at 60 Hz has to be greater than X + 20 dB. In some countries, especially in Europe, 50 Hz is used for power transmission, and therefore, the measurements have to be conducted at 50 Hz in those countries even though it is not stated.

Off-Hook Amplitude Tracking. It ensures transmission gain flatness for different amplitudes at 1004 Hz, which is similar to amplitude linearity at the selected frequency. Three different ranges are defined in TR-57. At lower signal levels, A-law and μ-law create more quantization noise, which makes gain deviation to increase below the −37-dBm input level. For inputs of greater than or equal to −37 dBm, maximum deviation is ±0.5 dB (average of ±0.25 dB). For inputs of −37 to −50 dBm, maximum deviation is ±1.0 dB. For inputs of −50 to −55 dBm, maximum deviation is ±3.0 dB, which implies that AGC cannot improve linearity for small signals. Most instruments can measure up to a minimum of −60 dBm. Test instruments use a C-message band-pass filter [URL (Sage935)] for the voice band. In some countries, psophometric (p)-message filters are used. The differences are outlined in the ideal channel noise section.

Overload Compression Tests. The A-law can accept 3.14 dBm and the μ-law can accept 3.17 dBm of a sine wave without distortion. With increased signal power, the sine wave is clipped on either end of the top to make it look like a trapezoidal waveform. Overload compression is for measuring distortions on overloaded power levels up to 9 dBm. These distortions are measured as loss in simple measurements. There are set limits to the maximum loss at different power levels of 0, +3, +6, and +9 dBm. If system loss is "X" dB at 0-dBm input, loss at +3 dBm has to be ≤X + 0.5. The loss at +6 dBm has to be ≤X + 1.8. The loss at the highest signal of +9 dBm has to be lower than ≤X + 4.5. The relations are good for no loss of X = 0 dB. When loss is significant, the measurements may show much better results. Assuming loss at an analog interface, even a higher power signal may pass without any distortions.

Off-Hook Ideal Channel Noise. In its simplest form, ideal channel noise is the noise heard when microphones are in mute. The noise can be heard in a silent room. Instruments in North America use a C-message filter for voice band and measure the noise power in dBrnc. Power in dBrnc (also referred as dBrnC) is C-message weighted noise power with a reference unit of 1 pw. The popular notation is dBrnc for noise power. For voice power, dBm units are used. In simple mapping, 90 dBrn = 0 dBm (1 mw), and with C-message weighting, the mapping is 88 dBrnc = 0 dBm. The psophometric weighting is used in Europe as 87.5 dBp = 0 dBm. The symbol "dBp" is used for dBrn with psophometric weighting. The variation is by 2 to 2.5 dB with the weighting windows. The ideal channel noise power has to be better than 18 dBrnc, which is equivalent to 18 − 88 = −70 dBm. An accepted noise power level is 20 dBrnc = −68 dBm. Long-distance analog telephony was accepting higher ideal channel noise [Bellamy (1991)]. After conversion to digital telephony with short analog loops and DLCs, ideal channel noise achieved better levels, as stated in this section.

SNR or Distortion Ratio. The SNR and distortion measure identifies the SNR limits at the signal level. Lower signal power levels and their SNR results are also important in transmission. A-law or μ-law compression is used in PSTN and VoIP. Both the highest and the lowest signal are coded in 4 bits with 3-bit level scaling and in 1 bit for sign (polarity of the signal), which limits the possible SNR to 38–41 dB with self-quantization effects. With low amplitudes, SNR starts degrading. It is measured with an input sine wave of 1004 to 1020 Hz, and signal power is varied from 0 to −45 dBm. For input of 0 to −30 dBm, the required SNR is more than 33 dB (usually >37 dB is achieved in most systems). For input of −30 to −40 dBm, SNR is >27 dB. For −40 to −45 dBm, SNR is >22 dB. When end-to-end losses are more, or ideal channel noise is not meeting the specifications, SNR will be lower for the same input signal levels. The signal-to-total-distortion (STD) ratio that takes into account total distortion, including second and third harmonics.

Impulse Noise. Impulse noise is the short duration noise power that exceeds the set thresholds. Impulse noise can create audible tick sounds, and fax/modems may fall back to a lower bit rate. Ideal channel noise is the steady low-level noise. Impulse noise durations are comparable with 10-ms duration. Signal loss from phase or amplitude distortion can be noticed during VoIP packet loss, coupling from adjacent channels, ringing, line reversals clock drifts, and a change in power supply load conditions in the hardware. Noise on the interfaces may occupy wider bandwidth. Impulse noise is heard with a C-message filter, which means it is limited to the voice band.

The number of impulse counts in 15 minutes should not exceed 15; therefore, one impulse count per minute is the maximum. The measurements are conducted on both ends of the system with tones and simple ideal channel configuration. In the no-tone case, impulses are detected with a threshold of

47 dBrnc (equivalent to −41 dBm). In the tone case, with the tone at 1004 Hz, −13 dBm is sent and impulse noise is measured with a threshold of 65 dBrnc (equal to −23 dBm). Several conditions with impulse noise exist as given in TR-57 [TR-NWT-000057 (1993)]. The impulse noise counts are also referred to as hits.

Intermodulation Distortion. This test makes use of four tones as per IEEE STD-743 (1995) with a total input level of −13 dBm. This test is mainly used for measuring the undesired nonlinearity. A pair of tones around 860 Hz and another pair of tones around 1380 Hz are used in this test. In this test, total second and third harmonic distortions are measured. The second harmonic has to be better than 43 dB, and the third harmonic has to be better than 44 dB.

Single-Frequency Distortion. It measures the distortions to the single pure tone. This distortion is separated into two frequency bands of 0–4 kHz and 0–12 kHz. In narrowband, the 1004–1020 Hz tone at 0 dBm is sent and the observed tones outside the sent tone have to be lower than −40 dBm in the frequency band of 0–4 kHz. In the wideband range, for any input at 0 dBm in the frequency range of 0–12 kHz, the observed output at any other frequency outside the sent tone frequency has to be less than −28 dBm in the 0–12-kHz frequency range. This type of test will require spectrum analyzers.

System Generated Tones. It measures any undesired system generated tones. Send and receive can be in either the terminated mode or the not terminated mode with proper impedance. Tones measured at any end have to be less than −50 dBm in the frequency range of 0–16 kHz.

PAR. PAR provides amplitude and phase distortion over time because of transmission impairments [URL (Sage-PAR)]. A PAR waveform is a complex signal consisting of about 16 non-harmonically related tones with a spectrum that approximates the modem signals in the voice frequency band. In general, PAR takes care of attenuation distortion, envelope distortion, four-tone modulation distortion, and phase distortion. PAR is measured on both ends. Most systems achieve a PAR of 94, and an acceptable PAR is 90. The PAR measurement cannot reveal what caused this distortion. In many types of equipment, PAR test is replaced with the better 23-tone test as per IEEE STD-743 (1995).

Clarification Note on PAR Measurement and PAR Speech Levels. In specifying the speech characteristics, peak-to-average levels are specified as 15.8–18 dB. The usual speech levels are of −16 dBm, and peak levels can go up to +2 dBm with an 18-dB speech-level PAR. The PAR specified in TR-57 and the measurements using instruments from [URL (Sage-PAR)] provide the distortion measure for voice band signals. A PAR of 100 is the best, and a PAR of 90 to 94 is the goal in digital-loop carrier [TR-NWT-000057 (1993)]. A PAR of 94 is 6% of overall transmission distortions.

Channel Crosstalk. This test is to ensure, end-to-end, that no significant coupling occurs in the adjacent channels. Therefore, 0-dBm power tones from 200 to 3400 Hz are sent and power is measured in adjacent channels. This crosstalk power has to be 65 dB lower or below –65 dBm of power. With this type of low coupling, crosstalk will submerge into ideal channel noise.

Frequency Offset. It measures the frequency drift from end to end, which happens mainly because of a sampling clock parts per million (PPM) difference between source and destination. The frequency drift has to be contained to within 0 ± 0.4 Hz at 1004 Hz. The instruments have to cater to a 0.1-Hz measurement resolution.

1.6.2   IEEE STD-743–Based Tests

IEEE STD-743 (1995) had many tests that overlapped with TR-57 measurements. Instrument manufacturers were following IEEE specifications. Several tests of TR-57 are contained in IEEE STD-743. The 23-tone test has been adapted in recent test equipment. It takes care of transmission and distortion tests. The 23-tone test signal consists of 23 equally spaced tones [URL (Sage935)] from 203.125 to 3640.625 Hz with known phase relationships. The tones are spaced at $8000/512 = 15.625$ Hz to allow easy fast Fourier transform (FFT) analysis with 512 and 256 points at 8000-Hz sampling. The signal repeats once every 64 ms ($1/15.625$ Hz = 64 ms). Frequency response is measured at 23 tones, and phase relations are observed at 22 frequencies.

1.6.3   Summary on Association of TR-57, IEEE, and TIA Standards

Many equipment manufacturers built instruments as per 1984 version of IEEE STD-743. This outlines the IEEE standard accuracy of various measurements and measurement procedures. It has many overlapping measurements of TR-57. IEEE STD-743–based instruments can perform TR-57 tests also. IEEE STD-743 specifications were revised in 1995. This revision had many tests suitable for TR-57. This standard became obsolete in 2004. Equipment manufacturers, however, still follow IEEE STD-743. Now the TIA-470 series is followed in many types of new equipment. TIA-470 has had several revisions and extensions for acoustic measurements, such as cordless phones and wireless systems that were not available in TR-57- and IEEE STD-743–based tests. TR-57 is limited to a narrowband of 0–4 kHz. IEEE STD-743 (1995) and TIA IEIA-470C (2003) take care of wideband voice tests. In general, all these standards have several matching and overlapping specifications that determine proper voice quality on PSTN-based systems and their extensions across many different types of telephone and fax equipment.

2

VoIP OVERVIEW AND INFRASTRUCTURE

This chapter presents an overview of VoIP and its supporting infrastructure. VoIP services operate on an Internet protocol (IP) to transmit compressed voice samples as frames and messages as a group of bytes over an IP data network. VoIP can be achieved on any data network that uses IP-based networks like the Internet, intranets, and local area networks (LANs). In residential applications, voice from front-end telephone lines is converted into a suitable signal level, digitized, compressed as voice payload, and sent as IP packets. Signaling protocols are used to set up and tear down calls. Signaling packets carry information required to locate users, negotiate capabilities, create call establishing, and set up call features. Fax calls also work similarly to voice calls. The payloads of packets are the fax-compressed bytes. Initially, a fax call operates like a voice call, on fax tone detection, switches to fax, and continues with fax messages. The popular VoIP signaling protocols are the Session Initiation Protocol (SIP), Media Gateway Control Protocol (MGCP), and ITU-T-H.323 (2006). Many initial deployments have used H.323. Other standards like MGCP, ITU-T-H.248 (2005)/Megaco, and SIP evolved after the release of H.323. SIP has been mainly adapted for recent deployments. Major extensions in VoIP signaling were happening on SIP signaling during writing of this book. SIP is relatively simple and is favored by many engineers and service providers. VoIP calls, various applications, infrastructure, and interfaces are presented in this chapter.

2.1 PSTN AND VoIP

In this section, public switched telephone network (PSTN) and VoIP mapping and different VoIP basic call combinations are given. PSTN-based voice calls and infrastructure have been familiar to users for several decades. Functionally, a user will be using PSTN and VoIP in a similar way. However, several differences between PSTN and VoIP exist in actual implementation and end-to-end transmission. In Table 2.1, some key aspects and functional differences

Table 2.1. PSTN and VoIP Comparison Overview

Attributes	PSTN	VoIP
Major mapping	Digital loop carrier (DLC) is the intermediate box between phone and CO and CO manages call routing	VoIP CPE provides functionality of DLC and some functions of PSTN CO
Phones analog termination	Phone is connected to DLC through analog two wires. After DLC, digital μ/A-law samples are sent as bits	CPE is kept within a few feet of the phone. Beyond CPE, IP packets are sent
RJ-11 phone connector	Analog wire of several 1000 feet from DLC is terminated on wall socket	Phone is directly connected to VoIP adapter/CPE
Distortions on analog line	Higher distortions due to long loop lengths and creates problem to fax	No analog transmission distortions due to short loop length, which helps fax calls
DLC output impedance	DLC may offer different impedance to match along with long lines	VoIP CPE provides matching impedance same as phone
Ring and drive strength	DLCs support three to five phones, under the supervision of CO	VoIP CPE drives usually two to three phones, sometimes under soft-switch supervision
Battery voltages	Sometimes higher voltages are provided to compensate for line voltage drops	Phone is directly going to adapter. Exact voltages are maintained
Protection on TIP-RING	Better protection for surges, short circuits, and lightening	This usually stays inside the customer location and a minimal protection is used
Line diagnostics	PSTN management system can conduct remote telephone diagnostics referred to as GR-909 in North America	A minimal diagnostics are built in to the VoIP CPE. Many CPEs do not implement this feature
Digit detection	Dialed digits from phone are detected on nearest central office	Dialed digits are detected on the VoIP CPE and or soft switch

Table 2.1. *Continued*

Attributes	PSTN	VoIP
Call routing, tones, and call features	Call routing and call progress is created through central office	The VoIP CPE and soft switches decides call routing; a direct CPE-to-CPE call is also possible
End-to-end voice transmission	Up to DLC, it is analog. From DLC, it is a digital μ/A-law sample. Digital samples are switched in multiple stages of PSTN on a time division multiplex network	Compressed voice in μ/A-law or low-bit rate codec samples are packetized and sent on IP. At destination the VoIP adapter converts packets to analog signal
End to end delays	Decided by the physical distance, intra-regional calls are of 16 ms	Under good network and implementations, 50 to 80 ms more than PSTN calls
Echo cancellation	Usually not required on regional calls, and for inter-regional calls, an echo canceller is used at selected central office nodes	Echo cancellation is always required due to long round-trip delays
Clock PPM	DLCs use stratum-3 or better clock precision	VoIP CPEs use 30 to 50 PPM. Some deployments use PSTN clock precision
Bandwidth on TDM networks	64 kbps	Varies from 16 to 128 kbps depending on the compression and network interfaces
MOS quality	PESQ-MOS 4.4	MOS 4.3 to 4.4 with G.711, high compression codecs give lower MOS

Abbreviations: MOS, mean opinion score; PPM, parts per million; PESQ, perceptual evaluation of speech quality; kbps, kilo bits per second.

are given. A functional representation of a PSTN-based voice call is given in Fig. 2.1(a). VoIP calls can be mapped very closely to PSTN on an IP network as shown in Fig. 2.1(b). VoIP customer premises equipment (CPE) resides close to the end user of the VoIP service.

2.1.1 CPE and Naming Clarifications of VoIP Systems in this Book

In this book, CPE, VoIP adapter, and VoIP gateway names are used frequently and interchangeably. A VoIP CPE name is generic in that it accommodates several end-user boxes such as VoIP adapters, integrated access devices (IADs), residential gateways (RGs), home gateways (HGs), terminal adapters (same

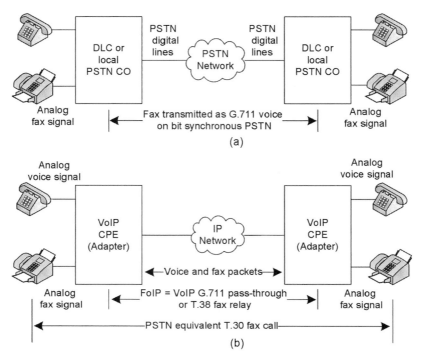

Figure 2.1. (a) PSTN voice and fax call; (b) representation of VoIP voice and fax call between two gateways.

as VoIP adapters), wireless fidelity (WiFi) IP phones, IP phones, VoIP soft phones on computers, and any new systems that are positioned directly at the end user. CPE terminals are available to the end user for directly making voice and fax calls.

Some confusion may happen while reading this book with usage of the names "VoIP gateway" and "voice gateway." In the early products, the VoIP gateway name was used for a high-channel VoIP–PSTN bridging gateway. Cisco products refer "gateway" to represent a high-channel system that usually resides at the subscriber enterprise or service provider [Donohue et al. (2006)] location. This shared naming occurs because residential gateways and home gateways are end-user CPE boxes, and more often, they are called simply "gateways." Also, several CPE users are using the VoIP gateway name to refer their CPE box. The CPE terminals with foreign exchange subscriber (FXS), FXS foreign exchange office (FXO) interfaces are referred to as gateways, and stand-alone terminals like IP phones, WiFi handsets, and soft phones are not referred to as gateways.

In this book, the VoIP gateway name is also used for CPE that is used for FXS and FXO interfaces. Wherever it is required to convey a high-channel gateway of the enterprise or service provider, it is exclusively referred to as the VoIP–PSTN gateway or as the PSTN–VoIP gateway. The name "gateway" without the prefix of "VoIP–PSTN" or "PSTN–VoIP" is a CPE in this book.

2.1.2 VoIP End-User Call Combinations

VoIP end-user call combinations vary widely based on the end-user equipment, service provider features, and the operational knowledge of the user. At the functional level, VoIP residential services are used with the following major call combinations. The combinations listed here are time sensitive and may change over time. Figure 2.2 provides functional representations that do not represent a complete deployment infrastructure.

CPE-to-CPE Calls. A CPE-to-CPE VoIP call is shown in Fig. 2.2(a). A voice call originating from one CPE can reach another CPE globally on the VoIP network, assuming the IP network is allowing end-to-end networking functions. This type of call is supported in a controlled way under the supervision of the service provider infrastructure or direct peer-to-peer VoIP calls between two CPEs. A CPE-to-CPE call combination is a true VoIP, and it provides a major benefit in the migration to VoIP.

CPE-to-PSTN Calls. A VoIP CPE to the existing PSTN and cell phone calls are shown in Fig. 2.2(b). In Fig. 2.2(b), the function of the VoIP–PSTN gateway is to interface between VoIP and PSTN. This high-channel gateway [Donohue et al. (2006)] handles calls between VoIP and PSTN. The VoIP–PSTN gateway handles several thousand simultaneous calls between VoIP and PSTN. A voice call originating from the CPE can reach an existing PSTN phone and a cell phone through the nearest VoIP–PSTN gateway. The calls are established mainly under the supervision of the service provider infrastructure. On determining the final destination of the VoIP call as a PSTN call, the gateway routes the calls to the PSTN. Calls are also established in the reverse direction from PSTN to VoIP.

PSTN-to-PSTN Calls through VoIP. PSTN-to-PSTN long-distance calls through VoIP are shown in Fig. 2.2(c). A PSTN call can use the VoIP network to route a long-distance call. At both destinations, the local PSTN network is used to complete the call. As an example, to make a call from San Jose to New York using the PSTN and VoIP network, the San Jose PSTN network can reach the gateway closest to San Jose (West coast), and the VoIP call can terminate on the VoIP gateway positioned close (from an operational point of view) to New York (East coast). In general, the VoIP–PSTN gateway establishes connectivity with the existing PSTN network. It performs transcoding and call signaling translation to work between IP and PSTN. Both PSTN clouds are of local regions.

CPE-to-Local PSTN Calls. As shown in Fig. 2.2(d), CPE can provide an additional FXO called as a lifeline connection for local PSTN connectivity.

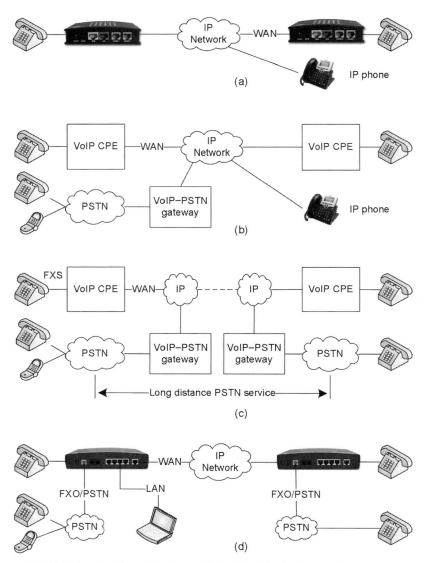

Figure 2.2. VoIP basic call combinations: (a) CPE-to-CPE VoIP call, (b) CPE-to-CPE and PSTN call through VoIP–PSTN gateway, (c) PSTN-to-PSTN call through VoIP–PSTN gateways at different regions, and (d) CPE calls to PSTN.

FXS can reach a local PSTN through the FXO directly at CPE. It facilitates additional local PSTN-specific services. CPE can also handle VoIP calls between PSTN and VoIP. The PSTN-supported CPEs with an FXO interface are also called residential gateways or home gateways.

CPE for Data Service. Data services are also supported on CPE through LAN interfaces in addition to voice. The interfaces include a wired universal serial bus (USB), Ethernet port, or wireless local area network (WLAN). The quality-of-service (QoS) implementation internal to the CPE provides the priority to voice packets to ensure low-delay, real-time operations for voice.

CPE as Mobile Device. A WiFi phone as a limited-range mobile device is in use now. Mobile convergence is now happening that allows VoIP CPE to possess mobile capabilities through various hand-over mechanisms among VoIP, the mobile network, and PSTN.

2.2   TYPICAL VoIP DEPLOYMENT EXAMPLE

A typical deployment scenario for VoIP is shown in Fig. 2.3. In VoIP deployment, various voice-interfacing CPEs, soft switches, and VoIP–PSTN gateways are associated to establish the VoIP call. The main functions of various supporting infrastructure boxes are indicated in this section. For voice and data applications, CPEs communicate to the IP network through LAN or wide area network (WAN) interfaces. The LAN and WAN interfaces include Ethernet, a digital subscriber line (DSL), a very high-speed DSL (VDSL), a WLAN, a passive optical network (PON), dial-up, cable interfaces, USB, and more.

The software running on computers and add-on cards with voice interfaces can also function like the CPE and communicate with the Internet through LAN interfaces. VoIP adapters have simple voice and LAN data functions, whereas IAD or RG will have a direct WAN interface with many data-centric LAN interfaces. The IP phone is a single-user CPE terminal used to make a VoIP call. CPE terminals are also called end points.

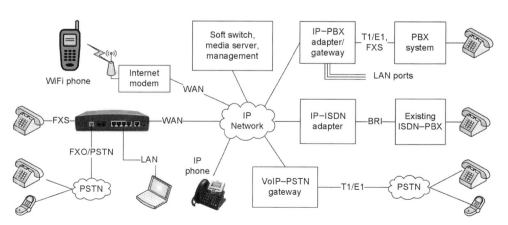

Figure 2.3. Typical VoIP deployment functions.

VoIP–PSTN gateways establish the routing of calls to the existing PSTN network. These gateways are connected to the PSTN network through T1/E1/ J1, ISDN, and FXO interfaces. Based on the signaling protocol used, the SIP server, the MGCP media gateway controller, or the H.323 gatekeeper establishes the VoIP calls. Most of the call establishing function is built into a combined unit, which is popularly known as the soft switch, that serves as the gatekeeper in many deployments [Donohue et al. (2006)] even though a gatekeeper is technically for the H.323 protocol. The soft switch is a scalable unit that controls 10,000 to 100,000 users and can accommodate many other functions, such as protocol translation (useful to make a call from one signaling CPE to another type of signaling CPE—an example SIP to H.323), billing, management of traffic, and coordination with media servers and the PSTN network. Media servers are closely associated with soft switches to support multiple voice features for conferencing calls and to provide conference bridge support. Private branch exchange (PBX) systems are associated closely with the PSTN network. Existing PBX systems can be migrated to VoIP with suitable adapters. The VoIP adapter in combination with the PBX system forms the IP–PBX adapter and is used to interface the existing PBX with the IP network. IP–ISDN is useful as an interface with the existing ISDN PBX system. The existing ISDN PBX or adapter for normal phones interfaces with the PSTN through the basic rate interface (BRI).

2.3 NETWORK AND ACOUSTIC INTERFACES FOR VoIP

Internet connectivity is the main requirement for VoIP. This connectivity is achieved through LAN or WAN interfaces. WAN interface directly goes to the Internet service provider. The popular WAN interfaces available to the residential user are DSL, cable, worldwide interoperability for microwave access (WiMAX), and direct dial-up. LAN interfaces are used with existing home or office network or with existing DSL and cable modems. Some of the popular LAN and WAN interfaces used with VoIP services are listed below:

1. Ethernet LAN interface available in residential, office, and public places
2. USB as a LAN interface
3. Bluetooth as a LAN interface
4. WLAN as a LAN interface
5. WiMAX as a direct WAN interface
6. Direct WAN interfaces used with DSL service
7. Cable modem interface used as WAN interface
8. Dial-up modem along with PC creates network connectivity as a low-bandwidth WAN

Acoustic interfaces for VoIP vary widely and are difficult to classify. VoIP service is used from multiple general-purpose computers, VoIP adapters, and residential gateways. Users will be hearing the voice through several acoustic interfaces. Some of the popularly used interfaces for user acoustics are follows:

1. Regular plain old telephone service (POTS) telephones connected to VoIP adapters
2. Fax machines connected to VoIP adapters also work for voice calls
3. IP phones connected on LAN interfaces
4. WiFi phones mainly used on a wireless LAN interface
5. USB handset phones connected to a USB interface of computers or CPE
6. Headsets/handsets to work with computers and soft phones
7. Digitally enhanced cordless telecommunications (DECT) phones used on a POTS interface or on a dedicated DECT wireless LAN interface
8. Bluetooth handset extension of base stations

PSTN interfaces of the T1/E1/J1 family are used on VoIP–PSTN gateways. Multiple FXO and ISDN BRI interfaces are used for a lower channel PSTN interface. The VoIP service provider mainly hosts PSTN connectivity; hence, the PSTN interface is optional on most CPE systems.

2.4 VoIP SYSTEMS WORKING PRINCIPLES

In this section, some high-level architectures and working principles are presented. VoIP supporting systems specific to the current context have to perform two different tasks—namely signal processing and network processing. While representing the architecture, two processors-based approach is also used in this book. A digital signal processor (DSP) is used for voice and fax processing, and a network processor is referred to as a host processor for networking functions. It is also common practice to use a single processor for the whole functionality. Processor architectures for VoIP and networking functions are presented in Chapter 19. The VoIP systems considered in this chapter are as follows:

1. VoIP adapter and extensions
2. RG or IAD
3. IP phones
4. WiFi phones
5. Personal computer (PC)-based VoIP softphones
6. VoIP-to-PSTN gateway

7. IP–PBX adapters to interface with existing PBX systems
8. Hosting VoIP through PSTN lines
9. Subscribed VoIP services

2.4.1   VoIP Adapter

VoIP adapters are used when IP service is available through another LAN interface in residential applications. The other names used for VoIP adapter are terminal adapter, CPE, and gateway. In this section, a VoIP adapter is presented as an end-user system with an accompaning example. Functional block-level representation, signal flow, and module level software blocks of the adapter are given in this section. A residential gateway can completely make use of the VoIP operation presented under the VoIP adapter, and RG additionally supports several data path applications and WAN connectivity.

User Box-Level Interfaces.   The popular network interface to the adapter is Ethernet. Some adapters are also made to work with USB, WLAN, and other LAN interfaces. The simplest adapter will have a single Ethernet marked as a WAN/LAN for the network interface. Several adapters in the market provide extra Ethernet ports for LAN extensions. When multiple interfaces are provided, one interface is usually marked as WAN and other interfaces as LAN. WAN is the main interface connected to the external Internet service provider's box, and LAN ports are used for data path applications such as computer and IP-enabled multimedia terminals. The VoIP adapter may be having RS-232 for optional configurations. In most boxes, this interface is mainly used at the development stage. In residential applications, the VoIP adapter supports one to two simultaneous phone connections, an Internet-connecting WAN, and one extra LAN interface. In small-office and home-office applications, up to four voice channels and one to four extra LAN interfaces are supported.

The VoIP adapter picture is shown in Fig. 2.4 that is used in the actual deployments. In Fig. 2.4(a), labels are visible that convey interface details. The labels appearing on the back panel are a 12-V DC power jack to provide external supply, RJ-45 connectors for WAN and LAN, Phones 1 and 2 for connecting normal POTS phones, and the reset button (RSTR) that is usually hidden but accessible on the back panel. Front-view indicators convey the operational status of the system and voice calls. More details on this product and extended VoIP products are available at http://www.innomedia.com.

Hardware Functional Description.   The VoIP adapter supports multiple TIP-RING two-wire telephone interfaces. It is the analog high-voltage FXS interface driven by the subscriber line interface circuit (SLIC) devices. The TIP-RING supplies DC voltages on the order of −48V, which is similar to the PSTN residential line interface. Voice as an analog signal from the phone is converted to a four-wire interface in the SLIC functional block, and it is

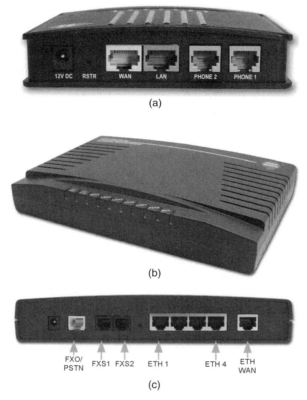

Figure 2.4. VoIP multimedia terminal adapters: (a) VoIP adapter view, (b) VoIP adapter with PSTN interface, and (c) VoIP adapter with PSTN interface panel view with connectors. [*Courtesy*: Pictures of InnoMedia's MTA 6328-2Re {Fig 2.4(a)}, and MTA 6338-2Re4S {Fig. 2.4(b), 2.4(c)} VoIP Multimedia Terminal Adapters printed with the permission from InnoMedia, Inc., USA, http://www.innomedia.com]

sampled in the hardware CODEC at 8 kHz. The hardware CODEC consists of minimum functions of analog-to-digital conversion (ADC) and digital-to-analog conversion (DAC), which is also referred to as a subscriber line access circuit (SLAC). When SLIC and CODEC are mentioned together, the SLAC name is used as SLIC–SLAC, and when SLAC is mentioned by itself, it is usually called a CODEC. These samples are interfaced to the processor through a pulse code modulation (PCM) interface. The PCM interface is a serial interface maintained at multiples of 64-kHz clock frequency as explained in Chapters 1 and 9. The hardware CODEC and SLIC communicate additional signaling and programming information through a serial peripheral interface (SPI). The SPI interface is mainly derived from the host network processor, and the PCM interface is derived from a digital signal processor. In Fig. 2.5(a), voice samples are compressed in DSP and communicated to the network

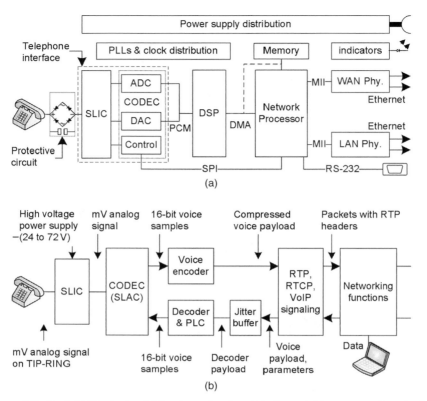

Figure 2.5. Basic VoIP adapter: (a) hardware functional blocks and (b) representation of voice flow in the VoIP adapter.

processor through established interprocessor communication (IPC) between the DSP and the network processor. The communication is through a direct memory access (DMA) operation through a parallel memory interface that gives high-speed access to the internal data memory (DM) and program memory (PM) of DSPs. The DSP may act as a slave and requires host support for booting and various configurations.

The network processor takes compressed payload from the DSP and formulates a complete packet after incorporating several packet headers of the real-time transport protocol (RTP), user datagram protocol (UDP), IP, and network interface (Ethernet, DSL, etc.) headers. End-to-end voice is delivered on the data network as IP packets through the LAN/WAN interface. Several hardware blocks will be used, such as memories, indicators, and protective devices; clock generation will occur through phase locked loops (PLLs) for processors and interfaces; and so on as indicated in Fig. 2.5(a).

As marked in Fig. 2.4(b) and (c), VoIP adapters are also made with a PSTN interface. This option gives more flexibility in the deployment to use lifeline in power faults and PSTN connectivity and to make VoIP calls bridge to PSTN. This interface is connected to the PSTN central office, and it can create similar functionality of the VoIP–PSTN gateway for one channel. More details on the products illustrated in Fig. 2.4 are available at InnoMedia Inc., USA (http://www.innomedia.com).

2.4.2   Voice Flow in the VoIP Adapter

Figure 2.5(b) illustrates voice flow in the main functional blocks of the adapter. The blocks shown in Fig. 2.5(b) are expanded in subsequent chapters of this book. Voice entering through the telephone TIP-RING interface is collected through SLIC and CODEC interfaces. SLIC will be supplying a −24- to −72-V DC battery on TIP-RING interfaces. The hardware CODEC samples the voice at 8 kHz. The analog signals at ADC analog input are in mV level as explained in Section 1.3. Voice samples are 16-bit numbers used as linear or coded in 8-bit A-law/μ-law format. These samples vary with selection of CODEC support and configuration. They are compressed in DSP and formatted as compressed payload. The host processor appends all required headers to the payload and routes them to the right destinations. Call establishing is mainly handled as signaling packets in the host processor. SLICs and CODECs report off-hook and on-hook signaling status to the host network processor. The host processor delivers voice and signaling packets on the Ethernet interface. The Ethernet interface joins with the service provider's network through a local modem, which resides at the customer premises. Modems have Ethernet on the one side and DSL or cable connectivity on the other side. In the return path, received voice packets and VoIP signaling information are analyzed and passed on for voice processing. In the receive path, voice packets may appear with network impediments. Jitter buffer in association with RTP, RTP control protocol (RTCP), regulates the packet flow to voice processing. Received voice samples are extracted from payload and passed to the decoder. These voice samples are delivered as an analog signal through a CODEC and SLIC interface.

2.4.3   Voice and Fax Software on VoIP Adapter

Basic VoIP software architecture is given in Fig. 2.6 for VoIP voice and fax support with block-level representation. Figure 2.6 represents the main functional modules. In general, several supporting modules are used that vary based on the deployment requirements. In this figure, VoIP voice and fax software is assumed to be distributed between the network processor and the DSP. In all adapters, IP network or router stack support is required to run VoIP applications. Network stack details are beyond the scope of this book. The following major modules run on the network processor for voice and fax over IP applications.

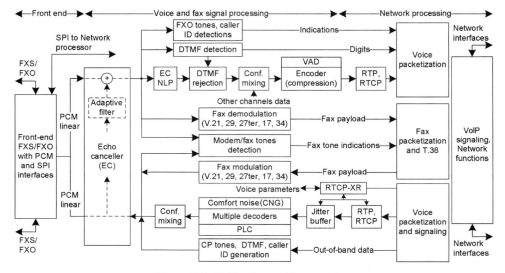

Figure 2.6. VoIP voice and fax modules.

- Networking (TCP/UDP/IP) stack with QoS
- VoIP signaling—H.323, SIP, MGCP and media gateway, and a media gateway controller (MEGACO) [ITU-T-H.248 (2005)].
- T.38 fax relay signaling
- Telephone call states and call feature functions
- Packetization, RTP, RTCP, and RTCP-XR
- Adaptive jitter buffer (AJB)
- PSTN state machine to handle FXO calls
- Communication interfaces for voice and other functions

The following voice software modules are applicable to DSP processing. These modules are explained separately in subsequent chapters of this book. In this section, a list of voice modules is provided in relation to Fig. 2.6.

- Narrowband compression codecs of G.711, G.729A, G.723.1, G.726, and iLBC, for voice samples compression [ITU-T-G.711 (1988), ITU-T-G.729 (1996), ITU-T-G.723.1 (2006), ITU-T-G.726 (1990), URL (iLBC)]
- Wideband compression codecs of G.722, G.722.2 and G.729.1, etc., [ITU-T-G.722 (1988), ITU-T-G.722.2 (2003), ITU-T-G.729.1 (2006)]
- G.168 [ITU-T-G.168 (2004)] echo canceller—tail span of 16 to 32 ms for CPE and more than 32 ms (48 to 128 ms) for VoIP–PSTN gateways
- Voice activity detection (VAD) and comfort noise generation (CNG)

- Packet loss concealment (PLC) for enhancing voice in packet drop situations [ITU-T-G.711 Appendix-I (1999)]
- Dual-tone multifrequency (DTMF) for generating tones, detecting digit tones, and rejection [ITU-T-Q.24 (1988)]
- Pulse dialing detections
- Call progress tone generation and detections
- On-hook and off-hook caller ID based on frequency shift keying (FSK) or DTMF and other call feature generation and detection
- Conferencing mixing operations
- Automatic gain control or automatic level control
- Fax and modem tone detections for discriminating voice, fax, and modem calls

The fax modulations and demodulations referred to as "data pump" are part of the functionality required for fax data transmission and reception. ITU-T proposed different standards for fax over IP such as T.37 and T.38. T.37 or T.38 resides on the network processor in most implementations. The main fax data pump modules involved in fax over IP are given below. Details on fax data pump modules and T.38 fax relay are given in Chapters 14–16.

- V.21 at 300-bps binary signaling modem [ITU-T-V.21 (1988)]
- High-level data link control (HDLC) framing and deframing [ITU-T-T.30 (2005)]
- V.27ter at 2400- and 4800-bps high-speed modems [ITU-T-V.27ter (1988)]
- V.29 at 7200- and 9600-bps high-speed modems [ITU-T-V.29 (1988)]
- V.17 at 7200-, 9600-, 12,000-, and 14,400-bps high-speed modems [ITU-T-V.17 (1991)]
- V.34 from 2400 to 33,600 bps in increments of 2400 bps for a total of 14 rates for super group-3 support [ITU-T-V.34 (1998)]
- Answering (ANS) or called terminal identification (CED), answering tone with phase and amplitude modulation (/ANSam) tone detection and generation
- Calling (CNG) tone detection and generation

2.4.4 Residential Gateway

The RG works similarly to the VoIP adapter for voice applications and supports multiple services with direct WAN connectivity. Separate DSL and cable modems are not required with RG. In the simplest form for voice applications, RG is treated as a higher end VoIP adapter with multiple network interfaces and multiple media services such as triple play (voice, video, and data). The

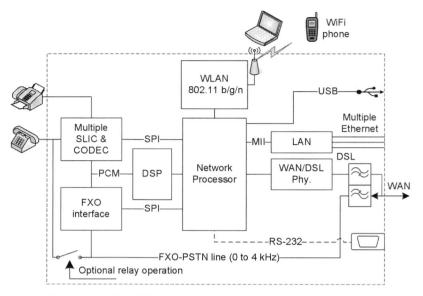

Figure 2.7. DSL WAN-based residential gateway functional blocks.

other popular names used for RG are IAD, HG, and next-generation gateway. The popular interfaces on RG are indicated in Fig. 2.7. The main RG features that relate to the VoIP adapter are given below.

- RG has a direct WAN interface such as DSL, cable, or fiber.
- The PSTN lifeline is one of the common features on RG with a DSL interface. Most RGs may use the DSL modem as part of the RG. DSL service is provided on the telephone TIP-RING interface. The same telephone interface is also used for making PSTN calls and lifeline emergency calls. The DSL and PSTN are isolated with the help of a DSL splitter that usually is part of the RG. Some RGs have a lifeline relay. In case of power faults, when RG is disconnected from the IP network, or during nonavailability of VoIP service, this relay can make a connection between the FXS and the FXO interface. It works as a lifeline even in power failure conditions. The FXO interface is used for a local PSTN call and for making PSTN-to-VoIP calls as represented in Fig. 2.2(d) call combinations. RGs may cater to secured voice and data applications.
- Additional LAN interfaces such as USB, WLAN, Bluetooth, and multiple Ethernet LAN ports are usually part of RG.
- WLAN functionality through the peripheral component interconnects (PCI) family and USB interfaces are commonly used. A WLAN card is usually plugged into a mini-PCI interface. The RG works as WLAN access point. Multiple devices such as WiFi phones, PCs, laptops, and wireless cameras can communicate with the RG using this WLAN interface.

- USB host and devices support is provided for communication with USB handset phones, IP phones, computers, audio, video appliances, and data storage devices.
- Multiple Ethernet interfaces are supported. One port may be marked as a WAN for network connectivity in non-DSL support. Several ports of LAN are provided for data distribution and for connecting multiple terminals such as IP phones, PCs, and IP-enabled multimedia terminals.
- RS-232 and joint test action group (JTAG) interfaces are used for configuration and initial debugging. In the stable products, configuration and debug requirements are handled through WAN and LAN ports. Hence, RS-232 and JTAG are not provided on the RG box.

2.4.5 Residential Gateway Example

F@st3500 family RG pictures are given in Fig. 2.8, which are popularly deployed in Europe. Figure 2.8(a) is the main RG system. Figure 2.8(b) is the rear view that shows the interfaces of multiple FXS, FXO/DSL, multiple Ethernet interfaces, USB slaves, WLAN antenna extensions, power supplies, and switches. Figure 2.8(c) is the side view that shows USB host interfaces, internal antenna pairing, and reset buttons. Some more details on residential gateway systems and applications are available at Sagem Communications, France [http://www.sagem-communications.com].

Residential gateways cater to higher processing, complex interfaces, higher data throughput, and multiple services. With several interfaces and higher performance of residential gateways, an IP converging digital home can be created in residential and small-office applications. The subscriber can use several computers, printing, game console, set-top boxes, and fixed or mobile phones simultaneously across interfaces of Ethernet, USB, USB host, WiFi, FXS, and FXO interfaces.

2.4.6 IP Phones

IP phones are used for making a direct VoIP call. Separate VoIP adapter and normal phones are not required with IP phones. IP phones are connected directly to the network interfaces. Normal analog phones send an analog signal on a two-wire TIP-RING interface. IP phones send packets directly on the IP network. On the IP phone, several additional keys are found compared with analog phone keys. These additional keys represent specific functions, and they are used for direct functionality. A processor inside the IP phone understands these digits and maps them to programmed features. In the IP phone, all messages, tones, commands, and compressed voice packets go as packets on the IP network. The physical interfaces such as Ethernet and USB carry these packets. Power to the IP phone is given either externally or through a powered Ethernet interface or USB.

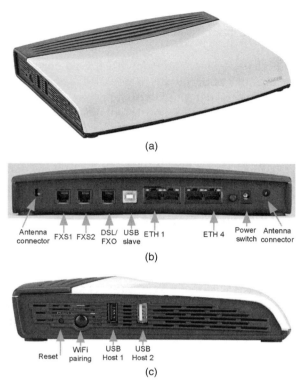

(a)

(b)

Antenna connector | FXS1 | FXS2 | DSL/ FXO | USB slave | ETH 1 | ETH 4 | Power switch | Antenna connector

(c)

Reset | WiFi pairing | USB Host 1 | USB Host 2

Figure 2.8. Residential gateway system with representation of multiple interfaces: (a) main RG system, (b) rear view with interface details, and (c) side view with interface details. [Courtesy: Sagem F@st™ 3504 Residential Gateway pictures in different views are printed with permissions from Sagem Communications, Paris, France; www.sagem-communications.com; Copyright "Le Square" agency.]

The IP phone has at least one network interface connector marked as a WAN or network port. Most phones use an Ethernet interface. WiFi and USB are also considered, but the usage is limited. There could be an extra one or two LAN connectors (usually of Ethernet) to use the phone as a switch. LAN ports are used to connect PCs and other data devices that work with the Internet. IP phones are bigger in size, which suits the desktop operation, and they require external power.

An IP phone screen is usually of four or more lines, which is helpful when viewing the network configuration menu and when editing the configuration settings with phone keys instead of connecting the phone to a computer for network configuration. IP phone chip manufacturers provide up to 7×7 keys. IP phones have several advanced features like connecting the headset directly to the phone base station, and several single-press hot keys programmed for dedicated purpose. A block-level representation of an IP phone is shown in Fig. 2.9(a). An IP phone chip and block diagram can be found at reference

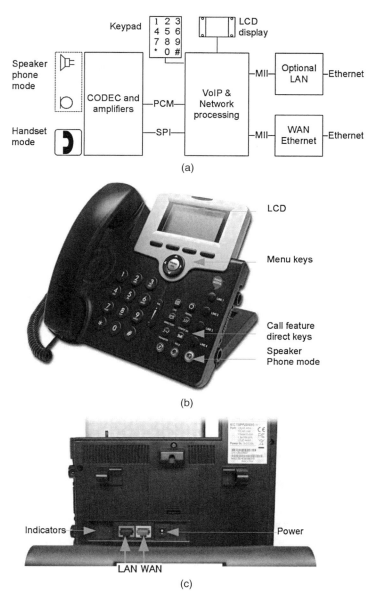

Figure 2.9. (a) IP phone functional blocks, (b) IP phone panel view, and (c) IP phone rear view. [Courtesy: Pictures of InnoMedia's IP phone model 6308M in different views are printed with permission from InnoMedia, Inc., USA, http://www.innomedia.com.]

[URL (Fijitsu-IP)]. A picture of an IP phone, interfaces, and some functions are marked in Fig. 2.9(b) and (c) for the Innomedia Model 6308M IP phone, and more details on IP phone features can be found at [http://www.innomedia. com].

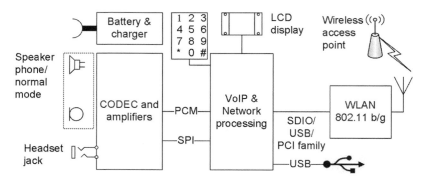

Figure 2.10. WiFi IP phone functional blocks.

2.4.7 Wireless LAN-Based IP Phone

WLAN-based IP phones are called WiFi phones or WiFi handsets. Wireless IP phones are compact in size with a look and feel similar to mobile phones, but they work more like IP phones with Internet connectivity. A functional block-level representation is given in Fig. 2.10. Wireless IP phones are connected through Wireless LAN IEEE 802.11 b/g interfaces. The family of these interfaces may change over time. A wireless router will establish Internet connectivity. The internal blocks of WiFi phones almost match with a regular IP phone. The differences are a stand-alone battery, one network interface or WLAN, and a compact size. Several IP phones and WiFi phone pictures and details are available at reference [URL (VoIP-Supply)]. The main interface for networking is WLAN, and it is interfaced to the main network processor through secured digital input out (SDIO), SPI, or USB. PCI family-based WLAN interfaces are most popularly used with VoIP residential gateways and IP phones. In IP and WLAN phones, the network processor alone will be performing total processing, including voice encoding and decoding.

2.4.8 VoIP Soft Phones on PC

A PC can also work as an IP phone. A IP phone on a PC will be a software application. Hence, PC-based IP phones are called soft phones. The hardware and physical interfaces of PC are reused for VoIP calls. PCs are also used for many other applications while performing a VoIP soft phone operation. A networked PC with an audio headset or with an internal microphone and speakers connected to the IP network is required for VoIP calls. Many PC-based VoIP applications are developed to emulate IP phones exactly. A computer keyboard and mouse are used to operate the virtual phone application running on a PC. While writing this book, Skype [URL (Skype)] and SJPhone [URL (SJ Labs)] were some popularly supported VoIP soft phones through a PC. For the PC, many third parties developed audio interfaces that are used to provide better comfort with dialing and acoustics. The popular interfaces that are headsets, USB-based phones that plug into a PC USB port, a Bluetooth-based handset, and WiFi extended handsets. Skype-supported IP phones are also available, which eliminates the requirement of a PC. The VoIP call operation falls under the IP phone category. Some main advantages of PC-based VoIP are the mobility and wideband audio support. The main disadvantage of a soft phone is that PCs have to be in ready state at both ends to send and receive calls. Applications on a PC are extended to video conferences. A PC screen works like a TV. A mini camera usually on a USB interface is used with a PC. PC-based soft phone and video phones on VoIP are widely used by travelers. Some more aspects of the soft phone are given in Chapter 19.

2.4.9 VoIP-to-PSTN Gateway

The function of the VoIP–PSTN gateway is to make voice and fax calls between the PSTN and the VoIP. These gateways support up to several thousand calls. The enterprises and VoIP service providers use these gateways. The VoIP–PSTN gateway is popularly known as the VoIP gateway, as the Trunking gateway, and sometimes as the PSTN gateway. In the situation of calls between VoIP and PSTN, a call that originates from the VoIP adapter reach the distant VoIP gateway (VoIP–PSTN gateway) and will be forwarded to the PSTN network. PSTN will use either µ-law or A-law. IP traffic will be of G.711, G.729AB, or G.723.1A. DSPs inside the VoIP–PSTN gateway will be working on transcoding the voice between VoIP and PSTN. The host processor will be working on translating the PSTN to IP and IP to PSTN signaling and on call management. Compared with RG, PSTN–VoIP gateways are simplified on networking functions. To cater to a larger number of voice and fax channels processing, more DSPs are used with a network processor. As shown in Fig. 2.11, these gateways are made with combinations of fixed and optional plug-ins with optional FXS, FXO, ISDN, and T1/E1 family interfaces. Gateways usually come with one WAN or two WAN and LAN Ethernet interfaces. In deployments, an Ethernet interface will join with another established high bandwidth

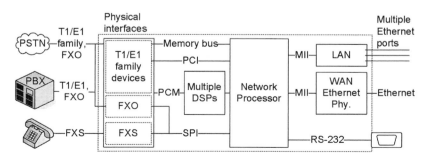

Figure 2.11. VoIP-to-PSTN gateway or PBX adapter.

connection. Several gateway pictures can be found at [URL (VoIP-Supply), Donohue et al. (2006)].

2.4.10   IP PBX Adapter

PSTN PBX systems are used in many offices. These PBX systems are connected to the telephone office or central office of the PSTN service provider through various telephone interfaces. The usual interfaces are T1, E1, BRI, PRI, and multiple FXOs. PBX adapter and VoIP–PSTN gateway will be having a similar architecture as illustrated in Fig. 2.11. The PBX adapter system has to cater to a less number of channels than the PSTN–VoIP gateway. FXS and FXO interface functions are explained in the previous sections and in Chapter 1. T1/E1 is a four-wire interface. The T1/E1 physical layer device delivers its output on a TDM interface. The TDM interface is joined to the PCM interface of the DSP processors. DSP retains the voice bytes, and signaling bytes are forwarded to the network processor. DSP can also perform time slot allocation or interchange (TSA/TSI) functionality for local switching of voice samples. Signaling information is also communicated from framer to processor through PCI or memory bus. The network processor manages the PSTN signaling. The T1/E1 framer and physical layer devices are programmed through either memory bus, PCI, or SPI. For more channels of voice processing, additional DSP processors are used along with the network processor.

2.4.11   Hosting Long-Distance VoIP through PSTN

Many international calling card services use hosted VoIP services. A user makes the VoIP calls through PSTN telephone lines. This type of PSTN to VoIP to PSTN calls representation is given in Fig. 2.2(c). In this service, a user will be dialing a local number or toll free number of the hosted service. This number is also for the VoIP service providing hosted VoIP. On receiving the call through the PSTN, the gateway analyzes the caller ID as a password or prompts with an interactive voice response (IVR) for entering a password. To continue the call, IVRs will continue prompting for destination number and other options. With country-specific dialing, a call is routed on the VoIP. The VoIP gateway takes the PSTN and routes a call to its destination on the IP. At the destination, a call is directly routed to the closest possible VoIP gateway. A VoIP-to-PSTN operation at the destination routes the local calls on the PSTN. Billing, accounting, and other management services monitor the call. Overall, users will be using a local PSTN, and a service provider will use VoIP to complete inter-regional calls, as represented in Fig. 2.2(c).

2.4.12   Subscribed VoIP Services

To use this service, a user is given a CPE, usually a VoIP adapter, IP phone, or higher end residential gateway. On connecting the VoIP adapter to the IP network through DSL, cable modem, or any established network, supporting infrastructure allows the VoIP adapter to be registered based on the configuration of the VoIP adapter box. The examples in Fig. 2.2(a) to 2.2(d) closely match for the following operation modes.

For VoIP CPE to another VoIP CPE, voice calls are established directly in VoIP. A service provider will not incur any extra cost to the PSTN service. For VoIP to local PSTN, a call is established with the nearest VoIP–PSTN gateway. The call is completed on VoIP and PSTN even though it is a local call. For a VoIP-to-long-distance operation, a voice call is established from one part of the country to another. If the call is required to reach another VoIP adapter of the same service provider, the call is routed as VoIP without any restrictions on the distance. If the termination is on PSTN, then it reaches through the closest local or inter-regional PSTN–VoIP gateway. Several deployment aspects on PSTN–VoIP gateways and call combinations are given in reference [Donohue et al. (2006)].

2.5   VoIP SIGNALING

VoIP signaling is essential to set up, tear down, and manage VoIP voice and fax calls, as well as various call features. Once a call is established, RTP/UDP packets of voice and fax are transmitted for the media. In this section, an overview on H.323, SIP, and MGCP signaling is presented. While writing this book, SIP signaling was popularly adapted by several deployments. SIP signaling is given in more detail to provide an immediate introductory-level explanation. Refer to [ITU-T-H.323 (2006), ITU-T-H.248 (2005), Andreasen and Foster (2003), Rosenberg et al. (2002), Sparks (2003)] for more details on VoIP signaling. SIP end points are called user agents (UAs), and they can work in the point-to-point mode without a SIP proxy. In a service provider environment, VoIP calls are made through SIP servers. MGCP end points are called media gateways (MGs). MGCP calls are made through the media gateway controller (MGC). MGCP customer premises equipment always requires MGC support to establish VoIP calls. H.323 end points are called terminals, and they can work in the point-to-point mode. In a service provider network, H.323 gatekeepers are used to establish VoIP calls.

2.5.1   VoIP–H.323 Overview

The ITU-T-H.323 standard (2006) was used in early packet media applications for the transmission of real-time audio/voice and video and data communications over packet-based networks. It specifies the components, protocols, and procedures providing multimedia communication over packet-based networks. The H.323 standard specifies four kinds of components to provide the point-to-point and point-to-multipoint multimedia communication services. These components are terminals, gateways, gatekeepers, and multipoint control units

(MCUs). H.323 can be applied in a variety of mechanisms of audio only (IP telephony), audio and video (video telephony), audio and data, as well as video and data. H.225 is a key protocol in the H.323 defined by the ITU-T [ITU-T-H.225.0 (2005)], and H.225 call control signaling is used to setup connections between two H.323 terminals. The registration admission and status (RAS) is used to perform registration, admission control, bandwidth changes, and status between end points and gatekeepers. H.245 is a control channel protocol [ITU-T-H.245 (2005)] used with H.323 sessions and capable of conveying information needed for multimedia communication as well as the opening and closing of logical channels used to carry media streams.

2.5.2 VoIP–MGCP Overview

MGCP is among the first few VoIP signaling protocols used in deployment. MGCP is defined in RFC3435 [Foster and Andreasen (2003)]. MGCP assumes a call control architecture where the call control "intelligence" is outside the gateways and handled by external call control elements called as MGCs or call agents (CAs). MGCP assumes that these call control elements known as CAs contain the intelligence to send commands to the gateways (end points) under their control to establish a call. MGCP is a master/slave protocol. The gateways are expected to execute commands sent by the call agents.

2.5.3 SIP Signaling

The SIP is a peer-to-peer protocol. SIP follows the client–server architecture. SIP is a text-based protocol. It can use UDP as well as the transmission control protocol (TCP) as the transport protocol. In SIP, user agents UAs and network servers will be working in coordination. Proxy, registrar, and redirect are the three types of network servers used in SIP deployment.

User Agent. The peers in a session are called UAs. The SIP user agent acts as both server and client. The UA that initiates a request is called a client or user agent client (UAC). The agent to which the request is destined and that which returns a response on behalf of a user is called a server or user agent server (UAS). The response accepts, rejects, or redirects the request. Two SIP user agents can communicate directly in a simple SIP-based VoIP voice call.

SIP Registrar. A SIP registrar is an entity where SIP users can be registered. A registrar imparts mobility to the SIP users. A SIP user can register with a registrar. If the user changes the location, the user agent has to register again with the registrar stating the latest contact information. Whenever a call has to be delivered to the user, the registrar can provide the information about the location where the user was active recently. A registrar is typically coexists with a proxy or redirect server and may offer location services.

Proxy Server. A SIP proxy receives a request, makes a determination about the next server to send it to, and forwards the request, possibly after modifying some header fields. As such, SIP requests can traverse through many servers on their way from UAC to UAS. Responses to a request travel along the same set of server but in reverse order.

Redirect Server. The redirect server does not forward requests to the next server. Instead, it sends a redirect response back to the client containing the address of the next server to contact. A redirect server is a server that accepts a SIP request, maps the address into zero or more new addresses, and returns these addresses to the client. Unlike a proxy server, it does not initiate its own SIP request.

2.5.4   SIP Call Flow

In this section, the call flow for an end-to-end VoIP voice communication using SIP signaling is described. SIP is an application-layer control protocol for creating, modifying, and terminating sessions with one or more participants. These sessions include Internet telephone calls, multimedia distribution, and multimedia conferences as given in RFC3261 [Rosenberg et al. (2002)]. SIP can invite parties to both unicast and multicast sessions. SIP can be used to initiate sessions as well as to invite members to sessions that have been advertised and established by other means. A SIP working group, within the Internet Engineering Task Force (IETF) developed SIP protocol RFC3261.

SIP Messages. Two kinds of SIP messages exist, requests and responses. User agent clients issue requests, and User agent servers answer with responses. These requests and responses include different headers to describe the details of the communication. To make SIP signaling more secure, encryption and authorization are used. Encryption prevents packet sniffers and other eavesdroppers from seeing who is calling whom. Authorization is used to prevent an active attacker from modifying and replaying SIP requests and responses.

The SIP messages are formatted according to RFC822 [URL (RFC822)]. All SIP messages are composed of three parts, namely a start line, one or more header fields, and an optional message body. Each line must end with a carriage return-line feed (CRLF).

Start Line. All SIP messages begin with a start line. The start line conveys the message type and protocol version. The start line can be either a request line (requests) or a status line (responses). The request line includes a request-URL, which indicates the user or service to which this request is being addressed. The status line holds the numeric status code and its associated textual phrase.

Requests. The request is characterized by the start line, called a request line. Requests start with a method token followed by a request-URI and the protocol version. Requests are referred to as methods. Some methods and their purposes are given below. Request-URI is a SIP URL or a general Uniform Resource Identifier (URI); this is the user or service to which this request is being addressed. The SIP version should be 2.0.

Request-Line = Method SP Request-URL SP SIP-Version CRLF

SP = Space, CRLF = Carriage Return Line Feed

INVITE—Initiates a call and changes call parameters. The INVITE method indicates that the user or service is being invited to participate in a session. For a two-party call, the caller indicates the type of media it can receive as well as their parameters such as a network destination.

ACK—ACK confirms a final response for INVITE. It may contain a message body with the final session description to be used by the callee. If the message body is empty, the callee uses the session description in the INVITE request. This method is only used with the INVITE request.

BYE—The BYE request terminates a call.

CANCEL—The CANCEL request cancels a pending request, but it does not terminate a call that has already been accepted. A request is considered completed if the server has returned a final response.

OPTIONS—Queries the capabilities of servers.

REGISTER—Registers the address listed in the To header field with a SIP server.

INFO—Sends mid-session information that does not modify the session state.

NOTIFY—Return current state information.

SUBSCRIBE—Used to request current state and state updates from other end points.

Responses. After receiving and interpreting a request message, the recipient responds with a SIP response message, which indicates the status of the server as success or failure. The response starts with a SIP version number followed by a response code (or status code) and a textual description of the status code. The responses can be of different kinds, and the type of response is identified by a status code, a three-digit integer. The first digit defines the class of the response. The other two digits have no categorization role. Two types of responses and six classes exist.

Status-Line = SIP-Version   SP Status-Code SP   Reason-Phase CRLF

Response Types. The server uses provisional (1xx class) responses to indicate progress, but they do not terminate a SIP transactions. Final (2xx, 3xx, 4xx, 5xx, 6xx classes) responses terminate a SIP request. A brief description and example responses are given below.

1. A 1xx provisional or informational request is for continuing to process the request. Example: 100 Trying and 180 Ringing.
2. 2xx is for indicating the success of an action as the action is successfully received, understood, and accepted. Example: 200 OK.
3. 3xx is redirection for additional action needs to be taken to complete the request. Example: 300 is multiple choice, and 301 is moved permanently.
4. 4xx is client error. The request contains bad syntax or cannot be fulfilled at this server. Example: 400 is a bad request, and 408 is a request time out.
5. 5xx is a server error that conveys that a server failed to fulfill an apparently valid request. Example: 500 is a server error.
6. 6xx is a global failure to convey that the request cannot be fulfilled at any server. Example: 600 is busy everywhere, 603 is declined, and 604 does not exist anywhere.

Header Fields. SIP messages use header fields to specify caller, callee, the path of the message, type, and length of message body. Some header fields are present in all messages, and the rest is used when appropriate. A SIP application does not need to understand all these headers, although it is desirable. The entity receiving headers may ignore headers that it does not understand. The order in which the headers appear is generally of no importance, except for the Via field and that hop-by-hop headers appear before end-to-end headers. The headers are similar in syntax and semantics to Hypertext Transfer Protocol (HTTP) header fields formatted as ⟨name⟩ : ⟨value⟩. Some header fields are as follows:

Call-ID—uniquely identifies a particular invitation or all registrations of a client

Contact—contains a location that is used different purposes depending on the message

Content length—indicates the message body length in bytes

Content type—indicates the media type of the message body

Command sequence (Cseq)—uniquely identifies a request within a Call-ID

From—indicates the initiator of the request

To—specifies the recipient of the request

Via—indicates the path taken by the request so far

Message Body. A message body is used to describe the session to be initiated, which includes audio codec types, telephone event types, fax parameters, a packetization period, and sampling rates in multimedia sessions. Message bodies can appear in both request and response messages. SIP makes a clear distinction between signaling information, which is conveyed in the SIP start line and headers, and session description information, which is outside the scope of SIP. The message body description of protocols is given as the Session Description Protocol (SDP) [Handley and Jacobson (1998)], which is used mostly in SIP signaling and multipurpose Internet mail extensions (MIMEs).

SIP Call Steps. SIP call steps between two user agents are given in this section.

Registration. On boot-up, VoIP adapter-A and adapter-B register with the registrar server with their SIP URIs and Contact addresses, as shown in Fig. 2.12. Both end points communicate with each other using a SIP URI similar to an e-mail address containing a username and a host name, such as sip:1111@ yahoo.com. The default SIP Port used along with the Host IP address is 5060 for TCP, and UDP. The FXS channels connected to both adapters register their user names with the registrar. If registration processing is successful, the reg-

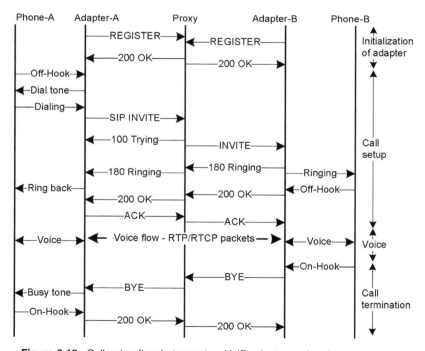

Figure 2.12. Call setup flow between two VoIP adapters using the proxy server.

istrar stores their user names and contact addresses for additional communication. The registrar sends 200 OK responses to adapter-A and adapter-B for successful registration. In the case of a failure, it issues a appropriate error code as a response to REGISTER request. The registration expiration time is proposed by adapter-A and adapter-B in the "Expires" header field or Contact header field. Proxy in response to the REGISTER request indicates the accepted value.

Call Establishment. Figure 2.12 shows a simple functional call setup between two VoIP adapters with the assistance of an intermediate proxy server. All three types of network servers such as proxy, registrar, and redirect are built into a single server in SIP deployment and are called SIP servers or SIP proxy. A list of a few SIP proxy servers that can be used in testing are given at references [URL (Radvision) URL (Brekeke)]:

- Phone-A connected to adapter-A goes off-hook, and adapter-A sends a dial tone to the phone-A.
- After dialing all the digits of the destination number of adapter-B, adapter-A composes the INVITE request with SDP information in its message body to the proxy with the "To" header field containing the address of adapter-B.
- The proxy server immediately responds with "100 Trying" provisional responses to adapter-A, which indicates that the message is received but not processed.
- Proxy searches the registration database in the location server to find the contact address corresponding to VoIP adapter-B. If it is not found, the error "404 Not Found" is returned to the VoIP adapter-A.
- If the contact address is found, then the proxy server decides to proxy the call and creates a new INVITE transaction based on the original INVITE request received from Adapter-A and forwards the INVITE request to the adapter-B.
- The moment the adapter-B receives the INVITE, it looks for the availability of the resources. If the resources are available and the VoIP adapter-B is ready to receive an incoming call, adapter-B responds with a "180 Ringing" response to the proxy. Adapter-B generates the ring to the phone connected at adapter-B. Otherwise, it returns an error with an appropriate error code (i.e., 486 "Busy Here").
- The proxy server forwards the 180 Ringing response back to the adapter-A. The response contains same To, From, Cseq, Call ID, and branch parameter in Via as in initial INVITE. In general, before forwarding any request, the proxy adds its own address in the Via Header. Changing characteristics of the active session can be done by sending re-INVITE. The message re-INVITE can be used to hold the call or for voice codec switching.

- The moment the adapter-A receives a 180 Ringing Response; it generates the ring back tone to the phone connected at adapter-A.
- When the phone at adapter-B goes off-hook, adapter-B sends 200 OK Response with SDP information that contains negotiated codecs, packetization periods, and RTP/RTCP transport addresses.
- The proxy forwards the 200 OK response to calling adapter-A. Additional SIP communications (ACK) may be sent directly to the peer adapter or via the proxy server depending on the record route based on the contact header information found in the 200 OK responses.
- Adapter-A generates ACK to the 200 OK response and is sent to adapter-B via the proxy server.
- Media flow starts between adapter-A and adapter-B. Media flow is routed depending on the behavior of the proxy while forwarding the SDP information.

Each proxy can independently decide to receive subsequent messages after a SIP Transaction (initial INVITE, 180 Ringing, and 200 OK), and those messages are passed through all proxies that elect to receive it [Rosenberg et al. (2002)]. In such a case, a proxy would add the Record-Route header field to the INVITE request before forwarding it to another proxy/end user. As shown in Fig. 2.12, all messages are exchanged through the proxy server assuming the proxy has added the Record-Route header field to the INVITE request.

Call Termination. Figure 2.12 shows call termination flow between two VoIP adapters with the assistance of an intermediate proxy server. The session termination proceeds when phone at adapter-A or adapter-B goes on-hook. When the phone at adapter-B goes on-hook, adapter-B sends a BYE request to adapter-A via proxy. After receiving the BYE request from the peer gateway, adapter-A generates the busy tone on phone-A. When phone-A goes on-hook, adapter-A sends 200 OK in response to the BYE request to adapter-B and no ACK is sent in response to 200 OK message in call termination.

3

VOICE COMPRESSION

Voice compression is essential for interactive voice communication systems of VoIP, mobile, public switched telephone network (PSTN), and satellite. Voice compression reduces the network bit rate or bandwidth on the communication channel. VoIP makes use of several compression codecs for minimizing the Internet bandwidth requirements. PSTN also makes use of voice compression except in small local and regional systems with analog switching. At a digital interface for PSTN, every 13- or 14-bit sample at 8-kHz sampling is converted to logarithmic 8-bit compression as per G.711 [ITU-T-G.711 (1988)] making it 64 kbps on a digital circuit switched network. G.711 is also used in VoIP. Voice with VoIP evolved to consider compression as one of the main parameters, and IP networks were considered band limited, at least in the early stage of VoIP considerations. Compression codecs such as G.729AB and G.723.1 are considered in VoIP to reduce network bandwidth. Wideband G.722 codecs are used to improve voice quality and to create better perception than PSTN. PSTN also makes use of higher compression of 16-, 24-, 32-, and 40-kbps adaptive differential pulse code modulation (ADPCM), which are usually referenced with an ITU codec name as G.726 [ITU-T-G.726 (1990)] to send more voice channels on the same digital network. In codec selection, the main parameters [Nikhil (2000)] considered are bit rate, quality, delays, and complexity (processing and memory) requirements. In this chapter, an overview on codecs, bit rate, quality aspects, complexity, and recommendations on choice of codecs for VoIP are presented.

VoIP Voice and Fax Signal Processing, by Sivannarayana Nagireddi
Copyright © 2008 by John Wiley & Sons, Inc.

3.1   COMPRESSION Codecs

Voice compression is broadly classified as waveform, vocoders, and hybrid codecs [Kondoz (1999)]. Waveform-based codecs of G.711, and G.726 encode–decode voice on an actual signal without making an assumption on speech models. They work on any input that is supported by sampling without any significant distortions.

A vocoder is a voice coder that makes use of a vocal tract voice production model. Vocal tract-based compression achieves a better compression ratio than waveform-based codecs. Vocoder-based codecs generate a set of parameters that represent the speech production models. On the receive side, voice is synthetically reproduced based on the parameters. These codecs take very low bandwidth (very high compression) and deliver lower quality. They are of a moderate complexity for processing. Linear predictive coding-10 (LPC-10) codec uses 2.4 and 4.8 kbps [Kondoz (1999)], which is much lower than waveform and hybrid codecs. The main benefit of vocoders is a lower bit rate.

Hybrid codecs achieve acceptable compression and quality. Compared with vocoders, hybrid codecs deliver better quality, and a wideband version of hybrid codecs can exceed waveform-based codec quality. Several extended techniques in hybrid codecs are beyond the techniques used in vocoders and waveform-based codecs. Hybrid codecs analyze the signal in frequency and time domain. Frequency-domain compressions are known as sub-band or transform-domain coding. Time-domain compression makes use of short-term redundancies through linear prediction, a long-term excitation rate through pitch analysis, and residual signals. Linear prediction is used in vocoders without several enhancements appearing in hybrid codecs. The waveform codecs of G.711 and G.726 and the hybrid codecs of the G.729 family are the most popular for VoIP applications.

The classification given here does not prevent any reuse of the benefits from one class of codecs to another. Waveform-based G.711 makes use of LPC for voice activity detection (VAD)/comfort noise generation (CNG) algorithms (discussed in Chapter 4), and pitch and linear prediction in packet loss concealment (PLC) operations (discussed in Chapter 5). Wideband codec G.722 is typically considered under waveform coding, and it makes use of ADPCM, but G.722 PLC extensions of the codec [ITU-T-G.722 Appendix III (2006)] make use of extended techniques of hybrid codecs. As a conclusive note, waveform and hybrid codecs are the most popular, and vocoders are not used in VoIP. In the naming conversion, hybrid codecs are known by name as low bit rate codecs, even though "low bit rate" does not convey any separate classification.

3.2   G.711 COMPRESSION

G.711 is an ITU-T-G.711 Recommendation (1988) pulse code modulation (PCM) logarithmic compression. The G.711 codec is an example of waveform-based compressing [URL (Cisco-coding)], which means that the decoder reconstructs the actual signal without making an assumption on any speech models. G.711 is supported in PSTN, hardware interfaces for voice samples, and VoIP deployments. PCMU is PCM using μ-law, which is popularly used in North America. PCMA is PCM compression using the A-law scheme used in Europe and some countries of Asia. PCMA and PCMU together are called as G.711-based compression scheme. G.711 takes a smaller code size of the order of 100 lines, and processing is insignificant compared with other voice processing operations. The basic difference between PCMU and PCMA [URL (PCM), URL (TI-PCM)] schemes are actual quantization, coding steps, dynamic range, and bit formats. Some more details on G.711 and power-level calculations are given in the subsequent sections.

3.2.1 μ-Law Compression of Analog Signal

In analog representation [Bellamy (1991), URL (Cisco-coding), URL (TI-PCM)] or for a continuous signal of μ-law compression, maximum input amplitude is mapped to normalized logarithmic output of ±1. For a given input x, the equation for μ-law encoding (compression) output y is given as follows:

$$y = F(x) = \text{sgn}(x)\frac{\ln(1+\mu|x|)}{\ln(1+\mu)} \quad -1 \le x \le 1$$

In the equation, $\mu = 255$ is the compression parameter, $\text{sgn}(x)$ is the sign of x, sgn (positive number) = 1, and sgn (negative number) = −1. Symbol "ln" is the natural logarithm; i.e., logarithm to the base $e = 2.71828$. A natural logarithm is used in applications of finding unknown variables, Taylor series expansion, differentiation, and quantization. Logarithm to the base 10 is popularly used in power calculations. A μ-law decoder (decompression) performs the reverse process of converting the compressed signal y back to the expanded analog signal $\hat{x}$. Expansion for the decoder input y is defined by the inverse equation [Bellamy (1991), URL (TI-PCM), URL (Cisco-coding)].

$$\hat{x} = F^{-1}(y) = \text{sgn}(y)\frac{\left|(1+\mu)^{|y|} - 1\right|}{\mu} \quad -1 \le y \le 1$$

Analog signals are usually represented as $x(t)$, $y(t)$, and $\hat{x}(t)$, with t as time. In the above equations, signals are denoted as x, y, and $\hat{x}$ for simplification in representation. In digital implementation, the linear input sample is converted to an 8-bit digital number after compression. Certain deviations of G.711 occur from continuous representation. Continuous representation gradually modifies step sizes, and G.711 follows finite segmented discrete steps.

3.2.2 PCMU for Digitized Signals

In G.711-based implementation, amplitude is split into eight segments for each polarity signal also called chords, and each segment has fixed quantization.

PCMU digital implementation takes a maximum linear sample sine wave amplitude of ±8159 equal to 3.17 dBm in power representation. The front-end analog circuit and ADCs are calibrated to take a 3.17-dBm sine wave and give ±8159 amplitude in digital number representation. The numbers 8159 and above are clipped to 8158 in the process of quantization. In the process of compression, the input range is broken into segments, and each segment will use different intervals. Most segments contain 16 intervals, and the interval size doubles from segment to segment as shown in Table 3.1. Large signals use a big quantization step and small signals use a small quantization step. This type of nonuniform quantization gives a signal-to-noise ratio (SNR) of 38 to 40 dB for most of the useful signal amplitude range and limits the significant signal dynamic range to 48 dB [Bellamy (1991)]. For full-scale amplitude of 3.17 dBm, signal to quantization is shown as 39.3 dB in reference [Bellamy (1991)].

A PCM 14-bit (±8159 amplitude) input is split into eight amplitude segments represented with 3 bits. Each segment is quantized uniformly into 16 levels using 4 bits. The polarity of the input is represented using 1 bit. A sign bit is represented as "1" for positive numbers and as "0" for negative numbers. In a two's complement number format, sign bit-1 is used for negative numbers. After framing 8 bits, the last 7 bits except the sign bit are inverted in PCMU.

This bit inversion increases the one's bit density in transmission systems that help timing and clock recovery circuits in the receiver. Ideal channel noise makes bits toggle between 01111111 (0x7F) and 11111111 (0xFF), which allows clock recovery to be better even under ideal channel conditions. Bit inversion is achieved by simple exclusive OR (XOR) operation of the encoded output with value 0x7F. This inversion is applied on both positive and negative values of input. At the decoder, the same operation of 7-bit inversion can retrieve the original decoded compressed byte. In Table 3.1, inverted encoder output is listed.

Interpretation of PCMU Coding.   Table 3.1 lists PCMU encoded and decoded output values for positive and negative numbers. In the table, positive values are presented first for all segments, and negative values are given for a few segments to limit the table size. Higher amplitude is split into 16 equal intervals of 256 size and is categorized as segment number 8[1] but coded in 3 bits as "111," which will occupy half (4063 to 8158) of the total amplitude. At the next level, 16 intervals of 128 are used, categorized as segment 7, and coded as "110," which occupies one quarter (2015 to 4062) of the total range. The interval keeps decreasing by a factor of two, causing a small amplitude to be quantized with small quantization steps. The last column of Table 3.1 is the decoder

---

[1]PCMU has eight segments numbered from 1 to 8. The segment number and interval size of that segment are related as Interval size = $2^{(\text{segment number})}$. In PCMU, the maximum interval size is 256, and the corresponding segment number is 8. However, encoded output has only 3 bits for representing the segment number, which means only numbers from 0 to 7 can be used for segment coding. Therefore, in PCMU, maximum and minimum segment numbers are coded as 111(7) and 000(0), respectively.

Table 3.1. PCMU Quantization Example for Positive Inputs and a Few Segments of Negative Inputs

Input Range	Segment in Bits	Interval Size	Step Number	Encoded Output After Bit Inversion	Decoder Output for Input Range
0	000	1	0	01111111(0x7F) 11111111(0xFF)	0
1–2	000	2	1	11111110(0xFE)	2
—	—	—	—	—	—
29–30	000	2	15	11110000(0xF0)	30
31–34	001	4	0	11101111(0xEF)	33
—	—	—	—	—	—
91–94	001	4	15	11100000(0xE0)	93
95–102	010	8	0	11011111(0xDF)	99
—	—	—	—	—	—
215–222	010	8	15	11010000(0xD0)	219
223–238	011	16	0	11001111(0xCF)	231
—	—	—	—	—	—
463–478	011	16	15	11000000(0xC0)	471
479–510	100	32	0	10111111(0xBF)	495
—	—	—	—	—	—
959–990	100	32	15	10110000(0xB0)	975
991–1054	101	64	0	10101111(0xAF)	1023
—	—	—	—	—	—
1951–2014	101	64	15	10100000(0xA0)	1983
2015–2142	110	128	0	10011111(0x9F)	2079
—	—	—	—	—	—
3935–4062	110	128	15	10010000(0x90)	3999
4063–4318	111	256	0	10001111(0x8F)	4191
—	—	—	—	—	—
7903–8158	111	256	15	10000000(0x80)	8031

For negative numbers, segments 0, 1, 6, and 7 are given to represent the beginning and the end of the range

−1 to −2	000	2	1	01111110(0x7E)	−2
—	—	—	—	—	—
−29 to −30	000	2	15	01110000(0x70)	−30
−31 to −34	001	4	0	01101111(0x6F)	−33
—	—	—	—	—	—
−91 to −94	001	4	15	01100000(0x60)	−93
—	—	—	—	—	—
−2015 to −2142	110	128	0	00011111(0x1F)	−2079
—	—	—	—	—	—
−3935 to −4062	110	128	15	00010000(0x10)	−3999
−4063 to −4318	111	256	0	00001111(0x0F)	−4191
—	—	—	—	—	—
−7903 to −8158	111	256	15	00000000(0x00)	−8031

output. For an input range of 7903 to 8158, output is one single value 8031. The value 8031 may be interpreted as the rounded mean value of the input range. To represent negative values, few selected ranges in the beginning and at the end are given.

3.2.3   PCMU Quantization Effects

Figure 3.1 has three graphs in three rows. The top graph is the encoder input and output relation for input values between 16 and 8158. The PCMU encoder output is given to the decoder, and the resulting absolute error is given in the middle graph. The middle graph shows a growing error with an increase in amplitude toward 8158. This error is happening because of the big step size at

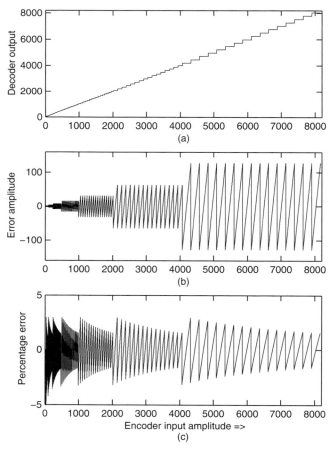

Figure 3.1. PCMU quantization. (a) input and output relation indicating bigger steps at high amplitude. (b) error between input and corresponding decoder output — more error at high amplitude. (c) percentage of error with input — error is relatively steady for a wide range of input signal levels.

higher amplitudes. At lower amplitude, error is very low. In the middle graph, it can be observed that half of the amplitudes between 4063 and 8158 are occupied in 16 intervals, and it is continuing to reduce by a factor of 2 for lower input amplitudes. The bottom graph is the percentage of error or relative error with reference to the original input. As observed, error is contained to approximately ±3%. The percentage error is remaining steady with growing amplitude. It shows that relative distortion or signal to quantization or SNR caused by logarithmic compression will remain almost steady for a wide range of input amplitudes. At lower amplitudes from 0 to 16, the percentage of error is more and it increases toward an amplitude of "0." In Table 3.1, for input–output relations of input "1," the cascaded decoder output is 2. Error here is "1," and percentage of error is 100%. In the bottom graph, it can be observed that percentage of error is growing toward an input amplitude of 16, which makes SNR caused by G.711 compression lower at lower amplitudes.

3.2.4 A-Law Compression for Analog Signals

In analog representation [ITU-T-G.711 (1988)] or a continuous signal of A-law compression, maximum input is mapped to a normalized logarithmic output of ±1. For a given input x, the equation for A-law encoding is defined by

$$y = F(x) = \text{sgn}(x)\frac{1 + \ln(A|x|)}{1 + \ln(A)}, \quad \frac{1}{A} \leq |x| \leq 1$$

$$= \frac{A|x|}{1 + \ln(A)}, \quad 0 \leq |x| \leq \frac{1}{A}$$

In the equations, x is the input to be compressed and $A = 87.6$ is the compression parameter. The A-law decoder performs a reverse process of expanding compressed samples to linear samples. A-law expansion is defined by a continuous inverse equation [Bellamy (1991), URL (TI-PCM)].

$$\hat{x} = F^{-1}(y) = \text{sgn}(y)\frac{|y|[1 + \ln(A)]}{A}, \quad 0 \leq |y| \leq \frac{1}{1 + \ln(A)}$$

$$= \text{sgn}(y)\frac{e^{(|y|[1+\ln(A)]-1)}}{A}\frac{1}{1 + \ln(A)}, \quad \frac{1}{1+\ln(A)} \leq |y| \leq 1$$

In digital implementation, a 16-bit linear input sample is converted to 8 bits after compression.

3.2.5 PCMA for Digitized Signals

PCMA takes a maximum input of ±4096 equal to 3.14 dBm in power representation. The front-end analog circuit and analog-to-digital converters (ADCs) are calibrated to take 3.14 dBm and to give ±4096 amplitude in digital number representation. The quantization procedure is the same as with PCMU with few deviations in quantization steps. The quantization steps of A-law use 4 bits for 16 quantization steps. The 16 intervals are of 128, 64, 32, 16, 8, 4, and 2 for two 16 intervals, which means the last two segments use the same interval 2. Of 16 total segments for positive and negative numbers, four segments from −63 to 63 use a step size of 2. Hence, PCMA is referred to as 13-segment quantization. PCMA encodes a 13-bit sample number to an 8-bit compressed sample. Similar to µ-law, there are certain deviations of PCMA from continuous A-law representation. Continuous representation gradually changes to step sizes. Table 3.2 shows A-law compressed and expanded output for both positive and negative values as per G.711.

In PCMU, the last 7 bits are inverted, and in A-law, encoder output even bits inversion (EBI) is used. In the A-law tables, bits before inversion and after EBI are also included in separate columns. EBI is performed in software by XOR operation of the encoded output with 0x55. EBI is applied on both positive and negative values of input. At the decoder, the same operation of even bit inversion can retrieve the original decoded compressed byte. EBI in PCMA once again helps clock recovery mechanisms during ideal channel or near zero amplitude signals.

3.2.6   PCMA Quantization Effects

In Fig. 3.2, three graphs are given. This figure looks similar to Fig. 3.1 with PCMU, with the main difference in input range and quantization steps. The top graph is the input and output relation for input between 16 and 4096. The PCMA encoder output is given to the decoder, and the error is provided in the middle graph. The middle graph shows a growing error with an increase in amplitude toward 4096. At lower amplitude, the error is lower. In the middle graph, it can be observed that half of the amplitudes from 2048 to 4095 are occupying 16 intervals and that they are continuing to reduce by a factor of 2.

The bottom graph is the percentage of error or relative error with reference to the original input. Here error is contained to approximately ±3%. The percentage error remains steady with growing amplitude. This graph shows that relative distortion, signal to quantization, or SNR caused by logarithmic compression will remain almost steady at different amplitudes. At lower amplitude (0 to 16), the percentage of error is more, which makes the SNR lower at very small amplitudes.

3.2.7   Power Levels in PCMU/PCMA and SNR

PCMU maximum power is 3.17 dBm for sine wave tone and corresponding input decimal numbers of ±8159. Any numbers above ±8158 are clipped to ±8158. For a square wave input of ±8158 or above, PCMU power is 6.17 dBm, 3 dB more than 3.17 dBm. A sine wave or perfect square wave may be used for calibration purposes. In actual measurements, a square wave test may not

Table 3.2. PCMA Quantization Example for Positive Inputs and a Few Segments of Negative Inputs

Input Range	Segment in Bits	Interval Size	Step	Encoded Output Before EBI	Encoded Output After EBI	Decoder Output for the Same Input Range
3968–4095	111	128	15	11111111(0xFF)	10101010(0xAA)	4032
—	—	—	—	—	—	—
2048–2175	111	128	0	11110000(0xF0)	10100101(0xA5)	2112
1984–2047	110	64	15	11101111(0xEF)	10111010(0xBA)	2016
—	—	—	—	—	—	—
1024–1087	110	64	0	11100000(0xE0)	10110101(0xB5)	1056
992–1023	101	32	15	11011111(0xDF)	10001010(0x8A)	1008
—	—	—	—	—	—	—
512–543	101	32	0	11010000(0xD0)	10000101(0x85)	528
496–511	100	16	15	11001111(0xCF)	10011010(0x9A)	504
—	—	—	—	—	—	—
256–271	100	16	0	11000000(0xC0)	10010101(0x95)	264
248–255	011	8	15	10111111(0xBF)	11101010(0xEA)	252
—	—	—	—	—	—	—
128–135	011	8	0	10110000(0xB0)	11100101(0xE5)	132
124–127	010	4	15	10101111(0xAF)	11111010(0xFA)	126
—	—	—	—	—	—	—
64-67	010	4	0	10100000(0xA0)	11110101(0xF5)	66
62–63	001*	2	15	10011111(0x9F)	11001010(0xCA)	63
—	—	—	—	—	—	—
32–33	001*	2	0	10010000(0x90)	11000101(0xC5)	33
30–31	000*	2	15	10001111(0x8F)	11011010(0xDA)	31
—	—	—	—	—	—	—
2–3	000*	2	1	10000001(0x81)	11010100(0xD4)	3
0–1	000*	2	0	10000000(0x80)	11010101(0xD5)	1

Table 3.2. *Continued*

Input Range	Segment in Bits	Interval Size	Step	Encoded Output Before EBI	Encoded Output After EBI	Decoder Output for the Same Input Range
For negative numbers, segments 0, 1, 6, and 7 are given to represent the beginning and the end of the range.						
0 to –1	000*	2	0	00000000(0x00)	01010101(0x55)	–1
–2 to –3	000*	2	1	00000001(0x01)	01010100(0x54)	–3
—	—	—	—	—	—	—
–30 to –31	000*	2	15	00001111(0x0F)	01011010(0x5A)	–31
–32 to –33	001*	2	0	00010000(0x10)	01000101(0x45)	–33
—	—	—	—	—	—	—
–62 to –63	001*	2	15	00011111(0x1F)	01001010(0x4A)	–63
–64 to –67	010	4	0	00100000(0x20)	01110101(0x75)	–66
—	—	—	—	—	—	—
–992 to –1023	101	32	15	01011111(0x5F)	00001010(0x0A)	–1008
–1024 to –1087	110	64	0	01100000(0x60)	00110101(0x35)	–1056
—	—	—	—	—	—	—
–1984 to –2047	110	64	15	01101111(0x6F)	00111010(0x3A)	–2016
–2048 to –2175	111	128	0	01110000(0x70)	00100101(0x25)	–2112
—	—	—	—	—	—	—
–3968 to –4095	111	128	15	01111111(0x7F)	00101010(0x2A)	–4032

*Note: In the G.711 recommendation, the last segment is shown as a long segment with 32 values. Here it is shown as segment 1, but a segment field in the segment position of 8 bits was changed to "000." In actual coding, this results in a 3-bit segment value of "000." To avoid confusion, in this section, it is listed as "0" in the segment column instead of "1." It has to be noted that the interval is 2 for segments 0 and 1. The modified values are used in Table 3.2 for the last segment. While referring to multiple references, the difference in representation has to be noted.

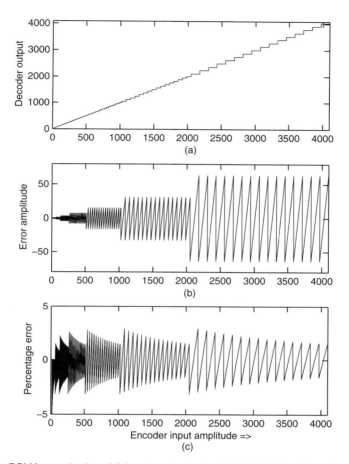

Figure 3.2. PCMA quantization: (a) input and output relation indicating bigger steps at high amplitude, (b) error between input and corresponding decoder output — more error at high amplitude, and (c) percentage of error with input — error is relatively steady for a wide range of input signal levels.

be used and is not accurate. The front-end telephone interfaces limit the square wave bandwidth, which causes rise and fall of the square wave. Hence, the measurements will not match with the mathematical representation of a square wave. In general, telephone measuring instruments use sine wave tone and a combination of sine wave tones for the tests.

In PCMA, maximum sine wave power is 3.14 dBm. For a square wave, this is 6.14 dBm. The same interpretation of PCMU is applicable for PCMA for amplitude limits at ±4096. The SNR of the PCMU and PCMA are in the range of 38 to 40 dB [URL (PCM)]. The SNR of PCMU is estimated as 39.3 dB [Bellamy (1991)] for a full-scale sine wave of 8159. The SNR varies based on the signal amplitude. To conduct measurements for the SNR or power-level calibration sine wave, a range of 1004 to 1020 Hz is used. The details on the acceptable SNR for different signal levels are given in TR-57 (1993). A tone of 1004 Hz is not harmonically related to 8000 Hz, and it exercises all encoder and decoder levels [Bellamy (1991)].

The quality difference between PCMU and PCMA is minor in nature. PCMA has a higher dynamic range, but PCMU signal quality is better than PCMA. For small signals, PCMA works as 2/4096 quantization, and PCMU is 2/8159; therefore, the signal quality from PCMU is better. Ideal channel noise is quantized better in PCMU. PCMA does not define zero-level output for the first quantization interval; hence, PCMA is a mid-riser quantizer. The quality degradation with this type of quantization difference is imperceptible [Bellamy (1991)].

3.3 SPEECH REDUNDANCIES AND COMPRESSION

G.711 provides pure quantized logarithmic compression on a sample basis. The basics on speech redundancies are given in [Bellamy (1991), Nikhil (2000), Kondoz (1999)]. Speech signals have several low-level signals and pauses, which allow for quantization of low-level signals with a few bits, whereas operations like VAD make use of the pause periods to improve compression by about 40% to 60%. In speech, sample-to-sample correlations are very high, leading to use of adaptive differential PCM (ADPCM) based compression. At a given time, a speech signal is concentrated in a few frequencies, which results in short time periodicity, and LPC makes use of this short time periodicity. The rate of excitation in the vocal tract is pitch. Pitch periods have long-term correlation that results in more compression. Many compression techniques operate in time domain. Some compression techniques use frequency-domain analysis, which is usually referred to as transform-domain compression.

3.4 G.726 OR ADPCM COMPRESSION

G.726 is an ITU-T-G.726 recommendation (1990) known as ADPCM. This waveform-based compression does not assume any vocal tract models [URL (Cisco-coding)]. It works on a sample basis by making use of a short-term correlation of samples. ADPCM is used in PSTN for overloaded digital circuit multiplication equipment (DCME) channels. In DCME, more T- or E-carrier channels are multiplexed on the same digital channels. Cordless and DECT phones use ADPCM for communication between the base station and multiple cordless units. A modified version of ADPCM is used in wideband G.722 [ITU-T-G.722 (1988)] with two sub-bands that deliver a higher quality than PSTN. In packet voice, G.726 was initially included in voice over asynchronous transfer mode (VoATM). In the migration of VoATM to VoIP, several compression codecs were considered that provide comparable quality at a lower bit rate than G.726. Hence, G.726 is not preferred in many VoIP deployments. In some VoIP-to-PSTN voice gateways, G.726 is supported for better interoperability.

ADPCM is easy to implement on dedicated chips. The processing is mainly through arithmetic and logical operations than through multiplications. The codec takes about 7 to 8 million cycles per second (MCPS) processing on a single multiplier accumulator-based generic digital signal processor (DSP) like TI-54x [URL (TI-54x)] and ADSP-218x [URL (ADI-218x)], which occupies much lower memory when compared with low-bit-rate codecs.

3.4.1 G.726 Encoder and Decoder

G.726 supports four rates of 16, 24, 32, and 40 kbps. Depending on the selected rate, the encoder gives 2, 3, 4, or 5 bits of output at a 8-kHz sampling rate. The principal application of 40-kbps channels is to carry data modem signals especially for modems operating at greater than 4800 bps. The most popularly used compression rate is 32 kbps for packet voice applications. For VoIP, a frame of 5- to 10-ms samples is grouped together as one frame. The ADPCM encoder and decoder operate independently at any rate between 16 and 40 kbps.

Encoder and Decoder. The encoder processes on linear samples. Some early versions of ADPCM had A- or µ-law 8-bit input, and an encoder was built to cater to conversion from an 8-bit to a 16-bit linear sample. In G.726, for every input sample, a difference signal is generated between input and estimate of the input signal. An adaptive 31-, 15-, 7-, or 4-level quantizer is used to assign five, four, three, or two binary bits, respectively, to the value of the difference signal.

An inverse quantizer produces a quantized difference signal from these same five, four, three, or two binary bits. The signal estimated is added to this difference signal to produce the reconstructed version of the input signal. An adaptive predictor operates on the reconstructed and the quantized difference signals. An adaptive predictor produces the estimate of the input signal, thereby completing the feedback loop. The feedback path of the encoder is a replica of the decoder. The decoder includes a structure identical to the feedback portion of the encoder, together with the optional linear to A/µ-law conversion. More details on G.726 and implementation are available in [ITU-T-G.726 (1990), ITU-T-G.191 (2005)].

Compression Rate Clarification. Original 32-kbps ADPCM was introduced by ITU recommendation G.721. Later on, G.723 was introduced that supports 24- and 40-kbps rates. In G.726, an additional 16-kbps rate is included in addition to combining 32 kbps from G.721 and 24/40 kbps from G.723. G.726 is a superset of stand-alone ADPCM codecs. More details are given in the G.191 manual [ITU-T-G.191 STL (2005)].

3.5 WIDEBAND VOICE

VoIP wideband voice is possible if an end-to-end call is only VoIP and not routed through PSTN. It has to be a direct VoIP call between wideband voice interfaces, including the acoustics. In wideband voice, acoustic interfaces and processing the signal can support frequencies from 50 to 7000 Hz sampled at 16 kHz. Traditional PSTN and VoIP service limits the analog speech bandwidth to 300 to 3400 Hz. Lower frequency enhancement (50–300 Hz) provides greater naturalness, presence, and comfortable listening. High-frequency enhancement (3400 to 7000 Hz) provides better intelligibility and fricative differentiation (e.g., s versus f). The implementation aspects of a wideband VoIP voice call are given in Chapter 9.

3.5.1 G.722 Codec

G.722 [ITU-T-G.722 (1988)] is a time-domain wideband-waveform-based codec that can work on a minimum of a block of two samples. In actual implementations, a block of 5 or 10 ms (160 samples) is used for processing. G.722 uses sub-band adaptive differential pulse code modulation (SB-ADPCM) that splits the frequency band into two sub-bands (0 to 4000 Hz and 4000 to 8000 Hz) and applies ADPCM to sub-bands independently. G.722 operates in three compression rates. The recommended default bit rate of 64 kbps gives the highest quality. A compression rate of 56 kbps allows for an auxiliary data channel of 8 kbps, and the 48-kbps mode allows for an auxiliary data channel of 16 kbps. Because of the availability of high-throughput independent data channels, the G.722-64-kbps mode is most relevant and preferred for deployments.

As shown in Fig. 3.3, the SB-ADPCM encoder splits signal into low- and high-frequency bands. The splitting is performed using quadrature mirror filtering (QMF). The QMF filter has 24 coefficients, this filtering process introduces a delay of 3 ms. The QMF filter also downsamples the each sub-band by a factor of two before encoding. Each sub-band is coded with an independent ADPCM coder. The lower band is coded with 6 bits per each sample at 8 kHz, which creates 48 kbps of compressed data. Note that the original ADPCM had a maximum of 5 bits per sample, and the G.722 low band uses 6 bits of ADPCM. A higher sub-band gives 2 bits per sample to create 16 kbps of data. The higher band has relatively low-amplitude signals that can be coded with 2 bits per sample. The ADPCM bits from low- and high-band channels can be multiplexed with auxiliary channel data up to 16 kbps. With auxiliary data, low-band bits are reduced to make a total bit rate of 64 kbps.

At the receiver, these data are decoded in low- and high-band channels separately. Auxiliary data status is indicated to the low-band ADPCM decoder to decode data at one of 32-, 40-, and 48-kbps rates depending on the auxiliary data rate. The decoded samples at 8 kHz are recombined in QMF. In the receiving channel, the QMF operation upsamples signals, performs filtering, and

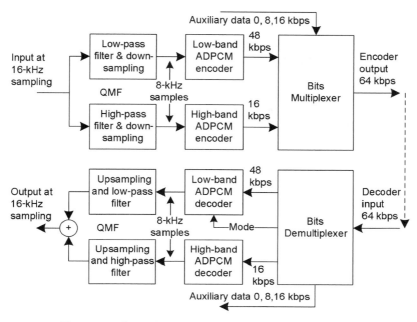

Figure 3.3. G.722 Sub-band ADPCM encoding and decoding.

combines both signals. Various signals, bit rates, and sampling are indicated in Fig. 3.3.

Practical Aspects of G.722 Codec. G.722 is a low-complexity codec. It takes less memory and processing than the other low-compression codec G.729AB. On a single multiplier DSP processor, G.722 takes about 12 MCPS and 2000 words of program memory. In general, the ADPCM-equivalent first lower sub-band is a narrowband toll quality channel that is slightly lower quality than G.711. Two sub-bands and wide bandwidth in G.722 improve voice quality to exceed that of G.711. More details are provided in Section 3.6. Wideband codec G.729.1, in the G.729 family, had backward compatibility with the narrowband codecs of G.729. Once an end-to-end system supports wide-band, G.722 may now coexist because of its low complexity even though it is not backward compatible to G.729 and other G.722 codecs (G. 722.1, G. 722.2).

3.6   G.729 FAMILY OF LOW-BIT-RATE Codecs

The G.729 family of codecs are popularly used in most VoIP deployments. G.729 [ITU-T-G.729 (1996)] is an ITU-T G.729 recommendation for a conjugate-structure algebraic-code-excited linear-prediction (CS-ACELP) speech compression algorithm. G.729 makes use of human vocal tract models suitable for voice signals unlike the G.711 and G.726 codecs, which use waveform-based compression. Basic G.729 compresses voice to 8 kbps, which provides eight times more compression than G.711 and provides good voice quality.

G.729 Annex A (G.729A) is the reduced complexity version of the G.729 recommendation, and it compresses voice to the same 8 kbps [ITU-T-G.729A (1996)] with the tradeoff of a slight loss of quality in comparison with G.729. This version is developed mainly for multimedia simultaneous voice and data applications, although the use of the codec is not limited to these applications. G.729A is bit stream interoperable with the full version of G.729. This codec has built-in packet loss concealment. The codec with suffix B denoted as G.729B or G.729AB supports VAD in the encoder and CNG in the decoder [ITU-T-G.729B (1996)]. G.729 has a wide family of codecs, and some of the popular codecs are listed in Table 3.3. In VoIP deployment, G.729AB is popularly used. In the wideband upgrade, these codecs may be replaced with G.729.1. These wideband codecs are interoperable with narrowband versions. Hence, a wideband codec alone should be sufficient for both wideband and 8-kbps narrowband support. Except G.722, other wideband codecs compress voice to bit rates lower than G.711. In the following section, a high-level overview on G.729AB is given. Codec G.729AB makes use of G.729 as a baseline standard. For better understanding of G.729AB, refer to the G.729, G.729B, and G.729A [ITU-T-G.729 (1996), ITU-T-G.729A (1996), ITU-T-G.729B (1996)] recommendations.

Table 3.3. ITU-T G.729 Family of Narrowbond and Wideband Speech Codecs

Codec Name	Description	Application
G.729	Basic 8-kbps CS-ACELP speech codec	Supported on VoIP voice gateways
G.729B	G.729 with silence compression	
G.729A	Reduced complexity 8-kbps CS-ACELP speech codec	Popular on VoIP adapters and IP phones, compatible with G.729, and G.729.1
G.729AB	G.729A with silence compression	compatible with G.729, G729B, G729.1
G.729E (G.729 Annex E)	11.8-kbps CS-ACELP speech coding algorithm	Not popular in VoIP, works better for music and background and is used in video applications
G.729.1 wideband and narrowband	Embedded variable 8–32 kbps, bits interoperable with G.729, G.729A	Wideband voice

3.6.1 G.729 Codec

The G.729 codec consists of a separate encoder and decoder. The codec compresses speech samples as frames using a CS-ACELP analysis-by-synthesis procedure. The codec operates with 10-ms frames with a look ahead of 5 ms, which results in total algorithmic delay of 15 ms. An overview on the encoder and decoder is given in relation to Fig. 3.4.

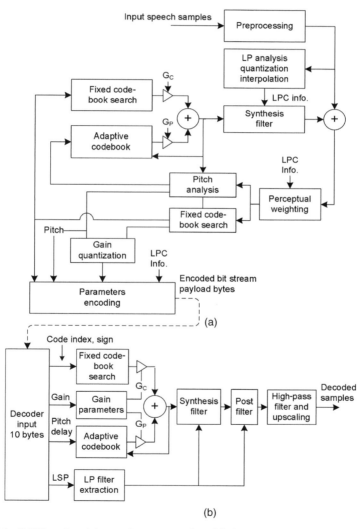

Figure 3.4. G.729 codec. (a) encoder-compression. (b) decoder-decompression from ITU-T-G.729 — redrawn with some simplification. [Courtesy: Reproduced with the kind permission of ITU; International Telecommunication Union, Geneva, www.itu.int).]

G.729 Encoder. The G.729 coder is based on the code-excited linear-prediction (CELP) coding model. For every 10-ms frame, the speech signal is analyzed to extract the parameters of the CELP model. The parameters are linear-prediction filter coefficients coded as line spectral pairs, adaptive and fixed-codebook indices, and gains. These parameters are encoded and transmitted as payload to VoIP application. As shown in Fig. 3.4, preprocessed speech is analyzed for LP filter coefficients. These coefficients are converted to line spectrum pairs (LSP) and are quantized using predictive two-stage vector quantization (VQ). The open-loop pitch estimation is computed for every 10-ms frame based on a perceptually weighted speech signal. Closed-loop pitch analysis is performed using the target signal and impulse response by searching around the value of the open-loop pitch delay. The new target signal is computed and used in the fixed codebook search to arrive at optimum excitation. The gains of adaptive and fixed codebook contributions are vector quantized. Finally, the filter memories are updated using the determined excitation signal. For every 80 samples of input, the encoder gives 10 bytes of compressed output making the total bit rate 8 kbps. These 10 bytes consist of several parameters as listed in Table 3.4. From the table, it can be observed that G.729 coding splits parameters into several classes with each of them having a few bits. It is entirely different from G.711 and G.726 compression. For more details on coding and parameters referenced in Table 3.4 refer to [ITU-T-G.729 (1996), ITU-T-G.729A (1996), ITU-T-G.729B (1996)].

G.729 Decoder. The G.729 decoder is illustrated in Fig. 3.4(b). The decoder generates 80 samples of 16-bit linear PCM values for every 80 bits (10 bytes) of data. The input parameters for the decoder are LSP coefficients, the two fractional pitch delays, two fixed-codebook vectors, and the two sets of adaptive and fixed codebook gains. Initially, the LSP coefficients are interpolated and converted to LP filter coefficients for each subframe. For each 5-ms subframe, the excitation is constructed by adding the adaptive and fixed codebook vectors.

Table 3.4. G.729 Encoder Parameters and Compressed Bits Allocation for 10-ms Frame

Parameter Name	Codeword	Subframe 1	Subframe 2	Total Bits per Frame
Line spectrum pairs (LSPs)	L0, L1, L2, L3			18
Adaptive codebook delay	P1, P2	8	5	13
Pitch delay parity	P0	1		1
Fixed codebook index	C1, C2	13	13	26
Fixed codebook sign	S1, S2	4	4	8
Codebook gains at stage-1	GA1, GA2	3	3	6
Codebook gains at stage-2	GB1, GB2	4	4	8
Total bits in 10-ms frame				80

Speech is reconstructed by filtering the excitation through the LP synthesis filter. The reconstructed speech is processed through the postprocessing stage, which includes an adaptive post-filter based on the long-term and short-term synthesis filters followed by a high-pass filter and scaling operation.

In addition to an algorithmic delay of 15 ms, delays may be possible because of the execution time of the encoder and decoder algorithms in the processor. Depending on the implementation, while processing several channels on one processor, delay increases for the last processed channel.

3.7 MISCELLANEOUS NARROW AND WIDEBAND Codecs

In the previous sections, codecs of G.711, G.726, and G.729AB are considered for narrowband voice, and G.722 is considered for wideband voice. Several other codecs are presented in [Kondoz (1999), Goldberg et al. (2000), Alexander (2006), Hersent et al. (2005)], but each regional VoIP deployment may confine to a finite set of codecs. Some retail market products incorporate more types of codecs to cater to deployments in more regions and to cater to retail market options. An overview on some popular narrowbond and wideband codecs considered for VoIP applications are given in this section.

3.7.1 Narrowband Codecs

G.723.1. It is a 8-kHz sampling ITU-T Codec. G.723.1 [ITU-T-G.723.1 (2006)] is one of the popular codecs supported in many VoIP retail market products. G.723.1A is having built-in VAD/CNG functions. G.723.1 compression rates are 5.3 and 6.3 kbps. The compression at 5.3 kbps makes use of ACELP and 6.3-kbps multipulse maximum likelihood quantization (MP-MLQ). The frames are of 30 ms giving 20 or 24 bytes in 30 ms, and total algorithmic delay is 37.5 ms, including the look-ahead delay of 7.5 ms. With G.723.1, end-to-end delays also increase that result in more degradation of voice quality. Note that G.723 is an ADPCM codec for 24 and 40 kbps and is not part of G.723.1 or G.723.1A. With the incorporation of G.726 with 4-rates, G.723 became part of G.726.

G.723.1 is more attractive from a compression point of view, but G.723.1 has lower quality than G.729A. G.723.1 takes about 40% more processing than G.729A. For the same 30-ms frame of G.723.1, G.729A gives 30 bytes of payload. Because of its small frames, G.729A is used for different packet sizes, and this helps in reducing end-to-end delay and improved packet loss concealment. The benefit of higher compression in G.723.1 is negligible compared with big IP headers and network interface overhead. Overall, the G.729 family is better adapted in VoIP compared with G.723.1. The G.729 family has a compatible wideband voice codec.

G.729E. It is a 8-kHz sampling narrowband ITU-T codec in the G.729 family. G.729E [ITU-T-G.729E (1996)] at 11.8 kbps works on the same 10-ms frame of G.729. The major benefit of G.729E is better voice quality even in the presence of music and background noise. G.729E is considered for the voice part of video conferences. This codec is not interoperable with G.729 or G.729A. Hence, it is not considered for VoIP applications.

G.728. It is a 8-kHz sampling ITU-T codec [ITU-T-G.728 (1992)]. It is a low-delay code-excited linear prediction (LD-CELP) codec giving a compression rate of 16 kbps. The major benefit of G.728 is low latency, and it can work with smaller frames of five samples. It has better quality than G.729/729A for VoIP applications and can support music and background noise. VoIP packetization makes use of several five-sample frames to increase the significance of voice payload against IP packet headers. G.728 is about three-times more complex in processing than G.729A, but takes lower memory. This codec is not popular in practical VoIP deployments because of higher processing requirements.

AMR Narrowband Codec. The adaptive multirate (AMR) codec is mainly used in wireless mobile applications. It was standardized by the European Telecommunications Standards Institute (ETSI) [ETSI EN 301 703 (1999)] and adapted by a third-generation partnership project (3GPP). The codec makes use of a multirate ACELP supporting eight compression rates of 4.75, 5.15, 5.90, 6.70, 7.4, 7.95, 10.2, and 12.2 kbps. This codec has a built-in VAD and operates on a 20-ms frame at 8-kHz sampling. The bit rate can be changed on a 20-ms boundary. The main motivation of AMR for VoIP is to make direct VoIP calls to a mobile infrastructure with minimal transcoding operations that helps to improve voice quality. Transcoding always degrades voice quality as explained in Chapter 20. For this mode of operation, the cell phone infrastructure has to support interworking directly with VoIP. A VoIP call can directly negotiate using an AMR codec for making an end-to-end call as AMR. Several other converging voice technologies such as IP multimedia system (IMS), fixed mobile convergence (FMC), and Femto cells can make use of this type of direct call with minimal transcoding. The AMR codec requires two times more processing than G.729AB.

iLBC. It is an Internet low-bit-rate codec (iLBC) at 8-kHz sampling [URL (iLBC)]. This is one of the most popular codecs outside ITU-T recommendations designed by global IP sound (GIPS). It is a free codec used mainly in dedicated deployments and PSTN–VoIP gateways for interoperability. The PESQ MOS (explained in Chapter 20) of iLBC in most of the good network conditions is slightly exceeding or matching G.729. Under higher packet drop, iLBC performs better. Compared with G.729A, iLBC takes less time to settle from PLC, making iLBC perform better for more packet drops of 15% to 20%. The iLBC codec supports 13.3 kbps at a 30-ms frame and 15.2 kbps at a 20-ms frame.

3.7.2 Wideband Codecs

G.729.1 Wideband Codec. It is the scalable narrowbond and wideband codec [ITU-T-G.729.1 (2006)]. This codec is also called G.729EV, where EV stands for embedded variable. The narrowband part of payload is backward compatible for the G.729 narrowband family of codecs. This is the main advantage of a G.729.1 wideband codec. This codec can adjust rates from 8 to 32 kbps. The narrowband rates are 8 and 12 kbps. Wideband rates are from 14–32 kbps in steps of 2 kbps. This codec works on a 20-ms superframe, while maintaining compatibility with the G.729 family of codecs for the same 10-ms frame and 5-ms subframe. This codec has an algorithmic delay of 48.9375 ms (783 samples at 16-kHz sampling). A narrowband codec can extract a 20-ms payload from the wideband payload for interoperability without calling for out-of-band signaling and renegotiations. This codec uses three different major coding algorithms:

1. Embedded CELP coding of a low band, which is compatible to narrowband coding. It generates 8-kbps G.729 family-compatible bits and an extra 4 kbps as narrowband enhancement.
2. Parametric coding of the higher band (4000 to 7000 Hz) is used by a time-domain bandwidth extension (TD-BWE) that generates an additional 2 kbps to make the rate total of 14 kbps. It is the main wideband extension on top of 12 kbps. Wideband can start working in 2-kbps steps up to a total of 32 kbps.
3. Enhancement of the full band (50 to 7000 Hz) by a predictive transform coding technique is referred to as time-domain alias cancellation (TDAC). This takes error from the CELP coding and signal directly from higher band signal. TDAC makes use of modified discrete cosine transform (MDCT).

G.722.2 (AMR-WB) Wideband Codec. It is the adaptive multirate wideband (AMR-WB) codec for wideband speech signals [ITU-T-G.722.2 (2003)]. It is the same as 3GPP AMR-WB planned for enhancements in mobile phone wireless communications and audio streaming. The applications of the codec include VoIP, mobile communications, ISDN wideband telephony, video conferencing, and audio streaming to mobile devices. Most wireless applications currently use narrowband AMR. Certain later enhancements and streaming to mobile applications are expected to adapt to wideband mode. AMR-WB operates at nine rates of 6.6, 8.85, 12.65, 14.25, 15.85, 18.25, 19.85, 23.05, and 23.85 kbps. This codec operates on a 20-ms frame at 16-kHz sampling, which corresponds to 320 samples in a frame and the bit rate can be changed on a 20-ms boundary. The codec makes use of a multirate wideband ACELP with built-in voice activity detection. ACELP is also used in G.729 codecs. AMR-WB is more complex in processing. It takes about four to five times more processing than G.722 and G.729A codecs.

G.722.2 is not interoperarable with other codecs mentioned in this chapter. The main advantages of G.722.2 are better wideband quality with a low bit rate and the bit rate is scalable. It is useful in making direct wideband VoIP calls. When mobile infrastructure is upgraded to wideband, VoIP calls can be made to mobile phones in wideband. For this mode of operation, mobile phone infrastructure has to support interworking with VoIP. PSTN does not have this wideband option; hence, an end-to-end call will go through narrowband transcoding.

iSAC. It is an Internet speech audio codec (iSAC), the wideband codec popularized through PC-based wideband VoIP calls of Skype service. GIPS own this codec [URL (iSAC)]. It is also packet drop friendly and claims to work better with a higher packet drop. Narrowband iLBC from GIPS is a free codec, but iSAC is not a free codec. iSAC operates on 16-kHz sampling, which is similar to other wideband codecs. iSAC adjusts rates from 10 to 32 kbps that work on packet sizes from 30 to 60 ms. Algorithmic delay is 3 ms plus the selected frame size. In most commercial deployments, G.722 or G.729.1 may be preferred compared with iSAC because of better interoperability. The complexity of iSAC is stated as comparable with the G.722.2 wideband codec.

3.8 Codecs AND OVERLOAD LEVELS

Relations between overload and maximum signal levels are given in the G.191 manual [ITU-T-G.191 STL (2005)]. As explained in the G.711 codec, G.711 accepts 3.17 dBm for μ-law and 3.14 dBm for A-law as an undistorted sine wave. In the case of G.722, it can accept 6 dB more than G.711, up to 9 dBm. In practice, ADC and digital-to-analog converter (DAC) devices come with some of the bit selection options in quantization. When such options are available, instead of ±8159 as full-scale quantization, two times ±8159 can be considered as the main input. Engineers may also shift data by 1 bit to the left to make it achieve 6 dB of more quantization. The better option would be directly getting required full-scale amplitude from ADC/DAC and scaling down to feed it to other modules. Some codecs accept 15-dBm levels as input. The meaning here is they accept signal levels ±32767, i.e., four times (12 dB) more than usual G.711 levels at their input. These levels are also required to be considered in association with other module operations as some of them may consider G.711 levels as the highest levels.

3.9 VOICE QUALITY OF Codecs

Mean opinion score (MOS) is used to measure the voice quality performance of the compression scheme or end-to-end voice call. Different methods like

E-model [ITU-T-G.107 (2005)], perception based on listening, and perceptual evaluation of speech quality (PESQ) are used for MOS reporting [URL (DSLAII), URL (Opticom-PESQ)]. To present MOS in this chapter, PESQ and rating (R) factor are used. Some more details on voice quality measuring procedures are given in Chapter 20.

PESQ is an active measurement conducted mainly on instruments. In a PESQ-based measurement, the instrument sends a reference waveform from one end and compares the distortions at the other end with the original. PESQ is an external measure. The PESQ and R-factor are mapped to MOS. PESQ-based MOS differs from R-factor MOS.

MOS is presented in the range of 1 to 5, but it is limited to 4.5 in actual usage. The PESQ-based MOS of G.711 is 4.3 to 4.4. Worldwide G.711 in PSTN is treated as delivering toll voice quality. MOS of greater than or equal to 4.0 is considered as toll voice quality. To achieve toll quality in VoIP, G.711 and G.726 at higher bit rates and wideband codecs have to be used.

In the narrowband case, G.729 and G.729A are popularly used. G.729A gives a PESQ MOS of 3.75 to 3.8. The main G.729 codec gives better quality at about 3.9, but it requires more processing. These are not toll (MOS more than 4.0) quality codecs, but they are accepted worldwide with voice quality broadly rated as very close to satisfactory and acceptable levels. Codec G.729.1 is for wideband voice applications. The listening perceptions of wideband codecs exceed G.711 PSTN quality. Higher quality than ideal G.711 may be limited to a MOS of 4.5. Wideband MOS extensions to represent beyond 4.5 are not documented clearly at this stage. Refer to the latest ITU recommendations for the updates on wideband MOS.

An overview of codec quality comparisons with an R-factor of the E-model and corresponding calculated MOS is given here. The R-factor connects quantitatively many parameters that influence voice quality. The R-factor calculation makes use of several parameters broadly classified under delays, echo, noise, phone characteristics, packet, and signal transmission characteristics. In this section, the R-factor is considered purely from the codec-dependent impairment factor (I_e); other dependencies are given in Chapter 20. For narrowband speech, the R-factor ranges from 0 to 100, with 100 being the MOS equivalent of 4.5. In telecommunications, a G.711-based full digital ISDN system gives a highest R of 93.2 (initially it was 94.2 and was amended to 93.2) and the corresponding MOS is approximately 4.4. Linear 16-bit samples give a MOS of 4.5. User satisfaction levels with R-factor and MOS are given in Table 3.5. MOS is calculated from the R-factor as

$$MOS = 1 + 0.035\,R + R(R-60)(100-R)(7\times10^{-6}), \text{ for } R = 1 \text{ to } 100,$$

$$MOS = 4.5, \text{ for } R \geq 100, MOS = 1 \text{ for } R \leq 0$$

Considering only codec compression, under ideal conditions R-factor is expressed as $R = R_0 - I_e$, where $R_0 = 93.2$ and I_e is the codec specific impair-

Table 3.5. Relation among R-Factor, MOS, and User Satisfaction

				R-Factor and MOS for R = 70 to 99						User Satisfaction Level
R	*99*	*98*	*97*	*96*	*95*	*94*	93	92	91	90	R of 90 to 93
MOS	*4.49*	*4.48*	*4.47*	*4.46*	*4.44*	*4.42*	4.41	4.38	4.36	4.34	is very satisfactory PSTN quality; *R of ≥94 is better than PSTN quality*
R	89	88	87	86	85	84	83	82	81	80	Satisfied; R ≥
MOS	4.31	4.29	4.26	4.23	4.2	4.17	4.13	4.1	4.06	4.02	80 and MOS ≥ 4.0 is referred to as toll quality
R	79	78	77	76	75	74	73	72	71	70	Some users
MOS	3.99	3.95	3.91	3.86	3.82	3.78	3.73	3.69	3.64	3.6	dissatisfied

ment factor. Different codecs I_e, R and corresponding R-based MOS mapping is given in Table 3.6. The quality of wideband codec exceeds that of G.711, and it has a rating of more than 93.2. Hence, the R-factor scale is extended to a maximum value of 129 that takes care of mapping both narrowband and wideband on the same scale. For this reason, a different R_0 of 129 is used [ITU-T-G.107 (2006), ITU-T-G.113 (2006), Alexander (2006)]. An R-factor of 129 is obtained with direct wideband 16-bit linear samples at 16-kHz sampling. To get the same R-factor for a narrowband codec on a wideband R scale, a new equipment impairment parameter I_e is introduced with name $I_{e,wb}$. The impairment parameter $I_{e,wb}$ values are modified for narrowband codecs to give the same narrowband results. In the original narrowband R-factor, G.711 has Ie = 0. On the wideband scale, $I_{e,wb}$ for the same G.711 is 35.8. The resulting R in both cases is the same (93.2 is the same as wideband 129 − 35.8 = 93.2). It is also expressed as $I_{e,wb} = I_e + 35.8$.

Combined narrowband and wideband R-values are listed in Table 3.6. In Table 3.6, both narrowband and wideband values are given in different columns for clarity in presentation. The result of the R-factor is the same for narrowband codecs. From the table, it is clear that wideband performs better than narrowband. In narrowband, G.711 gives the best quality. Based on the R-factor comparison, the wideband G.722 gives an R-factor of 116. These values represent much higher quality than G.711 at an R-factor of 93.2. Refer to later revisions of recommendations/standards for possible updates on the MOS

Table 3.6. Narrowband and Wideband R-Values for Different Codecs

Codec	Bit Rate (kbps)	MOS from R	Narrowband ($R_0 = 93.2$)		Wideband (R_0 wideband = 129)		Remarks
			I_e	$R = 93.2 - I_e$	$I_{e,wb}$	$R_{Nb,wb} = 129 - I_{e,wb}$	
G.711	64	4.4	0	93.2	36	93.2	PSTN quality
G.729E	11.8	4.3	4	89.2	40	89.2	
G.726	32	4.23	7	86.2	43	86.2	
G.728	16	4.23	7	86.2	43	86.2	
G.729	8	4.13	10	83.2	46	83.2	
G.729A	8	4.1	11	82.2	47	82.2	Commonly used in VoIP
G.723.1MP-MLQ	6.3	3.95	15	78.2	51	78.2	
G.722	64	4.5*			13	116*	Exceeding G.711 rating

*While preparing this book, MOS mapping was not available for R > 100; hence, the highest MOS given is (4.5).

scale for wideband and on proper connectivity in MOS mapping. In Table 3.6, a MOS of 4.1 appears against the G.729A codec. After considering several end-to-end transmission contributions, this level is below the toll quality. A PESQ-G.729A MOS is 3.75 to 3.85, which is below the toll quality, but it is widely accepted in deployments.

3.9.1   Discussion on Wideband Codecs Voice Quality

The ITU wideband codecs considered at this stage for VoIP are G.722, G.729.1, and G.722.2. Wideband codecs are rated on an extended R-factor scale of 129. In Table 3.6, a G.722 rating is given as 116. The rating of G.722.2 is stated as 128 based on [ITU-T-G.107 (2006), ITU-T-G.113 (2006)]. With the highest rating being 129, G.722.2 at provide 128 seems to higher quality. G.729.1 is still under study, and it is expected to appear in later revisions of ITU documents. While writing this book, the ITU study group 12 (SG12) was actively evaluating wideband codecs and arriving at various $I_{e,wb}$, R-factors and MOS mapping. At this stage, the results from auditory and instrument measures from the modified WB-PESQ vary in estimating $I_{e,wb}$. Among codecs G.722, G.722.2, and G.729.1, several combination results are noticed based on the SG12 results. Most tests convey that G.722.2 at 23.05 kbps are higher quality than G.722 and G.729.1. The results from G.722 and G.729.1 are very close. Some tests reveal that G.722 performs better than G.729.1, and some combinations reveal the opposite. In general, G.722.2 requires five times more processing than does G.722, and G.729.1 requires three to four times more processing than does G.722. Hence, G.722 is used in most early wideband products. To maintain backward compatibility with the narrowband, G.729.1 is considered. Codec G.722.2 may be considered to interoperate with wireless infrastructure enhancements and media convergence as well as to present a higher quality supported product.

3.10 C-SOURCE CODE FOR Codecs

The following summary gives an overview on the 16-bit fixed-point code. The source code for codecs is available as a 16-bit fixed-point code. The floating-point code is also available for some codecs, but most processors used for voice chain processing are of fixed-point processors. Hence, by default, voice codecs software is referred to as fixed-point code. The c-code is ANSI-compatible and compiles with most processor tools after making minor corrections to file interfaces.

G.711 and G.726 are available in ITU-T-G.191 (2005), and implementation guidelines are also available in the G.191 manual. ITU-T-G.729, G.729B, G.729A, G.729E, G.729.1 (G.729EV), G.723.1, G.722, and G.722.2 source code is available from the ITU. The source code for G.728 code is not directly available with the recommendation. Users make their own code based on the pseudo-code for functional blocks, guidelines, and test vectors given by the ITU recommendations. Several third parties provide optimized or processor-optimized code at cost. iLBC floating point code is available as a free download. Users have to derive fixed-point c-code. Third parties and GIPS also license fixed-point C-code at some expense. iSAC-optimized C-code is licensed from GIPS. The AMR codec code is mainly supported by ETSI and 3GPP, and some codecs are regulated through the ITU. Refer to Chapter 19 for software, operating system, and processor features.

3.11 Codecs IN VoIP DEPLOYMENT

VoIP deployments use multiple codecs. It is difficult to arrive at one generic conclusion on the selection of codecs. Codec selection will consider several aspects of quality, compression, packet loss performance, frame size and end-to-end delays, processing, memory requirements, any backward compatibility, roadmap and life of codec, ability to cater to multiple deployments, and finally acceptance in the market. An overview on this selection is provided in this section.

As mentioned, retail market products keep several codecs in the VoIP box to cater to wider requirements. VoIP-to-PSTN high-channel gateways like

CISCO gateways Donohue et al. (2006) keep several codecs to take care of wider market requirements. VoIP customer premises equipments (CPEs) mainly work with a particular deployment. The requirements vary with the region of interest and the service provider, but the requirements will narrow down to two to three codecs from actually working deployment. The service provider controls these parameters.

As a common practice, G.711 is used in all VoIP deployments. It is the best narrowband codec. It is helpful to use G.711 to get better branding on voice quality. Japan VoIP deployments use mainly G.711 because of higher Internet bandwidth support in most deployments and as it provides better quality. G.729A is supported on most VoIP CPEs, but G.729 is not used in actual VoIP calls. Europe is using G.711 and G.729AB, and some deployments are using G.722 for wideband. In the case of G.729AB, VAD is not preferred in deployments. VAD/CNG is kept to avoid any interoperability issues. When the destination gateway is sending VAD packets, CNG will continue to create comfort noise. Usually gateways are disabled for VAD, but CNG will always be enabled for complete interoperability.

North America uses mixed combinations. Some deployments go by G.711 and G.729AB; and some customers insist on G.723.1 and iLBC support. G.711, G.729AB, and iLBC is a good combination for North America. No extra benefit is earned with G.723.1 support when compared with G.729A. Wideband G.729.1 is still a new standard and requires some time for adaption by the deployments. Many existing boxes may not be able to accommodate this wideband as an upgrade. In general, European customers prefer wideband voice at this stage. Europe may adapt to both G.722 and G.729.1 in addition to G.711 and G.729AB. When G.729.1 is supported, G.729AB is not required on the same box because of backward compatibility. The preferred combination for Europe would be G.711, G.722, and G.729.1. Codec G.722.2 helps in achieving media convergence on multiple handheld terminals, and it may be marketed as a higher quality supported voice product. Uncontrolled deployments like Skype [URL (Skype)], and Google talk use wideband iSAC in addition to the many codecs like G.711 and iLBC. In general, codec G.711 in narrowband and G.722 in wideband are expected to remain in use for a long time. The G.722 highest rate of 64 kbps matches that of G.711. The G.722 recommendation is enhanced for packet loss concealment, and this is an added advantage. Studies are ongoing on for the G.711WB codec by an ITU study group. Refer to the latest updates and revisions from the ITU for more information on this codec.

4

GENERIC VAD/CNG FOR WAVEFORM CODECS

The voice activity detection (VAD) algorithm classifies the input signal as active speech or silence. The purpose of VAD is to reduce the number of packets during inactive speech and to send all available packets during active speech. Comfort noise generation (CNG) recreates a suitable background as a continuity of the inactive region. For complete end-to-end operation, VAD is required on the sending side and CNG is required on the receiving side. VAD is for detecting and giving a decision between speech and silence. Silence insertion description (SID) is the payload name or silence packet created from the sender. The packet transmission from the sender side is controlled by a discontinuous transmission (DTX) algorithm, which resides on the VAD operation/module. For sending a SID packet from the sender, information from both the VAD and the DTX controls are involved. The next transmission of SID packets will happen during a change of power or spectral characteristics. In actual usage, VAD, SID payload formulation, and DTX functions are referred to by the single name of VAD. The name "VAD packet" is commonly used for representing the "SID packet." In this book, the VAD packet is also used at several places as this quickly conveys a VAD operation, but the correct name is "SID packet." The combination of transmitter and receiver is referred to as VAD/CNG. CNG is for comfort noise creation at the receiver based on received SID packets. CNG uses the SID packet of the last transmitted frame to generate comfort noise, and it continues until receiving the next SID packet. A SID packet from the transmitting side

VoIP Voice and Fax Signal Processing, by Sivannarayana Nagireddi
Copyright © 2008 by John Wiley & Sons, Inc.

is interpreted at the receiving side to generate comfort noise with the same spectral characteristics indicated in the SID packet. VAD/CNG is built inside several codecs such as G.729AB, G.723.1, adaptive multirate (AMR) narrowband (AMR-NB), and AMR wideband (AMR-WB). Waveform-based codecs such as G.711, and G.726, and the low-bit-rate codec (LBC) G.728 are not having a built-in VAD/CNG operation. This chapter is presented for the VAD/CNG scheme used with waveform-based G.711/726 codecs and provides some overview on the VAD/CNG principle of operation for other low-bit-rate codecs.

4.1   VAD/CNG AND Codecs

The speech is a half-duplex under normal conversation, meaning one person at a time speaks into the phone. In a telephone conversation between terminals (or users) A–B, speech from A to B or B to A is present at a given time. In the middle of the same person's speech, significant durations of inactive (silence) zones occur. VAD/CNG exploits inactive zones and eliminates the need for sending voice packets to the network, thus saving about 40% to 50% of Internet bandwidth. The bandwidth saving happens with a certain degradation of voice quality, even though it is not a significant degradation. As a general recommendation, the use of VAD/CNG has to be eliminated on availability of sufficient network bandwidth for a VoIP voice call.

VAD/CNG is embedded with some compression codecs such as G.729AB and G.723.1A. These codecs work based on the vocal tract model. VAD and the comfort noise created during silence closely match and continue to generate the original background modeled through the vocal tract. The codecs G.711, G.726, and G.728 do not have a built-in VAD/CNG scheme. These codecs are used with an external VAD/CNG scheme. Two popular VAD/CNG schemes for G.711 and G.726 exist.

Power- or energy-based VAD/CNG is the simplest VAD/CNG scheme used for G.711 and G.726 codecs. The names "power based" or "energy based" are used for representing this scheme. In this chapter, the power-based VAD/CNG name is used. Some engineers also call this power-based VAD/CNG a VAD/CNG-I or VAD-I. Power-based VAD/CNG sends 1 byte of payload of power based on the RTP payload for comfort noise (CN) as given in RFC3389 [Zopf (2002)].

G.711-Appendix-II VAD/CNG [ITU-T-G.711 (2000)] is another VAD/CNG scheme used for both G.711 and G.726 that is derived based on G.729B. G.711-Appendix-II VAD/CNG gives 11 bytes. Out of 11 bytes, 1 byte is power and 10 bytes are quantized reflection coefficients derived in the process of linear predictive coding (LPC). The VAD scheme is referred to as VAD/CNG-II, VAD-II, G.711 VAD, or Appendix-II VAD. VAD-II creates a spectral envelope that matches the background, and it is preferred because of better voice quality. Many VoIP systems were built before the G.711 Appendix-II VAD recommendation. To maintain interoperability, both power- and Appendix-II based VAD/CNG schemes are included in the VoIP implementation.

The DTX algorithm determines the frequency of SID frame transmission during periods of inactive speech. The DTX algorithm analyzes the input signal and transmits only when a significant change in ambient noise characteristics is detected. Some more details on VAD/CNG are given in subsequent sections. More details on VAD/CNG can be found at references [Kondoz (1999), Zopf (2002), Schulzrinne et al. (2003), ITU-T-G.711 (2000)].

4.2   GENERIC VAD/CNG FUNCTIONALITY

Generic VAD/CNG functionality is illustrated in Fig. 4.1(a) for the G.729AB encoder and decoder. For simplification of representation, only the encoder and decoder are shown in the voice call, and voice is shown in one direction in Fig. 4.1(a). G.729AB has a built-in VAD/CNG function. In the VAD disable mode, G.729AB generates voice packets for every 10 ms or multiples of 10 ms. The packets are marked on the top row as "S S S" in boxes. In the VAD-enabled mode, speech packets marked as "S" and the VAD–SID packet marked as "V" are sent on the network. The packets shown without any marking (empty boxes), are of no packet or a null packet. For completeness of representation, Fig. 4.1(a) and (b) are shown with an empty box for no packets. The empty boxes are the bandwidth-saving operations of not sending any packet on the network. At the destination, the decoder decompresses speech and VAD packets. The VAD packet helps in continuing the generation of comfortable noise during no packets.

The G.711/726-based VAD/CNG is shown in Fig. 4.1(b). In this scheme, external VAD and CNG functions are required. For simplification, the VAD, SID packet, and DTX are referred to as VAD. The VAD module generates detection of a decision regarding speech or silence and a decision regarding when to send an updated packet, which is controlled by the DTX operation. The VAD module provides a required selection to choose between speech or VAD packets. With reference to Fig. 4.1(b), encoder packets marked with "S" and a VAD module packet marked with "V" are sent on the network. At the destination, the status of the packet is retrieved and payload is sent to the normal decoder function or CNG decoder based on received packet type.

4.3   COMFORT NOISE PAYLOAD FORMAT

As explained in the previous section, two popular types of VAD/CNG schemes are included in G.711. The noise power levels in both VAD/CNG methods are expressed in −dBov to interoperate with each other. The unit "dBov" is the dB level relative to the overload of the system. It is not a Volts reference for

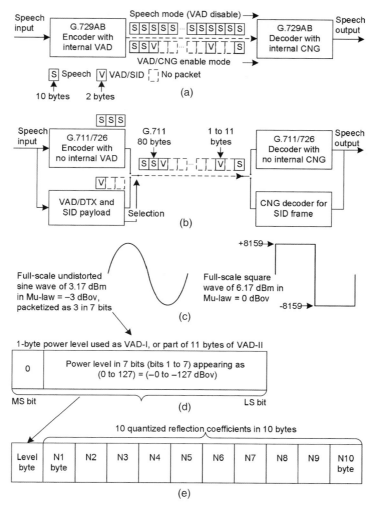

Figure 4.1. VAD/CNG representations. (a) VAD/CNG operation with G.729AB built-in VAD/CNG. (b) VAD/CNG with external VAD/CNG. (c) Power-level mapping with sine and square wave. (d) 1-byte power payload. (e) 11 bytes of payload.

dBV. For example, in the case of μ-law-based coding, the maximum sine wave signal power without distortion is 3.17 dBm with amplitude ±8159. Square wave power is 3 dB more than sine wave power. The maximum possible power of signal from a square wave with an amplitude of ±8159 is 0 dBov as reference, which corresponds to a 6.17-dBm power level in the μ-law system. Hence, 0 dBov = 6.17 dBm is used in μ-law system. Sine and square wave power levels and signal amplitudes are shown in Fig. 4.1(c). Representation relative to the overload point of a system is particularly useful for digital implementations, because one does not need to know the relative calibration of the analog circuitry.

Noise power levels in VAD are comparable from −40 to −60 dBm. The noise power is represented as −dBov with input power ranging from 0 to −127 dBov. In packet format, the absolute value of −dBov is with values between 0 and 127 that fits in 7 bits of number format. A power level of −40 dBov is packed as a numerical value of 40 in the payload. The noise level is packed with the most significant (MS) bit always set to 0, and the remaining 7 bits represent a value from 0 to 127. One byte of power-level payload is represented in Fig. 4.1(d). A payload value of 127 is the least power of −127 dBov (i.e., −120.83 dBm (−127 + 6.17 = −120.83 dBm). A good voice system works up to idle channel noise of −80 to −70 dBm as explained in Chapter 1. Hence, the power levels can be clipped to represent a system's idle channel noise.

The VAD-II SID frame is 11 bytes in length. The power or energy is in 1 byte, and the reflection coefficients representing the spectral envelope are quantized in 10 bytes. Quantized reflection coefficients are packed in subsequent bytes in ascending order as shown in Fig. 4.1(e). The bytes are denoted with the symbols N1, N2 ..., and N10. In the absence of spectral information, these coefficients are filled with a value of zero. As with the speech packet, 1 byte or 11 bytes of SID payload is packetized with Real-Time Transport Protocol (RTP), User Datagram Protocol (UDP), and Internet Protocol (IP) headers.

4.4 G.711 APPENDIX-II VAD/CNG ALGORITHM

G.729B is the VAD/CNG recommendation that works with the G.729 and G.729A codecs. G.711 Appendix-II (referred to as VAD-II) also uses a part of the G.729B with slight deviations in some parameter calculations. In G.729AB, various modules and parameters are shared between VAD and regular speech compression operations. In VAD-II, the operations of VAD/CNG and the G.711 codec are independent. A detailed description is available in [ITU-T-G.711 (2000), ITU-T-G.729B (1996), Kondoz (1999), Goldberg et al. (2000)]. VAD detection goes through several algorithmic steps as well as the decision process. The main operations summary is given here.

1. Preprocessing: The VAD algorithm takes the signal through preprocessing. The input signal is preprocessed by a first-order high-pass infinite impulse response (IIR) filter to remove an unwanted low-frequency component and any impulse noise spikes.

2. Autocorrelation: Short-term prediction or LPC of the speech signal is performed once per speech frame using autocorrelation with a 25-ms asymmetric window. The analysis window consists of two parts. The first part is half of a Hamming window, and the second part is a quarter of a cosine function cycle. The analysis window applies a 20-ms duration

of previous samples and a 5-ms duration of look-ahead samples, which maintains compatibility with G.729 implementations that account for the algorithmic delay of 5 ms at the encoder stage. A set of 11 coefficients are computed in autocorrelation. The first coefficient $R(0)$ is used to derive the first byte of power in the SID payload.

3. Linear prediction (LP) computation: The autocorrelation coefficients are used to obtain the LP filter coefficients. Reflection coefficients are also derived in the computation of LP coefficients. Reflection coefficients are used as payload after making a VAD decision. Normalized reflection coefficients are used in generating 10 bytes of SID packet payload.

4. A required SID packet payload can be generated from autocorrelation and reflection coefficients derived in the process of LPC parameters. To arrive at a silence-or-speech decision, and to help the DTX algorithm decide on when to send the next SID packet, it is essential to perform several operations.

5. LP-to-line-spectral-pairs (LSP) conversion: The LP filter coefficients are converted to LSP coefficients for quantization and interpolation purposes. LSP coefficients are found by evaluating the polynomials created with LP parameters as coefficients.

6. LSP-to-line-spectral-frequencies (LSF) conversion: The LSP coefficients are normalized in the frequency domain $[0,\pi]$ and are mapped to LSF.

7. In G.729B, LSF coefficients are used with a codebook that allows the table index to convey required information. G.729B or G.729AB sends a total of 2 bytes as a SID frame. In G.711, the reflection coefficients are sent in 10 bytes without coding.

8. The VAD decision process makes use of the following parameters:
   a. Spectral distortion computed based on LSF coefficients.
   b. Full-band energy estimation based on autocorrelation coefficients.
   c. Low-band energy estimation based on autocorrelation coefficients.
   d. Zero crossings based on sign change of the input signal.

9. Using the above set of parameters, difference parameters are computed. The set of difference parameters are spectral distortion, energy difference, low-band energy difference, and zero-crossing difference. The differences are calculated between the instantaneous frame parameters and running averages. The difference parameters are required only for decision making in VAD-II and are not used for actual SID frame payload generation.

10. The initial VAD decision using a multiboundary is made based on 14 conditions applied on the above-named differential parameters.

11. The algorithm goes through several checks mainly to avoid declaring a wrong decision such as speech as VAD. The next step is to give a final

VAD decision using smoothing and hangover to reflect the long-term stationary nature of the speech signal. It is done in four stages as described in G.729B. For a smooth decision, energy consideration together with neighboring past frame decisions are used.

12. Once VAD decision is made, power (or energy) level and reflection coefficients are used as payload with proper scaling and normalization. Reflection coefficients are normalized to occupy an 8-bit number with values from 0 to ±254. The value 255 is kept as reserved in the payload. The mean square error between the instantaneous and averaged normalized autocorrelation coefficients are computed to decide whether to use the instantaneous or averaged normalized autocorrelation coefficients for the calculation of reflection coefficients.

13. The frame energy is computed from the first normalized autocorrelation coefficient $R(0)$, where the frame energy is expressed in a base-2 logarithm. The running average is used to avoid any sudden frame energy changes in the speech or background noise. The final value of frame energy is quantized and packed as described in the previous section.

4.4.1 DTX Conditions

The VAD algorithm determines whether the current packet is speech or silence. For each nonactive voice frame, the DTX module decides on nonactive voice update parameters by measuring the changes in the nonactive voice signal. Absolute and adaptive thresholds on the frame energy and the spectral distortion measure are used to obtain the update decision. If an update is needed, the nonactive voice encoder sends the information needed to generate a signal, which is perceptually similar to the original nonactive voice signal. This information comprises an energy level and a description of the spectral envelope. If no update is needed, then the nonactive voice encoder sends a NULL (or no) packet. The nonactive voice signal is generated by the nonactive decoder according to the last received frame energy and spectral shape information of a nonactive voice frame. For each frame, the DTX decision is output as one of the three frame types. This convention may vary in different implementations.

Frame type = 1 Speech frame, control passes to speech encoder

Frame type = 2 SID frame, control passes to VAD encoder

Frame type = 0 Untransmitted frame or NULL packet

To consider sending a new SID frame, a significant change in energy of the spectral envelope between two consecutive SID frames and a minimum time gap of two 10-ms frames are considered. The power-level difference is checked to differ by at least 2 dB.

4.4.2 CNG Algorithm

CNG provides a representation of the background noise at the listening end. The CNG algorithm uses a random number generator to create a white noise source. The generation adaptively modifies the random samples to indicate the background noise at the transmitting side of the system. The decoder produces comfort noise by passing a scaled white noise excitation through a linear prediction synthesis filter. The summary steps of the CNG algorithm are given here and the actual steps are given in [ITU-T-G.711 (2000), ITU-T-G.729B (1996)].

- In parameters update, the reflection coefficients from the last frame are used. The power is used after smoothing. Let LE_{SID} be the received power converted from dBov to base-2 logarithm, and the smoothed power is estimated as $LE(i) = 0.9\,LE(i-1) + 0.1\,LE_{SID}$. This smoothing procedure is done to avoid any abrupt changes in signal energy in the comfort noise.
- Excitation generation is used for generating the random noise. A random number generator with Gaussian distribution is used to create noise source, and this noise is scaled to the correct energy by a factor generated from the smoothed frame energy.
- LP synthesis produces noise matching in the spectrum to the original input signal. The reflection coefficients are used to generate the linear prediction synthesis filter coefficients. The scaled excitation is passed through the filter to produce final comfort noise. The length of excitation is generally equal to frame length. In the first frame of SID (i.e., voice to VAD transition), additional samples are generated equal to the model order ($M = 10$) of coefficients. The first M samples are discarded for proper synthesis.

4.5 POWER-BASED VAD/CNG

Energy- or power-based VAD makes use of 1 byte of payload. The other 10 bytes of G.711 Appendix-II are not available in this energy-based VAD. To maintain the compatibility of the payload, 11 bytes are also used with zero filling of the last 10 bytes. Input to the VAD module is a speech signal. The output from the VAD signifies either the presence or the absence of the speech and 1 byte of power payload at a suitable frame. At the receiver, CNG is generated based on power payload in the VAD packet. The same functional block diagram of Fig. 4.1(e) of VAD/CNG functionality can be referred for power-based VAD/CNG. The power on a frame of samples can be calculated in several ways. Some options used in this book are given here.

The formula given in G.168 [ITU-T-G.168 (2004)] is one of the options to estimate voice power over a block or frame of samples. It is given in Eq. (1.1) in Chapter 1. In VAD-II, the first autocorrelation coefficient R(0) represents power. The parameter R(0) can be calculated easily as sum of squares of amplitudes "e_i" with normalization $R(0) = \dfrac{1}{n} \sum\limits_{i=k}^{k+n-1} e_i^2$, where "$i$" is the index, "$k$" is the starting sample index of frame, and "n" is the frame length in samples. The power payload has to be suitably normalized to match the calibration of 6.17 dBm matching to 0 dBov.

In practice, VAD detection is set when the VAD power level falls below the -42 ± 2-dBm threshold, and it varies in different designs based on user configurations, loss planning, and so on. This threshold is also provided to the user for fixing the required threshold. In any frame, when the signal goes below this power threshold, the signal is declared as silence. The DTX algorithm can apply the logic used in VAD-II on top of power-based detections. At the receiver based on the power value, required noise power is generated. There is no spectral shaping to the random noise in power-based VAD. It is very simple and gives reasonably acceptable quality. VAD-II is always the preferred choice because of spectral shaping and a better way of discrimination between speech and nonspeech.

4.5.1 Signal-Level Mapping Differences

In the literature [Zopf (2002)], amplitude 8031 is taken as the highest amplitude value to generate a 0-dBov level. In this book, 8159 is taken as the maximum value for μ-law scaled maximum amplitude. In a cascaded operation of pulse code modulation μ-law (PCMU), an input range of 7903–8158 is quantized as one particular 8-bit compressed word. As shown in Table 3.1 of Chapter 3, the decoder extracts this as 8031 for any value of encoder input between 7903 and 8158. Assuming the hardware interface is already using a μ-law interface, the processor then has to convert the μ-law input to linear before sending the samples to VAD and other voice modules. Hence, the maximum overload point has to be with 8031. When the interface used is linear, an amplitude of 8159 will propagate to the VAD module calibration. Many implementations use a linear interface to get better quality. The difference in 8031 and 8159 in dB scale $20\log_{10}(8159/8031)$ is 0.13 dB. The dBov values are quantized to the 1-dB level. Hence, using any of these numbers will not create any significant difference. In general, VAD algorithms send a new update on change of power by 2 to 3 dB and the difference of 0.13 dB is negligible. For clarity, interpretation of these two dBov representations with 8031- and 8159-based relations is given. In the literature, both numbers are used for normalizing to 0 dBov.

In PCM A-law (PCMA)-based interface, amplitude range 3968–4095 is mapped to 4032 in decoder output as shown in Table 3.2 of Chapter 3. The usual practice in voice and fax processing is to convert 4095 or 4032 maximum amplitude to μ-law 8159 or 8031 as maximum amplitude. In practice, engineers multiply A-law amplitude by factor of 2. The upscaled A-law amplitude also fits into μ-law with insignificant error.

4.6 VAD/CNG IN LOW-BIT-RATE Codecs

In this section, an overview on low-bit-rate codecs VAD/CNG is given. Low-bit-rate codecs are given in Chapter 3. The codecs such as Internet low bit rate codec (iLBC) [URL (iLBC)], G.722 [ITU-T-G.722 (1988)], and G.728 [ITU-T-G.728 (1992)] do not have built-in VAD/CNG support. Some users considered G.711-Appendix-II as a VAD/CNG scheme with non-VAD/CNG codecs. Several low-bit-rate codecs do support VAD/CNG. Low-bit-rate codecs provide better compression. Because of the availability of higher network bandwidth from service providers, VAD/CNG is not used in deployments, even though VAD/CNG is supported in the codec and as a product feature for maintaining better interoperability. Some features of VAD/CNG operations in low-bit-rate codecs are given here.

G.729AB and G.729B [ITU-T-G.729B (1996)] support VAD/CNG, and the suffix B conveys VAD/CNG support. VAD packets are of 2 bytes with parameters, namely 1 bit for a switched predictor index of the LSF quantizer, 5 bits for the first-stage vector of the LSF quantizer, 4 bits of the second-stage vector of the LSF quantizer, and 5 bits of gain (energy). It has close relevance with G.711 VAD-II. In G.711 VAD-II, all 10 bytes are directly sent, and in G.729AB, these are coded with a table index. Hence, the payload during VAD/CNG reduces to 2 bytes in G.729AB.

G.723.1A supports VAD/CNG with 4 bytes of payload, and G.723.1 is without VAD/CNG support. At the encoder, for each SID frame, the algorithm computes a set of LPC parameters and quantizes the corresponding LSPs to 24 bits using the encoder's LSP quantizer. The VAD operation also evaluates the excitation energy and quantizes it with 6 bits. This yields encoded SID frames of 4 bytes, including the 2 bits for bit rate and DTX information. VAD/CNG basic operations are the same as in G.729AB/G.729B.

Wideband codec G.729.1 does not have a VAD [ITU-T-G729.1 (2006)], which means it cannot send a silence region as separate packets. However, it accepts CNG similar to G.729B and G.729AB. The codec G.729.1 is backward compatible with a narrowband 8-kbps G.729 family. Hence, it is required to support CNG as part of G.729.1 to ensure interoperability between G.729.1-capable end points and G.729B end points. The only constraint is that, G.729.1 has a 20-ms frame duration instead of 10 ms for narrowband G.729.

The AMR codec has a built-in VAD/CNG, but it works differently than G.729B and G.729AB, and it works on a 20-ms frame. The VAD algorithm indicates the type of signal transmitted (i.e., speech, music, or information tones during that frame). Samples of the input frame are divided into nine sub-bands (0–4000 Hz) to arrive at a signal's sub-band energy. AMR VAD [ETSI EN 301 708 (1999), ETSI EN 301 706 (1999)] is generally more complex, and it makes use of the signal level in a sub-band, pitch detection, tone detection that indicates the presence of an information tone, and complex signal detection that indicates presence of a correlated complex signal such as music. The VAD decision function also makes use of estimates from background noise levels. The comfort noise evaluation algorithm uses the averaged LSF parameter vector and averaged logarithmic frame energy of the eight most recent frames before the VAD frame. The SID frame consists of 35 bits.

4.7 MISCELLANEOUS ASPECTS OF VAD/CNG

In this section, practical aspects of VAD/CNG packetization, interoperability, bandwidth saving, and testing aspects are given.

4.7.1 RTP Packetization of VAD/CNG Packets

VAD/CNG is packetized similar to any other voice payload with some minor exceptions. The deviations are in payload type, marker bit setting, and multiple frames handling. VAD algorithms work in a frame of 10 ms or with the basic frame of the codec. With an implementation of 5 ms for the smaller frames, these algorithms may be made to operate on two 5-ms frames. Simple power-based VAD-I can be made to operate with flexible frame sizes of 5, 10, and 20 ms.

The VAD packet can appear immediately after the speech packet, which coincides with speech-to-silence transition. Immediately after the VAD packet, for at least two more voice frames duration, VAD packets are not delivered on the network, which is similar to hangover operation, and after completing VAD hangover time, VAD packets can be sent on the network. The DTX algorithm will decide on when to send the next updated VAD packet. A speech packet can be an adjacent packet to the VAD packet, which means the speech packet can precede or succeed the VAD packet.

Voice solutions use multiple frames up to 80 ms in RTP payload duration. At the input of RTP, several frames are collected to form packets with a required duration. In multiframe packetization, to send any VAD packet, RTP has to release the available voice frames without waiting for required packetization payload period. RTP may send voice frames in combination or separately from available VAD frame. On getting an isolated VAD frame as the update, RTP has to release the isolated VAD packet without waiting for any other frames. It is possible that the next update of VAD or speech may happen after several 100-ms duration and holding of VAD packet may prevent comfort noise updates.

In RTP, payload type 13 is used for CN. The G.711 main codec will use a PT of "0" for PCMU and "8" for PCMA. RTP packets switch Payload type (PT) between 13 and 0 or 8 for sending VAD and speech packets. In some RFC drafts, a PT of 19 is also considered, and subsequently, a PT of 19 was updated with 13.

4.7.2   VAD Duplicate Packets

The speech packets can precede VAD packets. A loss of VAD packet in end-to-end delivery may force the decoder to continue in speech and PLC will try to fill a speech extension for about 50 ms. Several logic conditions for VAD and PLC combinations are given in [ITU-T-G.729B (1996)]. Even though it is not in the recommendation of ITU, it is a common practice to create some duplicate packets of VAD to counter packet impediments. It is always helpful to create two to three duplicate packets at the source for the first VAD packet after the speech packet. At the destination, RTP or jitter buffer discards unwanted duplicate packets.

A new VAD packet is sent on change of background characteristics. In practical implementations, previous VAD packets are repeated at regular intervals of 100 to 1000 ms without waiting for the next packet. This type of timed duplicate packet is helpful in situations of packet impediments and eliminates the disconnection of voice sessions in long silences. In some implementations, RTP sessions disconnect voice when packets are not present for 3 to 30 minutes. This process varies by implementation.

4.7.3   VAD/CNG Interoperability

In VoIP call establishment, VAD/CNG may be negotiated by PT as comfort noise (CN). If CN is not negotiated, the RTP allows discontinuous transmission on any audio payload format [Zopf (2002)]. In G.711 VAD, the payload of 1 or 11 bytes is not known in advance. In practical implementations, PT also differs among multiple VoIP systems. These combinations can create interoperability issues. To improve on the interoperability, it is required to keep flexible implementation. Any one of the payload types of 0, 8, 13, and 19 have to be accepted as VAD. A payload length of 1 or 11 bytes has to be accepted for G.711. In a generic way, the implementations have to cater to any payload size from 1 to 11 bytes irrespective of the available VAD module. If the actual implementation is VAD-I, and received SID is of 11 bytes, then CNG has to discard spectral coefficients. If the implementation is VAD-II, and received SID is of 1 byte, then the spectral coefficients have to be made as zeros in the CNG. Some existing systems in the deployment are not supporting the required logic. VAD-I always interoperates with VAD-I and VAD-II SID packets. Hence, many new VoIP systems use a VAD-I power-based scheme as the default VAD/CNG scheme. On activating specific configurations, VAD-II may be enabled.

G.729AB is the VAD-supported version of G.729A. The suffix B indicates VAD support. A payload type (PT) of 18 is used for generic narrowband G.729 and 98 for G.729.1. The VAD packets from the G.729B codec may select a PT of 18 or 13, and a VAD payload size with G.729 is 2 bytes. In the deployments, it is also observed that a G.729 voice call sends 1 byte of payload with payload type 13, which disturbs the G.729 decoding. G.729 VAD expects 2 bytes of payload with parameters. A payload of 1 byte creates disturbance to G.729 CNG decoding. The 1 byte is power payload. Hence, VAD-I has to operate and decode this even if PT is not matching. Similar situations may happen with other codecs. It is essential to keep flexible implementation that validates PT, presents an active call codec, payload length, and the ability to use G.711 VAD/CNG for low-bit-rate codecs.

CNG has to be supported at all times for better interoperability. The preferred options would be to disable VAD in the send path and keep the CNG module active all the time. On getting any SID packet from destination, the CNG decoder can continue comfort noise generation for proper interoperation.

4.7.4   Network Bandwidth Saving

It is difficult to predict bandwidth savings with VAD/CNG. In qualitative terms, a VAD/CNG operation can save as much as 40% to 60% of network bandwidth. In the ITU handbook [ITU-Handbook (1992)], active speech is given as occupying 41% in some experiments. It will give about 59% in savings during VAD/CNG operations. The bandwidth is dependent on the input signal to the encoder and VAD module used. In case of tone tests with significant power, no VAD silence detections will occur. In the presence of a loud, continuous, and disturbing background, several VAD packets are generated. In recent deployments, bandwidth availability is much higher than the savings from VAD/CNG operation. Higher bandwidth allows voice calls to continue in the VAD-disabled mode.

4.7.5   VAD/CNG Testing

In the previous sections, emphasis is given for power-based VAD and ITU-T G.711-Appendix-II-based VAD. These modules do not have standard test vectors. One good option for VAD-II testing is to use G.729B test vectors. The results of VAD-II can be compared with G.729B VAD detections. The payload cannot be directly compared, but active and nonactive regions closely match within a frame. G.729B gives 2 bytes during VAD, and VAD-II gives 11 bytes. Low-bit-rate codecs have a separate test vector for VAD/CNG testing. A user will be using the bit exact implementation.

Several instruments also support VAD/CNG testing. Instruments [URL (DSLAII)] measuring mean opinion score (MOS) [for example, perceptual evaluation of speech quality (PESQ)] also support VAD analysis. In these instruments, a reference waveform is sent from the encoder side and a received waveform is compared for VAD clippings and any wrong decisions of VAD. It is suggested to select instruments with such VAD/CNG analysis. No stated benchmarks are used on MOS degradation with VAD/CNG. In practical PESQ Listening Quality (PESQ-LQ) measurements, a MOS degradation of about 0.05 to 0.1 in VAD/CNG mode is observed. In general, the use of VAD/CNG creates a certain amount of voice quality degradation.

4.7.6 VAD Clippings

Voice clippings are more common with VAD/CNG. Clipping is the wrong detection of useful speech as silence because it sends it as a VAD packet. Speech clippings happen at silence-to-speech and speech-to-silence transitions. The clippings have to be contained to less than 0.2% to 0.5% of the active speech [ITU-T-G.116 (1999)]. Typical talk-spurts (concatenated speech) are of 300- to 500-ms active speech. The quality goal of 0.5% in 500 ms is 2.5 ms, which is smaller than the usual VAD/CNG frame of 10 ms. Hence, enough care has to be taken in meeting the clipping specifications.

Misclassifying inactive speech as speech results in an increase of the transmission rate, but speech quality is unaffected. Misclassifying active speech as inactive speech causes the speech signal to be clipped, and the speech quality degrades. Most DTX algorithms employ a hangover period during transition from active speech to inactive speech, which minimizes clippings at the tail end of the active speech, but the problems of silence-to-speech clippings remain. The implementations with G.711 VAD-II and low-bit-rate bit exact implementations take care of these requirements, but power-based VAD-I may create some clippings based on the signal characteristics.

4.8 SUMMARY ON VAD/CNG

For waveform-based narrowband codecs such as G.711 and G.726, two popular VAD/CNG algorithms are used. The simplest VAD/CNG works on power levels. In this scheme, VAD detection is made when power level falls below a certain threshold, usually of −42 to −44 dBm represented in dBov units. After sending one VAD packet, another VAD packet will be sent after holding for two frames and power-level changes by more than 2 dB. In the receiver, VAD packets will go through AJB, and this can create a jitter buffer under run. It is suggested to keep memory of the AJB statistics in the transitions of silence and speech. On getting speech packets, established jitter buffer characteristics can be reimposed. In some designs, VAD packets are sent at regular intervals between 100 and 1000 ms to ensure that VAD/CNG works even under packet losses and avoids RTP session disconnects in long silences.

In payload, all negative values of dBov will appear as positive numbers in the range of 0 to 127 dBov. It will correspond to power levels of 6.17 to

−120.83 dBm. Ideal channel noise of −70 dBm is the minimum signal in the system. Hence, catering or clipping at −70 dBm or 76 dBov will be sufficient in deciding the power levels. In case of a sudden power change within the VAD power limits, it is required to limit the rate of a VAD packet update. As per G.169, a signal gain change of 10 dB per second is suggested. In the case of VAD-I, the interpretation is to send packets once in 200 ms at a 2-dB gain change that amounts to a 10-dB change per second to match the ITU-T-G.169 (1999) recommendation.

VAD packets are of 10 ms in packetization. In VoIP, packets can take 10 to 80 ms. For long packetization, VAD packets should not be grouped for long packets. When retrieving the VAD frame, whatever be the current accumulated packet size, it has to be released end-to-end. As an example, in 20-ms packetization of G.711, two 10-ms frames are included in one RTP packet. When retrieving the VAD packet, an RTP packet has to be sent as one 10-ms packet even if 20 ms of payload is not ready.

VAD-II makes use of a power and vocal tract model, and VAD packets are sent on change of characteristics. It is a more complex process than power change in VAD-I. VAD-II gives better quality because of spectral matching of background in silence.

Many low-bit-rate codecs have a built-in VAD. Because of the coding operation of LPC parameters in low-bit-rate codecs, payload bytes are shorter in size than waveform-based VAD/CNG. Because of higher compression in low-bit-rate codecs, VAD/CNG is not preferred, but CNG is always enabled to maintain proper interoperability with any other VoIP system that is sending VAD packets. Overall VAD is not preferred when sufficient network bandwidth is available to the VoIP call, but CNG is kept in active mode for maintaining better interoperability. In case of network bandwidth restrictions, instead of going for low-bit-rate codecs, G.711 with VAD/CNG can provide better quality.

5

PACKET LOSS CONCEALMENT TECHNIQUES

In VoIP network packet drops, packet delay, delay variations, errors, and fragmentation are very common and may degrade voice quality significantly. Packet loss concealment (PLC), also known as packet erasure concealment, synthesizes the missing voice samples during the packet erasures without creating the noticeable artifacts. This chapter on PLC is presented for improving the voice quality in the presence of packet drops. PLC techniques are categorized as transmitter–receiver- and receiver-only based.

Transmitter–receiver-based techniques are simply referred to as transmitter-based techniques and as encoder–decoder-based techniques. In this method, receiver support is required and both transmitter and receiver use the same technique. Transmitter-based techniques are commonly applicable to all compression codecs. These techniques are based on standard methods of redundancy and forward error correction. Receiver-only methods are known as decoder-based techniques. Receiver-based techniques are based on the selected codec and can accommodate nonstandard and proprietary techniques. In this chapter, generic transmitter-based methods and waveform-coding-based receiver methods are given.

5.1 PACKET LOSS CONCEALMENT OVERVIEW

VoIP voice is based on real-time. Internet Protocol (IP) packet delivery. Reducing end-to-end delay for IP packets delivery is important for voice

VoIP Voice and Fax Signal Processing, by Sivannarayana Nagireddi
Copyright © 2008 by John Wiley & Sons, Inc.

quality. Hence, any loss of packet cannot wait for retransmission. Packet impediments are caused by several factors, namely quality-of-service (QoS) actions on the sending side, network impediments, drops in downstream, jitter buffers, and so on. Details on IP packet impediments are given in Chapter 10. These net results of packet drop can result in nonavailability of voice frame for voice decoding. A PLC operation details how to synthesize voice during a resultant packet drop.

Voice codecs based on ITU G.711, G.726, G.729AB, and G.723.1A are popularly used in VoIP applications. The vocal-tract-based codecs G.729AB and G.723.1A have built-in packet loss concealment operations and operate on a vocal tract model for packet loss concealment. These codec models have a long memory, and, hence, it takes more time to recover from lost packets. G.711 and G.726 are waveform-based codecs, which recover quickly from loss conditions to correct speech. G.711 is a sample-based codec that can recover the packet loss state to the correct state instantaneously. G.726 maintains a short memory of previous conditions, but it recovers more quickly than G.729 and G.723.1. Because of relatively quick transitions in G.726, the techniques used for G.711 are applicable for G.726. In this chapter, PLC techniques are described under G.711, but they are also applicable to G.726.

Speech signals are locally stationary. If the erasure is short, it is possible to extract missing speech with local stationary characteristics of previous speech samples. PLC algorithms [ITU-T-G.711 (1999), Funkai and Nakamua (2005), Mahfuz (2001), Elsabrouty et al. (2004)] have to maintain some of the following good properties:

- A synthesized signal should have closely matching pitch and spectral envelopes.
- Should not create un-natural artifacts.
- A minimum possible algorithmic delay.
- Long bursts of erasures should decay to silence zones.
- Should not extend or reduce the original PLC duration.
- The recovery from PLC should be smooth without any artifacts and should recover as quickly as possible to retrieve good, valid packets.

PLC algorithms may not perform well if erasure falls in rapidly changing transitions, which implies that speech-to-silence and silence-to-speech transitions of erasure may not be reproduced properly. Long erasures cannot be recovered, and most algorithms will make long erasures as silences in the process of minimizing the artifacts of speech.

5.2 PACKET LOSS CONCEALMENT TECHNIQUES

The broad classification of transmitter-based and waveform-coding receiver-based PLC techniques is given in Fig. 5.1. In the terminology usage, trans-

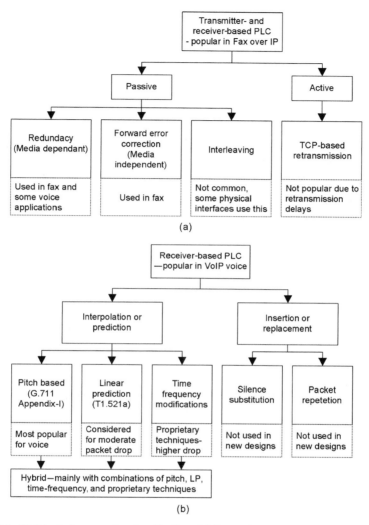

Figure 5.1. PLC techniques major classification. (a) Transmitter–receivers-based techniques. (b) Receiver-only based techniques.

mitter, sender, and encoder are used for the sending side of packets. The keywords "receiver", "receiving", and "decoder" represent the receiving side of the voice packets. The combination of these words also appears in the literature. Transmitter–receiver or transmitter-based techniques are the same. It is implied that transmitter-based techniques use receiver support to decode. Hence, these techniques are simply referred to as transmitter based. The classification is based on major techniques given in [Perkins et al. (1998)]. Several proprietary extensions are available mainly for decoder-based techniques on top of the classification shown in Fig. 5.1(a).

Transmitter-based techniques are classified as active and passive as shown in Fig. 5.1 (a). The techniques use mainly redundancy as per RFC2198 [Perkins et al. (1997)] and forward error correction (FEC) as per RFC2733 [Rosenberg and Schulzrinne (1999), Perkins and Hodson (1998)] for erasure concealment. Interleaving and Transmission Control Protocol (TCP)-based retransmissions are not used in voice applications, because of delay concerns.

In the receiver- or decoder-based techniques, implementers usually combine several features and arrive at their own techniques. In decoder-only based implementation, a transmitter does not need to know the decoder implementation method. The derived techniques are made to work based on several conditions and on a combination of packet loss characteristics. Hence, several proprietary, as well as a combination of, techniques in implementation are available beyond what is reported in the literature. Decoder-only supported techniques are represented in Fig. 5.1(b). Simple decoder-based techniques are based on silence insertion or replacement. Some decoder-only based techniques are as follows:

- Silence substitution, which means mute or noise insertion.
- Packet repetition or selected substitution from previous packets.
- Interpolation-based techniques.
- Pitch-based PLC as per ITU-T G.711 Appendix-I [ITU-T-G.711 (1999)].
- Linear prediction (LP)-based PLC as per T1.521a-2000 Appendix-B [T1-521a (2000)].
- Proprietary techniques of time-frequency modification and GIPS NetEQ [URL (Gips-G711)] proprietary techniques and linear prediction variants [Gunduzhan and Momtahan (2001)].
- Hybrid techniques make use of several combinations of techniques like pitch, LP, interpolation, and time frequency modifications [Perkins et al. (1998b), Funkai and Nakamoya (2005)].

5.3   TRANSMITTER- AND RECEIVER-BASED TECHNIQUES

In the transmitter-based scheme, the transmitter assumes that a certain packet drop could occur in the end-to-end operation. The transmitter sends additional payloads or modifications that help the receiver (decoder/destination) to recover the lost packets. The popular options for these payloads are redundancy and FEC. Retransmission with the TCP-based method is possible for listening and for the broadcast mode of voice communication, but it is not used in interactive voice conversations.

5.3.1   Retransmission or the TCP-Based Method

Retransmission is a TCP-based method. In principle, retransmission-based methods work for any end-to-end media, signaling, or data packets. A lost packet is identified at the destination, and a request is initiated to the transmitter for retransmission of the lost packet. The longest delays of 300 to 400 ms will happen with inter-regional voice calls. As per calculations given in RFC0793, RFC2988 [Postel (1981), Paxson and Allman (2000)], inter-regional calls create a retransmission time-out of 1.0 to 1.6 seconds. Interactive audio applications have tight latency (delay) bounds. End-to-end delays need to be as low as possible and preferably in the upper limit range of 150 to 250 ms with the exception of intermediate satellite links and hard-to-reach areas. For this reason, interactive voice applications typically do not employ retransmission-based recovery for lost packets. Fax tolerates more end-to-end delay up to 3 seconds. Hence, the TCP-based method is possible for fax transmission, even though other techniques are most popular with fax. In practice, the TCP method is used for VoIP voice and fax signaling packets. Retransmission-based schemes work better when loss rates are relatively low, and they take lower bandwidth than other transmitter-based methods of FEC and redundancy. As loss rates increases, the overhead from retransmission and bandwidth requirements increases. At higher packet loss, redundancy or FEC becomes more effective for bandwidth utilization.

5.3.2 FEC

Several FEC techniques have been developed to repair losses of data or media during transmission. These schemes rely on sending additional FEC packets in addition to regular Real-time Transport Protocol (RTP) voice packets. At the receiver, a lost packet is recovered from the regular packets and from the FEC packets. FEC techniques are broadly classified as media-independent FEC, and the media-dependent technique is usually known by the name "redundancy" [Rosenberg and Scholzrinne (1999)]. In practice, media independent is the default referenced name for FEC. Media-independent FEC is given in this section, and redundancy is explained in the next section.

FEC schemes were popularly used for bit error recovery. In digital communication, many media-independent FEC techniques use block or algebraic codes to produce additional packets in transmission to aid the correction of lost packets. In a generic notation, FEC takes k data packets (with each packet of several bytes) and generates $n - k$ additional FEC check packets for the transmission of n packets over the network. Multiple FEC scheme options are given in RFC2733 that were considered for VoIP voice applications. FEC takes care of bursty or more packet losses based on the employed FEC scheme. In the process of an FEC operation, a certain amount of extra information is sent to the transmitted packets. The transmitter adds error correction payload, and the receiver tries to recover the missing data from extra packets or bytes. This operation adds extra overhead to the transmitted bytes and increases bandwidth.

To make use of FEC benefits, the receiver has to hold the received packets in a buffer, which increases the end-to-end delay by one to four packet intervals based on the selected FEC scheme. The complexity of FEC varies based on the selected implementation. For RTP payloads, parity-based algorithms are popularly used. These parity-based algorithms use the exclusive OR (XOR) operation. Most processors support a 16- or 32-bit XOR operation in their general arithmetic. Compared with other voice processing algorithms, FEC processing is insignificant because of XOR and packet loss detection logic operations.

Forward error correction is used in many digital communication systems. It is also useful in T.38 [ITU-T-T.38 (2005)] based fax calls as given in Chapter 16. A FEC technique also helps to encrypt the transmitted bits that help in sending secured VoIP voice and fax. Some FEC-based techniques are reported as patented for VoIP as a part of a secured VoIP communication.

The example used in RFC2733 [Rosenberg and Schulzrinne (1999)] is taken here. For the flow of packets a, b, c, d, and so on, FEC packets f(a,b) and f(c,d) are generated. The operation f(a,b) is for bitwise XOR logic operation of total payloads a and b. The example given below creates two FEC packets for every four original packets, which makes a total of six-packets of transmission for every four packets. For a, b, c, and d, the packets generated by the sender are given here:

$$a \quad b \quad c \quad d \qquad \Rightarrow \text{Original RTP media stream}$$
$$f(a, b) \qquad f(c, d) \Rightarrow \text{Addition FEC stream}$$

where time progresses to the right. In this example, the error correction scheme introduces 50% of overhead. In this example, if packet b is lost, a and f(a,b) can be used to recover b.

Scheme-1 of RFC 2733. This scheme creates double the packets and caters to the bursts of two consecutive packet losses. The packets generated by the sender are given here:

$$a \quad b \quad c \quad d \quad e \quad \Rightarrow \text{Original RTP media stream}$$
$$f(a, b) \; f(b, c) \; f(c, d) \; f(d, e) \Rightarrow \text{Additional FEC stream}$$

Scheme-2 of RFC2733. In this scheme, original packets are not sent. All packets are of FEC. This scheme allows for recovery of all single packet losses and some of the consecutive packet losses but with slightly less overhead than scheme-1. The packets generated by the sender are given here:

$$f(a, b) \; f(a, c) \; f(a, b, c) \; f(c, d) \; f(c, e) \; f(c, d, e) \Rightarrow$$
$$\text{only FEC stream (no original RTP stream)}$$

Scheme-3 of RFC2733. This scheme requires the receiver to wait an additional four-packet interval to recover the original media packets. However, it can recover from one, two, or three consecutive packet losses. The packets generated by the sender are given here:

$$a \quad b \quad c \qquad d \quad \Rightarrow \text{Original RTP media stream}$$
$$f(a, b, c) \quad f(a, c, d) \quad f(a, b, d) \Rightarrow \text{Additional FEC stream}$$

The advantages of the FEC scheme are that its media-independent operation does not depend on the contents of the packet. Recovery is an exact replacement for the lost packet and takes relatively less computation. FEC is simple to implement compared with the voice or fax algorithms. The disadvantage of the FEC scheme is increased delay and bandwidth, but the increase in bandwidth in FEC is relatively less compared with redundancy.

5.3.3 Redundancy

In redundancy-based scheme, packets are sent with a concatenation of payloads with the procedures given in RFC2198 [Perkins et al. (1997)]. Redundancy techniques are popularly used in fax pass-through and T.38-based fax transmission. Fax signals are composed of phase and frequency modulations. In fax transmission, decoder-only based concealment techniques are not used because of the limitations in PLC implementations to recover phase and frequency modulations. In redundancy, primary payload is appended with a required number of previous payloads. This is a simple operation of concatenating the required number of previous basic payloads. Based on the selected media and transport mechanisms, different packet headers are used. RTP-based voice and fax payloads follow implementations as per RFC2198. UDPTL-based fax payloads implement different headers as given in Chapter 16.

Figure 5.2 shows the basic principle of redundancy. In this example, each frame is considered for a duration of 10 ms. In row 1 of the figure, frames marked as 10, 11, 12, 13, and 14, are transmitted for every 10 ms from the transmitter side to the IP network. In row 2, packets with redundancy = 1 are shown. Here the current 10-ms frame and previous 10-ms frame are grouped together to send as one packet. In row 3, the previous two frames and current frame are transmitted as one packet for redundancy = 2.

Redundancy = 1 takes care of one packet loss. As an example, if packet with frames of {13, 14} are lost, packets with frames {12, 13} and {14, 15} can make it to recover all the required frames of a lost packet with {13, 14} frames. If packet drop exceeds more than one packet, redundancy = 2 is required. Redundancy is advantageous for single packet drops or drops not exceeding the redundancy level.

Redundancy = 2 with two packet drops is shown in the fourth row. As an example, if packets with frames of {12, 13, 14} and {13, 14, 15} are lost, the previous and next packets are used to recover the required frames. To get the

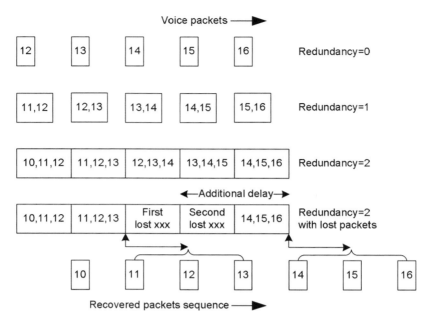

Figure 5.2. Frames recovery using redundancy. [Payloads in a packet are marked as {10, 11, 12} for simple interpretation. Actual order is {12, 11, 10}.]

benefit of this redundancy, the receiver has to wait for extra packet delay. This delay should not vary on a packet basis. Once redundancy is selected, the buffering delay has to hold the required number of packets. In redundancy = 2, the buffer has to hold two more packets to get the benefit out of redundancy.

Bandwidth Considerations for FEC and Redundancy. A redundancy scheme duplicates the payloads in successive packets based on the redundancy level. Redundancy increases the bandwidth in proportion (as approximation) to the redundancy level. To cater for redundancy = 1, bandwidth requirements increase by two times (at least payload). In FEC, one packet loss can be managed with about 50% more bandwidth. In practice, engineers show more favor to redundancy because of simple interpretation. However, FEC gives more benefits because of the availability of direct RTP packets along with FEC packets. The receiver that does not support FEC can simply continue the call with main non-FEC packets. FEC also scales better for security features of voice and fax packets.

5.3.4 Interleaving

Interleaving is used in higher throughput data networks. Interleaving disperses the small units of payloads. Hence, this technique is useful for distributing bursty erasures of packets or payloads into short-duration losses. In digital subscriber line (DSL), very high-speed DSL (VDSL), wireless, and passive optical networks interleaved techniques are used at the physical layer to protect data from burst errors in physical transmission. The interleaver works by segmenting the payload into units. Several units are combined to form one packet. The units are distributed in several packets. At the destination, it requires several packets to retrieve the units. A loss of a single packet from an interleaved stream results in multiple small gaps in the reconstructed stream, as opposed to the single large gap that would occur in a noninterleaved stream. As an example for voice, a 20-ms payload of G.711 can be fragmented into four small payloads of 5 ms. The four small 5-ms frame units can be dispersed. If some packet is lost, the receiver will see only 5 ms of loss and not the whole 20 ms of consecutive data. A short-duration packet drop is easier to recover for PLC algorithms. In practice, the interleaver and FEC are used in combination. The interleaver disperses the burst errors, and FEC takes care of dispersed packet drops or errors.

The main advantage of interleaving is that it does not increase the bandwidth requirements when compared with redundancy and FEC. The interleaver complexity also changes in different applications. The main disadvantages of the interleaver are interoperability, increase in delay compared with FEC and redundancy, more processing, and extra memory to hold several small units of packets. For voice payloads, redundancy is most popularly used. FEC is used as the next option. The interleaver is not popular on direct voice and fax payloads.

5.4 DECODER-ONLY BASED PLC TECHNIQUES

Speech signals are known to contain significant short-duration redundant information, and human perception can allow certain artificially created short-duration voice signals in place of lost signals. In the decoder, the PLC algorithm produces these signals by making use of the history of the samples. During lost packets, the PLC algorithm extrapolates the missing voice samples, and on receiving a good packet, it interpolates for smooth continuity. The PLC algorithm stores a history of the previous speech samples to the extent of 50 ms to regenerated missing speech samples. Figure 5.3 shows a packet drop in the middle of good packets.

In Fig. 5.3, packets 17 and 18 are lost together because of packet drop or they are delayed beyond current acceptable conditions. These packets are assumed to be of 10-ms duration. The PLC algorithm creates synthetic extrapolated speech based on packets 12–16. Interpolation-based algorithms use good packets like 19 and beyond to arrive at the interpolated speech. Interpolation delays the packets and introduces more delay than extrapolation. Different receiver-based recovery techniques are classified as illustrated in Fig. 5.1. A pictorial example of the techniques is given in Fig. 5.3. From the figure, the following observations can be made.

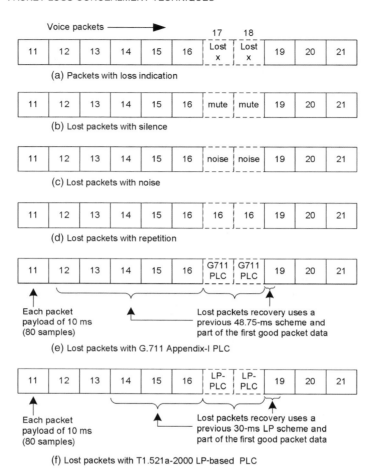

Figure 5.3. Decoder-based PLC techniques. (a) Packet flow with two packets dropped. (b) Lost packet replacement with silence. (c) Lost packet replacement with noise. (d) Lost packets with repetition of previous packets. (e) PLC with G.711 Appendix-I. (f) PLC with LP based on T1.521a-2000.

Insertion-based scheme repairs packet losses by inserting fill-in packets [Goodman et al. (1986), Wasem et al. (1988)]. The fill-in is usually of simple silence, noise, or repetition of the previous packet. These techniques are easy to implement, and at the same time, they do not provide any significant improvement. Packet filling examples are marked in Fig. 5.3(b), (c), and (d). These techniques are not discussed in more detail here because of limited use in the current implementations. In the next section, the other techniques of G.711 Appendix-I, T1.521a linear prediction-(LP)-based, and hybrid techniques are presented. G.711 Appendix-I based scheme is the most popular and makes use of a previous 48.75 ms of speech as marked in Fig. 5.3(e). The

LP-based T1.521a scheme is also found to perform better. The LP-based scheme works on a previous 30 ms of speech as marked in Fig. 5.3(f).

5.5 PLC TECHNIQUES DESCRIPTION

In this section, the G.711 pitch-based technique, the T1.521a LP technique, and an overview on hybrid techniques are presented for performing a PLC operation.

5.5.1 PLC Based on G.711 Appendix-I

The PLC scheme as per G.711 Appendix-I recommendation [ITU-T-G.711 (1999)] is the most popular packet loss concealment scheme to use with G.711 and G.726 waveform-based codecs. The G.711 Appendix-I PLC is known as the PLC scheme based on Annex A. The G.711 Appendix-I scheme gives reasonably better and acceptable quality at low complexity of computation. Unlike CELP-based coders such as G.729, G.723.1, and G.728, the codec G.711 has no model of speech production. With CELP-based coders, the decoder's state variables take time to recover after an erasure. PLC in G.711 has the ability to recover rapidly to the original signal after an erasure is over. This PLC scheme is well characterized and widely used in deployments. The results are available in the published literature for different conditions [Britt (2007), ITU-T-G.107 (2005), ITU-T-G.113 (2002), ITU-T-G.113 (2007), TIA-EIA-810A (2000), TIA-EIA-116A (2006)].

The G.711 Appendix-I scheme is shown in Fig. 5.4. This figure is a simplified version to represent the principle of the scheme. In this decoder path, the jitter buffer provides voice packets to the decoder. The packet type information is analyzed by the control block. This block generates required control and timing information to various internal blocks of PLC. PLC has major operations of identifying the pitch from previous samples and applying overlap and add, scaling to minimize speech artifacts. The G.711 PLC operation with reference to good and lost data is shown in Fig. 5.4(b) and (c). Referring to Fig. 5.4(b), the lost packets use a previous 48.75 ms of data and a small part of the first good packet data. For better smoothing effect, overlap add, and scaling operations are performed on the synthesized speech. In the process of correlation and pitch detection, a part of previous good samples are used as marked in Fig. 5.4(c). On enabling PLC-operation, either decoder output or PLC output is selected based on the control block information. In the PLC-enabled mode, voice is delayed by 3.75 ms. Operations of various functional blocks are described in more detail in the subsequent part of this section. Refer to G.711 Appendix-I [ITU-T-G.711 (1999)] and G.191 [ITU-T-G.191 (2005)] for details, waveforms, and pseudo-code of PLC operations.

During Good Frames. During good frames, the algorithm stores a previous 390 samples (48.75 ms) of the decoded speech in a history buffer. It is marked

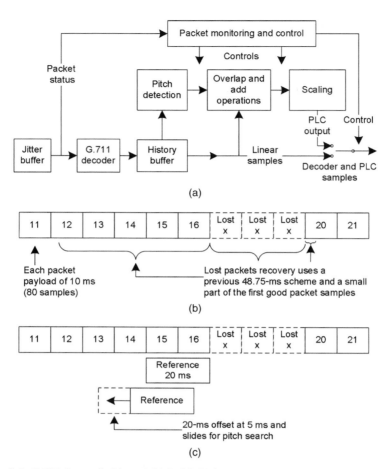

(a)

(b)

(c)

Figure 5.4. G.711 Appendix-I-based PLC. (a) PLC scheme based on G.711 Appendix-I. (b) Diagram is shown for three packets loss and the dependencies on previous and future samples. (c) Representation of correlation windows for pitch detection.

in Fig. 5.4(b). This history buffer is used to calculate the current pitch period and to extract the waveform during an erasure. The storing of samples in the history buffer does not introduce any delay into the output signal. As per the representation of Fig. 5.4(b), this algorithm requires the previous packets 12–16 10-ms frames each to start working for the missing data in packet 17. The output is delayed by 3.75 ms, i.e., of 30 samples. This algorithm delay is required for overlap-add (OLA) and to create smooth transitions between the real and the packet loss concealed by the synthesized signal.

During First Bad Frame. The main basis for this algorithm is pitch detection on the previous 48.75-ms data. The pitch-aligned backward signal from the previous 48.75 ms is replicated in the lost interval. An additional copy of the most recent one-quarter pitch period is also made in case the erasure is longer than 10 ms. The pitch period calculation is done only during the first bad frame.

Pitch Detection. PLC is based on pitch alignment of the waveform. In this scheme, the pitch period is calculated based on a peak location of the normalized cross-correlation of the most recent 20 ms of speech in the history buffer with speech at taps from 5 to 15 ms with reference to most recent speech in the history buffer. The durations of correlation windows are marked in Fig. 5.4(c), which gives a value of pitch in 5 to 15 ms (40 to 120 samples). The pitch range is chosen based on a range used by the G.728's post-filter. Although G.728 uses a lower bound of 2.5 ms (20 samples), here it is increased to 5 ms (40 samples) so that the same pitch period is not repeated more than twice in a single 10-ms erased frame. Pitch is extracted in two steps. The coarse search is on a 2:1 decimated signal. Fine estimation is performed on the localized zone of the coarse correlation peak. To save computation, coarse resolution can also be used directly with slight degradation in the quality. The pitch peak location helps in identifying the alignment for replaying the speech after additional processing steps. As shown in Fig. 5.4(c), the most recent 20 ms of speech is used before the erasure of packet 17. This 20 ms of speech is used as a reference in generating the synthetic speech signal. The extension of the 20 ms is for the correlation window span. The window slides back at taps of 40 to 120 samples. The normalized cross-correlation between these two windows gives the pitch period. This pitch period is used in identifying the synthetic speech signal of appropriate pitch length.

Synthetic Signal Generation for First 10 ms. During the first 10 ms of an erasure, the synthetic signal is generated from the last pitch period with no attenuation. The most recent 1.25 pitch periods of the history buffer are used during the first 10 ms. An OLA is performed using a triangular window on one quarter of the pitch period between the last and the next-to-last period. The OLA process ensures that a smooth transition occurs between real and synthetic speech. This operation creates a smooth transition if the pitch period is repeated multiple times. For one-quarter wavelength, the signal starting at 1.25 pitch periods from the end of the history buffer is multiplied by an upsloping ramp and is added to the 0.25 pitch period in the additional one-quarter pitch period buffer multiplied by a downsloping ramp.

The result of the OLA replaces the tail of the history buffer. It is also output by the receiver during the tail of the last good frame, replacing the original signal. This process introduces the algorithm delay—the tail of the last frame cannot be output until it is known whether the next frame is erased. If an erasure occurs, OLA modifies the signal in the tail of the last good frame to provide a smooth transition to the synthesized signal. The synthesized signal for the 10 ms during the erasure is generated by copying the most recent pitch period samples to the output. If the pitch period is less than 10 ms, the last pitch period in the history buffer is repeated multiple times during the 10-ms erasure.

For Long Erasures. If the erasure interval is more than 10 ms, pitch is extended backward in the waveform. Although repeating a single pitch period works well for short erasures like 10 ms, on long erasures, it introduces unnatural harmonic artifacts, which is especially noticeable if the erasure lands in an unvoiced region of speech or in a region of rapid transition. These artifacts are reduced by increasing the number of pitch periods used to synthesize the signal as the erasure progresses. Playing more pitch periods increases the variation in the signal. If the erasure is 20 ms long, the number of pitch periods used to synthesize the speech is increased to two, and if erasure is 30 ms long, a third pitch is added. Beyond 30 ms of erasure, no changes are made to the history buffer. Transition in the synthesized signal should be smooth as the number of pitch periods used in the history buffer is increased. It is accomplished by continuing the output of the existing history buffer for one quarter of a pitch period at the start of the second and third erased frame, updating the history buffer, keeping the buffer pointer synchronized with the correct phase, and then doing an OLA with the output from the new history buffer.

The history buffer is updated exactly as during the first erased frame, except that the number of pitch periods is increased. For example, at the start of the second erased frame, for one-quarter wavelength, the signal starting at 2.25 pitch periods from the end of the history buffer is multiplied by an upsloping ramp and is added to the one-quarter wavelength in the additional copy of the most recent one-quarter pitch period buffer multiplied by a downsloping ramp. The result of the OLA replaces the last one-quarter wavelength in the history buffer. To maintain the phase of the synthesized signal, pitch periods are subtracted from the pointer until it is in the first pitch period used.

As the erasure gets longer, the synthesized signal more likely diverges from the real signal. Without attenuation, strange artifacts are created by holding certain types of sounds too long. The replicated signal is applied with attenuation at the rate of 20% per every 10 ms. For the first 10 ms of erasure, the signal is not attenuated. From the start of the second erasure, the synthesized signal is linearly attenuated with a ramp at a rate of 20% per 10 ms. By the end of the sixth erasure, the synthesized signal is zero.

First Good Frame After an Erasure. At the first good frame after an erasure, smooth transition is needed from synthesized speech to the real signal. This transition is achieved by continuing the synthetic speech beyond the end of the erasure, and it is mixed with the real signal using OLA. The overlap and add window length depends on both the pitch period and the length of the erasure. For 10 ms of erasure, one quarter of the pitch period is used as a window. For longer erasures, the OLA window is increased at a rate of 4 ms per 10 ms of erasure, up to a maximum of 10 ms.

As mentioned, the G.711 Appendix-I-based PLC algorithm has an algorithm delay of 3.75 ms. To minimize the computational delay, the PLC algorithm can be executed after every good frame, before it is known if the next frame is erased at a slight cost in million instructions per second (MIPS) and memory. If the next frame is erased, the synthetic signal is available immediately. If the next frame is not erased, the synthetic signal is just discarded.

5.5.2 LP-Based PLC

LP conveys prediction of speech that helps with packet loss concealment. LPC is linear prediction coding conveying coding or compression. The usage of LPC is more common than LP. In practice, engineers use the LPC-based PLC name even though the LP-based PLC name is more correct. LP-based schemes are based on T1.521a (2000). This standard is also referred to as Annex B.

In many compression codecs, such as G.729 and G.723.1, waveforms are modeled through LPC. In compression codecs, about 10 to 12 LPC coefficients are derived. Compressed versions of the coefficients are sent to the destination along with other parameters. When noise or derived excitation is passed through these coefficients, the LPC filtered signal closely matches with original speech. A similar technique is extended for packet loss concealment. In a decoder-based PLC scheme, instead of using LPC coefficients for compression, they are used for prediction of speech.

Speech consists of the predictable part of the vocal tract model as well as the residual part that contains excitation. The goal is to derive model and excitation for the missing part of speech based on the history of the speech. To predict the lost speech, the history of the speech is preserved. On detecting a lost packet, 20 LPC coefficients are derived based on the history of the speech. These coefficients are valid for a short duration to reconstruct the missing part of the speech.

As shown in Fig. 5.5, the main blocks of the LP-based scheme are history buffer, LP analysis and filter, pitch detector, excitation generator, inverse LP filter, OLA unit, and scaling unit. LP and inverse LP coefficients are derived based on the previous speech buffer samples. The LP filter is derived based on autocorrelation and on the Levinson–Durbin recursive scheme. Once filter coefficients are derived, they are used during the remaining part of the packet erasure. Passing the speech through the LP filter produces a residual signal. Pitch is extracted on the residual signal. The pitch period and residual signal are used in deriving the excitation. Excitation when passed through an inverse LP filter can recover the reconstructed speech, which is the predicted speech to be used for replacing the erased part of the speech. The speech is applied with time-dependent OLA and attenuation in a scaling operation, which helps reduce the artifacts in a long erasure of frames.

The basic features of LP-based PLC are as follows:

- This scheme has a 5-ms algorithmic delay. The history buffer stores 25 ms of previously played out speech and 5 ms of speech that have been received but not yet played, which produces a 5-ms delay. The overlap buffer contains a 5-ms extension of the generated signal. It is used for overlap and

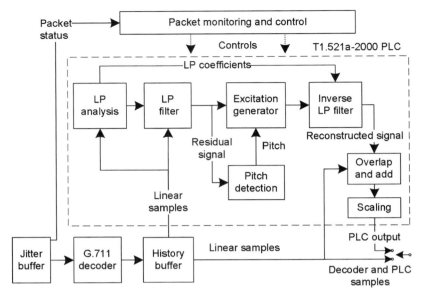

Figure 5.5. LP-based PLC scheme as per T1.521a-2000. [Courtesy: Reprinted from T1.521a-2000 (Supplement to T1.521-1999, Packet Loss Concealment for use with ITU-T Recommendation G.711) © 2001 Alliance for Telecommunications Industry Solutions. For more information about this document, please see the ATIS document center at https://www.atis.org/docstore/.]

adding with the first good packet that comes after the lost segment. Most computations are done at the first packet of the lost segment.

- The last 15 ms of the history buffer is used to compute 20 LP coefficients. If the signal energy is too low, LP coefficients are not computed and an all-zero vector of LP coefficients is passed to the LP filter and to the inverse LP filter.
- The pitch period of the previous speech frames is estimated by searching for the peak locations in the normalized autocorrelation function of the residual signal. Pitch periods ranging from 2.5 ms to 15 ms are searched at a resolution of one sample. Samples of the pitch period are passed to the excitation generator. These are also stored and used in the case of a consecutive packet loss.
- The residual signal and the computed pitch period of the previous speech frames are used to generate an excitation signal. The size of the excitation signal is of 120 samples (15 ms). It is also stored for future use in case the next packet is lost.
- The excitation signal is filtered by the inverse LP filter to add the vocal tract information. The output signal is 120 samples long for the packet size of 80 samples. The first 80 samples are used to replace the lost segment, and the last 40 samples are used for OLA with the next segment.

- The OLA unit uses the reconstructed signal and the samples from the speech buffer to generate the output frame. The two 5-ms segments at the beginning and the end of the reconstructed signal (output signal from the inverse LP filter block) are used for OLA operations. The 5-ms (40 samples) segment at the beginning of the reconstructed signal (size of 120 samples) and the 5-ms (40 samples) segment at the end of the speech buffer (size of 240 samples) are weighted by a triangular window and summed. This resulting (windowed and summed) signal replaces the 5-ms (40 samples) segment at the end of the speech buffer. The 5-ms (40 samples) segment at the end of the reconstructed signal is copied into the overlap buffer. It is used for OLA with the next packet if the next packet is not lost.

- The output frame is scaled down before it is played-out to the speaker by multiplying each sample by the current value of the scale. The scale is set to 1.0 when the algorithm is initialized. Starting from its value at the beginning of the current output frame, it is decreased at each sample with a slope of 0.054 per 10-ms packet, and it is multiplied by that sample. This process continues up to the last sample of the output frame or for 20 ms— whichever is first. After 20 ms, the slope is increased to 0.222 per 10-ms packet. The scale is zero after 60 ms of consecutive packet loss. The output reconstructed frame is completely muted after the scale is zero. The value of the scale at the end of the output frame is retained, and it is used at the following packets.

- If the current frame is lost and the previous frame indicator is also set as lost, then consecutive packet loss is declared. In this case, a new excitation signal is generated using the previously computed excitation signal and pitch period information for the first lost packet. The excitation signal generator block, the inverse LP filter, OLA block, and scaling block perform the same operations as before. LP analysis and pitch detection operations are not used in consecutive packet drop.

5.5.3 Hybrid Methods

Some hybrid methods make use of the combined benefits of ITU-T G.711 Appendix-I and T1.521a-2000. The technique in [Funkai and NaKamura (2005), Gunduzhan and Momtahan (2001)] is published as a hybrid method. Some main features of the hybrid technique are given here. Hybrid techniques use a combination of G711 PLC, LPC, and some proprietary algorithms.

In one hybrid technique, a part of the G.711 pitch-based PLC signal is used to excite the LPC-based system. The outputs from both LPC based and G.711 PLC based are combined with scaling to create a signal on a relative scale.

The hybrid techniques reported by Global IP sound [URL (Gips-G711)] are held confidential by the company and only made available to their customers along with other modules. These techniques are reported to be working much better than the G.711 Appendix-I for higher packet drops of 10% to 20%. These techniques are much more complex and take a lot of MIPS and memory to implement on processors. These techniques are found to produce a good perception mean opinion score (MOS). Voice quality with Perceptual evaluation of speech quality (PESQ) [ITU-T-P.862 (2001)] may be sensitive to this type of measure.

In our internal evaluation of the hybrid technique, LPC, G.711 PLC schemes have produced similar results with up to 5% packet drops. Beyond 7% packet drops, hybrid scheme seems to be performing slightly better. While writing this chapter, many deployments are targeting for less than 1%, and this guideline of 1% is also listed in TIA/EIA-116A (2006). Hence, ITU-G.711 Appendix-I PLC or T1.521a PLC may be sufficient for practical applications of recent deployments.

5.6   PLC FOR LOW-BIT-RATE Codecs

In the previous sections, PLC is given for the G.711 and G.726 codecs that makes use of previous speech to extrapolate the lost packets. Waveform-based codecs store a large amount of previous samples. As such, previous parameters are not important with waveform codecs. In low-bit-rate codecs of G.729, G.723.1, and Internet low bit rate codec (iLBC), coded parameters of LPC, pitch, and excitation are used. The decoder keeps preserving the parameters during good frames, and these parameters from good frames are reused for predicting the lost speech. During long erasures, both waveform and low-bit-rate codecs (LBCs) reduce their amplitude and make it merge with the silence level. The transmitter–receiver-based approaches are common in all codecs. High-level applications outside the speech codec manage operations of FEC and redundancy. Even after applying FEC and redundancy, some loss of packet could occur based on the impediment conditions. In such situations, a decoder-based PLC has to continue improving on packet losses.

In summary, waveform codecs preserve previous speech, and low-bit-rate codecs preserve previous parameters like LPC, pitch, and excitation. PLC algorithms for low-bit-rate codecs are embedded with a decoder of the main codec. A user may not make any changes to the PLC operation in low-bit-rate codecs and may modify certain application-level conditions outside the decoder to decide on packet loss conditions. The test vectors used with the decoder ensure that these PLC modules are well tested. For completeness, summary points on a low compression codec's PLC schemes are given here. High-level codec details are given in Chapter 3. Refer to a codec's main recommendation documents for more details.

PLC for G.729AB.   The PLC in G.729 is operational on any erased frame of 10 ms. The concealment strategy has to reconstruct the current frame based on previously received information. A synthesis filter in an erased frame uses the

LP parameters of the last good frame. These parameters are repeated during frame erasures. Generation of the replacement excitation depends on the periodicity classification of the previous frame as periodic or nonperiodic. The method replaces the missing excitation signal with one of similar characteristics, while gradually decaying its energy to silence during long erasures.

PLC for G.723.1. The codec G.723.1 supports the compression rates of 5.3 and 6.3 kbps with a 30-ms frame. If a frame erasure occurs, the decoder switches from regular decoding to frame erasure concealment mode. The frame interpolation procedure is performed independently for the line spectral pair (LSP) coefficients and the residual signal. It works similar to G.729/G.729A codecs with different frame sizes.

PLC for iLBC. The codec iLBC has a built-in decoder/receiver-based PLC. The scheme in iLBC works like a hybrid scheme of PLC and caters to higher packet drop compared with G.729 family of codecs. iLBC makes use of 20- and 30-ms frames depending on the bit rate. It operates on LP filters, pitch, and excitation signals. During good frames, the decoder state structure preserves information of the current block, LP filter coefficients for each sub-block, and the entire decoded excitation signal. This information is used on packet loss.

PLC for G.722 Wideband codec. The Wideband codec G.722 PLC works as a waveform-based PLC. G.711/G.726 both use a waveform-based PLC. The G.722 algorithm operates on frame size of 10 ms. During packet loss, past buffered output speech of G.722 is provided to the wideband pulse code modulation (WB PCM) PLC. The WB PCM PLC is based on periodic waveform extrapolation (PWE), and pitch estimation is an important component of the WB PCM PLC. The WB PCM PLC output is a linear combination of the periodically extrapolated waveform and noise shaped by LPC. For extended erasures, the output waveform is gradually made into silence (mute). The muting starts after 20 ms of frame loss and becomes complete after 60 ms of loss. There are six types of frame erasures conditions with G.722 PLC, and each of these categories follows extended procedures as given in [ITU-T-G.722 (2007)].

PLC for AMR. In adaptive multirate codecs, the network shall indicate erroneous/lost speech or lost Silence Insertion Description (SID) frames by setting speech bad, speech lost, and SID bad flags. If these flags are set, the speech decoder performs parameter substitution to conceal errors. To mask the effect of an isolated lost frame, a predicted frame based on previous frames substitutes the lost speech frame. For several subsequent lost frames, a muting technique is used to indicate to the listener that transmission has been interrupted.

PLC for G.728. In G.728, Annex I describes the modifications of the G.728 decoder to make it suitable for handling frame erasures in the received bit stream. The concealment operation of erased frames includes extrapolation of the excitation and the LPC filter coefficients, continuation of certain backward adaptation operations for LPC and gain predictors, and limiting the rate of growth for the gain after frame erasures.

PLC for G.729.1. The codec G.729.1 is backward compatible to narrowband ITU-T-G.729 codecs. G.729.1 operates on 20-ms frames referred to as a super-frames, whereas narrowband 10-ms frames and the 5-ms frames are called sub-frames. Because of wide combinations with G.729.1 and backward com-patibility, frame erase concealment in the G.729.1 codec makes use of signal classification. The speech signal can be roughly classified as voiced, unvoiced, and pauses. Voiced speech contains an important amount of periodic compo-nents and is divided into voiced onsets, voiced segments, voiced transitions, and voiced offsets. In general, G.729.1 is much more involved in PLC steps than regular G.729.

A voiced onset is defined as a beginning of a voiced speech segment after a pause or an unvoiced segment. During voiced segments, the speech signal parameters (spectral envelope, pitch period, ratio of periodic and nonperiodic components, energy) vary slowly from super-frame to super-frame. A voiced transition is characterized by rapid variations of a voiced speech, such as a transition between vowels. Voiced offsets are characterized by a gradual decrease of energy and voicing at the end of voiced segments.

5.7 PLC TESTING

The PLC requirement arises with loss of packet in the network. In the deploy-ment, it may not be possible to monitor the PLC performance. PLC testing may not be straightforward and simple. It requires a lot more interpretations of conditions and results. While making the modules and product, PLC has to be evaluated thoroughly. To create packet drop, several approaches can be used depending on the available instrumentation and product cycle. To test PLC performance at the module level,

1. It is required to drop packets at the input of the decoder based on the required loss pattern. Stand-alone PESQ software may be used that compares both reference input and degraded output (PLC-enhanced output signal) for various PLC and decoder operations. The results from this type of test may be used as reference for system-level tests.

2. The next level of module-level integration, jitter buffer input, can be used to erase the packets. Achieving similar results of the previous module-level test is the desired goal. Sometimes the jitter buffer may influence the packets, mainly in disturbing the packet loss duration adjustments. Increase of actual loss duration or reduction of loss duration in jitter buffer or at any other stage in the path is more harmful to the PLC operation.

3. The third levels of tests are on IP network with the test setup shown in Chapter 13. IP impediment generators are used on the IP network that creates several impediments, and they can be controlled for the required conditions. In this setup, actual instruments are used that measure the score to check the PLC performance.

4. While making the measurements and comparing them with published results, it is essential to know the various conditions applied in the test. When PLC is enabled, some differences can always exist among MOS reported by subjective methods and objective measurements as listed in Chapter 20.

5.8 PLC SUMMARY AND DISCUSSION

Packet loss is the major contributor of voice quality degradation as given in this chapter and in Chapter 20. Packet losses are unavoidable at least in certain conditions. The first priority should be to avoid end-to-end packet drops, and that may be achieved at the cost of actions that may increase end-to-end delay. In most situations, increasing delay is more acceptable than packet drop and PLC action. Packet drop contributions are usually apportioned to the IP network, but in practice, major contributions happen with Internet service provider interfaces, terminating routers, and end VoIP systems. IP backbone networks provide much better performance in transporting IP packets. In practice, the user will have a tendency to increase data applications that quickly consume the available bandwidth. Activating QoS mechanisms and ensuring effective functioning of these in real deployments, reserving enough bandwidth for voice, monitoring, and dynamically incorporating feedback, can improve on packet delivery. When more bandwidth is available for VoIP, using lower compression codecs such as G.711 and G.722 (in wideband) is more beneficial to counter packet losses. Applying transmitter–receiver-based FEC, redundancies are preferred, but in the case of bandwidth constraints, these techniques degrade packet delivery. Hence, it is essential to operate with codec and packetization combinations that allow some reasonable margin within the available bandwidth.

Packetization size choice is also important for better PLC performance. If the packet frame size is larger, then encountered burst loss is more. For better PLC performance, a smaller packet size is preferred. Ensuring end-to-end no-packet drop is one of the major deployment success parameters to provide voice quality. Ideally, a VoIP call should not call for a PLC operation.

RTP Control Protocol (RTCP) payloads, E-model-based voice quality, and voice quality monitoring techniques like VQmon can provide feedback to reduce the packet drop. Using real-time feedback from these modules and applying corrections for improvement helps in reducing the demand for long-duration PLC. In VoIP, when fax transmissions are happening in T.38 or fax pass-through is enabled, receiver-based PLC is kept in the disabled state. Adaptive jitter buffers have to be fine-tuned to minimize the packet drops outside the network impediments. In the early stages of VoIP evolution, packet drops exceeding 5% to 10% were considered. In recent deployments, networks and bandwidth is managed better. Many times, no packet drops occur. If the packet drops are contained in less than 5%, G.711 PLC as per ITU or T1.521a-2000 is sufficient. If packet drop exceeds 5% to 7%, some hybrid techniques are more helpful. Low-bit-rate codecs have a built-in PLC, and the designers can stick to the characterized performance with low-bit-rate codecs. Trade-off in choice of algorithm is also dependent on the networks, Internet service provider (ISP), and country of deployment. In general, ITU- and LP-based PLC techniques are presented clearly with pseudo-code in the main documents. The other techniques listed in the literature require a separate effort and a lot of tuning and experimentation as well as measurements. Hence, some users may stick to G.711 Appendix-I or to T1.521a PLC.

6

ECHO CANCELLATION

In telephonic voice conversation, echo is the return of a person's speech with delay, with reduced or modified sound level, and with a certain amount of distortions. In voice communication, echoes are created as acoustic and electrical. Acoustic echo is generated through an acoustic medium of a speakerphone or hands-free phone functionality. Electrical echoes are created in the two-to-four-wire telephone conversion hardware hybrids. Electrical echo is also called line echo.

In public switched telephone network (PSTN) local service, analog private branch exchange (PBX) systems, and VoIP, echoes are created inside the telephone and at two-to-four-wire telephone interface circuits. Echo in isolation is not a troubling signal in PSTN or VoIP. The increase in end-to-end delay makes echo perceivable and more annoying. In VoIP, end-to-end delay is caused by packet-based transmission. End-to-end delays also increase with higher packetization intervals and IP impediments to voice packets. The delays in PSTN voice calls are low even for local long-distance calls. Lower delay helps to tolerate echo as a sidetone. Sidetone is an attenuated version of the microphone signal fed back to the speaker. Sidetone returns immediately within 1-ms [ITU-T-G.107 (2005)] delay and for delays above 2 to 4 ms, a sidetone may be perceived as echo. In inter-regional PSTN calls, delays are in the order of 100 to 300 ms [ITU-T-G.114 (2003a)], and it becomes mandatory to incorporate echo cancellers (ECs) at the long-distance PSTN terminations. By the very nature of packet-based transmission, VoIP calls encounter

VoIP Voice and Fax Signal Processing, by Sivannarayana Nagireddi
Copyright © 2008 by John Wiley & Sons, Inc.

typically 50 to 80 ms more delay compared with a PSTN call for the same distance. Local PSTN calls do not need an echo canceller, but all VoIP calls will need an echo canceller because of 50- to 100-ms delay, even for local calls.

In the early operation of echo cancellation, echo suppressors [ITU-T-G.164 (1988)] were used. An echo suppressor works as a switch, and it removes echo based on signal power and timing. Echo suppressors are not suitable for good-quality voice. Echo suppressors are not in use with the availability of sophisticated adaptive algorithms and processing hardware. In the current context, echo cancellers are used as part of the inter-regional PSTN or any VoIP system to reduce the echo level. An echo canceller removes echoes by adapting to the varying echo path conditions. Echo cancellation is one of the major voice quality-enhancing operations in VoIP and long-distance PSTN service.

Echo cancellers are classified as acoustic echo canceller (AEC) and line echo canceller (LEC). LEC is required with calls that use two-to-four-wire conversion or interfaces. Analog phones used with VoIP adapters and foreign exchange office (FXO) interfaces use a two-wire TIP-RING interface. These interfaces and phones use hybrids that create electrical echo. The electrical echo canceller when referenced with packet-based calls is called a network echo canceller.

AEC is applicable to all hands-free interactive voice applications. Acoustic echo is much stronger than electrical echo and more dynamic in echo path characteristics. In relation to VoIP, many Internet Protocol (IP) and wireless fidelity (WiFi) phones use speakerphone mode. IP and WiFi phones use a four-wire interface; hence, in non-speakerphone mode, no echo from these devices occurs. AEC is an essential requirement in speakerphone mode. Both LEC and AEC share several common techniques, but AEC will use additional control operations to cater to rapid change in signal dynamics.

An echo canceller in a total VoIP voice chain is represented in Fig. 2.6 of Chapter 2. Several other voice quality aspects and echo canceller contributions on overall voice quality are given in Chapter 20. In this chapter, LEC, AEC, basics on adaptation algorithms, control plane, and echo canceller measurements are given.

6.1 TALKER AND LISTENER ECHO IN PSTN VOICE CALL

Figure 6.1(a) represents functional PSTN-based voice conversation. In the figure, a voice conversation is happening between two phones—A and B. In a PSTN call, the telephone interface is terminated at the central office (CO) or digital loop carrier (DLC) through a two-wire TIP-RING interface. The DLC hybrid as part of a subscriber line interface circuit (SLIC) creates a two-to-four-wire conversion. Analog-to-digital converter (ADC) and digital-to-analog converter (DAC) are referred to as a subscriber line access circuit (SLAC), also known as hardware CODEC. The DLC-to-PSTN CO and CO-to-CO interfaces work as four-wire conversions with separate send and receive paths. In the local PSTN call, the CO completes the call directly or through another DLC. In the long-distance call, the PSTN CO will send voice and signaling bits to the destination CO. The destination CO completes the call through its local DLC and telephone interfaces. The interface between the DLC and the telephone is a two-wire TIP-RING in most countries. Inside the telephone, a handset works equivalent to a four-wire interface, but the telephone interface going to the DLC is of two wires. A hybrid inside the telephone creates a two-to-four-wire conversion. Echo is created at every two-to-four-wire hybrid of DLC and telephone. A telephone hybrid echo level of 6 to 14 dB is more dominant than a DLC electronic circuit hybrid echo of 18 to 24 dB. In Fig. 6.1(a), echo is created at the telephone hybrid. For simplification in representation, the echo path through the DLC hybrid path is not shown in Fig. 6.1(a), but it is marked in Fig. 6.1(b) on the phone-B side.

Talker echo is explained with an example call between A and B. In the example shown in Fig. 6.1(a), the person at phone-A is speaking and is called the "talker." The person at phone-B is the "listener." Voice from A goes up to B, and part of A comes back to A along with voice originating from B. Person at phone-A will be hearing his own voice called "talker echo" with round-trip delay. The talker listening to his own voice after delay is the talker echo. Round-trip delay is the transmission time for voice to travel from A to B and returning from B to A as echo. The delay from either A-to-B or B-to-A is called one-way delay. Average one-way delay is calculated as half of round-trip delay. While A is speaking, the presence of echo with increased round-trip delay limits the conversation comfort for person-A. The person at A slows

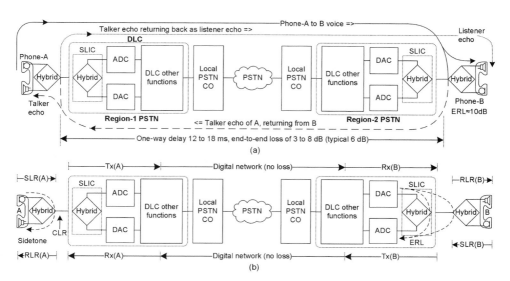

Figure 6.1. Echo representations in PSTN. (a) Talker and listener echo functional representation. (b) End-to-end losses and echo representation.

down conversation with long pause periods and lowers the speaking sound level to minimize the echo annoyance. The same interpretation is applicable to B. When B is speaking, B will also be getting his own voice (talker echo) with round-trip delay.

Echo cancellation requirements are given in G.168 [ITU-T-G.168 (2004)]. G.131 [ITU-T-G.131 (2003)] provides recommendations for acceptable talker echo in relation to one-way delays [ITU-T-G.114 (2003a), ITU-T-G.114 (2003b)]. At lower delays, some amount of talker echo is not a problem. The annoyance of echo keeps growing with an increase in the end-to-end delay.

It is also possible to see multiple circulations of echo, creating listener echo. Listener echo characteristics are given in the G.126 recommendation [ITU-T-G.126 (1993)]. Similar to talker echo, listener echo is also sensitive to one-way and round-trip delays. A's voice after returning back from B as talker echo may leak back to the phone-A hybrid and go back up to B. This process makes B to hear the same speech of person-A's voice a second time or multiple times. The second time voice of A is the "listener echo"—the echo of the talker's voice heard by the listener a second time. In both listener and talker echo, B is listening to A's voice. Listener echo is mainly created when both A and B are not keeping any echo cancellers or minimum losses in the system. VoIP systems use an echo canceller as part of voice processing algorithms. Hence, listener echo is not applicable in VoIP voice calls with minimal echo cancellation operation. Listener echo is more delayed than talker echo and more annoying than talker echo. The presence of listener echo indicates instability in overall transmission that may be resulting from the improper gain/loss planning in the overall system and the characteristics of phones used. The incorporation of talker echo cancellation can prevent listener echo.

6.1.1 Echo and Loudness Ratings

In this section, SRL, RLR, CLR, transmit (Tx)/receive (Rx) losses, LR, TELR, ERL, TCLw, mean one-way delay, and sidetone are presented in relation to markings made in Fig. 6.1(b). These parameters play a major role in echo characterization and perception. The values in this section with dB scale are loss parameters, and a high value indicates a higher amount of loss.

SLR is the send loudness rating. It is the ratio represented in dB scale of the sound pressure produced by a talking person to the voltage produced by the telephone on the TIP-RING interface. Analog telephones give an SLR of about 7 dB [ITU-T-P.79 (1999)] and vary by country requirements. In Chapter 1 under TR-57 tests, another parameter SRL is given. This SRL is the singing return loss, which is not the same as this SLR. This SRL is an electrical reflection loss that is closely associated with ERL.

RLR is the receive loudness rating. It is the ratio represented in dB scale of the telephone electrical signal to the acoustic signal conversion [ITU-T-P.79 (1999)]. Analog telephones give an RLR of about 3 dB and vary by country requirements. RLR and SLR vary based on the country-specific phone. The combined value of SLR and RLR should be 10 dB for a handset of nominal performance. The SLR minimum is 3 dB, and the RLR minimum is 1 dB. The minimum values will make the voice too loud and create degradation of voice quality.

CLR is the circuit loudness rating decided by the two-wire lines from PSTN DLC to the telephone. In PSTN, long lines will go from DLC/CO to the end-user telephone. CLR will also contribute to the losses, but it is not significant. In VoIP service, CLR is not accounted for because of the short lines from the VoIP system to the user telephone.

Tx and Rx losses are the intentional losses kept in the system by design. Tx loss is associated with the sender. Rx loss is at the receiving end. The losses are marked as Tx(A) and Rx(B) in Fig. 6.1(b). These losses are also called padding (meaning attenuation) losses. These losses are called send (top) path and receive (bottom) path losses. The same losses are also applicable in each side of the VoIP system. There are no losses in four-wire digital PSTN and on IP network.

ERL is the echo return loss. ERL [ITU-T-G.168 (2004)] is attenuation of a signal from the receive-out (R_{out}) port to the send-in port (S_{in}) of an echo canceller. Hybrids inside SLICs and telephones create echo. In practical systems, phone ERL (6 to 14 dB) is the more dominant contributor than SLIC ERL (18 to 24 dB). For echo cancellation, the resultant ERL marked as passing through a SLIC and telephone hybrid in Fig. 6.1(b) is considered. The singing return loss explained in Chapter 1 is related to ERL. ERL has to be as high as possible. Good phones with matching impedance of phones and interfaces offer an ERL of 24 dB.

TCLw is the weighted terminal coupling loss of the telephone sets. TCLw is applicable to digital handsets and to IP phones in non-speakerphone mode. Digital phones are used with PBX and integrated services digital network (ISDN) service, and IP phones directly use a four-wire interface. Electrical echo is not present on four-wire interface phones. A small part of acoustic echo couples through handsets, but the coupling loss in TCLw is of 45 dB [ITU-T-G.131 (2003)]. This loss is sufficient to use four-wire equivalent phones without echo cancellation. TCLw is measured as attenuation from the digital input to the digital output at the 14 1/3-octave bands between 200 Hz and 4 kHz [ITU-T-P.341 (2005)]. TCLw is measured when no signal occurs from the local user speech in the send path.

LR is the loudness rating. It is the loss from the mouth of A to the ear of B, similar to end-to-end transmission loss taking into account both end phones [ITU-T-P.79 (1999), URL (Cisco-EC), TIA/EIA-912 (2002)]. The loudness rating value consists of losses in the system—namely the SLR, CLR of lines, RLR of the receiving system, and end-to-end any padding losses. For sending voice from A to B, EC at A will not introduce any loss. CLR is usually a small value. In relation to markings made in Fig. 6.1(b), LR from A to B is given as follows.

$$LR(A \text{ to } B) = SLR(A) + CLR(A) + Tx(A) + Rx(B) + CLR(B) + RLR(B)$$
$$= SLR(A) + \text{send path loss} + RLR(B)$$

$$LR(B \text{ to } A) = SLR(B) + CLR(B) + Tx(B) + Rx(A) + CLR(A) + RLR(A)$$

The LR suggested value is from 8 to 12 dB [ITU-T-G.111 (1993)], and the preferred value is 10 dB. LR outside the range of 8 to 12 dB degrades the quality. A higher value of LR is attenuated voice, and a lower value is loud voice. Attenuated or loud voice degrades voice quality. This parameter is also called the overall loudness rating (OLR).

TELR is the talker echo loudness rating. It is the level difference between original acoustic levels of voice at microphone to the received acoustic echo power [ITU-T-G.131 (2003)] on the same phone speaker. TELR is the round-trip acoustic and electrical combined loss. A higher value of TELR is desired, and it improves voice quality by decreasing the echo level. With reference to Fig. 6.1(b), TELR is signal loss from the mouth of A to the ear of A after returning as echo. In TELR calculations, voice makes one round trip by going up to B and coming back to A. It is not sidetone. By considering CLR and digital network (PSTN or VoIP) losses as insignificant, TELR(A) as seen at phone-A and TELR(B) as seen from B are given below. All parameters associated here are expressed in dB scale. When echo cancellers are used, echo canceller rejection to echo in dB also contributes to TELR. TELR has to be as high as possible. The TELR requirements are given in relation to mean one-way delay. For higher delays, TELR has to be more than 65 to 75 dB. A higher value of TELR is achieved by employing an echo canceller.

$$TELR(A) = SLR(A) + Tx(A) + Rx(B) + ERL(B) + Tx(B) +$$
$$(EC \text{ rejection at } B) + Rx(A) + RLR(A)$$

$$TELR(B) = SLR(B) + Tx(B) + Rx(A) + ERL(A) + Tx(A) +$$
$$(EC \text{ rejection at } A) + Rx(B) + RLR(B)$$

In the above formulation, ERL(A) and ERL(B) include the combined influence of the SLIC hybrid and the phone. While viewing from phone-A, Tx(A) + Rx(B) is considered as send path loss, and Tx(B) + (EC rejection) + Rx(A) is considered as receive path loss. Receive path includes echo canceller rejection. TELR (A) is also expressed as

$$TELR(A) = SLR(A) + \text{send path loss} + ERL(B) + \text{receive path loss} + RLR(A)$$

TELR(A) will not go through SLR(B), RLR(B) of phone-B. It will only go through an equivalent ERL of hybrids at B. Additional details on TELR are given in G.122, G.131, P.310, G.107 [ITU-T-G.122 (1993), ITU-T-G.131 (2003), ITU-T-P.310 (2003), ITU-T-G.107 (2005)], and TIA/EIA-116A (2006).

Sidetone is a small part of the microphone electrical signal that is fed back immediately as leakage to the same phone speaker. Sidetone is also similar to talker echo, but sidetone arrives within imperceptible delay of 1 ms [ITU-T-G.107 (2005)], and for delays above 2 to 4 ms, sidetone may be perceived as echo. Sidetone creation happens as part of the phone circuitry as intentionally created leakage or from the phone hybrid. Sidetone marked in Fig. 6.1(b) on the phone-A side is essential to create a balanced sound level from the talker. In the absence of sidetone or very low-level sidetone, the talker will get the perception of hollowness (silence) and start speaking loudly. Too much of sidetone creates annoyance, and the talker will have tendency of reducing the speaking sound level. Sidetone mask rating (STMR) levels of 7 to 12 dB and a maximum of 20 dB [ITU-T-G.121 (1993)] is considered in the early evaluations as required for VoIP communication. In recent revisions of G.107 [ITU-T-G.107 (2005)], sidetone is considered as 10 to 20 dB with a default value of 15 dB.

Mean one-way delay is the half of the round-trip delay. Round-trip delay considers a talker's echo reaching the ear after completing a round trip from the destination. This delay has to be as low as possible. One-way delays up to 150 ms are treated as comfortable for VoIP voice communication [ITU-T-G.114 (2003b), ITU-T-Y.1541 (2006)(2006)], even though lower delay improves voice quality. In inter-regional PSTN and VoIP calls, one-way delay increases up to 300 ms [ITU-T-G.114 (2003b)]. In usage of terminology, end-to-end delays and one-ways delays are used without much distinction.

6.2   NAMING CONVENTIONS IN ECHO CANCELLER

Naming conventions of the echo canceller are marked in Fig. 6.2. The echo canceller is represented as a four-terminal device connected in the receive and send voice path. Echo cancellation blocks and literature uses the names of R_{in}, S_{in}, L_{RES}, S_{out}, and S_{gen} for representing the signals [ITU-T-G.168 (2004), ITU-T-G.161 (2004)]. Signal R_{in} is the reference signal coming from the far end. In the receive path, no modification for this signal R_{in} occurs. The signal coming out of EC in the receive path is marked as R_{out}. This signal is the same as R_{in}. The signal R_{out} is played on the speaker. Echo is created based on the R_{out} signal. The near-end signal is S_{gen}. During the near-end signal, the echo canceller will work as simply pass-through. The combined signal of echo and S_{gen} is the input S_{in} to the echo canceller. The echo canceller uses R_{in} as a reference and cancels echo present in S_{in}. The echo canceller first level removes the linear part at the summing junction and gives the nonlinear part and small uncancelled linear part as residue represented as L_{RES}. The suffix "RES" denotes residue. In this book, S_{RES} is used in place of L_{RES} for easy identification of the send path. This residue is processed in nonlinear processor (NLP) to obtain a nonlinear echo removed signal denoted as S_{out} or L_{RET}. The signal at S_{out} is expected to be combination of imperceptible echo and S_{gen} without any significant distortions. The returned echo level at S_{out} is the level of the signal at the send-out port that is returning to the talker.

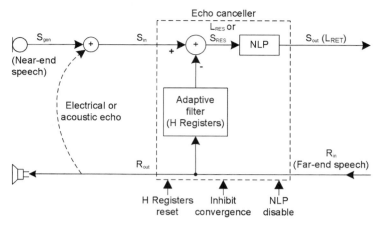

Figure 6.2. Echo canceller representation and naming conventions.

Echo return loss enhancement (ERLE) is the attenuation of the echo signal [ITU-T-G.168 (2004)] as it passes through the summing junction of the echo canceller. ERLE is the linear part of improvement without including the non-linear processing. ERLE = $S_{RES} - S_{in}$. The nonlinear enhancement is $S_{out} - S_{RES}$. Echo canceller total enhancement or cancellation is $S_{out} - S_{in}$. The signals R_{in}, S_{in}, L_{RES}, S_{out}, and S_{gen} are estimated in dB scale as dBm units. The losses or enhancements are expressed as addition and subtraction operations on these dB-scale converted signals. Echo power calculation and block sizes for echo estimation are given in the G.168 recommendation.

Three controls shown in Fig. 6.2 to the EC block are H-register reset, inhibit convergence, and NLP disable. H-registers within the echo canceller store the impulse response model of the echo path. These are the adaptive filter coefficients. The H-register is externally controlled for resetting and holding the contents based on the echo canceller conditions. The holding of coefficients is the inhibit convergence operation for freezing the filter updates. NLP control is an independent control to enable or disable NLP operation. The three controls are used while conducting the tests with simulated signals and with instruments.

6.3 LINE AND ACOUSTIC ECHO CANCELLER

Echo cancellers for voice processing are classified as LECs and AECs based on the source of echo generation. Several techniques and requirements overlap between these two echo cancellers. In general, acoustic echo canceller can be reused as line echo cancellers. Line echo canceller may not cater to all requirements of the acoustic echo cancellation. The main differences of the line and the acoustic echo canceller are listed in Table. 6.1.

Table 6.1. AEC and LEC Functional Differences

Parameter	Line Echo Canceller	Acoustic Echo Canceller
Echo source	Electrical echo with wired connections	Acoustic echo with speakerphone mode and mechanical body coupling
Applications	Used in VoIP CPEs with FXS, FXO, and VoIP-to-PSTN interfaces	Speakerphone mode of operation for IP phones, WiFi handsets, and soft phones
Echo span	Typical 8–32 ms, as high as 128 ms	Typical 32–128 ms, as high as 256 ms
Echo characteristics	Slowly varying, sparse, typical tail span ≤12 ms	Rapidly varying, dispersive tail span, but decays rapidly with echo tail span
Echo return loss	ERL > 6 dB	Echo can be stronger and varying in characteristics
Processing complexity	Depends on design and tail length	In addition to design and tail length, also depends on sampling rate variations
Voice quality enhancements	Adaptive level control	Adaptive level control and background noise reduction

As shown in Fig. 6.3(a), acoustic echo is the coupling/leakage of voice through acoustic media, usually through air as media in most telephone applications. Acoustic echo is created while using phones in speakerphone mode. Several phones (mobile phones, IP phones, WiFi phones, car phone adapters) are supporting speakerphone mode. With speakerphone, acoustic waves are stronger in strength to reach several feet distance. The strong acoustic waves from the speaker will also be reaching the microphone creating acoustic echoes. The signal level caused by acoustic coupling is much stronger than the usual speech level reaching the microphone. The phones with a speakerphone option come with a built-in acoustic echo canceller. The acoustic echo canceller is mainly governed by P.340 [ITU-T-P.340 (2000)] recommendations. Many proprietary acoustic echo cancellation techniques [Gay and Benesty (2000), Gustafsson (2001), Borys (2001), Benesty et al. (2001)] are also used based on the actual application and implementation. In a normal handset-based phone, some amount of acoustic coupling will be happening through air and the mechanical body structure of the handset. This echo coupling is 45 to 50 dB lower and merges with the ideal channel noise. Hence, in the handset mode of operation, the effect of acoustic echo is less.

Line echo cancellers are related to coupling of electrical signals (no acoustic coupling) as shown in Fig. 6.3(b). Talker echo and listener echo indicated in Fig. 6.1(a) are related to the line echo canceller. Line echo cancellers are used with foreign exchange subscriber (FXS), and FXO electrical interfaces. In a simple a VoIP adapter with an FXS interface, echo is generated inside the phone and at the SLIC–CODEC telephone hybrid. Line echo cancellation is

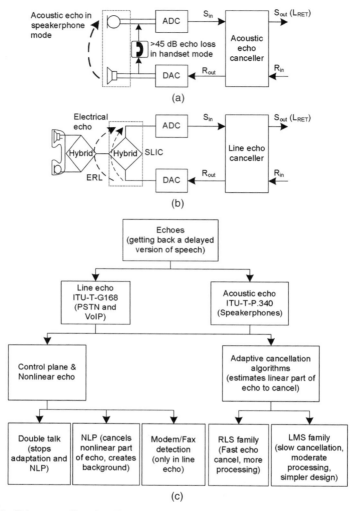

Figure 6.3. Echo canceller classification. (a) Acoustic echo creation. (b) Electrical echo creation. © Echo canceller classification and functional blocks.

mainly governed by G.168, G.131, and G.169 [ITU-T-G.168 (2004), ITU-T-G.131 (2003), ITU-T-G.169 (1999)] recommendations. The other names used for the line echo canceller are the electrical echo canceller, network echo canceller, or simply echo canceller referring to the context of the application.

AEC is not always associated with speakerphone mode. Even without a hands-free mode of operation, acoustic echo can be noticed because of mechanical coupling though miniature form factor handsets such as mobile and WiFi phones. As shown in Fig. 6.3(a) and (b), AEC or LEC removes echo in one direction at the near end (close to echo creation). In the communication between mobile phones and PSTN, the central communication nodes keep both the near-end and the far-end echo canceller. This type of back-to-back echo canceller operation can help remove echoes from the mobile and the PSTN side. The echo canceller on the mobile phone side has to cater to low-level acoustic echo. PSTN side can take care of echo problems, even if mobile phones are not incorporated with an echo canceller and end-to-end delays are more between mobile and PSTN calls.

Figure 6.3(c) is presented for broad classification of the echo canceller functional blocks and their relation. The functional blocks are common to LEC and AEC. Modem and fax tone detection are not applicable in AEC. Some blocks indicated in the figure are explained in next sections of this chapter. The main functional blocks of the echo canceller are as follows:

1. Adaptive filter—adapts to the echo path and estimates the closely matching signal to cancel the echo.
2. Double talk detection—indicates the presence of simultaneous conversation from both ends also referred as presence of near end speech and echo.
3. Nonlinear processor (NLP)—removes the nonlinear and residual part of the echo and optionally inserts matching background as comfort noise.
4. Modem answering and fax tone detection—detect the presence of modem/fax tones to disable certain functions of the echo canceller. This is presented in Chapter 14.
5. Monitoring—residual echo estimators, echo residual monitoring, as well as echo and signal parameters to voice quality monitoring during an active call. Configuration has to ensure that echo canceller parameters are properly set for various conditions.

In this chapter, echo cancellation operations are given in relation to LEC. The operations and algorithms of LEC are also useful for AEC.

6.4   TALKER ECHO LEVELS AND DELAY

This section is mainly relevant for line echo cancellation. The talker echo loudness rating and the delay relation are shown in Fig. 6.4 in relation to G.131 [ITU-T-G.131 (2003)] and G.107 [ITU-T-G.107 (2005)] recommendations. In PSTN, delays are lower than VoIP, and TELR requirements will be lower even for inter-regional calls. Several examples of delay and TELR combinations are given in G.131. Some of these examples with echo cancellation requirements are given in Table 6.2(a). The first four rows of the table are for without the echo canceller, marked as a "None" improvement under the echo canceller. The last five rows are with the echo canceller operation. For the delays of 150,

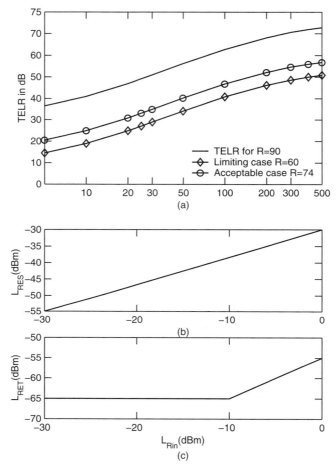

Figure 6.4. TELR in relation to ITU-T-G.131 and echo cancellation requirements as per ITU-T-G.168. (a) G.131 TELR graph for limiting, acceptable, and for R-factor 90. (b) G.168 linear part echo cancellation. © G.168 linear and nonlinear part of echo cancellation. [Courtesy: Reproduced with the kind permission of ITU; International Telecommunication Union, Geneva, www.itu.int).]

200, 300, and 500 ms, the corresponding TELR and the echo canceller enhancement are given in Table 6.2.

In relating TELR from G.131 with G.107 R-factor, the TELR limiting and acceptable cases are given in G.131 as first-level considerations. The limiting case is the echo level for which 10% of the listeners reporting echo problems. The acceptable level is 1% of the users reporting echo. Echo complaints are quality degradations captured in the E-model through R-factor estimation. Narrowband digital phones with a G.711 voice call can achieve a maximum R-factor of 93.2. TELR of the limiting case establishes R-60, and the accept-

Table 6.2. Talker Echo Requirements. (a) TELR Requirements for R = 60 to 90, (b) linear and NonLinear Part Cancellation Requirements as per G.168

ERL	SLR	RLR	Send + Receive Path Loss	EC Rejection dB	TELR	One-Way Delay for R-factor
17	7	3	6	None	33	25 ms for R-74
14	7	3	6	None	30	18 ms for R-74
8	7	3	6	None	24	9 ms for R-74
8	2 (min)	1 (min)	6	None	17	7 ms for R-60
10	7	3	6	24	50	150 ms for R-74
10	7	3	6	26	52	200 ms for R-74
10	7	3	6	28	54	300 ms for R-74
10	7	3	6	45	71	300 ms for R-90
10	7	3	6	48	74	500 ms for R-90

R_{in} dBm	S_{in} with Phone ERL of 6 dB	S_{RES} as per G.168	ERLE- Minimum Linear Part	S_{out} as per G.168 Requirement	NLP Improvements in dB
−0	−6 dBm	−30 dBm	24 dB	−55 dBm	25 dB
−10	−16 dBm	−38 dBm	22 dB	−65 dBm	27 dB
−20	−26 dBm	−47 dBm	21 dB	−65 dBm	18 dB
−30	−36 dBm	−55 dBm	19 dB	−65 dBm	10 dB

able case R-74, which thus degrades voice quality. Maintaining the acceptable condition as a goal is not sufficient because of a lower mean opinion score (MOS) with an R of 74. Hence, the goal has to be to achieve more than the acceptable case of the G.131 recommendation.

Additional graphs on TELR are given in G.131 for different TELR levels and the R-factor. In this section, TELR specific to R of 90 from G.131 is considered the recommended value. In this book, TELR of R 90 is referred to as TELR(For R-factor 90). Refer to the latest revisions of ITU recommendations for any regularized naming convention for TELR exceeding the acceptable case. TELR with limiting, acceptable, and R-factor 90 are given in Fig. 6.4. Refer to Chapter 20 for more details on other parameters in arriving at the R-factor. TELR can be noted from the graphs given in the G.131 recommendation or from the following equations. For mean one-way delay in ms "t", TELR for these three categories are calculated as [ITU-T-G.131 (2003), ITU-T-G.107 (2005)] given below. The results from these equations are established as graphs in Fig. 6.4(a). It is noted by visual inspection that about a ±1-dB difference is observed between the equation estimates and the G.131 graphs. This deviation is insignificant compared with the TELR values of 65 to 75 dB.

$$\text{TELR(limiting)} = 8 + 40\log\left(\frac{1+t/10}{1+t/150}\right) - 6e^{-0.3t^2} \quad \text{gives R} = 60 \quad (6.1)$$

$$\text{TELR(acceptable)} = 6 + \text{TELR(limiting)} \quad \text{gives R} = 74 \quad (6.2)$$

$$\text{TELR}(\text{R-factor 90}) = 16 + \text{TELR}(\text{acceptable})$$
$$= 30 + 40 \log\left(\frac{1 + t/10}{1 + t/150}\right) - 6 e^{-0.3t^2} \tag{6.3}$$

In Fig. 6.4, TELR limiting, acceptable, and recommended (R-factor 90) are given. From the G.131 graphs, it is observed that a TELR of R 90 is 15 to 17 dB higher than TELR acceptable values. A typical value of 16 dB higher than are acceptable case values is considered in this book for presenting R of 90. For the R-factor 90 condition, a TELR requirement at 300-ms one-way delay is 72 dB. It is recommended to achieve higher TELR in practical deployments beyond values marked for an R-factor of 90.

6.4.1   Relating TELR and G.168 Recommendations

The echo canceller based on G.168 considers electrical input signal power to decide on required echo cancellation. TELR considers one-way delay and various losses from mouth to ear. Mean one way-delay used with TELR is not clearly connected with G.168 requirements. TELR is not relating input power. It is useful to connect signal power, one-way delay, and echo rejection requirements.

As per G.168 recommendations, echo canceller requirements are given in Table 6.2(b) through notation of R_{in}, S_{in}, S_{RES}, and S_{out}. The main goal as per the G.168 recommendations is to maintain echo levels at lower than −65 dBm. By considering padding losses (send and receive path loss) of 3 dB in each path, the talker echo level merges with ideal channel noise of −68 dBm. Details on ideal channel noise are given in Chapter 1. From the Table 6.2(b) [ITU-T-G.168 (2004)], it can be noted that the linear part of the EC has to improve by 19 to 24 dB and that the nonlinear part has to improve by 10 to 27 dB depending on the input reference (R_{in}) level. Many EC implementations exceed this requirement. The linear part is cancelled by 24 to 35 dB, which reduces the NLP requirements to 6 to 18 dB. Meeting G.168 requirements and achieving echo reduction to −65 to −68 dBm closely matches with the R-factor of 90 requirements at delays of 300 to 400 ms. At a 300-ms delay, the TELR requirement is 72 dB for the R-factor 90 condition. This requirement takes care of meeting −68 dBm even for strong signals of the order of 0 dBm. In summary, to maintain good quality from echo cancellation considerations, it is essential to exceed the requirements of the G.131 TELR acceptable condition. The echo level should submerge with ideal channel noise to minimize dependencies on delay and power levels.

6.4.2   Convergence Time

Convergence time [ITU-T-P.340 (2000)] is the time interval between the instant when a specified test signal is applied to the R_{in} port of the terminal after all functions of the echo canceller have been reset and then enabled, and the instant when the returned echo signal at the S_{out} port is attenuated by at least a predefined amount. During convergence local user speech S_{gen} is inactive.

As per G.168, rejected level at L_{RES} in 1 s is 20 dB (including the phone ERL of 6 dB) below R_{in} level. The S_{out} listed in Table 6.2(b) are steady state values. Steady state is 10-s duration from the initial reset conditions. In practice, echo cancellers are designed to converge faster than the G.168 requirements. In practical implementations, echo cancellers are implemented with two to five times faster convergence than G.168 recommendations. Fast convergence can also lead to fast divergence in bad conditions. The improvement in convergence has to be considered in relation to the robustness of the control function during nonadapting conditions.

6.5   ECHO CANCELLATION IN VoIP ADAPTERS

Echo canceller functioning between two VoIP adapters is represented in Fig. 6.5. For easy representation, two phones (phone-A and phone-B) connected to two adapters are marked as A and B. These phones are shown to generate two different types of signals. Phone-A is voice from continuous reading of letters "abcd," simply marked as **abcd..**, and phone-B voice is also called voice-B or B-voice. In Fig. 6.5, SLIC, hardware CODEC (ADC, DAC), echo canceller, voice encoder, and decoder are shown as part of each VoIP adapter. The other voice functions of dual-tone multifrequency (DTMF), tones, packet loss concealment (PLC), other packetization, and VoIP signaling are not shown. Refer to Chapter 2 for details on complete VoIP adapter blocks and on an end-to-end VoIP voice call.

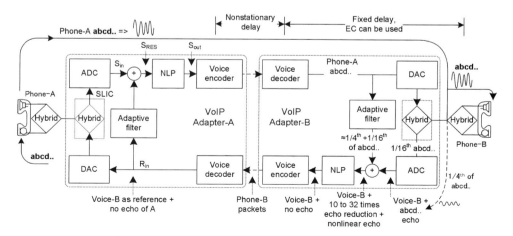

Figure 6.5. Echo cancellation between two VoIP adapters.

The VoIP adapter will interface with the telephone through an FXS TIP-RING interface. Hybrid inside the phone converts a two-to-four-wire interface for connecting voice to the handset microphone and speaker. In the VoIP adapter, hybrid converts the two-wire TIP-RING interface to the four-wire interface for ADC and DAC signals. In actual implementation, hybrid is part of the SLIC. ADC and DAC are part of the hardware CODEC chip also called the SLAC. In recent devices, SLIC and hardware CODEC are manufactured as a single device with few passive components located outside the chip [URL (Si3015)]. The echo canceller is a software module working on a processor along with DTMF, voice compression, voice activity detection (VAD)/comfort noise generation (CNG), PLC, and call progress tones.

End-to-end echo cancellation steps are given as follows:

- In this echo cancellation example of Fig. 6.5, phone-A is the talker and B is the listener. Echo is generated at B and comes back to A as talker echo.
- When person at phone-A speaks (say "**abcd..**"), voice is fed into the phone-A hybrid and is converted on to a two-wire telephone interface between phone-A and the VoIP adapter. In the adapter-A, the hybrid converts the two-wire signal into a four-wire interface. Voice from phone-A goes through ADC, echo canceller, voice compression path of VoIP adapter, and finally reaches adapter-B as IP packets. At destination VoIP adapter-B, voice packets are decompressed and sent to the telephone through SLAC and SLIC hybrid interfaces.
- In the end-to-end voice call between two phones, voice goes through a total of four hybrids. All hybrids have a certain amount of leakage. This leakage is referred to generically as ERL. The phone hybrid is the most dominant echo creator. The ERL for a good VoIP adapter SLIC hybrid is of the order of 18 to 24 dB. In voltage scale, about one sixteenth (for 24-dB ERL) of the voltage of DAC output directly enters the ADC path as leakage without going through the phone. This leakage of one sixteenth of abcd.. is marked in the path from ADC–SLIC–DAC in adapter-B.
- The main contributor for echo beyond adapter-B SLIC is the phone-B connected at the listener. Phone-B will convert a two-wire TIP-RING to a four-wire (microphone and speaker) headset. A typical phone with an ERL of 12 dB returns one quarter (25%) of the received signal as echo. In Fig. 6.5, signal one quarter of abcd.. voice is shown with dotted lines returning from the phone hybrid.

As shown in Fig 6.5, a total of one quarter A's voice from the phone hybrid and one sixteenth of A's voice from the SLIC hybrid will be fed back to the ADC of VoIP adapter-B. This return signal will be heard as echo at phone-A. To cancel this echo, VoIP adapter-B will need an echo canceller in the B-to-A path. In general, this type of near-end cancellation in the VoIP adapter will benefit the other end. Overall, VoIP calls based on the VoIP adapter will need line echo cancellers. The echo cancellation requirements vary based on end-to-end delays, phone characteristics, interfaces, loss planning of country, and quality goals. Echo canceller implementations take care of these requirements.

6.5.1 Fixed and Nonstationary Delays

In Fig. 6.5, fixed and nonstationary delays are marked on the upper part of the figure. In the voice call between two gateways, telephone, telephone interface, and blocks on the telephone side offer stationary/fixed delays. The echo canceller can adapt to the fixed delays. In this path of fixed delay, the delay is not fixed on a permanent basis. A user may connect different telephones or another parallel telephone, or conference in or out of the call, which changes the echo path in the middle of the conversation. These operations offer slowly varying delay/echo path models. In Fig. 6.5, between two adapters, delay is not stationary because of IP network and packet impediments. Continuous and sudden delay changes are not suitable for echo cancellation. Hence, far-end echo cancellation is difficult with VoIP solutions. In far-end echo cancellation, adapter-A has to cancel echo in its decoder path after the A-voice completes the round trip. Adapter-B will not be canceling the echo. Round-trip delay is not stationary in VoIP. Far-end echo cancellation is avoided in deployments, but such operation is possible with PSTN because of fixed delays on the established voice call. To get the best out of echo cancellation, the echo canceller has to be mounted close to the echo path and must avoid nonlinearities and nonstationary echo paths.

6.5.2 Automatic Level Control with Echo Cancellers

Automatic level control (ALC), also known as automatic gain control (AGC), is more commonly used with acoustic echo cancellers. In AEC, level control is used after S_{RES}, as given in reference [ITU-T-P.340 (2000)]. The positioning of ALC is very important. In summary, the gain control should not appear between R_{in} and S_{RES} of the echo canceller. ALC can appear before giving the input R_{in} or at the output of S_{RES} or S_{out}. The echo canceller as a four-terminal block should not see gain variations because of ALC. Otherwise, ALC in the echo path destabilizes the adapted filter coefficients, and steady state echo cancellation will reduce significantly. Echo path variations can create signal-level variations, and this is more common with acoustic echo. This echo path variation is unavoidable, and the echo canceller is supposed to takes care of this. Some characteristics of gain control and its relations can be found in references [ITU-T-G.169 (1999), ITU-T-P.340 (2000)]. As per the G.169 recommendations, the gain range is usually limited to 15 dB, and gain is controlled at the rate of 10 dB per second; typically, slower control is better.

6.5.3 Linear and Nonlinear Echo with Example

Echo cancellation happens by locally estimating the replica of the echo in the adaptive filter. With reference to Fig. 6.5, the filter adapts and provides an equivalent echo of "abcd.." that will appear as a delayed and attenuated (reduced in amplitude) version of the original. In the Fig. 6.5 example, about one quarter + one sixteenth of the original signal is given at the summing junction at adapter-B. The summing junction works for reducing the echo by 10 to 30 times (in dB scale, 20 to 30 dB). This improvement is the ERLE. The summing junction output is reduced in level in comparison with the original signal ("abcd.."). Practically, the SLIC hybrid and phone hybrid will have certain nonlinear parts in their electrical path model. The echo canceller adaptive filter adapts to the linear part of the SLIC and telephone hybrids. The nonlinear part appears as the residue at the output of the summing junction. The NLP block is used to remove the residue and to create comfortable noise in place of nonlinear echo.

For most voice power levels of −10 to −30 dBm, echo has to be removed to the −65-dBm level ITU-T-G.131 (2000)]. In practical echo cancellers, linear echo cancellation removes echo by 24 to 35 dB. An additional linear part of echo residue and nonlinear part has to be improved by 12 to 24 dB to make a total echo level of −65 dBm. This process calls for nonlinear echo cancellation. In lower end-to-end delays of the order of 50 ms, a linear echo cancellation of 24 to 35 dB would be sufficient, assuming the requirements of TELR(for R-factor 90) are met.

Echo residue is processed in NLP. NLP works when it senses the presence of small residue echo when compared with the main reference signal. The NLP has to operate only when NLP input is a small residue. If phone-B voice also appears at NLP input along with residue, then NLP will not operate. The simplest function of NLP is to replace residue with zero signals during echo residue. When voice from phone-B appears, then NLP will pass the signal without introducing any distortions. When people at phone-A and phone-B speak simultaneously, then it is referred to as double talk (DT). During double talk, NLP will be disabled to avoid distorting the phone-B voice. In double talk, echo residue is not removed. When residue of "abcd.." and voice from B present in the S_{RES}, then NLP will not operate. Hence, the echo residue will pass to the NLP output as is without any degradation.

6.5.4 Linear Echo Improvement with 16-Bit Samples

As illustrated in Fig. 6.5, in the communication between telephone and echo canceller, ADC and DAC are present that interface between analog signals and digital samples through a pulse code modulation (PCM) interface. This PCM interface can use 16-bit linear or 8-bit A/μ-law. The 8-bit format has a limitation of a signal-to-noise or quantization ratio of only 38 to 40 dB, and for low amplitude signals, this will reduce to 27 dB. Hence, this quantization effect limits the possible linear part of the echo cancellation. By making use of the 16-bit linear format in the communication between ADC/DAC and the processor, the linear part of the echo cancellation can be improved. The linear part of cancellation is also limited by the linear part of echo at the phone and by the capabilities of the echo cancellation algorithms. In summary, when ADC/DAC limits the samples to 8-bit, linear echo cancellation dB levels may be lower. Linear 16-bit samples also improve voice quality compared with 8-bit A/μ-law. As explained in Chapter 9, wideband voice communication makes use of 16-bit linear samples to achieve better voice quality. A low-bit-rate codec such as G.729AB is also found to provide slightly better objective measurements for a MOS with a linear mode of direct input.

6.6   ECHO PATH

Voice is an analog signal transmitted from the telephone to the VoIP adapter's ADC/DAC interface. Digital samples enter DAC and give analog signals from the SLIC hybrid. The SLIC hybrid introduces certain leaks and sends back part of the signal to the ADC path. The analog signal entering the TIP-RING line goes through distortions. Multiple telephones connected on the TIP-RING interface and their telephone hybrids and impedance mismatches create additional echo. The combined influence of SLIC hybrid, TIP-RING, and telephone is treated as a filter. It is also called by different names as echo impulse response, echo path, echo span, echo model, echo spread, and echo tail. The maximum spread of the significant response of the echo path filter is the tail length. G.168 provides eight echo path models to conduct tests as per G.168 recommendation.

For analog phones used with VoIP adapters, an echo canceller with a 12- to 16-ms span or tail length is sufficient. On the commercial side, many boxes are marketed as supporting 32 ms. For FXO side interfaces, echo span varies based on the termination point of the destination phone. A typical span of 48 ms would be sufficient for more regions. Some PSTN-to-VoIP gateways use an echo canceller span of 128 ms mainly to cater to the local long-distance PSTN side influences. Independent echo canceller systems are built up to a 512-ms echo span as given in [URL (Orion-EC), URL (Ditech)].

6.6.1   Delay Offset and Tail-Free Operations to Reduce Echo Span

In Fig. 6.5, an echo canceller is present between R_{in} and S_{in}. The signals S_{in} and R_{in} are interfaced to the voice compression (encoder and decoder). Voice codecs operate on a block of samples. In practical implementations, echo cancellers are also implemented on a block basis. This type of block processing and accumulation of voice samples into multiple frames creates fixed delay between echo canceller R_{in} and S_{in} reference points. This echo path has two main components for delay, the slowly varying part caused by the echo path and the fixed part usually caused by buffering of digital samples at ADC output and DAC input or fixed echo path delay. Sometimes, these delays are significant compared with echo path span. In the absence of any correction, this delay has to be catered in the echo path. To use the echo span to its best and to save on computation and memory in the processor, an offset is created in the echo path modeling. The easy way to remove the echo fixed delay is to introduce the matching delay in the receive (R_{in}) path. The fixed delay is usually called bulk delay [URL (Ditech), Dyba et al. (2004)], fixed delay, or delay offset. The delay is usually kept as programmable from 0 to a few tens of milliseconds to cater to various situations. For a particular product and deployment conditions, this can be set as a preset value. To use the echo span to the optimum, far-end echo cancellers used in PSTN will keep a minimum offset delay as a programmable value from 0 to 512 ms and echo span. A commercial product weblink is given in reference [URL (Orion-EC), URL (Ditech)] that accounts for such situations. A good representation of delay offset is given in [Dyba et al. (2004)]. Some specification of the echo canceller may be listed as higher echo span. The echo span specification has to be analyzed for fixed offsets and real span.

Tail-Free Echo Cancellation. The popular concept in echo cancellation in reducing the echo span is the tail-free or floating-tail echo cancellation. In tail-free echo cancellation, echo adaptive filter is capable of analyzing the dominant echo zones [URL (Ditech)]. These zones are spread across wide time zones. Typically, four short windows of 8 ms are identified within the stated span of say 128 to 512 ms. Pictorial representation of this tail-free echo cancellation is given in reference [Dyba et al. (2004)]. The algorithm will be trying to identify the zones and will keep adapting for the identified zones. These algorithms also keep identifying for varying conditions of change in echo path location. This type of scheme demands lower processing in the steady state operation of echo canceller and offers the benefits of a long tail length.

6.7   ADAPTATION FILTERING ALGORITHMS

The echo path consists of a linear and a nonlinear part. The linear part is modeled as a finite impulse response (FIR) or a moving average (MA) filter. G.168 [ITU-T-G.168 (2004)] has eight echo path models that work exactly as FIR filters. The nonlinear part requires higher order filter models. Volterra filters [(Mathews et al. (2000), Borys (2001)] model higher order nonlinear terms. In the literature, the applicability of Volterra filters were proved to be suitable for echo cancellation [Borys (2001)], but the utility is limited because of the higher memory and processing requirements of the Voltera filters. Nonlinear echo cancellation with higher order filters is beyond the scope of this book.

Linear part adaptation is estimated using popular least mean square (LMS) [Haykin and Widrow (2003), Haykin (1996)] and recursive least squares (RLS) based algorithms. LMS was widely adapted in the early designs of the adaptive filters. LMS algorithms are simple and stable as well as consume less memory and processing. These filters are updated in either the frequency domain or the time domain. If the filter span is longer, frequency-domain techniques are advantageous. The main disadvantage of LMS algorithms are that they adapt slowly. However, these algorithms were able to achieve the G.168 requirements. LMS algorithm adaptation further slows down with low signal levels. Normalized LMS (NLMS) normalizes signals that improve adaptation even for low-volume signals. For every sample of new input, these adaptive filters can be updated. To save computation, a block of samples is used to update the adaptive filter. This block operation is known by name block NLMS. Several other LMS derivatives work in both time and frequency domains. In wideband applications and acoustic echo cancellation, total signal frequency bandwidth is split into lower and upper bands or multiple bands. The purpose is to create an adaptive algorithm in multiple sub-bands. It saves processing time and allows certain processing scaling to consider only required sub-bands for updates. Similar concepts of LMS, blocks, and sub-bands are applicable to the other adaptive algorithms such as affine projections and RLS.

6.7.1 Adaptive Transversal Filter with LMS

In a digital echo canceller, both the reference and the echo signal are available as digital samples. Therefore, the overall influence of echo path impulse response is represented in digital form by $h_k[\ ]$ as represented in Fig. 6.6(a). In the absence of speech from near-end phone-A, the input signal S_{in} is equal to the echo signal r(i) that has to be cancelled. The reference signal $R_{in}(i)$ is denoted as $y(i)$ in the following equations:

$$r(i) = \sum_{k=0}^{N-1} h_k\, y(i-k) \qquad (6.4)$$

where $r(i)$ is the echo signal at sample index "i" and "N" denotes the number of taps in the filter as marked in Fig. 6.6(b).

The adaptive filter can cater to only the linear part. The echo path impulse response of a finite "N" nonzero coefficients linear system is represented as [Haykin (1996)]

$$\hat{r}(i) = \sum_{k=0}^{N-1} a_k\, y(i-k) \qquad (6.5)$$

The goal is to adjust a_k such that $\hat{r}(i) = r(i)$. When $a_k = h_k$, for $k = 0, 1 \ldots N-1$ the returned and estimated echoes are identical, which results in cancellation of the linear part of echo. The echo path error is given as $e(i) = [r(i) - \hat{r}(i)]$.

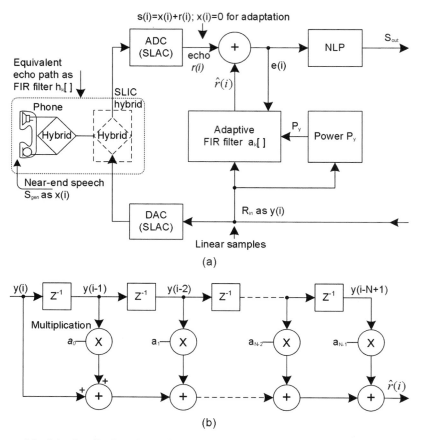

Figure 6.6. Adaptive filtering. (a) LMS-based filtering. (b) Basic echo path modeled as FIR filter.

The coefficients of the traversal filter are updated to match the slowly time-varying echo path impulse response by minimizing the mean-squared residual error $e^2(i) = [r(i) - \hat{r}(i)]^2$. The update of the coefficients at each iteration i is controlled by step size β. The typical value of β is 2^{-8}. Here iteration is the sample index for every new sample.

$$a_k(i+1) = a_k(i) + 2\beta E[e(i)y(i-k)] \tag{6.6}$$

where i = current sample index, and filter taps $k = 0, 1, 2 \ldots N - 1$.

The expectation of error and input signal requires prior knowledge of the reference signal probability distribution. Common practice is to use an unbiased estimate of the gradient, which is based on time-averaged correlation error. Thus, replacing the expectation operator with short time average gives [URL (SPRA129), URL (SPRA188)]:

$$a_k(i+1) = a_k(i) + \frac{2\beta}{M} \sum_{m=0}^{M-1} e(i-m)y(i-m-k) \qquad (6.7)$$

When $M = 1$, Eq. (6.7) is simplified as

$$a_k(i+1) = a_k(i) + 2\beta e(i)y(i-k) \qquad (6.8)$$

The current error $e(i)$ is correlated with $y(i)$, and up to $y(i - N - 1)$ to update coefficients a_k. It is called the standard LMS algorithm or the stochastic gradient algorithm, which works better with noisy conditions. The convergence time of the algorithm is determined by step size β and the power of the far end signal $y(i)$. In general, making β a higher value speeds up the convergence, whereas a smaller β reduces the asymptotic cancellation error. The convergence time constant is inversely proportion to the power of $y(i)$. The algorithm will converge very slowly for low signal levels. To overcome this situation, the loop gain is normalized by an estimate of the far-end signal power. This type of normalization is known as NLMS. The normalization is performed on step size β.

$$2\beta = 2\beta(i) = \beta_1/P_y(i) \qquad (6.9)$$

In the equation, β_1 is the compromise value of the step size as $0 < $ step size $\beta_1 < 2$/input power. Power $P_y(i)$ is an estimate of the average power of $y(i)$ at sample index i. The algorithm converges quickly with the larger step size, but the steady state error level will be more. If step size is small, the steady state error will be less, but its convergence is slow. A typical value of β_1 is 2^{-8}. This method optimizes the convergence time constant. Power is estimated as

$$P_y(i) = [L_y(i)]^2 \text{ and } L_y(i) = (1-\rho)L_y(i-1) + \rho|y(i)| \qquad (6.10)$$

This type of power estimate is used because the calculation of the exact average power is computationally expensive. The recommended value of ρ is 2^{-7} [URL (SPRA129)].

Block NLMS Algorithm. In the block NLMS algorithm, instead of updating all the filter coefficients for each input sample, a set of filter coefficients is updated for each sample. After some blocks of input samples, all filter coefficients are updated. In the block update approach, the coefficients may be updated less frequently within a thinning ratio of up to M called the block NLMS algorithm. The block NLMS algorithm performs better than the standard LMS algorithm with noise and speech signals and is computationally faster [URL (SPRA129), URL (SPRA188)].

$$a_k(i+M+1) = a_k(i) + 2\beta \sum_{m=0}^{M-1} e(i+M-m)y(i+M-m-k) \qquad (6.11)$$

Generally, a smaller block size is chosen for the implementation because it takes advantage of the coefficient multiply and accumulate capabilities of the processor. For the block size $M = 16$ and filter coefficients $N = 256$ for 32 ms, during each sample period of 125 µs, 16 of 256 coefficients are updated using correlation of the 16 past error and signal values. After a block of 16 samples, all filter coefficients are updated. The NLMS algorithm uses a more accurate estimate of the gradient vector because of the time averaging. Estimation accuracy increases with block size. This scheme allows an echo tail programming resolution of 2 ms for a block length of 16 samples.

Leaky LMS. The possibility of sensitivity to round off errors and other disturbances exists because the LMS update equation is essentially an integrator. The introduction of the small leakage parameter can guard against such numerical problems. The formulation for leaky LMS is shown below:

$$a_k(i+M+1) = \frac{lm-1}{lm}a_k(i) + 2\beta \sum_{m=0}^{M-1} e(i+M-m)y(i+M-m-k) \qquad (6.12)$$

Additionally, if the reference signal is a pure sinusoid, then the adaptive filter will diverge and then become unstable. The inclusion of the leakage factor $[(lm-1)/lm]$ in Eq. (6.12) has the equivalent effect of adding a white-noise sequence of zero mean and variance $1/lm$ to the input processes $y(i)$. As a result of this operation, filter stability will improve for tones and will prevent overflow [Haykin (1996)]. The value of this leak depends on the level of cancellation required. The usual value of lm is less than or equal to 32767. In actual implementation $(lm-1)/lm$ is made equal to 32767 in a 16-bit numbering representation [ITU-T- G.191 STL (2005)]. The leak factor should be bigger to avoid rapid decay of filter coefficients.

As per G.168, the adaptation has to be inhibited once the echo canceller is fully converged. Then after two minutes, the output residual echo level should not decrease more than 10 dB with respect to the steady state value. If the leak factor is applied for each block, the leak will not support the two-minute test. To achieve this, the leak factor is to be introduced for every fourth to tenth block instead of for every block.

6.7.2   Overview on Adaptive Filtering with RLS and Affine Projections

Two other algorithms popularly considered are the affine projection algorithm and RLS. An overview on these two algorithms is given here.

The RLS algorithm is the method of least squares that will be used to derive a recursive algorithm for automatically adjusting the coefficients of a tapped-delay-line filter, without invoking assumptions on the statistics of the input signals. The RLS algorithm estimate is based on minimizing the sum of the squares of instantaneous error values. The procedure, in the RLS algorithm, is capable of realizing a rate of convergence that is much faster than the LMS algorithm, because the RLS algorithm uses all the information contained in the input data from the start of the adaptation up to the present.

In the LMS algorithm the correction filter coefficients are based on the instantaneous sample value of the tap-input vector and the error signal. On the other hand, in the RLS algorithm, the computation of this correction uses all the past available information. The RLS algorithm has the effect of decorrelating the successive tap inputs, thereby making the RLS algorithm self-orthogonalizing. RLS algorithms are suitable for very large scale integration (VLSI) implementation. RLS is always preferred to make more channel echo cancellers with a single dedicated chip.

The LMS algorithm requires approximately $20N$ iterations to converge in mean square, where N is the number of tap coefficients contained in the adaptive filter. On the other hand, the RLS algorithm converges in mean square within less than $2N$ iterations. The rate of convergence of the RLS algorithm is, therefore, in general, faster than that of the LMS algorithm by an order of magnitude.

RLS Disadvantages. The RLS algorithm demands much higher computation and memory. The RLS algorithm requires a total of $3N (3 + N)/2$ multiplications, which increases as the square of N, the number of filter coefficients. LMS algorithm requires $2N + 1$ multiplication, increasing linearly with N. Numerical instability is also one of the major concerns with RLS algorithms. In general, RLS is stated as a higher computation operation, but fast algorithms such as fast RLS, QR decomposition RLS (QRD-RLS) are used in RLS that reduce computation requirements.

Affine Projection Algorithm. Affine projection algorithms are a family of LMS algorithms with higher order. These algorithms provide added benefit of catering to wideband voice and acoustic echo cancellation. Under least order, affine projection algorithms work like LMS algorithms. Affine with higher order gives faster convergence and low steady state echo residue compared with LMS algorithms.

6.8 ECHO CANCELLER CONTROL FUNCTIONS

Adaptive filters use an FIR-equivalent filter for forming the echo path. The echo path is updated with different algorithms of LMS and RLS. In recent implementations, the adaptive filter is split into two parts. In Fig. 6.7, the filters are marked as adaptive as well as the hold filter. The adaptive filter keeps adapting as per the previously explained algorithms. The hold filter gets an update from the adaptive filter in good conditions. Although the adaptive filter

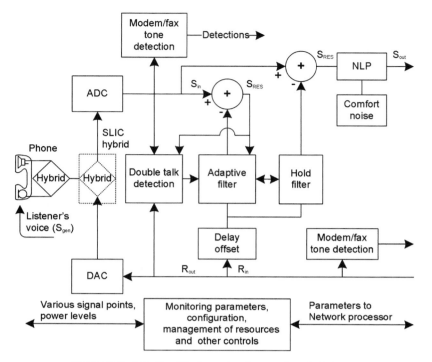

Figure 6.7. Adaptive filtering extensions and control plane.

is settling, the hold filter keeps taking update. The hold filter will not have any filter coefficient adaptation. The adaptive filter summing junction output is mainly used for the adaptive filter closed-loop adaptation. The hold filter summing junction output is the actual output used with an NLP operation. When the adaptive filter is disturbed, it can reload good coefficients from the hold filter. When there is a near-end signal or a double talk condition, the adaptive filter is not updated. This type of double-filtering scheme minimizes the disturbances during undesired conditions [URL (Cisco-G168)]. The quality of echo removal will be better and stable with this scheme. This type of scheme requires slightly higher memory and processing mainly to validate the updates as well as an additional filtering operation. More aspects of the hold filter in associating with double talk are given in the subsequent part of this chapter.

In the echo canceller, the control plane includes the double talk detection, nonlinear processor, and modem/fax tone detectors. Double talk detection is the main control that influences the adaptation. Modem and fax detections are mainly single events that decide on enabling adaptation and NLP enable/disable of echo cancellation operations. The NLP operation is used to remove the echo residue.

6.8.1 Double Talk Detection

DT is the simultaneous presence of the near-end and far-end speech. The main purpose of the DT detector is to avoid adaptation whenever the double talk signal is present. Double talk detection has to be declared with any significant presence of near-end signal (nonecho) irrespective of the far-end signal and echo. The double talk detector should also detect near-end low-level signals, music, and strong background noise conditions. Hence, the DT detector is also called the near-end speech detector or near-end voice activity detection. DT detection freezes the adaptive filter and disables the NLP operation. However, the linear part of the echo cancellation will occur with the previously adapted filter. With reference to Fig. 6.6(a) and 6.7, DT detection passes near-end speech S_{gen} directly to S_{out} without any distortion while cancelling the linear part of the echo.

As shown in Fig. 6.7, the main inputs to the DT are the far-end signal (R_{in}) and the S_{in} signal that includes echo and the near-end signal. Echo is also a strong signal present in the S_{in}. Echo alone should not detect DT without the presence of actual near-end speech, even with low ERL. Low ERL means a strong echo in S_{in}. Some popular DT detection approaches are given below.

Geigel Algorithm. In the early implementations of double talk, the Geigel algorithm was used. In recent times, this is supplemented with several other signal-processing techniques. The Geigel algorithm [URL (SPRA129)] compares near-end speech S_{in} with a short-term history of R_{in}. In the following equations, signal $S_{in}(i)$ is referred to as $s(i)$ which is a combination of near-end speech $x(i)$ and echo $r(i)$. $R_{in}(i)$ is referred to as $y(i)$ to maintain consistency with Fig. 6.6(a). The Geigel algorithm detects the presence of DT when the following condition is satisfied:

$$|s(i)| = |x(i) + r(i)| \geq \frac{1}{2} \max\{|y(i)|, |y(i-1)|, \ldots, |y(i-N)|\} \qquad (6.13)$$

where N is the FIR adaptive filter order. The factor of one half is based on the hypothesis that the echo path loss known as ERL through a hybrid is at least 6 dB. For different ERL requirements, this threshold has to be modified. Keeping one quarter in place of one half will cater up to a 12-dB ERL. The selected threshold, and the variations in ERL under multiple usage scenarios should not create double talk detections from self-echo. The preferred condition is $|x(i)| \gg |r(i)|$ for better DT detection. In Eq. (6.13), N samples are compared for every new input sample. In actual implementation to reduce computation, this will be optimized as partial maxima over a few small blocks and not all previous $N-1$ samples are required in each step.

The more robust version of the Geigel algorithm uses the short-term power estimate, $\hat{s}(i)$ and $\hat{y}(i)$, for the power estimates of the recent past of the far-end signal and near-end hybrid signal $s(i)$. These estimates are computed recursively by the equations.

$$\hat{s}(i+1) = (1-\alpha)\hat{s}(i) + \alpha|s(i)| \tag{6.14}$$

$$\hat{y}(i+1) = (1-\alpha)\hat{y}(i) + \alpha|y(i)| \tag{6.15}$$

In the equations, the filter gain $\alpha = 2^{-5}$. The near-end speech or double talk is detected whenever

$$|\hat{s}(i)| \geq (1/2)\max\{\hat{y}(i), \hat{y}(i-1), \ldots, \hat{y}(i-N)\}$$

As the near-end speech detector algorithm detects short-term peaks, it is desirable to continue declaring near-end speech for some hangover time after initial detection [URL (SPRA129)]. During hangover time, the previous conditions are continued. Once hangover time is over, the adaptation of filter coefficients is allowed and NLP is enabled.

Correlation-Based DT Detections. In correlation-based detections, correlation is computed on reference $y(i)$, and near-end speech with echo $s(i)$. It is expected that actual echo $r(i)$ will correlate better with $y(i)$ and that near-end speech $x(i)$ will have less correlation with $y(i)$. These correlations are validated with thresholds and absolute power levels. There are several extensions to this basic correlation-based method. Assuming that the echo canceller is in an adapted state, in relation to Fig. 6.6(a), $\hat{r}(i) \approx r(i)$. During double talk, $r(i)$ will be growing more quickly than $\hat{r}(i)$. This measure is helpful to build extra logic for DT detection.

ERL and ERLE-Based DT Detection. The echo adapting filter is a linear part of the hybrid ERL contribution. To make use of ERL for DT detection, ERL is monitored continuously. This process may require normalization of filter coefficients and converting ERL to the dB scale. During the beginning phase of DT, the adaptive filter will keep adapting and ERL will keep decreasing. It is one of the early indications of double talk, which will happen because of filter coefficients quickly trying to adapt to a strong near-end signal even though it is not echo. This procedure also goes by the name "filter disturbance detection." As a continuity of the previous condition, a reduction of ERLE at the summing junctions is also treated as the likely condition of DT.

Double Filter as a Helpful Option to DT. In recent implementations of the adaptive algorithm, double filters are used. As shown in Fig. 6.7, one filter marked as the hold filter will preserve the good version of adapted coefficients. The incorporation of the hold filter gives relaxed conditions for DT detection [URL (Cisco-G168)]. Good versions of the adaptive coefficients are always preserved in the hold filter. If the adaptive filter is disturbed, the hold filter can upload the good coefficients to the adapting filter. The hold filter will take a copy from the adaptive filter, when all the conditions are favorable. The double talk detection is based on hold filter echo cancellation. This double filter scheme cannot detect any double talk detection, but it is helpful to use with other detection methods.

Power-Based Normalization. The LMS algorithm given in Section 6.7 adapts slowly for lower powers. The technique indicated in [Al-Naimi (2003)] makes use of slowing down adaptation by controlling the resultant step. During a strong $s(i)$ signal, the adaptation step is reduced so that it slows down adaptation. The goal is to make the adaptation too slow on any likely conditions of DT. Once it comes out of the DT condition, the adaptive filter will continue to adapt at the set rate. This type of approach is usually attempted with a single adaptive filter scheme. No separate DT detector is used in this approach.

6.8.2 NonLinear Processing

An example of a device that reduces or cancels small echo signals by nonlinear operation on the samples of the transmitted audio signal is a center clipper ITU-T-P.340 (2000)]. Nonlinearities in the echo path of the telephone circuit, uncorrelated near-end speech, speech clippings, and quantization effects [URL (SPRA129)] of codecs limit the amount of achievable echo rejection in the adaptive filter to 19–35 dB. In most situations of VoIP, echo has to be removed to −65 dBm [ITU-T-G.168 (2004)]. Test equipment terminated on the VoIP system may behave as equivalent to a linear FIR filter creating the possibility to reject most of the echo as linear. With practical phones, the adaptive FIR filter will not remove echo to this level, which calls for residual echo suppression algorithms. When the near-end signal is present, the residual echo suppressor must pass the signal without any noticeable distortions. A suppression algorithm detects when to operate the NLP to remove the residue.

In simple suppression control, a decision is made when the error signal falls below a certain level in relation to the reference signal. The residue is eliminated by turning the signal into complete silence. In some implementation, instead of complete silence, attenuation of usually 12–24 dB is applied. Sudden application of silence or attenuation will create an annoyance to the perception. Hence, time-dependent attenuation shaping is applied in the NLP window of operation. This process will help to reduce annoyance, but it will not eliminate it. To help improve perception, comfort noise (CN) is created during the NLP region [URL (Cisco-G168)]. As explained in Chapter 4, VAD can be power based and match the spectral envelope. The G.168 requirements mainly talk about power-based tracking. Several advanced implementations [Bourget (2003)] create a pleasant background that completely eliminates the hollowness. The creation of comfort noise based on power or spectral shaping has to track the background. Some of these techniques may not take care of low-level near-end speech or music. It is essential to detect these sounds to avoid removing these along with echo residue. To create improved perception, it is also essential to send the background speech or music directly without treating it through comfort noise. Taking care of these conditions will help improve the perception [Bourget (2003)].

Once a major part of the echo is rejected in the adaptive filter, major voice quality enhancement or degradation happens in the NLP block. Echo residue extends in time. Hence, certain hangover time is created that mainly coincides with the end of the speech utterance. This time varies based on signal conditions. Typical hangovers are from 50 to 120 ms [ITU-T-G.168 (2004)]. Higher hangover is undesirable, but this has to be sufficient enough to create rejection and better perception. Simpler algorithms operate on power levels. To arrive at optimal hangover, signal analysis is also performed.

Power-Based NLP Detection. Simple power-based detection [URL (SPRA129)] is given here. In this formulation, the echo residue is shown as $e(i)$. The signal power is estimated as

$$\hat{e}(i) = (1-\rho)\hat{e}(i-1) + \rho|e(i)|$$

The reference $y(i)$ power is estimated as

$$\hat{y}(i) = (1-\rho)\hat{y}(i-1) + \rho|y(i)|$$

Suppression is enabled on the transmitted $S_{out}(i)$ to zero whenever

$$\hat{e}(i)/\hat{y}(i) \leq \frac{1}{16}$$

The threshold of one sixteenth corresponds to 24 dB. A recommended value of ρ for this operation is 2^{-7}. Recently, implementation caters to higher end signal processing techniques that analyze the residual signal and apply required rejection. During double talk mode and modem/fax 2100-Hz tone detection, NLP-based residue detection and removal is disabled completely.

NonLinear Filtering. In theory, echo residue can be decomposed as a small linear part and as a nonlinear part. The framework of nonlinear filtering is given in [Mathews et al. (2000), Borys (2001)]. A nonlinear filter in a simple way is a higher order filter with several product terms. The higher order terms help in modeling the nonlinearities of echo. While writing this book, the nonlinear filter was not widely adapted in echo cancellers implementation because of several limitations and constraints.

Nonlinear filters are more complex by design. They take more processing. Numerical precision for higher order product terms is also important. Memory requirements will grow with the order of the filtering. Fixed nonlinear filtering is reasonably easy to establish. Adaptive nonlinear filtering complicates the operation. To establish a reasonable balance of complexity, the models are truncated in order. The truncated order will limit the cancellation. Hence, another level of cancellation is required even with nonlinear filtering.

In future designs, new designs may adapt nonlinear filtering because of unlimited (relative to the nonlinear filtering) availability of processing with higher numerical precision and memory. The end-to-end delays may reduce over time because of improved network and bandwidth conditions. As a result, the requirements of nonlinear echo rejection will reduce, which provides the possibility to manage nonlinear filtering with lower order.

NLP and Quality Issues in Relation to Codecs and VAD/CNG. NLP

removes or reduces echo residue that consists of a linear and a nonlinear part. Depending on the NLP implementation, the NLP output can be silence, power-matched background noise, spectral matching comfort noise with additional checks on background voice and music, and so on. Some voice quality aspects of NLP are given in reference [Bourget (2003)]. In the case of simple NLP such as silence or power-level-based comfort noise, voice codecs and VAD go through distortions as illustrated in Fig. 6.8.

In VAD disable mode, all samples are compressed as speech frames. Imperfections in NLP as well as the inability to reproduce the exact background at NLP output can disturb the parameters mainly with code excited linear prediction (CELP) codecs like G.729A. At the decoder, audible artifacts (sudden ticks or hits) are produced from disturbed parameters from the encoder. These artifacts are of very low power, but they will be audible in a clean acoustic environment. In waveform-based codecs, NLP imperfections are transparent through the encoder and decoder. As explained in Chapter 4, VAD modules used with waveform codecs use simple power-based and CELP techniques. Power-based VADs will not create any disturbance with the imperfections of the NLP operation. In CELP-based VAD/CNGs, artifacts can be observed with the imperfections of the NLP operation.

In general, NLP has to be made perfect to use any codec with or without VAD/CNG mode. When end-to-end delays are lower, it is possible to disable NLP, but end-to-end delays may vary in different call combinations. Hence,

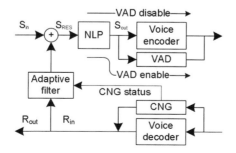

Figure 6.8. NLP, codecs and VAD/CNG relation.

perfecting NLP is essential to maintain good voice quality. In the receive path, when CNG is operating, the R_{in} signal level will be very low of the order of −40 to −60 dBm. It is good to disable adaptation during comfort noise generation to avoid disturbance to the adapted coefficients.

In general, the NLP operation influences codecs, and VAD/CNG operations and these disturbances are audible in a reasonably silent environment and degrade quality in multiparty conferences.

6.8.3 Monitoring and Configuration

Real-line Transport Protocal (RTP) transports voice packets and RTP Control Protocal (RTCP) transports packet statistics. RTCP extended reports (RTCP-XR) is the RTCP extension that monitors several voice quality parameters. RTCP-XR, which is discussed in Chapter 20, makes use of parameters residual echo return loss, signal level, and noise level. It is possible to derive these parameters from echo canceller functional blocks. In addition to RTCP-XR parameters, echo canceller monitoring can be extended for validating adaptive filter convergence, double talk detector, NLP, tone detectors, coordination of control plane, and processing resources utilization. The echo canceller will require configurations for several modes of operations. The important configurations are echo tail length, stress on adaptation, NLP, DT controls, hangover, reset states, and echo canceller external testing controls.

6.9 ECHO CANCELLATION IN MULTIPLE VOIP TERMINALS

As explained in Chapter 2, VoIP deployment makes use of several user interface devices such as IP phones, WiFi phones, personal computers (PCs) as soft IP phones, residential gateways with FXO interfaces, and VoIP-to-PSTN gateways. The echo cancellation for these VoIP systems is given in this section.

6.9.1 Echo Cancellation in IP and WiFi Handsets

The IP phone works as a four-wire interface without using a two-to-four wire hybrid. The microphone and speaker are used as electrically separate units. In a normal analog phone, sidetone is generated in the design of hybrids and coupling at two-to-four-wire interfaces. In the IP phone, sidetone has to be created as an internal electrical signal provided from the microphone path to the speaker path, and this function will be limited as sidetone and will not be continued as electrical echo. In summary, the IP phone works as a four-wire digital phone in regular handset mode. In digital phones, TCLw from the handset is 45 dB. This level is sufficient for VoIP calls up to delays of 400 ms. IP phone acoustics are used as a normal headset, speakerphone, and headset interface. The normal headset mode works as a four-wire digital phone. In speakerphone mode, acoustic echo is created and an acoustic echo canceller is used to cancel it. IP phone acoustic interfaces are extended through headsets and Bluetooth terminals. In general, these interfaces work as four-wire interfaces. A small acoustic echo could occur based on the acoustics and mechanical structures.

WiFi phones also use three types of interfaces: acoustics on the main handset, speakerphone mode, and acoustics extended through the headset. In wireless phones, TCLw may be lower and nonlinear echo is present because of the size of the mechanical structure and the short distance between the microphone and speaker. In general, acoustic echo cancellation is required with normal mode or speakerphone mode.

6.9.2   Echo in VoIP–PSTN Gateways

VoIP-to-PSTN gateways communicate with the local PSTN network and IP network. Electrical echo is created on the PSTN side, and this echo has to be rejected on the PSTN side of the gateway. In a VoIP adapter with a normal plain old telephone service (POTS) phone, echo path delays are minimal. In the PSTN-to-VoIP gateway, distance to the gateways and to the user may be more. Hence, additional echo span is created. The actual tail span may be lower but offset at some delay. For this mode of operation, tail-free or bulk delay techniques are used to reduce required tail span.

6.9.3   Echo in PC-Based Softphones

PC-based softphones are used with several user interface options. When the PC is used in speakerphone mode, it is essential that the PC support acoustic echo cancellation. In the absence of AEC support, the destination listener will report loud echo, and it is too difficult to continue a VoIP call with softphones. When a user is not sure of AEC functioning in the soft phone, it has to be validated with the destination listener. If the destination listener is reporting echo, then the soft phone does not have a proper echo canceller. The alternative here is to use a headset or handsets with a PC.

6.10   ECHO CANCELLER TESTING

As per G.168 requirements [ITU-T-G.168 (2004)(2004)], the tests recommended are given in Table 6.3. These tests are identified with a test number. The numbers remained the same across several revisions of echo canceller recommendations, and the results are usually referenced with the test number. In the table test number, the test name and a brief description are given. For more details on these tests, it is suggested to refer to the G.168 recommendation [ITU-T-G.168 (2004)]. These tests ensure time-dependent echo attenuation [ITU-T-P.340 (2000)] for different power levels, NLP tracking and time responses, double talk conditions, echo attenuation over frequency, and characteristics of fax/modem signals and tones.

Table 6.3. Echo Canceller Tests as per G.168 Recommendation

Test No.	Test Name	Brief Description
Test 1	Steady state residual and returned echo level test (deleted in revisions)	Test 1 is deleted and incorporated in Test 2
Test 2A	Convergence and reconvergence with NLP enabled	To ensure that the echo canceller converges rapidly
Test 2B	Convergence and reconvergence with NLP disabled	This test case is the same as Test 2A with NLP disabled
Test 2C	Convergence test in the presence of background noise	To ensure EC converges in the presence of background noise
Test 3A	Double talk convergence test with low canceller-end levels	To ensure that the double talk detection is not sensitive to low-level near-end speech
Test 3B	Double talk stability test	To ensure that DT detection is sensitive
Test 3C	Double talk under simulated conversation	To ensure EC does not produce undesirable artifacts during and after double talk
Test 4	Leak rate test	To ensure that the leak time is not too fast, filters do not go to zero too rapidly in 2 minutes
Test 5	Infinite return loss convergence test	To ensure EC has to prevent the unwanted generation of echo
Test 6	Nondivergence on narrowband signals	To verify EC remains converged to narrowband signals after converging on wideband signals
Test 7	Stability test	To verify that the echo canceller will remain stable for narrowband signals
Test 8	Nonconvergence of echo cancellers on tones (optional)	To verify EC operation with mono or bi-frequency tones
Test 9	Comfort noise test	Comfort noise matches the near-end background
Test 10	Facsimile test	To ensure EC converges rapidly on initial fax handshaking sequences and to prevent the unwanted generation of echo by these signals
Test 10A	EC operation on calling station side	Adaptation with CNG, CED, and CSI/DIS
Test 10B	EC operation on called station side	Adaptation while TSI, DCS are applied
Test 10C	EC operation on calling side during page breaks (further study)	This is for V.29 modem, under study for V.27ter and V.17

Table 6.3. *Continued*

Test No.	Test Name	Brief Description
Test 11	Tandem echo canceller test	To ensure EC operating in tandem is not degrading the quality of a call
Test 12	Residual acoustic echo test (further study)	To check the performance in the presence of residual acoustic echo
Test 13	Performance with low-bit-rate coders in echo path (optional)	To ensure on performance with low-bit-rate codec in echo path
Test 14	Performance with V.series low-speed data modems	To ensure no impairments to data modems with EC enabled
Test 15	PCM offset test (optional)	To ensure EC works with PCM offset
Fax tone	Modem and fax tone disabler tests	To ensure fax and modem are detected

6.10.1   Simulated Tests

Echo canceller test signal generation, testing procedure, and timing sequence are given in G.168. The test signal consists of synthetic speech, pseudo-random noise, and silence zones. Developers/implementers also generate additional test vectors to create various corner cases and actual conversation speech files. Such an extended suit is used for internal testing. Most of these tests are useful for actual software development and a simulation environment. These test vectors are used in evaluating the software in Matlab, C, and processor-specific assembly. For actual testing in an integrated product, echo canceller test equipment is used. Equipment manufacturers use some G.168 test signals and several proprietary signals to help simplify the measurements in a known way.

6.10.2   Instrument-Based Tests

The echo canceller is a four-terminal block with R_{in}, R_{out}, S_{in}, and S_{out}. In the echo canceller, R_{in} and R_{out} are exactly the same. As shown in the previous sections, the major part of the electrical echo is created at the two-to-four wire hybrids of SLIC interface and telephones. Most users will be using a two-wire telephone interface and analog telephones. The measurement of echo will have several limitations on the two-wire interface; therefore, four-wire interface is more suitable.

Basic Instrument-Based Two-Wire Tests. Some instruments [URL (Sage935)] perform basic steady state echo cancellation tests on a two-wire interface. Sage instrument-based basic tests are called echo sounder and responder tests. The echo sounder generates a signal and measures the echo. The echo responder or echo generator creates the echo. On a two-wire interface, it is difficult to isolate the leakage. Hence, proprietary test signals are sent in the beginning to perform end-to-end hybrid calibration before creating the echo test. This is found to be useful to evaluate basic echo cancellation and to measure steady state cancellation. Some more test details are available at [URL (Sage935)].

Basic Instrument-Based Four-Wire Tests. Four-wire tests are represented in Fig. 6.9. Four-wire-based tests allow several combinations of echo canceller tests given in Table 6.3. In Fig. 6.9(a), two VoIP adapters A and B are used for conducting the tests. In the EC tests of Fig. 6.9(a), the adapter-A interface works for sending the signal and for measuring the echo. It also works for sending the talker voice and for measuring the talker echo. The other side of adapter-B is the echo creator. In this measurement, EC of adapter-B is under evaluation. For measuring EC performance, it is required to activate all hybrids, phones, and so on. Hence, it is useful to create echo on a two-wire interface of adapter-B. It is also possible to create echo on a four-wire interface as shown with dotted lines on adapter-B of Fig. 6.9(a). In general, the adapter-B side can be used with two or four wire. Instruments also provide various controls and configurations to the echo canceller in maintaining timing relations. An

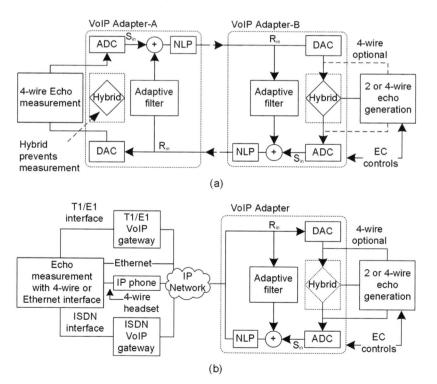

Figure 6.9. Echo measurements with VoIP systems. (a) Echo measurement on VoIP adapters. (b) Echo measurement through four-wire or Ethernet options.

RS-232 serial interface is used in most instruments. As shown in Fig. 6.2, there are three controls to the EC block: H-register, inhibit convergence, and NLP disable. These controls are mapped through the communication interface from the instrument or computer controlling the instrument.

With reference to Fig. 6.9(b), the measurement on adapter-A should not see a two-to-four-wires hybrid. It is essential to provide the direct interface from the ADC and the DAC. The SLIC hybrid on adapter-A is shown as isolated. This type of arrangement is possible in the development phase of the product. In a finished product, it is not possible to extract the four-wire interface. Alternatively, four-wire interface-capable VoIP systems can be used as shown in Fig. 6.9(b). In Fig. 6.9(b), the VoIP gateway with a T1/E1 or ISDN interface is used. In these four-wire interfaces, the hybrid is not present. Another simpler option is to use the IP phone. The IP phone does not have any hybrid. The IP phone internal architecture is given in Chapter 2. The IP phone ADC, DAC path signals are directly interfaced to the headset microphone and speaker. This headset interface will go directly to the instrument. In Fig 6.9, two instruments are shown for distinction of echo generation and measurements. In practice, both sides will be part of the same instrument. Several details on these tests can be found at equipment supplier websites [URL (Sage935), URL (GL)].

Echo Measurement through Network Interface. In Fig. 6.9(b), the Ethernet interface is shown to be going to the instrument. In general, the Ethernet interface is used with several instruments for network measurements. The purpose of the Ethernet interface shown here is for echo measurement. The instrument on the adapter B-side sends the talker signal as RTP packets as if they are delivered from the gateway. Internal representation blocks of this type of measurement are given in [URL (GL)]. The received IP packets (RTP packets) will also be captured directly from the network side of adapter-A. Echo is created on adapter-B side. It is more easy and accurate to measure echo with ideal four-wire interfaces. It is expected that such echo measurements will be made on a dedicated local area network. Hence, IP impediment contributors will not be perceived. The system may work like a fixed round-trip delay system for the purpose of measurement.

6.10.3 Perception-Based Tests

Simulation and instrument-based echo cancellation tests reveal most of the performance issues of the echo cancellers. Certain impediments of NLP, pleasantness in DT, interactive voice calls with delays, voice calls under background noise, and so on, are evaluated with perception tests. These tests are conducted with audio interfaces. Audio interfaces and listeners are positioned in acoustic silence chambers. A listener observes various conditions of the echo cancellation. The perception tests have to be conducted after thorough evaluation and meeting the echo cancellation specifications in simulated and instrument-based tests. Echo cancellers certified by agencies such as ATT-USA and British Telecom exact-UK validate the carrier-grade credentials of the product. These tests include both measurement and perception-based tests. This type of certification is essential for ensuring better voice quality. The echo cancellers mainly aimed at PSTN–VoIP gateways and long tail-length operations are subjected mandatorily to certification and subjective testing. Certified echo cancellers are widely used and accepted in inter-regional calls with end-to-end delays up to 400 ms.

7

DTMF DETECTION, GENERATION, AND REJECTION

DTMF is a dual-tone multifrequency signaling tone generated while pressing the telephone keypad numbers in tone mode. It is known by several names, including multiple frequency push button (MFPB) and digit tones, but the most well known is DTMF. When a key is pressed on a telephone keypad, the telephone typically generates a dual-frequency tone composed of a low-frequency component and a high-frequency component, which is recognized by equipment connected to the phone. This method of signaling replaces an older method known as pulse dialing where the number entered is represented by a series of line make-and-break contacts that result in electrical pulses. In pulse dialing, a single pulse represents one, and two pulses represent two, and ten pulses represent zero. Often modern phones support both methods, typically by setting a pulse/tone switch on the phone. Currently, DTMF is the most popular dialing in use.

DTMF-based signaling is used in many telecommunication applications, including public switched telephone network (PSTN) telephony, VoIP, answering machines, Internet Protocol Private branch exchange (IP PBX), telephone exchanges, video conferencing, video phones, IP phones, voice messaging systems, wireless telephony, and interactive voice message announcement systems. The telephone keypad has keys with the digits 0 through 9 and the symbols hash (#), star (*), and other feature [ITU-T-Q.23 (1988), TIA/EIA-464C (2001)] keys known as A, B, C, and D digits. The total 16 DTMF digits are mapped as digital numbers inside the processor memory usually as 0 to 15

VoIP Voice and Fax Signal Processing, by Sivannarayana Nagireddi
Copyright © 2008 by John Wiley & Sons, Inc.

Table 7.1. DTMF Digits and Tone Frequencies

Row Frequencies	Column Frequencies and Digits			
	1209 Hz	1336 Hz	1477 Hz	1633 Hz
697 Hz	1	2	3	A
770 Hz	4	5	6	B
852 Hz	7	8	9	C
941 Hz	*	0	#	D

[Schulzrinne and Petrack (2000)] or as 1 to 16 [Pessoa et al. (2004)]. To avoid confusion with the alphabet letters A to Z marked on top of the telephone 2 to 9 keys along with numbers, the A, B, C, and D keys (if used) are marked with applicable functionality. In pulse dialing mode, only 10 pulses are counted between 0 and 9, and the remaining *, #, A, B, C, and D buttons are not functional. Modern phones typically have a hook flash key to interrupt the line for about 200 to 400 ms. For phones without such a button, the same function can be invoked by briefly pressing the receiver hook switch. The hook flash key is used along with some call features such as call waiting to signal the equipment connected to the phone that some operation is required, such as switching to a second line.

As indicated in Table 7.1, DTMF digits are mapped to a combination of four rows and four column frequencies. The row frequencies 697, 770, 852, and 941 Hz are in low band, i.e., below 1000 Hz. The column frequencies 1209, 1336, 1477, and 1633 Hz are in the high band between 1000 and 2000 Hz.

7.1 SPECIFICATIONS OF DTMF TONES

The characteristics of DTMF or MFPB tones are given in Q.24 recommendation [ITU-T-Q.24 (1988)]. Different telecommunication administrations may have their own specifications. A few regional specifications are listed in Table 7.2 for NTT, AT&T, and Danish administration with reference to the specifications given in Q.24. It is also suggested that the reader refer to local country PSTN standards directly wherever relevant.

7.2 DTMF TONES GENERATION

Telephones generate DTMF tones, and these tones reach the PSTN local switch or VoIP gateway in the microphone path. The foreign exchange subscriber (FXS) interface on the VoIP adapter also generates DTMF tones to the telephone during DTMF caller ID and call waiting caller ID. As shown in Fig. 7.1, the foreign exchange office (FXO) is equivalent to the telephone resident inside the VoIP gateway. FXO interfaces also generate or accept DTMF

Table 7.2. Values of Multifrequency Push Button Receiving Parameters for Multiple Administrations from ITU-T-Q.24

Parameters		Parameter Values		
		NTT (Japan)	AT & T (N. America)	Danish Administration
Signal frequencies	Low group	697, 770, 852, 941 Hz	697, 770, 852, 941 Hz	697, 770, 852, 941 Hz
	High group	1209, 1336, 1477, 1633 Hz	1209, 1336, 1477, 1633 Hz	1209, 1336, 1477, 1633 Hz
Frequency tolerance $|\Delta f|$	Operation	≤1.8%	≤1.5%	≤(1.5% + 2 Hz)
	Nonoperation	≥3.0%	≥3.5%	
Power levels per frequency	Operation	−3 to −24 dBm	0 to −25 dBm	−2 to −27 dBm
	Nonoperation	Max. −29 dBm	Max. −55 dBm	Max. −36 dBm
Power level difference between frequencies (Twist)		Max. 5 dB	+4 to −8 dB**	Max. 6 dB
Signal reception timing	Signal operation	Min. 40 ms	Min. 40 ms	Min. 40 ms
	Signal nonoperation	Max. 24 ms	Max. 23 ms	Max. 20 ms
	Pause duration	Min. 30 ms	Min. 40 ms	Min. 40 ms
	Signal interruption	Max. 10 ms	Max. 10 ms	Max. 20 ms
	Signalling velocity	Min. 120 ms/digit	Min. 93 ms/digit	Min. 100 ms/digit
Signal simulation by speech		6 false/46 hours for speech with a mean level of −15 dBm	For codes 0–9, 1—false/3000 calls; For codes 0–9, *, #, 1—false/2000 calls; 0–9, *, #, A-D 1—false/1500 calls	46 false/100 hours for speech with a mean level of −12 dBm
Interference by echoes			Should tolerate echoes delayed up to 20 ms and at least 10 dB down	

Courtesy: Reproduced with the kind permission of ITU; International Telecommunication Union, Geneva, www.itu.int.
**High-frequency group power levels may be up to 4 dB more or 8 dB lower than the low-frequency group.

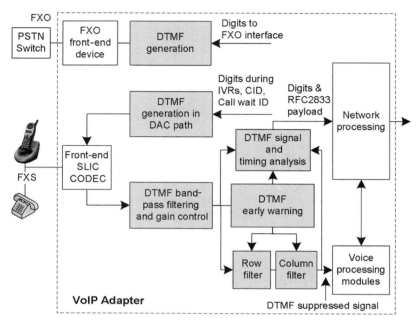

Figure 7.1. DTMF functional representations in the VoIP Adapter.

tones. Almost all modern telephones have a set of mechanical switches (often arranged as a 4 × 4 matrix) interfaced to a microcontroller that scans and debounces these contact closures; however, the microcontroller will have many other functions in addition to DTMF. When a mechanical switch is pressed, it typically "bounces," which generates a series of open and closed states before settling to the final state. This process typically occurs within 16 to 100 ms of when the button is pressed and needs to be debouched by the microprocessor so that it is not viewed as multiple button presses.

DTMF tones are sent on the phone line TIP-RING interface. A DTMF tone is generated as a combination of two frequencies as per Table 7.1. When the digit key is held down continuously, most phones generate a continuous DTMF tone. To avoid making multiple detections of the same digit, PSTN or VoIP gateways decide on the digit after completion of a long duration tone. Several specifications for DTMF tone parameters are given in Q.24 and listed in Table 7.2.

As listed in Table 7.2, DTMF digit tones have a higher dynamic range in power levels, but the generation operates at nominal power levels of −12 to −3 dBm. Usually the lower part of the signal power is not used for DTMF digit generation as this can become sensitive to various disturbances and losses. Some DTMF tone generation aspects are presented in [ADI Vol-1 (1990), Schmer (2000)]. In the generation of DTMF, two tones are generated with a required frequency and normalized amplitude. These two tones are combined with required amplitude scaling to generate tones with matching power levels. Simpler phones may use active circuit-based oscillators for DTMF generation. In processor-based implementations, sine computation with Taylor and Maclaurin series or digital oscillator method is typically used. Processor-based methods generate tones typically with a 0.003% precision level as per reference [ADI Vol-1 (1990)]. The reference clock precision used in the tone generation also limits the precision.

7.2.1   Sine Wave Computation in the Processor

This routine evaluates the sine function to 16 significant bits using fifth-order polynomial expansion. To improve accuracy with limited order, the polynomial suggested in ADI Vol-1 (1990) is used here.

$$\sin(\theta) = 3.1406250\theta + 0.0202263670\theta^2 - 5.3251960\theta^3 + 0.54467780\theta^4 + 1.8002930\theta^5 \tag{7.1}$$

In the above formulation, θ is in radians, with suitable mapping to a digital number of processor arithmetic. Valid values of θ in radians is from 0 to $\pi/2$ (first quadrant, representing 90 degrees) with $\pi/2$ equated to 0.5 in mapping of θ. For the remaining three quadrants, $\sin(\theta)$ may be computed by the simple mappings:

$$\sin(\theta) = \sin(\pi - \theta) \quad \text{(Second quadrant)}$$

$$\sin(-\theta) = -\sin(\theta) \quad \text{(Third and fourth quadrant)}$$

A fifth-order polynomial implementation with 16-bit processor arithmetic provides a tone generation precision of 0.003% as given in reference [ADI Vol-1 (1990)]. For each sample of sinetone, the phase is supplied to the polynomial. Phase increments or phase advances are fixed for subsequent sample generation. Consider an implementation in which the phase is represented by a 16-bit integer, with zero degrees as #0000, 90 degrees as #4000, and so on. Note that 360 degrees cannot be represented in this way, but it is equivalent to zero degrees and can therefore be represented by the value #0000. It effectively works as a modulus operation with a period of 2π radians or 360 degrees. The polynomial evaluation is performed using an accumulated phase. Assuming a sample rate of 8000 Hz, the phase change per step is given in reference [ADI Vol-1 (1990)].

$$\partial\theta = \frac{65536}{8000}(f) = 8.192(f) \tag{7.2}$$

Tone frequency f is in Hertz. The phase accumulator value θ_{acc} is the phase supplied to the polynomial. For every subsequent sample after required initialization, phase is updated as $\theta_{acc}(i) = \theta_{acc}(i-1) + \partial\theta$ and applied in Eq. (7.1) for sine wave generation. The initial value of $\theta_{acc} = 0$, or it is initialized with the required initial phase of the sine wave. The phase supplied to polynomial expansion is the initial phase advanced accumulated value. On accumulation exceeding h#FFFF, the 16-bit number is folded back to zero. In sine computation, the first quarter of the cycle phase is computed. The remaining three quadrants are derived using the mapping with the required sign (making it positive or negative) change. In this scheme, the sine wave is better approximated with negligible (0.003%) error. One drawback with this method is it demands computation of the polynomial and phase accumulations, but this is insignificant compared with processing in other voice modules of voice compression and with echo cancellation.

7.2.2 Digital Sinusoidal Oscillator Method

A digital sinusoidal oscillator may be viewed as a form of a two-pole resonator for which the complex conjugate poles lie on the unit circle. A second-order system transfer function for oscillator is given by

$$y(i) = \text{Coef}_{\cos} y(i-1) - y(i-2) \tag{7.3}$$

$$\text{Coef}_{\cos} = 2\cos(2\pi f/f_s) \quad \text{and} \quad \text{Coef}_{\sin} = \sin(2\pi f/f_s) \tag{7.4}$$

In the equation, $y(-1) = 0$ and $y(-2) = -A\text{Coef}_{\sin}$. The parameter $\text{Coef}_{\sin}$ is the frequency-dependent sine coefficient, and $\text{Coef}_{\cos}$ is the cosine coefficient, f_s is the sampling frequency, f is the DTMF tone frequency in Hz, and A is the required individual sine wave amplitude. Keeping constant $A = 1291$ for amplitude gives -10 dBm of two-tone combination as per μ-law power mapping. In some implementations, $\text{Coef}_{\cos}$ is used as $\cos(2\pi f/f_s)$ and oscillator is represented as $y(i) = 2\text{Coef}_{\cos}(k)y(i-1) - y(i-2)$. Both representations generate the same sine wave. For DTMF tones, two oscillator outputs have to be combined into a single composite waveform. The oscillator frequency is controlled by the coefficients, and a list of coefficients for different row and column frequencies is given in Table 7.3. DTMF digit-1 waveforms are given in Fig. 7.2 that consist of 667- and 1209-Hz tones. The combined tone is illustrated by the third graph of Fig. 7.2.

7.3 DTMF DETECTION

DTMF detection identifies the digit from a two-tone combination. As per specifications given in Section 7.1, DTMF detection has to take care of several constraints in the process of digit detection. The main constraints from the signal parameters are as follows:

- Low- and high-frequency tones are designed as f_L and f_H and their corresponding amplitudes as A_L and A_H. The amplitudes levels of A_L and A_H

Table 7.3. DTMF Tones and Cos and Sin Coefficients Used in Tone Generation

Tone frequency (f)	$\text{Coef}_{cos} = 2\cos(2\pi f/f_s)$	$\text{Coef}_{sin} = \sin(2\pi f/f_s)$
697	1.7077	0.5205
770	1.6453	0.5686
852	1.5687	0.6203
941	1.4782	0.6736
1209	1.1641	0.8132
1336	0.9964	0.8671
1477	0.7986	0.9168
1633	0.5685	0.9587

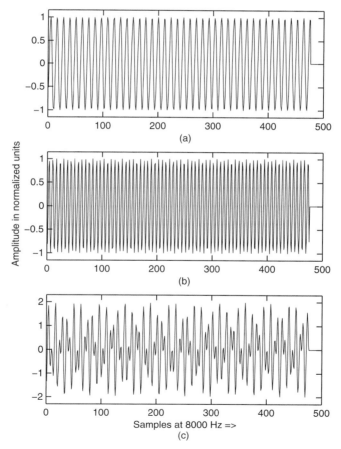

Figure 7.2. Digit-1 tone generation example for 320 samples, 40 ms. (a) Row tone at 697 Hz at unity amplitude. (b) Column tone at 1209 Hz at unity amplitude. (c) Combined tones for digit-1.

are the same in an ideal case, but some phones may generate tones with twist or asymmetry, which means that the amplitudes of f_L and f_H are not the same. If the amplitude of f_L is greater than f_H, it is called a reverse or negative twist. If the amplitude of f_H is greater than f_L, it is called a forward twist or positive twist.

- Frequency drift creates different filter responses. Frequency drift along with twist creates additional issues for correct tone detection.
- The DTMF digit minimum duration is greater than 23 ms (typical 50 ms), and the maximum can last for several seconds if the key is held for a long duration.
- Interruptions in the signal up to 10 ms should be tolerated, and a total minimum signal should be greater than 23 ms.
- DTMF tones have to be detected in the presence of speech, music, and background activity or noise. It is important to ensure that the speech and background music is not interpreted as a valid DTMF tone pair.
- DTMF processing has to cater to a wide signal dynamic range of DTMF tones.
- Detection processing has to tolerate echoes while extracting the tones.

Detection of a DTMF digit requires both a valid tone pair and the correct timing intervals while discriminating against different background activity. Tone detection involves mathematically transforming the time-domain samples into frequency domain. On the magnitude of tones, several signal-level and duration analysis steps are performed to determine the correct digit. The detection schemes are commonly based on:

- Goertzel algorithm
- Teager and Kaiser (TK) algorithms
- Frequency analysis with Fourier transforms

7.4 GOERTZEL FILTERING WITH LINEAR FILTERING

Linear filtering is based on a set of basic resonators in which each resonator selects one frequency. This approach is specified in references [ADI Vol-1 (1990), Schmer (2000)]. The Goertzel algorithm efficiently estimates the Fourier transform of the signal at specific frequencies. This process is similar to computing the discrete Fourier transform (DFT) at any one of the selected frequencies and for any arbitrary block of samples. DTMF has a limited set of frequency tone combinations; hence, the Goertzel method is more efficient than the DFT-based method. The Goertzel algorithm may be considered as a second-order infinite impulse response (IIR) band-pass filter and work similarly to the DTMF tone generation resonator, with the main difference that filtering operates on actual input and the oscillator requires initialization. The Goertzel filter output $y_k(n)$ for input $x(n)$ is given by the difference equation.

$$y_k(n) = x(n) + \text{Coef}(k) y_k(n-1) - y_k(n-2) \tag{7.5}$$

In this equation, $y_k(-1) = 0$; $y_k(-2) = 0$, $\text{Coef}(k) = 2\cos(2\pi k/N) = 2\cos(2\pi f/f_s)$, where f is the tone frequency of interest (697 Hz for the first row bin); f_s is the sampling frequency usually at 8000 Hz; $n = 0, 1, \ldots N - 1$; and N is the DTMF processing block size in samples. In practice, the frequency index $k = Nf/f_s$ is rounded to the integer value. It is suggested that the rounding operation on $\text{Coef}(k)$ be applied after normalizing to a required precision and 16-bit numbering format. Note that $\text{Coef}(k)$ is not dependent on N—only the intermediate value of k is dependent on N. The DFT-equivalent response at frequency bin k is given by:

$$Y_k(n) = y_k(n) - e^{j2\pi k/N} y_k(n-1) \tag{7.6}$$

In a block of samples from 0 to $N - 1$, $Y_k(N - 1)$ is the complex amplitude at the end of $N - 1$ samples given by:

$$Y_k(N-1) = y_k(N-1) - e^{j2\pi k/N} y_k(N-2) \tag{7.7}$$

$$Y_k(N-1) = \left[y_k(N-1) - \cos\left(\frac{2\pi k}{N}\right) y_k(N-2) \right] - j\left[\sin\left(\frac{2\pi k}{N}\right) y_k(N-2) \right] \tag{7.8}$$

In these equations, $j = \sqrt{-1}$ is for representing the imaginary part of a complex number. It gives real and imaginary parts similar to DFT. Some literature may use the representation $Y_k(N - 1)$ as $Y_k(N)$ to represent amplitude calculated at the end of the N-samples. $Y_k(N)$ representation is more convenient for a quick understanding result after N samples, and then the index can be thought of as being in the range of 1 to N.

In DTMF tone analysis, phase is not important so the computation is simplified by computing only the magnitude. It is called the modified Goertzel algorithm. The main difference of modified algorithm is the last step of magnitude computation.

$$[Y_k(N-1)]^2 = [y_k(N-1)]^2 + [y_k(N-2)]^2 - \text{Coef}(k)[y_k(N-1)][y_k(N-2)] \tag{7.9}$$

7.4.1 Selection of Frequency Bins

DTMF has four-row tones and four-column tones. To distinguish between speech and tones, the second harmonic of each tone is analyzed. For harmonic calculation, Eq. (7.5) and Eq. (7.9) are used on the same input $x(n)$. The values $\text{Coef}(k)$ vary depending on the harmonic frequency. The values for index k

Table 7.4. DTMF Frequencies, Coefficients, and Frequency Gap Measure

Frequency classification	DTMF tone and harmonics in Hz	$\text{Coef}(k) = 2\cos(2\pi f/f_s)$	Close neighbor frequency Hz	Frequency difference with closest neighbor in Hz
Row	697	1.7077	770	73
frequencies	770	1.6453	697	73
1, 2, 3, 4	852	1.5687	770	82
	941	1.4782	852	89
Column	1209	1.1641	1336	127
frequencies	1336	0.9964	1394	127 fundamental; 58 with harmonic
5, 6, 7, 8	1477	0.7986	1540	141 fundamental; 63 with harmonic
	1633	0.5685	1704	156 fundamental; 71 with harmonic
Second	1394	0.9164	1336	58
harmonic	1540	0.7069	1477	63
of row	1704	0.4608	1633	71
frequencies	1882	0.1851	1704	178
9, 10, 11, 12				
Second	2418	−0.6449	2672	254
harmonic	2672	−1.0072	2418	254
of column	2954	−1.3622	2672	282
frequencies	3266	−1.6768	2954	312
13, 14,				
15, 16				

and $\text{Coef}(k)$ for different DTMF frequencies are listed in Table 7.4. Note that in Table 7.4, there are extended column entries "close neighbor frequency" and "frequency difference with close neighbor." The details on these two columns are given in later sections.

7.4.2 Goertzel Filtering Example

Goertzel filtering and associated waveforms are illustrated in Figs 7.3 and 7.4 for digit-1 tones. The tone input in Fig. 7.3 is 667 Hz. The filter output increases with time when the input frequency of tone matches with the filter-designed frequency. In Fig. 7.3, the input is interrupted for 80 samples (samples 320 to 399 inclusive). During interruption of the input signal, the filter maintains the same amplitude to behave as an oscillator. The amplitude keeps increasing again after the interruption (samples 400 to 479 inclusive). From sample index 480 onward, it continues as an oscillator in the absence of input. This feature also allows a block-based zooming operation with proper phase corrections to each block of Goertzel filtering. The third row graph of Fig. 7.3

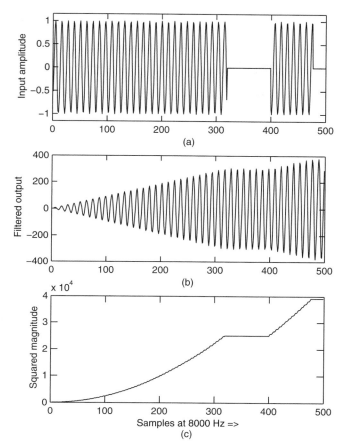

Figure 7.3. Goertzel filter response for 697 Hz tone. (a) Input tone at 697 Hz. (b) Goertzel filter output. (c) Magnitude response with time.

is a squared magnitude computation. The amplitude keeps tracking on a relative scale.

Figure 7.4 shows a 1209-Hz tone. For the same input reference, an amplitude of ±1 of the middle plot of the Goertzel filter output does not match the amplitude indicated in Fig. 7.3 for 667 Hz, but the final squared magnitude computation is the same as in Fig. 7.3. The final magnitude computation of a 1209-Hz tone uses the tone-dependent Coef(k) to achieve amplitude normalization. In summary, the Goertzel filter output is not the same in amplitude for the same matching center frequency input, but squared magnitude is same.

7.4.3 Goertzel Filtering in the Presence of Frequency Drifted Tones

When the input deviates from center frequency, the filter output appears as an amplitude envelope with the envelope matching the drifted frequency. A

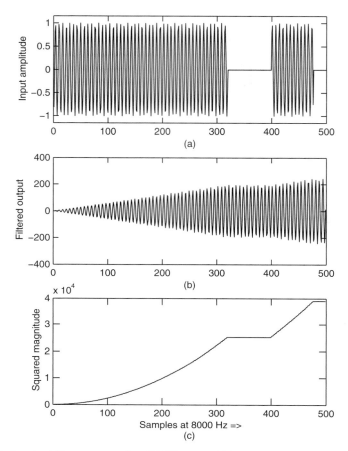

Figure 7.4. Goertzel filter response for 1209 Hz tone. (a) Input tone at 1209 Hz. (b) Goertzel filter output. (c) Magnitude response with time.

set of drift influences is illustrated in Fig. 7.5. This figure consists of three rows of graphs. The first row is the DTMF tone at 1227 Hz. The middle is the Goertzel filter output with the filter positioned at 1209 Hz. The third graph is the magnitude envelope.

The tone input of 1227 Hz in this example is 1.5% positive drifted frequency (i.e., 1209 + 18 = 1227 Hz), and this is within the acceptable range for a DTMF tone. From Fig. 7.5, it may be observed that the filter output is increasing for some samples, and after this, the amplitude is reducing with a sample index. For example, at the end of 80 samples, the amplitude is good and it keeps increasing until sample 222. After this point, the amplitude keeps falling and reaches zero at the 444th sample. Samples of 444 correspond to one eighteenth (1/18th) of a second in duration and exactly match with a drift frequency of 18 Hz calculated as 8000/18 = 444 samples. Once output reaches minimum

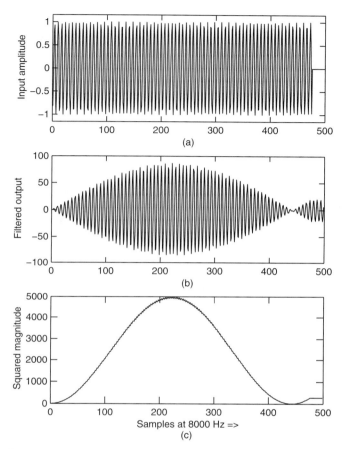

Figure 7.5. Goertzel filter influences for drift. (a) DTMF tone at 1227.135 Hz (i.e., digit-1 tone of 1209 Hz drifted by 1.5% in frequency). (b) Goertzel filter output. (c) Magnitude response with time.

amplitude, the filter output once again increases. The amplitude envelope exactly matches the drift frequency of 18 Hz. A DTMF digit duration is typically not long enough to study several cycles of a 18-Hz envelope. Hence, it is necessary to implement several trade-offs and zooming techniques to cater to multiple specifications. In the third row of the graph, it may be observed that the Goertzel filter output is continuing as an oscillator from the last sample of input excitation at a 480-sample index.

7.4.4 Frequency Drift Trade-Offs for the Highest DTMF Tone

Drift in the tone frequencies have to be analyzed to cater to the deviations in DTMF tone tolerances. Tone tolerances for detection are of 1.8% on either side of the center frequency. Drift percentage mainly influences the column frequencies. A drift of 1.8% on the highest column frequency of 1633 Hz is 30 Hz, so the DTMF highest tone can appear anywhere between 1603 and 1663 Hz, which creates a 60-Hz span. To contain loss to less than 3 to 4 dB (effectively turning the 1603- and 1633-Hz points into 3-dB filter bandwidth points), the processing samples window in the samples has to be less than 8000/60 = 133 samples. With 266 samples processing, the drifted column frequency is eliminated as a side lobe. Hence, it is important to consider analyzing with a smaller window of 80 to 102 samples at the first level. Once the possible tone is identified, additional frequency analysis around 1633 Hz can cater up to 30 Hz on either side. In practical implementations, this process will mean keeping one filter at 1633 Hz and adding two additional filters at approximately 1.2% away from the 1633-Hz tone center frequency. Initial analysis is then performed with 80 to 102 samples. On identifying the 1633-Hz tone as valid, two additional filters are created. In row frequencies, a 1.8% deviation can be accommodated in 240 samples processing; hence, row processing may be performed without additional adjacent filters.

7.4.5 Frequency Spacing and Processing Duration Trade-Offs

As listed in Table 7.3, in the beginning of row frequencies, tones are spaced at a gap of 73 Hz. This gap increases with column frequencies. In the last column, the gap is 156 Hz and this gap value closely matches the geometric progression with a center frequency of row and column frequencies. In a geometric progression, when center frequency doubles, the frequency gap also doubles. This result is referred to as constant-Q in resonance circuits and band-pass filters. In resonance circuits, Q = bandwidth/center frequency, and with constant Q, the relative performance across multiple tones remains the same. For DTMF tones, a similar interpretation can be applied as tone spacing to tone center frequency.

Row frequency spacing of 73 Hz demands a minimum analysis time of 1/73 seconds = 13.7 ms, which implies 110 samples for processing. Some previous published algorithms used 102- and 204-sample processing [ADI Vol-1(1990), URL (Julian)]. Processing of 102 samples is close to the 78-Hz mean value of the row spacing. The mean of row frequencies spacings (73, 73, 82, and 89 from the Table 7.3) is 79.25 Hz. With 102-sample processing, each row is distinguished by a difference of 3 to 4 dB. It is sufficient for initial identification of likely tone but not suitable for a complete decision. Hence, it is necessary to use 204-sample processing.

Overall, 204 samples with a sampling frequency of 8000 Hz represents a filter response time of 25.5 ms. This response is close to a rectangular window with side lobes matching with minimum signal power. In general, a DTMF tone typically lasts for 50 to 70 ms to meet most country-specific DTMF tone length specifications. In most implementations, VoIP packets are aligned to multiples of 10 ms (i.e., 80 samples). The signal of 10-ms duration creates resolution of 100 Hz. This resolution is sufficient for initial likely decision. Later proposals of DTMF processing consider 80 samples for coarse identification and frequency zooming to 160 and 240 samples. Processing and decision making with 240 samples is better than with 204 samples in PSTN or VoIP and matches better with voice frame processing of 80 samples.

7.4.6 Frequency Twist Influences

Twist is the difference in amplitude of row frequency f_L to column frequency f_H amplitude. If amplitude of f_L is greater than f_H, it is the reverse twist. For amplitude f_L less than f_H, it is called forward or standard twist. Reverse twist is given as given as -8 dB, and forward twist is given as $+4$ dB for AT&T [ITU-T-Q.24 (1988)]. In most cases, reverse twist is more common (i.e., f_L amplitude is higher than f_H amplitude). As per the Q.24 recommendation [ITU-T-Q.24 (1988)], AT&T North America has both forward and reverse twist, but several other administrations such as Japan NTT, Australia, Danish, and Brazil use the same value of forward and reverse twist.

The frequency spacing of rows, columns, and their harmonics play a major role in DTMF analysis. Some frequencies are very close to each other, and their frequency pattern is fixed. From the listings in Table 7.3, it can be observed that the minimum gap of columns and row second harmonics are 58 Hz. To get better differentiation of these two frequencies, it is necessary to process 275 samples to maintain the rectangular window side lobes. The processing of 240 samples can closely meet this requirement.

7.4.7 Overall DTMF Processing with Goertzel Filtering

In the previous sections, several trade-offs are considered concerning frequency, drift, power levels, twist, frame durations, and timing for processing. In this section, an overview on Goertzel-filtering-based digit detection is presented in relation to previous sections and Fig. 7.6.

At the end of frame processing, row and column frequency energies are computed. After performing sanity checking for valid bins and verifying for minimum signal power and first-level twist, more analysis on the identified bins and frequency zooming can be attempted.

Row and column filtering and the power estimation is an integral part of the filtering. These levels may be relative or normalized to dBm scale. DTMF processing starts with identifying the row and column frequencies with the highest power levels. More details on the implementation are given in [ADI Vol-1 (1990), Emiya et al. (2004)]. As a first step, these power levels are checked against minimum valid row–column powers and coarse-level twist. To get better analysis in the presence of drift, it is necessary to analyze more column frequencies, which is referred to in this book as column frequency zooming. The block marked with first-level power checking keeps analyzing for row column as well as column zooming frequencies and power levels.

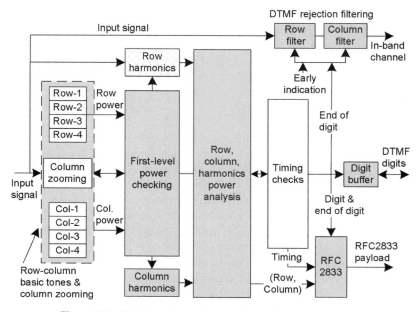

Figure 7.6. Digit detection using the Goertzel filtering scheme.

In the next step, harmonics are analyzed to check for speech, as speech produces more harmonic power compared with DTMF digits. Harmonic analysis is usually conducted on the selected row–column frequencies of interest. Finally, a timing check block analyzes the time envelope of the bins to assess the duration of specific tones, breaks, and total duration as well as to arrive at the digit detection. VoIP signaling uses the collected digits from the digit buffer. Implementations preserve previous digits in buffer for any retransmission or total dial plan analysis.

The RFC2833 operation explained in the later part of this chapter makes use of the digit information, row–column frequencies, and power levels. DTMF rejection requires an early indication of row–column frequencies and an end of the digit to stop the rejection operation. If this block is sufficiently intelligent, it can also consider row–column frequency power levels and drifts to arrive at optimal filter selection as well as rejection levels. In Fig. 7.6, several arrows are shown with signal parameter names. In general, several parameters and analysis state decision blocks are present in the DTMF detection. Automatic gain control (AGC) may be used on the input side, but this is mainly determined by the processor's arithmetic precision. For 16-bit processors performing an AGC operation helps; however, with 32-bit processing, it is not necessary. The processing steps of DTMF vary depending on the processing approach of Goertzel, fast Fourier transform (FFT), or Teager–Kaiser (TK).

7.5   TONE DETECTION USING TEAGER AND KAISER ENERGY OPERATOR

The TK energy operator is used for generic tone detection. TK is an efficient method that provides faster detection than traditional methods based on Fourier transforms and on the Goertzel algorithm. The TK operator is a frequency modulation (FM)–amplitude modulation (AM) demodulation technique for detecting the presence of a tone with arbitrary frequency and amplitude. It is mainly useful as a DTMF early warning detector to help with DTMF rejection. The design of Teager and Kaiser is given in [Pessoa et al. (2004), Emiya et al. (2004), Kaiser (1990), Fiebrink (2004)]. The TK algorithm is based on a differentiating technique and is hence sensitive to speech and noisy conditions. Some important features of the TK detector relevant to DTMF detection are as follows:

- The TK operator can extract localized frequency of the tone with few samples.
- It can work as a FM demodulator. It can extract localized amplitude envelope and can simultaneously work as an AM–FM detector.
- It works on every new sample.
- It can start working on a minimum of three samples of data. It is a differentiator type of operation. Hence, it cannot add any improved signal-to-distortion discrimination. If the tone is presented along with background noise or disturbance, the analysis is degraded because of the differentiation operation of TK detectors.
- It cannot discriminate between multiple tones in the original form. If two or more tones are presented, it gives a combined response.

Figure 7.7 shows, at a highlevel, TK-based DTMF processing. In TK-based processing, it is important to separate row and column frequency bands. Lowpass or band-pass filters are used for row frequency separation. The column filter can be a high-pass or a band-pass filter. Because of the band-limited nature of row and column filter frequencies, it is possible to reduce the sampling frequency at the output of the filters. In a TK energy operation, higher sampling helps to provide better numerical precision for trigonometric terms associated with amplitude and frequency estimation. Additionally, TK processing is of much lower computation compared with other voice functions; hence, 8000-Hz sampling may be retained at the end of filtering.

The row and column filters are usually realized as quadrature mirror filters (QMFs) [Vidyanathan (1992)]. From the first-level calculation of TK operations, frequency $K_\Omega = \cos^2(\Omega)$ and amplitude as $K_A = A^2$ are available. The DTMF state machine operates on amplitude and frequency values. As given in Starcore implementations [Emiya et al. (2004)], the state machine can be made to work directly on $K_A = A^2$ and $K_\Omega = \cos^2(\Omega)$, thus saving computation

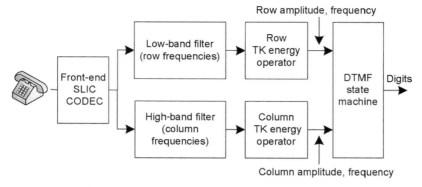

Figure 7.7. TK basic operations for DTMF processing.

and simplifying the DTMF state machine. Most voice processing is performed with 16-bit arithmetic, and operating directly on power levels, it requires that some arithmetic operations implementation in 32-bit numbering format cater to the wide range of arithmetic operations with power levels.

The TK energy operator is mainly useful for sinusoidal signals. The formulation below is given for a single sinusoidal tone. The energy of the sampled signal $x(n)$ is given by following two equations, each of which yields the same result:

$$\Psi[x(n)] = x^2(n) - x(n+1)x(n-1) \tag{7.10}$$

or

$$\Psi[x(n)] = x^2(n-1) - x(n)x(n-2) \tag{7.11}$$

Based on the formulation given in [Kaiser (1990)], signal $x(n)$ is considered a sinusoid. Let $x(n) = A\cos(\Omega n + \phi)$, where Ω is the frequency in radians and ϕ is the initial phase in radians.

$$x(n+1) = A\cos(\Omega(n+1) + \phi) \tag{7.12}$$

$$x(n-1) = A\cos(\Omega(n-1) + \phi) \tag{7.13}$$

$$x^2(n) = A^2\cos^2(\Omega n + \phi) \tag{7.14}$$

and

$$x(n+1)x(n-1) = A^2\cos(\Omega(n+1) + \phi)\cos(\Omega(n-1) + \phi) \tag{7.15}$$

Using the basic trigonometric identity operations:

$$\cos(\alpha+\beta)\cos(\alpha-\beta) = (1/2)[\cos(2\alpha)+\cos(2\beta)] \quad \text{and}$$
$$\cos(2\alpha) = 2\cos^2(\alpha) - 1 = 1 - 2\sin^2(\alpha)$$

Eq. (7.15) can be simplified as

$$x(n+1)x(n-1) = \frac{A^2}{2}[\cos(2\Omega n + 2\phi) + \cos(2\Omega)]$$
$$= A^2 \cos^2(\Omega n + \phi) - A^2 \sin^2(\Omega) \qquad (7.16)$$
$$= x^2(n) - A^2 \sin^2(\Omega)$$

Therefore, the energy of a signal is given as

$$\Psi[x(n)] = x^2(n) - x(n+1)x(n-1) = A^2 \sin^2(\Omega) \qquad (7.17)$$

For small values of Ω, the approximation $\sin(\Omega) \approx \Omega$ can be used. By limiting tone frequency to one eighth of the sampling frequency, the energy computation in Eq. (7.17) can be approximated as

$$A^2\Omega^2 \approx x^2(n) - x(n+1)x(n-1) \qquad (7.18)$$

This energy value is a combination of amplitude and frequency terms. This formulation of approximation gives a maximum error of 11%. This type of error is acceptable for the amplitude or estimation of the tone signal, but it is not acceptable for DTMF frequency estimation. Beyond 1.8% frequency drift, DTMF digits are rejected as an invalid digit or may be misidentified as a different digit. It is important to estimate frequency accurately in DTMF tone detections. The methods used for separating amplitude and frequencies are given under the name discrete energy separation algorithm (DESA). Two algorithms are referred to as DESA-1 and DESA-2 with minor differences in arithmetic in calculating the energy operators. These algorithms provide a simple and elegant method of estimating the amplitude and frequency of a single sinusoid. The equations given in this section are based on DESA-2 for frequency and amplitude estimations as per references [Pessoa et al. (2004), Emiya et al. (2004)].

Let $y(n) = [x(n+1) + x(n-1)]/2$ be the mean value of two adjacent samples. If an average value is not used, a multiplication factor of four appears in the result. The use of an average removes the scaling factor, and the results are simpler for presentation. The energy operation on $y(n)$ is given as

$$\Psi[y(n)] = y^2(n-1) - y(n)y(n-2) \qquad (7.19)$$

On simplification of the equation with a trigonometric identities energy operator is given as

$$\Psi[y(n)] = A^2 \cos^2(\Omega) \sin^2(\Omega) \tag{7.20}$$

Eq. (7.20) is derived with the following simplifications and trigonometric identities:

$$x(n+1) + x(n-1) = A\cos(\Omega(n+1)) + A\cos(\Omega(n-1)) \tag{7.21}$$

From trigonometric identity of $\cos(\alpha) + \cos(\beta) = 2\cos[(\alpha + \beta)/2]\cos[(\alpha - \beta)/2]$,

$$y(n) = [x(n+1) + x(n-1)]/2 = A\cos[\Omega n]\cos[\Omega] \tag{7.22}$$

$$y(n-1) = A\cos[\Omega(n-1)]\cos[\Omega]$$
$$y(n+1) = A\cos[\Omega(n+1)]\cos[\Omega] \tag{7.23}$$

$$y(n-1)\,y(n+1) = A^2\cos^2\Omega[\cos(\Omega(n-1))\cos(\Omega(n+1))]$$

and

$$[\cos(\Omega(n-1))\cos(\Omega(n+1))] = [\cos(2\Omega n) + \cos(2\Omega)]/2$$

From trigonometric identities of

$$\cos(2\Omega n) = 2\cos^2(\Omega n) - 1 \ \& \ \cos(2\Omega) = 1 - 2\sin^2(\Omega)$$
$$A^2\cos^2\Omega[\cos(2\Omega n) + \cos(2\Omega)]/2 = A^2\cos^2\Omega[\cos^2(\Omega n) - \sin^2(\Omega)] \tag{7.24}$$

$$y(n-1)\,y(n+1) = A^2\cos^2\Omega[\cos^2(\Omega n) - \sin^2(\Omega)]$$
$$y(n-1)\,y(n+1) = A^2\cos^2(\Omega)\cos^2(\Omega n) - A^2\cos^2(\Omega)\sin^2(\Omega)$$
$$y(n-1)\,y(n+1) = y(n)^2 - A^2\cos^2(\Omega)\sin^2(\Omega)$$
$$\Psi[y(n)] = A^2\cos^2(\Omega)\sin^2(\Omega) \tag{7.25}$$

The ratio of energy operators on signals $y(n)$, $x(n)$ and arithmetic simplifications results in the separation of frequency K_Ω and amplitude K_A terms.

$$K_\Omega = \frac{\Psi[y(n)]}{\Psi[x(n)]} = \cos^2(\Omega) \tag{7.26}$$

$$K_A = \frac{\psi[x(n)]}{1 - K_\Omega} = A^2 \tag{7.27}$$

These terms are then used to compute actual amplitude and frequency. To calculate Ω from $\cos^2(\omega)$, several approximations and polynomials are used. In DTMF, the interest is mainly to compare the tone frequency and to validate whether the tone is falling within the specified maximum deviation for the tone frequency. In DTMF implementations, $\cos^2(\Omega)$ is directly useful for tone frequency analysis. In the case of absolute value requirements of frequency, frequency f in Hz is calculated as $f = f_s\Omega/2\pi \ or \ \Omega = 2\pi f/f_s$. Here, f_s is the sampling frequency, usually of 8000 Hz. To arrive at the formulations of TK energy, amplitude, and frequency information, several authors have used different formulations. These formulations use several trigonometric simplifications and approximations. Each type of simplification may help particular application and/or work efficiently with the arithmetic supported by a particular signal processing implementation on a processor.

7.6 DFT OR FFT PROCESSING

The FFT efficiently calculates all possible frequency points in the DFT for an N-point DFT as given in reference [Madisetti and Williams (1998)]. A generic DFT calculation is given by

$$X(k) = \sum_{n=0}^{N-1} x(n) W_N^{nk}$$

In the above equation, $k = 0, 1, 2 \ldots N - 1$ is the frequency index; $W_N^{nk} = e^{-j\frac{2\pi nk}{N}}$; and $j = \sqrt{-1}$ is for complex numbering representation. To get actual frequency mapping, frequency index "k" has to be multiplied by frequency resolution f_s/N, where f_s is the sampling frequency usually of 8000 Hz. For an 80 sample block (N) at an 8000-Hz sampling rate, the frequency resolution of analysis is $8000/80 = 100$ Hz. For DTMF analysis, only selected DFT frequency bin values are required. The frequencies of interest are at four row values, four column values, and total of eight second harmonics of row and column frequencies. For selected frequency bins of computation, the DFT algorithm is more efficient than the FFT-based algorithm. A typical DTMF signal varies from 25 to 50 ms in duration, which gives few small blocks (about 10 ms) of DFT processing similar to short-time Fourier transform (STFT). DTMF tones do not appear at integer multiples of f_s/N, which calls for frequency-domain interpolation by padding several zeros before computing DFT. Although this procedure is theoretically correct, it is computationally involved when compared with the Goertzel method. When processing and available memory for the DTMF processing module is not a constraint, long zero-padded FFT is used to get better localization of peaks. Once the dominant frequency components have been recognized, DTMF processing can be performed similar to Goertzel-algorithm-based detection.

7.7 DTMF REJECTION

During a voice call, DTMF tones are sent to the destination to allow various interactive operations. In PSTN calls, DTMF is not distorted as compared with VoIP. In VoIP, an end-to-end call may use a compression codec such as G729AB, which results in distortion of the DTMF signal. In packet delay or lost situations, DTMF transmission is disturbed and may not necessarily be detected as a valid digit at the destination. Packet loss creates discontinuity in DTMF transmission even with G.711-based VoIP calls.

The preferred approach in VoIP is to send DTMF as out-of-band packets with redundancy. The digit value, and its power, duration, digit starting, and digit ending information are sent as message packets to the destination. RFC2833 [Schulzrinne and Petrack (2000)] mainly provides guidelines for sending digits as packets. Digits sent this way are called out-of-band packets or out-of-band digits. At the destination, these packets are used to regenerate DTMF digits with right parameters. Before this method can be used, initial negotiation is needed to establish that both VoIP end terminals support out-of-band digit transport through RFC2833.

Additional difficulties with out-of-band support exist. At the source, DTMF detection takes several tens of milliseconds, and during this period, a significant part of the DTMF tones have already transmitted in RTP packets. These in-band tones (inside the IP voice packets) can reach the destination with possible distortion depending on the codec and IP network impediment characteristics. Sometimes this in-band signal as voice is detected as a DTMF digit at destination, which creates the possibility of sending the same digit in in-band and out of band. To avoid this problem, the in-band DTMF signal is removed at the source. In the DTMF specifications, a minimum DTMF tone used for detection is of 23 ms. If the tone transmitted in in-band is comparable with 23 ms, then the destination may start detecting it as a valid digit. Hence, it is required to begin removal of the DTMF tone within 23 ms as the upper limit. Two popular options for DTMF rejection are as follows:

1. Based on a DTMF early warning
2. Delaying all samples at the source to the extent of DTMF normal detection time, usually of greater than 23 ms

In the previous sections, it is noted that DTMF initial decisions could be arrived with the first 80- to 102-sample data. However, it cannot confirm the digit, but it indicates the likely presence of DTMF tones and can be used to begin DTMF rejection of row and column frequencies.

The removal process may be performed using one of two basic methods. The simplest way is to remove the samples completely and to send silence or comfort noise, but this can be annoying in many situations. A better option would be to pass the signal through row and column frequency notch filters. The filter rejection bandwidth is usually of 20% of the row and column center frequency. It is required to minimize the amount of non-DTMF tone information removal and at the same time ensure that DTMF is removed even if it has drifted by 1.8%. As an example, drifted 1633 Hz can appear anywhere between 1603 and 1663 Hz. Frequencies in between this range have to be removed by the notch filters by 24 to 27 dB. Typically, DTMF tones are generated at −12 to −3 dBm. These tones have to be removed to lower than −27 dBm of the residual DTMF signal. It demands notch filter rejection of about 24 dB in the frequencies between 1603 and 1663 Hz. If the DTMF is of a lower power (e.g., −12 dBm), rejection requirements are of 15 dB to make tone power to below −27 dBm. It is rare, but some phones are generating stronger DTMF tones of the order of 0 dBm, and this demands a 27-dB rejection across the band of center frequency with drift.

A fourth-order IIR notch filter with 20% bandwidth can satisfactorily address most rejection requirements. A user may fine-tune the order of filter and bandwidth based on the rejection requirements. A program such as Matlab generates these coefficients with built-in functions for frequency analysis.

Filter rejection output gives the same duration output as the original DTMF tones. Assuming notch filter bandwidth is of 140 Hz, the notch filter introduces an approximately 7-ms delay at the output. In general, DTMF early warning detection also introduces an additional delay of 5 ms. The combined delay is of 12 ms. In most implementations, the combined delay is achieved as 10 to 12 ms to create sufficient margin with an upper allowed limit of 23 ms.

An in-band tone of less than 23 ms creates a tone tick followed by tone discontinuity. It can create a perception of poor quality for the end user or listener on the line. Many deployments insist on removal of initial DTMF tones even though they are not detected as DTMF tones. To make listening comfortable, two main popular options are as follows:

1. Delaying the transmission of the in-band tones by 10 to 20 ms. In this scheme, the DTMF algorithm operates on the current samples. These samples are delayed in sending a path by about 20 ms. DTMF detection information is used either to reject tones or to send the original signal. It is the simple way of implementing the DTMF rejection. The main disadvantage of this scheme is of increased end-to-end delay that causes degradation of voice quality.

2. Adding additional DTMF early warning detector on a sample basis helps to reduce the delay. In sample-based early warning, decisions are validated on every sample without waiting for 80- or 102-sample processing. DTMF rejection can start immediately on getting an early warning. Assuming a DTMF early warning is indicated within 1 ms using a TK detector, and that DTMF rejection takes 7 ms to settle, the in-band signal is time limited to 8 ms. This type of early warning is useful to give reduced delay and a better perception of voice quality. The Teager and Kaiser detector can perform such a sample-based tone analysis for early warning of digits.

Figure 7.8 illustrates a DTMF rejection example. The top graph of Fig. 7.8 is a DTMF tone of 50 ms for digit-1. The tone is of 0-dBm power mapped to a μ-law calibration of ±8159 sine wave that corresponds to 3.17-dBm power. These tones are passed through cascaded row and column filters as illustrated in Fig. 7.1. The middle graph is for a DTMF rejected signal. The bottom graph is the logarithmic value of rejection power in comparison with the input.

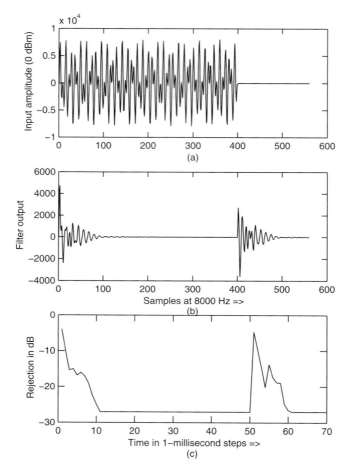

Figure 7.8. DTMF rejection example for digit-1 tones. (a) Digit-1 tone (667 Hz + 1209 Hz) at 0-dBm μ-law power mapping. (b) DTMF rejection output as in-band residue tone. (c) Rejection level in dB, low level clipped to required level of –27 dB.

In Fig. 7.8, when output goes below –27 dBm, it is shown clipped to –27 dBm for display purposes. By observing the figure, it may be observed that DTMF notch-filter-based rejection is taking approximately 10 ms to reach a –27 dBm output level for 0-dBm input. The tone power below –27 dBm is not detected as a valid tone by detection algorithms. For DTMF digit input of –10-dBm power, DTMF rejection takes approximately 5 ms to create 12-dB rejections.

7.8 DTMF RFC2833 PROCESSING

RFC2833 [Schulzrinne and Petrack (2000)] describes how to carry DTMF signaling, other tone signals, and telephony events in Reat Time Protocol

(RTP) packets. The RTP payload format for a DTMF event is designated as a "telephone-event," with the media type as an "audio/telephone-event" in negotiation. The RFC2833 payload format for telephone events does not have a static payload type number. The payload type number is established dynamically. The default clock frequency is 8000 Hz, but the clock frequency can be redefined when assigning the dynamic payload type. The named telephone events are carried as part of the voice stream and use the same sequence number and time-stamp base as the regular voice channel. RFC2833-based out-of-band telephone events are used for five different types of events/ signals—namely DTMF tones, fax-related tones, standard subscriber line tones, country-specific subscriber line tones, and trunk events.

The RTP header contains 12 bytes, and the details are given in Chapter 10. The use of the RTP header fields time stamp, sequence number, and marker bit are updated in the RFC2833 packet. Chapter 10 has some additional details on RTP header fields, which are common for both RFC2833 packets and voice packets. The RTP marker bit indicates the beginning of a new event. The marker bit should be set to "1" for the beginning of the DTMF packet and to "0" for all succeeding packets relating to the same event.

7.8.1 RTP Payload Format for Telephone Digits

The RTP payload format for telephone events is shown in Fig. 7.9. The RTP payload contains 4 bytes. The 4 bytes are distributed over several fields denoted as event, End bit, R-bit, volume, and duration. The use of these fields is given as follows:

Event is of (1-byte). The event field is a number between 0 and 255 identifying a specific telephony event. DTMF-related event codes within the telephone-event payload format are shown in Fig 7.9.

End (E) bit: If set to a value of one, the "end bit" indicates that packet contains the end of the event. The duration parameter with end bit set to "one" measures the complete duration of the event.

R-bit: This field is reserved for future use. The sender must set it to zero, and the receiver must ignore it.

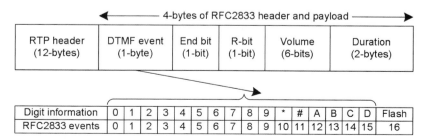

Figure 7.9. RTP payload format for DTMF telephone events.

Volume (6-bits): This field describes the power level of the event, expressed in dBm0 after discarding the sign. Power levels range from 0 to −63 dBm0. The range of a valid DTMF event is from 0 to −36 dBm0. Power levels of lower than −55 dBm0 are rejected as an invalid digit.

Duration: The duration field in time-stamp units indicate the duration of the event or reported segment. For a nonzero value, the event begins at the instant identified by the RTP time stamp and has so far lasted as long as indicated by this parameter. The duration value of zero is reserved to indicate that the event lasts forever, and it is considered effective until updated with new details.

The transmitter packetizes the event packet as soon as it recognizes an event and every packet interval thereafter. The packet interval can vary with previous voice RTP packet intervals. The beginning of the event is indicated with the RTP time-stamp value incremented by a minimum packet interval, and the RTP sequence number is incremented by one. The duration of the start event begins at the instant identified by the RTP time stamp. The marker bit field in a 12-byte RTP header should be set to "1" for the beginning of the DTMF packet.

If the DTMF event is pressed continuously, the RFC2833 encoder generates an RFC2833 packet with the same RTP time-stamp value corresponding to the beginning of the event and the RTP sequence number is incremented by one for each subsequent event packets. The duration of the event is increased by packet interval in time-stamp units.

If the DTMF event is released, the encoder generates the end of the event packet with the same RTP time-stamp value corresponding to the beginning of the DTMF event and the sequence number is incremented by one. The duration of the event corresponds to the total duration of the DTMF event that is pressed. The end bit field in RTP event payload is set to one to indicate the end of the event.

The start and end event packets can be retransmitted up to three times to ensure the duration of the event can be recognized correctly even if the packet is lost. The RTP sequence number and duration is not modified for duplicate or retransmitted packets. Under network impediments, there is a possibility that all end event packets, including duplicate packets, are lost. In this scenario, the receiver may play the tone continuously. To resolve the end bit miss case, tone is not extended by more than three packet intervals.

7.8.2 RFC2833 Telephone-Event Negotiation

The gateway can negotiate the RFC2833 telephone events using a session description protocol (SDP) during the call setup phase. This negotiation makes use of dynamic payload type since the "telephone-event" does not have a static payload type defined in the RTP header. The sender may send an initial tone packet as soon as a tone is detected, or it may wait until a prenegotiated packetization period has elapsed. The first RFC2833 RTP packet for a tone should have the marker bit set to 1. In the case of longer duration tones, the sender should generate multiple RFC2833 packets for the same tone instance. The RTP time stamp is updated for each new packet. Subsequent packets for the same tone should have the marker bit set to 0, and the RTP time stamp in each subsequent packet must equal the sum of the time stamp and the duration in the preceding packet. A final RTP packet is generated as soon as the end of the tone is detected without waiting for the latest packetization period to elapse. At the receiver, the RFC2833 packets are decoded and playout the event as received. To take care of lost packets, the events are played with a playout delay. The receiver should play the tone without change or a break when playing out contiguous RFC2833 events with marker bit zero.

7.9 DTMF TESTING

Even though DTMF is of a simple two-tone combination, it demands a significant amount of testing because of its interaction with voice and various line impairments and conditions. DTMF testing is typically conducted with the following combination of tests.

For testing Matlab, C, and assembly software implementations, engineers often create various test vector files for digit combinations and digit deviations/impediments. It is easy to create required tests for simulation purposes.

Once implementations are ready and working on the processors in real time, digits are fed to actual hardware through several telephones. A user may test with phones meeting the requirements of several countries, redial mode, digits of long duration, and short duration. It is common practice to test DTMF with Telcordia and Mitel test tapes as given in [FR-763-01 (2006), Mitel-tapes (1980)]. These tape tests are used for validating detection and for testing the robustness with several impediments.

For testing robustness, speech files are used that run for several hours. For these tapes, a false detection count benchmark is provided that is similar to the listed specifications in Table 7.2. Implementers always try to achieve a lower false digit count with a clear margin. However, it is difficult to ensure test tape signal levels and other playback parameters. The interfaces for connecting the tape players with telephone interfaces may create a mismatch in calibration of signal levels. It is necessary to ensure consistent calibration in interfacing tape tests. Note that initially tapes were used for these tests. Recently tapes were converted to compact disc (CD) [FR-763-01(2006)] format to improve on playout precision. The digitized versions are also useful with simulations without waiting for the actual hardware and telephone interfaces.

Several test-sets such as those from Sage Instruments [URL (Sage935)] also support basic DTMF tests. Signal and timing parameters are controlled accurately with these instruments. It is useful to test with instruments before subjecting to tape tests. It is also helpful to incorporate equipment as a telephone line tester to monitor various signal levels used in DTMF tests. One example of such monitoring instrument can be found at [URL (Advent-5120)]. Different countries may use target lines with different impedance characteristics. These instruments usually support one or two country impedances. Testing for additional country-line impedances can be an extra effort. These impedances mainly influence the performance of DTMF total power levels by up to 1 to 2 dB. Implementations typically cater to such small margins. In practice, users create their own test cases as either simple test vectors or test vectors played on telephone interfaces to ensure proper functioning, calibration, and robustness for various impediments.

DTMF rejection tests are based on both measurement and user perception of quality. In measurement-based tests, high-power DTMF digits are passed through DTMF rejection and they should not be detected through in-band packets. After ensuring the timing and signal levels, perception tests are important. Several digit combinations and impediments are generated while listening for DTMF rejection ticks.

7.10   SUMMARY AND DISCUSSIONS

A DTMF signal is a burst of two pure sinusoidal tones. DTMF generation is always very simple in implementation. DTMF detection and rejection operations work in close coordination. DTMF generation and detection are essential in FXS and FXO interfaced VoIP systems. DTMF rejection is optional in some implementations. In the process of analyzing DTMF detection and rejection implementation, engineers may come across several contradicting requirements that force incorporation of extra logic to tolerate deviations in algorithmic steps. The DTMF processing effort is low compared with voice compression and echo cancellations. Standard single multiplier accumulator (MAC)-based digital signal processors (DSPs) require 1 to 4 MIPs (million instructions per second) processing for DTMF detection and rejection functions. DTMF modules occupy about 1000 lines of code in the DSP assembly implementations. It is difficult to justify the best scheme for DTMF. Every scheme discussed in this chapter had certain merits and demerits. Implementers always tweak certain logic and algorithms to tune the algorithms. It becomes an implementation-specific proprietary effort to achieve good overall performance from the selected DTMF detection and rejection method.

8

CALLER ID FEATURES IN VoIP

Caller identity delivery (CID) is a member of the custom local-area signaling services (CLASS) family of telephone services offered by local telephone companies. CLASS services vary by the service provider. Caller ID is one of the most popular call services. Caller ID, which is also known as calling number delivery (CND), is a method of transmitting telephone caller information, such as telephone number and/or caller name. In a public switched telephone network (PSTN), the terminal receiving the call features is referred to as terminal equipment (TE). In most PSTN applications, the telephone is the TE. The customer premises equipment (CPE) name is also used in place of TE. In general, the call features description makes use of the names TE and CPE from the PSTN service. In VoIP, the CPE name is used for the VoIP adapter, Internet Protocol (IP) phone, wireless fidelity (WiFi) handset, and so on. In VoIP, deployment infrastructure, VoIP CPEs, and telephones used with CPE to establish call features. Two primary methods of caller ID delivery services are normal caller ID and call wait ID (CWCID). Normal caller ID happens during the on-hook phase of the phone. It is also known as Type-1 caller ID. Call wait ID or call wait caller ID is displayed in the middle of the on-going call. It happens in the off-hook state of the phone. Call wait caller ID is known as Type-2 caller ID. Depending on the country, governing local standards, and central office (CO) switch capabilities serving a given area, the caller ID data are transmitted in the voice band as frequency-shift keying (FSK) or dual-tone multifrequency (DTMF). FSK-based caller ID follows two types of

VoIP Voice and Fax Signal Processing, by Sivannarayana Nagireddi
Copyright © 2008 by John Wiley & Sons, Inc.

recommendations outlined by Telcordia [GR-30-CORE (1998), GR-31-CORE (2000)] and ETSI [ETSI ETS 300 659-1 (2001), ETSI ETS 300 659-2 (2001)]. DTMF caller ID makes use of DTMF digits. Initial versions of the caller ID documents are provided by Bellcore, but currently Bellcore is part of Telcordia. Therefore, some previous versions of documents may appear under Bellcore and later versions may appear under Telcordia [www.telcordia.com].

Caller ID and many other call features make use of direct current (DC) line conditions of off-hook, on-hook, flash-hook, open switch interval (OSI), and line reversal. In this chapter, different methods for caller ID, call wait ID in PSTN, and the mapping of these features to VoIP foreign exchange subscriber (FXS) and foreign exchange office (FXO) interfaces are presented. VoIP-based FXO caller ID requires additional call progress tone detections (CPTDs) and incoming caller ID detection. In PSTN, caller ID data are transmitted as a voice band signal from the PSTN central office. In VoIP, IP packets carry caller ID data and VoIP end gateways generate voice band caller ID signals. VoIP gateway emulates functions of the PSTN central office, and no distinction is made between the caller ID coming from PSTN or VoIP.

The caller ID parameters such as frequency, timing, and signal characteristics are country dependent. Country-dependent information is not provided in this chapter. For most countries, the caller ID service is based on either Telcordia or ETSI recommendations with slight deviations in signal characteristics and data formats. The TE alerting signal differs between the different recommendations. In this chapter, FSK and DTMF caller ID parameters, timing intervals, and FSK and alerting signal characteristics are presented. For converging on the overview on caller ID, some relevant parameters are included with tolerances from the referenced ETSI and Telcordia documents. For complete information on caller ID service and signaling specifications, refer to the relevant ETSI and Telcordia recommendations and to the country-specific local PSTN standard documents.

8.1   FSK CALLER ID ON PSTN

The popular methods for sending Type-1 FSK caller ID information are as follows:

1. Data transmission between first ring and second ring
2. Data transmission after alert but before actual first ring
3. Data transmission after an OSI

All these implementations use the same basic data transmission sequence with minor variations in FSK generation among ETSI and Telcordia implementations.

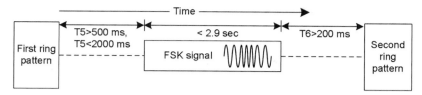

Figure 8.1. FSK Caller ID between two rings. [Courtesy: Figure printed from ETSI 300 659-1 V1.3.1 (2001-01). "© European Telecommunications Standards Institute 2001. Further use, modification, redistribution is strictly prohibited. ETSI standards are available from http://pda. etsi.org/pda/".]

Case-1 Data Transmission Between First Ring and Second Ring. It is commonly used in several countries such as North America, Canada, and so on. In this signaling method, FSK modulated data are sent during the silence periods between the first and the second ring. This on-hook service provides both number and name or only name as caller ID. The ring pattern is called cadence and varies with the country requirements. Data transmission shall occur during the first long silent period between ringing patterns as shown in Fig. 8.1. The first long silent period shall be of sufficient duration for transmitting FSK data. The initial application of the ringing provides an alert signal to the TE (phone) indicating the expected data transmission. The timing for FSK and ring pattern is shown in Fig. 8.1, and this should not take more than 2.9 seconds to transmit the entire data [ETSI ETS 300 659-1 (2001)]. Usually FSK data transmitted at 1200 bps takes a very short duration compared with the available 2.9 s.

Case-2: Data Transmission After Alert but Before Actual First Ring. This type of alert-based caller ID is used in countries such as France and Japan. In this case, the data transmission occurs before the first ring. As shown in Fig. 8.2, an additional TE alerting signal (TAS) is sent before the data transmission [GR-31-CORE (2000)]. TAS signal characteristics differ between the different recommendations. In Fig. 8.2, timing intervals are marked as T, T0, T1, T2, T3, and T4, and these marking names may vary in country-specific documents. The three popular alerting signals used are dual-tone alerting signal (DT-AS), ringing pulse alerting signal (RP-AS), and line reversal followed by DT-AS [ETSI ETS 300 659-1 (2001), GR-30-CORE (1998)]. Japan uses line reversal followed by receiving terminal activation signal (CAR) as the alerting signal for on-hook caller ID display. The CAR signal is the receiving terminal activation signal or signal receiver seizing signal and is a short ring signal of 15–20 Hz. Refer to the NTT recommendation [URL (NTT-E)] for more details on Japan caller ID. To make caller ID work, any one of these alerting signals shall precede FSK modulation transmission within set time limits. The data transmission before the first ring using a dual-tone alerting signal is shown in Fig. 8.2(a). DT-AS shall precede FSK modulation transmission as $45 \leq T4 \leq 500$ milliseconds (ms). The alerting signal DT-AS is also referred to as the CPE

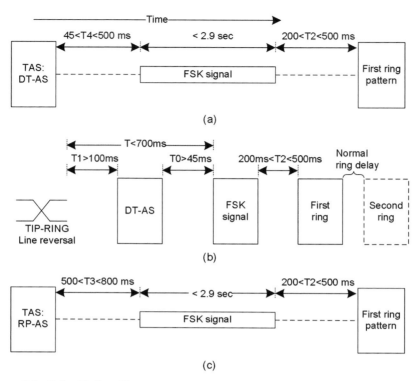

Figure 8.2. Caller ID Data Transmission Before the First Ring Using (a) DT-AS, (b) DT-AS With Line Reversal, and (c) RP-AS. [Courtesy: Figure printed from ETSI 300 659-1 V1.3.1 (2001-01). "© European Telecommunications Standards Institute 2001. Further use, modification, redistribution is strictly prohibited. ETSI standards are available from http://pda.etsi. org/pda/".]

alerting signal (CAS) in some places in the document. Both names convey the same alerting signal. The data transmission before the first ring using RP-AS is shown in Fig. 8.2(c). RP-AS shall precede FSK modulation transmission as $500 \leq T3 \leq 800\,$ms. The line reversal followed by DT-AS shall precede FSK modulation transmission by not less than 45 ms (T0 > 45 ms) as shown in Fig. 8.2(b). The total period between the line reversal and the start of FSK modulation is less than 700 ms marked as "T." In all three methods, the application of ringing current shall start not less than 200 ms and not more than 500 ms (marked as T2) after FSK modulation transmission is stopped. The lower limits are required to enable TE to apply and remove appropriate impedance for data reception. In RP-AS, the time gap is more compared with DT-AS because of a high-voltage alert signal in RP-AS.

During caller ID, one of the three TAS signals is sent before data transmission. It varies by local PSTN standards. In this case, data transmission is associated with ringing and the FSK modulated caller ID message is followed by ring patterns. An appropriate idle condition can also be applied to the local loop after FSK transmission. The important characteristics of DT-AS are listed here for the ETSI recommendation [ETSI ETS 300 659-1 (2001)].

Frequencies: 2130 and 2750 Hz ± 0.5%

Duration: 100 ± 10 ms

Signal Level: −40 dBV to −9 dBV/tone

Signal purity: The total power of all extraneous signals in the voice band shall be at least 30 dB lower than the power of signal fundamental frequency measured at the point of application.

Case-3: Data Transmission After an OSI. The OSI is the time duration when DC voltages are removed from TIP-RING lines. OSI is used as an alert to send messages and indications to the TE while the line is in the on-hook mode. This method is also known as data transmission without power ringing. In this case, the FSK modulated data are sent following a time period after OSI [GR-30-CORE (1998)]. No ringing follows the FSK data. The duration of the OSI shall be 150 to 300 ms, and the data transmission happens between 300 and 500 ms after the end of the OSI. The default delay value is 500 ms [URL (Advent-CID1)].

8.2   FSK CALLER ID DATA TRANSPORT PROTOCOL

The basic data transport protocol for caller ID is divided into four layers, namely,

1. Physical layer
2. Data link layer
3. Presentation layer
4. Application layer

The first three layers provide the actual data transport, and the application layer is used for caller ID-specific data and signaling for alerting the TE.

8.2.1   Physical Layer

The physical layer provides the interface between the caller ID service and the analog line. The physical layer provides two main functions of data transmission of service-specific information and signaling mainly for alerting the TE.

The data transmission is performed using continuous-phase FSK modulation. Data is always sent as serial binary bits in simplex mode. Data transmission is continuous, and no carrier dropouts are allowed. The start of data transmission must not corrupt the first data bit. The data transmission is

Table 8.1. Data Transmission Characteristics

Parameter	ETSI-Based Modulations	Telcordia-Based Modulations
FSK major distinction	FSK as per ITU-T-V.23	FSK as per Bell 202
Mark (bit-1)	1300 Hz ± 1.5%	1200 ± 12 Hz
Space (bit-0)	2100 Hz ± 1.5%	2200 ± 22 Hz
Data rate	1200 bps ± 1%	1200 ± 12 bps
Message format	Type-1: MDMF	Type-1: SDMF, MDMF, GDMF
	Type-2: MDMF	Type-2: MDMF, GDMF
Signal generation levels	−13.5 dBm ± 1.5 dB, signal purity 30 dB	−13.5 dBm ± 1.5 dB, signal purity 30 dB
Receiver operating	−8 to −36 dBV (0 dBV = 2.2 dBm)	−12 dBm to −36 dBm (space) −12 dBm to −32 dBm (mark)

stopped immediately after the last bit of the data-link message. The FSK data is sent asynchronously at a signal level of −13.5 dBm in both ETSI and Telcordia recommendations as listed in Table 8.1. This power level is applicable at the central office. The FSK signal level may differ for each country, because of country-specific deviations of overall loudness rating (OLR) as well as because of send and receive gain/loss planning. To get a first-level understanding on ETSI and Telcordia basic specifications, a summary is given in Table 8.1. It is suggested to refer to the ETSI [ETSI ETS 300 659-1 (2001), ETSI ETS 300 778-1 (1997)] and Telcordia [GR-30-CORE (1998)] recommendations for more details on these specifications.

8.2.2 Data Link Layer

The data link layer is responsible for providing data byte framing (start/stop asynchronous, 8-bit data) and checksum. The data link layer is also responsible for the addition of the preamble, which comprises the channel seizure signal and the mark signal as well as padding of mark bits between bytes for data flow variation.

Note that the definition of the "data link layer" varies slightly between Telcordia [GR-30-CORE (1998)] and ETSI [ETSI ETS 300 778-1 (1997)-Annex E]. The ETSI treats the message type byte and message length byte as part of the data link layer, whereas Telcordia treats the message body as a presentation layer message. Telcordia incorporates the message header (type and length bytes) as part of the presentation layer. The Telcordia recommendation refers to the presentation layer as the message assembly layer. Review the complete specifications regarding these differences.

The format of the data sequence at the data link layer is shown in Fig. 8.3(b). The data transmission between rings is presented in Fig 8.1(a) with timing information for the ETSI recommendation [ETSI ETS 300 659-1 (2001)]. The

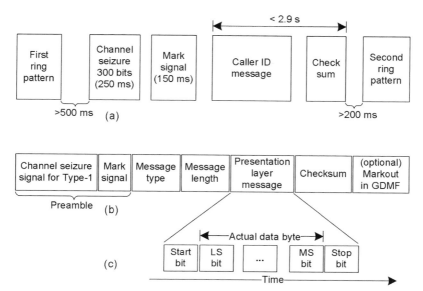

Figure 8.3. Data link layer format. (a) Details on data between two rings. (b) Message. (c) Actual data byte format. [Courtesy: Figure printed from ETSI 300 659-1 V1.3.1 (2001-01). "© European Telecommunications Standards Institute 2001. Further use, modification, redistribution is strictly prohibited. ETSI standards are available from http://pda.etsi.org/pda/".]

data consist of four functional blocks—channel seizure signal, mark signal, caller ID information, and the checksum. For both ETSI [ETSI ETS 300 659-1 (2001)] and Telcordia [GR-30-CORE (1998)], every byte in the sequence (excluding the contents of the preamble, but including the checksum) is framed in an asynchronous manner with each 8-bit word preceded by a start bit (space) and followed by a stop bit (mark). The seizure signal is sent first, followed by the mark signal, and then each byte is sent in sequential order, finally followed by the checksum. No parity is used in the data framing. In the bit transmission, start bit, LS bit ... , MS bit, and stop bit are sent as illustrated in Fig. 8.3(c). Note that actual byte is represented as stop bit, MS bit ... LS bit and as start bits.

Channel Seizure Signal. The channel seizure signal is used to signal the start of the data transmission to the CPE or TE to alert for an incoming message, and it conditions the receiver ready for the data. This signal is used to ensure that noise is not falsely identified as a carrier. If the seizure signal is not received, the TE must cease reception, and the next messages are ignored. The seizure signal consists of 300 continuous bits (duration of 250 ms at 1200 bps) of alternating "1" and "0." The first bit must be a "0," and the last must be a "1." The seizure signal is used only in Type-1 FSK-based on-hook caller ID and is not used in Type-2 call wait ID. The channel seizure bit size is different in ETSI and Telcordia. In some networks, the ETSI allows for a channel seizure length of 96 to 300 bits [ETSI ETS 300 778-1 (1997)].

Mark Signal. The second signal in the preamble is the mark signal. This signal consists of continuous mark bits. It is used to provide an idle period before the start of the message type byte to allow the receiver to identify the byte. For both ETSI and Telcordia, the same quantity of bits is sent as part of the mark signal. Type-1 is of 180 mark bits (duration of 150 ms at 1200 bps), and Type-2 is of 80 mark bits. Note that the ETSI allows the option of 80 mark bits for Type-1. The tolerances on the mark signal differ between the two standards. The ETSI allows ±25 bits, and Telcordia allows ±10 bits.

Checksum. The CID information is followed by a checksum for error detection. The checksum is a single byte value, which contains a two's complement of the modulo-256 sum of all bytes in the message starting from the message type byte up to the end of the message. The channel seizure and mark signals are not included in this checksum. When the TE receives the message, it checks for errors by taking the received checksum byte and adding the modulo-256 sum of all other bytes received in the message. The checksum is verified by summing (modulo-256) all bytes in the message between the message type byte and the checksum. If the contents are correct, the checksum is zero. If the checksum fails, the entire message is discarded without any error correction or message retransmission. The TE must not send any acknowledgment or response back to the stored program control signal (SPCS). The SPCS is part of the central office.

Additional Marks (Padding). To allow for some variations in the arrival of the data to be transmitted, and to ensure that the output sequence remains continuous, a small quantity of additional marks may be added between bytes. This quantity is treated differently between the ETSI and Telcordia. The ETSI allows for additional marks between any bytes within the message [ETSI ETS 300 778-1 (1997)-Annex E], whereas Telcordia allows for insertion of additional marks between selected bytes only [GR-30-CORE (1998)]. It is an objective of the Telcordia specification that no additional marks be inserted between bytes in a parameter message. For example, mark bits (0–10) can be transmitted between message type byte and message length byte, or between parameter type and parameter length. Inserting the additional marks between parameter messages bytes is not allowed.

The number of extra mark bits (not including the stop bit) should be between 0 and 10 bits. These bits are added after the stop bit and before the start bit of the next byte. Additional marks are not permitted in the generic data message format (GDMF) payload used by Telcordia. In GDMF, to prevent corruption of the final stop bit after the checksum or last byte of GDMF, and hence destruction of the checksum, additional marks (markout) may also be used following the checksum.

8.2.3 Presentation Layer

The presentation layer is responsible for the framing of the application-specific data for provision to the data link layer. Data can be organized according to three possible formats: single data message format (SDMF), multiple data message format (MDMF), and GDMF. The ETSI recommendation supports only the MDMF format for both Type-1 and Type-2 caller ID as indicated in the message formats of Table 8.1. The SDMF is almost obsolete, and Telcordia has made it an objective that the SDMF not be used [GR-30-CORE (1998)].

The SDMF contains the calling number, date, and time. A message in SDMF includes a message type byte, a message length byte, and the actual message bytes as shown in Fig. 8.4(a). The MDMF supports multiple messages such as calling number, caller name, date and time, and other optional information. A message in MDMF also includes a message type byte, a message length byte, actual message bytes, and a parameter type and parameter length bytes as shown in Fig. 8.4(b). At certain points within these messages, up to 10 mark bits may be inserted to allow for equipment delays in the central office [GR-30-CORE (1998)]. The message type byte defines whether the message is in SDMF or MDMF. The message length byte is the binary representation of the number of bytes in the message, not including the message type, and the checksum bytes. It supports message payload lengths of up to 255 bytes. In MDMF, the parameter type byte identifies the type and format of the parameter message. The parameter length byte contains the binary representation of the number of bytes contained within the parameter message, excluding the parameter header. The maximum length of the parameter payload is 253 bytes since the maximum message length is 255 bytes, of which the parameters header consumes 2 bytes. The parameter type byte has its unique value for

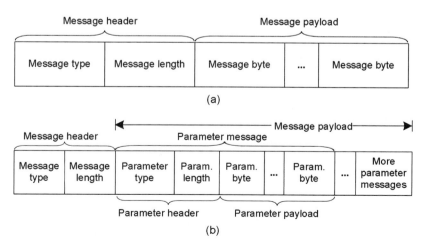

Figure 8.4. Presentation layer data formats. (a) SDMF. (b) MDMF (note: The Param. acronym is used for Parameters in the diagram).

each feature in MDMF. Refer to the ETSI or Telcordia recommendations for more details on various parameter type code words.

8.2.4   Application Layer

The application layer is where the actual caller ID service is provided. This layer determines the actual content of the data sent to the presentation layer, to suit the information to send to the TE. The application layer is country-dependent and varies between ETSI and Telcordia as well as other specification's that perform all the necessary interpretation of the data being transported, select the method caller ID display, and provide necessary timing information that is largely dependent on the specific country.

8.3   DTMF-BASED CALLER ID

Most caller ID signals use FSK-based implementation. Some countries like Finland, Denmark, Iceland, the Netherlands, India, Belgium, Sweden, Brazil, Taiwan, Saudi Arabia, and Uruguay make use of DTMF-based caller ID. In practice, many DTMF-caller-ID-supported phones are also made to work with FSK-based called ID. The phone displays are capable of supporting the numbers and alphabets. In the case of DTMF caller ID, only a number display is supported.

8.3.1   DMTF Caller ID on PSTN

For the transfer of display information over analog subscriber lines, 16-DTMF digits are used in DTMF-based caller ID. Except for these special display procedures, the normal signaling procedures and physical properties for analog subscriber lines also apply to DTMF caller ID [ETSI ETS 300 659-1 (2001)]). In case of an incoming call for a subscriber, the exchange shall seize the corresponding subscriber line for the terminating call. The line seizure may be indicated to the subscriber line by means of a polarity reversal of two lines of TIP-RING or through ring alert. In line reversal, the return to the idle polarity takes place after the information transfer phase. Two different modes for the data transmission are possible with DTMF—namely data transmission before the first ring and data transmission between the first and the second ring.

Data Transmission Before the First Ringing.   In this mode, the data transmission shall occur before the beginning of the first ringing pattern. Two different procedures are possible to alert the TE namely subscriber line polarity reversal or ring alert. The subscriber line polarity is not applicable with ring alert. In both cases, the first ringing has to be within the time limits of 200 to 500 ms after the data transmission is stopped. In polarity reversal, the data transmission should not start sooner than 200 ms after the polarity reversal.

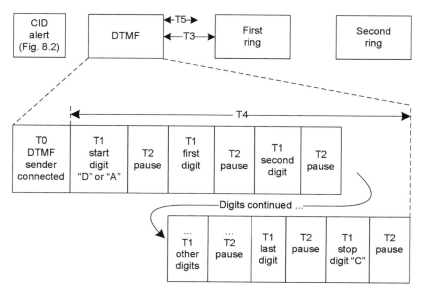

Figure 8.5. DTMF caller ID transmission.

The DTMF-based caller ID transmission method sends a series of DTMF digits before the first ringing cycle. In some implementations, alert signals are not used before data transmission in DTMF caller ID. In such cases, the central office provides a pause to connect the DTMF sender to the voice path [URL (Advent-CID2)]. As shown in Fig. 8.5, the calling number is sent after a "start" digit and ends when the "stop" digit is detected. Figure 8.5 represents the DTMF caller ID events before the first ring. The CID alert can be either ring pulse or polarity line reversal for DTMF-based caller ID. The transfer of number information is to be regarded as complete when the DTMF stop digit "C" is received or the ringing signal is detected. The following timing details of DTMF caller ID are marked in Fig. 8.5.

T0—Pause for the central office to connect the DTMF sender to the voice path when no alert signal is preceded with DTMF transmission. T0 is between 50ms and 400ms.

When line reversal is used as an alert, the delay from the end of the alert signal to the start of the data transmission is between 200ms and 500ms.

T1—Duration of DTMF digit, ≥50ms varies by country.

T2—Inter-digit pause, ≥50ms varies by country.

T3—Delay from end of digits to start of first ring, 200 to 500ms.

T4—Time required to send all DTMF digits, ≤3000ms.

T5—Return time to quiescent state after DTMF stop digit, ≤150ms.

Data Transmission Between First and Second Ring. In this mode, the data transmission shall occur during the first long silent period between two ringing patterns. The initial application of ringing is to provide an alert signal to the TE on possible data transmission. The data transmission should not start sooner than 500 ms after the first ringing pattern. At the end of caller ID data, the second ringing will start according to the normal ringing cadence. The timing of DTMF caller ID generation in the middle of the two rings will be similar to DTMF caller ID before the first ring of Fig. 8.5. In this method also, the second ringing has to be within the time limits of 200 to 500 ms after the data transmission is stopped. The caller ID number is sent after a "start" digit and ends with a "stop" digit. DTMF caller ID can display number and cannot display name.

Digits Data Coding. The DTMF transmission makes use of the 16-DTMF digit signals. Most telephones will use a keypad for the 0, 1, 2, 3, 4, 5, 6, 7, 8, 9, *, # digits. The DTMF-caller-ID-capable telephone can understand incoming A, B, C, and D digits for different interpretations. The interpretation of DTMF digits by the TE during caller ID is given here.

<A>   DTMF code "A" is used as a start code for the calling party number. The start and end codes are different for different countries. Refer to country-specific PSTN recommendations for exact details on the DTMF caller ID method and timing intervals. For example, the start code is digit "A" for Brazil and "D" for Taiwan.

<B>   DTMF code "B" is used as a start code for the special information concerning the nonavailability or restriction information of the calling party number. It specifies the user category.

<C>   DTMF code "C" is used as an end code for the information transfer.

<D>   DTMF code "D" is used as a start code for the diverting party number in case of call diversion.

<0 to 9>   DTMF codes "0 to 9" are used as number digits of calling/diverting party or special information code value.

<*, #>   DTMF digits *, and # are not used in DTMF caller ID.

DTMF digit details and tone generation are given in Chapter 7. The digits and pause periods vary depending on the country. It is required to refer to Q.24 [ITU-T-Q.24 (1998)] and to local country PSTN standards for the detailed specifications of DTMF digits and pause periods.

8.4 COUNTRY-SPECIFIC CALLER ID OVERVIEW

Caller ID standards for data transmission and signaling have been published by various international standards committees, regulatory agencies, and network carriers, and they fall into one of two different categories based on how caller ID data is transmitted. They are FSK modulation or DTMF. The main deviations for multiple countries are on network transmission interfaces (power level, impedances) and timing variations for each country. These deviations will be available in country-specific PSTN standards. The Telcordia recommendation is used in the United States, Canada, Australia, China, Hong Kong, New Zealand, and Singapore, among others [URL (CID-FAQ)] with different data formats and with some deviations in signal characteristics. In on-hook caller ID, data are sent after the first ring tone is used. This scheme uses a 1200 baud rate for data and uses Bell-202 for FSK modulations. The data may be sent in SDMF, which includes the date, time, and number, or as MDMF that adds a NAME field. The most common and default method used is MDMF.

BT and CCA recommendations are used in the United Kingdom. British Telecom [BT SIN 227 (2004)] developed a different recommendation, which uses FSK modulations for data transmission based on the ETSI specifications.

DTMF-based [(URL (Cisco-CID)] caller ID is used in Brazil, Belgium, Denmark, Holland, Finland, Iceland, India, the Netherlands, Saudi Arabia, Sweden, and Uruguay. These are based on ETSI recommendations [ETSI ETS 300 659-1 (2001)] using DTMF codes for data transmission with a slight deviation in start and stop digit usage and timing intervals for each of these countries.

FSK-based caller ID is used in Ireland, Germany, Norway, Italy, Spain, South Africa, and Turkey, among others. The FSK modulations and mode of data transmission are based on ETSI [ETSI ETS 300 648-01 (1997), ETSI ETS 300 659-1 (2001)] recommendations for these countries.

STI-4 recommendation [FT STI-4 (2004)] is used in France that uses FSK modulations based on the ESTI recommendation with slight deviations on power level. Taiwan follows the ETSI in permitting DTMF and FSK signals. The NTT in Japan has developed a V.23 [ITU-T-V.23 (1988)] FSK simplex system, wherein FSK data are transmitted after a short first ring. The framing of the application-specific data is used as defined in the NTT recommendation [URL (NTT-E)].

8.5 CALLER ID IN VoIP

VoIP enables many of the CLASS family of telephone services offered by telephone companies. Several VoIP call combinations and systems are given in Chapter 2. In this section, an overview on Type-1 on-hook caller ID generation in a VoIP system is given. For continuity of the description, a functional representation of VoIP call is shown in Fig. 8.6. Various voice interfacing VoIP adapters and soft switches are associated to establish the VoIP call. VoIP signaling protocols such as SIP, H.323, MGCP, and Megaco are used to set up,

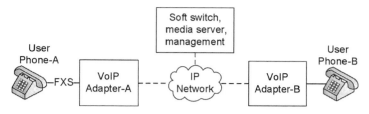

Figure 8.6. VoIP adapters and caller ID working functional representation.

tear down calls, carry user information required to locate users, and negotiate capabilities. Plain old telephone service (POTS) interfaces connected to analog interfaces on these VoIP systems are called FXS. In Fig. 8.6, FXS-A (referred as User phone-A) and FXS-B (referred as User phone-B) are end-point interfaces of different VoIP adapters connected to the IP network. The VoIP adapter samples voice from the telephone interface at 8 kHz, compresses samples, and generates IP packets with proper headers. The VoIP adapters are registered to the soft switch and subscribe to the specified call feature services. Soft switch and VoIP systems work together to emulate a PSTN central office for call feature establishing. Depending on the supported VoIP signaling, the soft switch will be working as a SIP—proxy server, MGCP—Media Gateway Controller, and H.323 Gatekeeper.

The VoIP adapter controls the caller ID service feature (i.e., enable or disable) through signaling negotiation by specifying message calling line identification presentation (CLIP) and caller ID restriction (CLIR) methods. The soft switch does not support any functionality of message framing and data transmission. Depending on the VoIP protocols selected, the soft switch enables proxy servers, gatekeepers, and media controllers to control the voice call and forward signaling packets. The soft switch modifies and rewrites a request message before forwarding to the end points. In practice, the proxy server name is used in place of the soft switch for SIP-based calls.

In VoIP deployment, the VoIP adapter is responsible for FSK generation to display caller ID. The VoIP adapter should also be capable of framing a caller ID message format using MDMF/SDMF. The signaling protocol in the VoIP adapter controls the call flow and the processor [usually digital signal processor (DSP) for tones and modulations] that resides in the VoIP adapter performs the generation of caller ID. The caller ID feature can be controlled for activation and deactivation in the VoIP adapter itself. The caller can restrict the caller ID supplementary service for a particular call or for a set of calls by sending the CLIR message to the proxy server.

The From header field in the SIP INVITE request indicates the initiator of the call. If caller ID service is restricted from the calling party at VoIP adapter-A, then VoIP adapter-A sends the CLIR message in the SIP INVITE request to the server proxy. The proxy server modifies and rewrites the request and forwards to the VoIP adapter-B. VoIP adapter-B restricts the caller ID message by keeping the display number as "P" private (Caller ID delivery is locked) or "O" unavailable in message framing. In this case, User-B gets unavailable, optional, or private on phone display. In some deployments, the VoIP adapter uses the display name as "Anonymous" in caller ID message framing when CLIR is enabled. In this case, the user receives Anonymous on the phone display.

If VoIP adapter-A is subscribed for caller ID service, then it enables the CLIP message in the signaling packet along with the display name of the calling party. VoIP adapter-B decodes the caller ID number from negotiation fields (From Field), and caller ID data are generated in VoIP adapter-B to display on phone-B. For example, using a SIP signaling call is established between two end-user agents A and B with the assistance of an intermediate soft switch or proxy server. If User-A has subscribed for caller ID service and caller ID display service is enabled at adapter side (User-A), the caller ID number "From" field value in the INVITE SIP Header is extracted to generate the caller ID data. The From header field in the INVITE request indicates the "display-name or number" of the initiator of the call [Rosenberg et al. (2002)].

8.6 CALL WAIT CALLER ID

Call wait caller ID (CWCID) or calling identity delivery on call waiting (CIDCW) includes all features of Type-1 caller ID plus the capability of call wait caller ID display in the off-hook state. CIDCW uses only MDMF, and this feature is also called Type-2 caller ID or off-hook Caller ID. CIDCW can happen when a call is already in progress. Ring is not used with CIDCW as alert. The channel seizure signal is not used with CIDCW. With CIDCW, the identification of the calling party can be seen without keeping the current call on hold or any additional operations. In this section, the FSK-based Type-2 caller ID method is described.

8.6.1 Call Wait ID Flow in PSTN

In this section, call wait caller ID functioning in PSTN is explained. A and B are the end telephones connected to the PSTN central office as indicated in Fig. 8.7(a). With the help of the central office, A and B will establish the call and continue the conversation. In the example, A receives another call from C through the central office. The basic sequence from the established call between A and B are described as follows:

- The voice call continues between A and B.
- C is dialing A, and the central office senses that A is busy.
- If subscriber A does not have the call wait service, the central office sends a busy tone to C and User-A cannot speak to User-C.

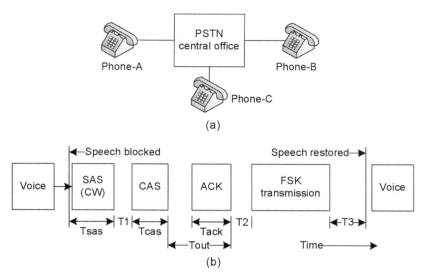

Figure 8.7. Call wait ID. (a) Functional representation of phones in call wait ID. (b) Call wait ID signal sequence.

- If User-A has subscribed for call wait service, the central office can play a call wait short beep tone to A by muting speech from B to A path. This tone is called the subscriber-alerting signal (SAS) or call waiting tone. The central office will keep playing a special ring-back tone to C if User-C has activated the call waiting/call forward service; otherwise, a normal ring-back tone is played to C.

- After hearing the call wait beep, User-A can switch to C by using flash-hook (or any other programmed digit). After flash-hook, User-B will be in wait mode, and at User-B, a music tone is played.

- After completing the conversation with C, A can switch back to B, or A can keep switching between B and C.

- If A has subscribed for call wait ID service, the central office will send CAS to phone-A immediately after the first call wait tone is played. The CPE refers to TE or the phone here. If User-A has a call-wait-ID-capable phone, it will keep the microphone path on mute and send an ACK (acknowledge) signal back to the central office. This ACK signal is either a DTMF "A" or "D." The DTMF digit "D" is the most common ACK signal.

- If the central office receives ACK from phone-A, it sends an FSK caller ID signal to phone-A. In the absence of ACK, the central office can terminate the call wait ID operation. ACK will not be delivered, if the phone is not supporting call wait ID service.

- Phone-A displays a call wait ID. On the sensing end of the FSK signal, phone-A comes out of microphone mute and voice can continue. If the central office does not send FSK before timeout, phone-A will un-mute on its own.

- After the call wait beep, phone-A can always establish a call with phone-C (using flash-hook) without waiting for call wait ID or at any time during this phase.
- The central office will be sending SAS, CAS, and FSK signals in place of speech during call wait ID. Speech is interrupted while playing these signals.

8.6.2 Call Wait ID Signals and Tones

Type-2 caller ID includes all the features of Type-1 caller ID for FSK generation plus the capability of CIDCW. The sequence of events for off-hook data transmission is the same for both Telcordia and ETSI [ETSI ETS 300 659-2 (2001)]. Refer to the recommendations for timing and timing deviations [ETSI ETS 300 659-2 (2001)], [GR-30-CORE (1998)]. The basic sequence of events at the SPCS of the central office is shown in Fig. 8.7(b). Typical call wait ID timing parameters of Fig. 8.7(b) are given below.

Tsas—SAS tone duration, 300 ± 50 ms

T1—Allowed time between SAS and CAS, 0 to 150 ms

Tcas—Duration of alerting signal duration, 75 to 85 ms

Tack—Duration of ACK detection, 55 to 65 ms

T2—Duration between ACK and the FSK data transmission, 50 to 200 ms

T3—Time to unmute the voice path, 40 to 120 ms

Tout—ACK timeout, 155 to 165 ms

The signals and timing referred in Fig. 8.7(b) is explained to some more details here.

SAS: The subscriber-alerting signal is a single frequency of 440 Hz that is applied for approximately 300 ms. This tone is heard when a call is in progress and call waiting beeps to indicate a second call. The SAS tone (also called a call waiting tone) is used to indicate the user on the second call. This tone is not required for the TE to receive the CID information.

CAS: A CPE alerting signal alerts the TE that it has CID information to send followed by the SAS tone. The CAS is also referred to as the DT-AS dual-tone–alerting signal with a dual-tone signal combination of 2130 Hz and 2750 Hz for 80 ms. The power levels are of -15 dBm ± 2 dB per tone and accept up to -32 dBm per tone. Once the TE hears the CAS, it mutes the handset of the telephone and returns an ACK signal to the central office. There are slight deviations in specifications among

Telcordia and ETSI [GR-30-CORE (1998), ETSI ETS 300 659-1 (2001)] for the CAS signal.

ACK: The ACK signal has a nominal tone duration of 60 ms and is either a DTMF digit "A" or "D." The digit "D" is the most common ACK signal, and it consists of the frequencies 941 Hz and 1633 Hz. Digit "A" consists of the frequencies 697 Hz and 1633 Hz. On receiving the ACK signal, the central office sends the CID information. The DTMF tone details are given in Chapter 7. There are slight deviations in specifications between the ETSI and Telcordia. The digit specifications and timing also vary by specific country requirements. For ETSI [ETSI ETS 300 659-2 (2001)], it is a network option to consider DTMF digits "A," "B," or "C" as a valid alternative ACK at the SPCS side. As per Telcordia, the SPCS allows the use of digit "A" as a valid alternative ACK. In either case, the DTMF tones must comply with the relevant specifications for DTMF transmission and reception.

FSK Transmission: FSK data have to be prepared in proper format for CIDCW operation. CIDCW uses only MDMF [GR-30-CORE (1998)]. The message type indicates only MDMF format for CIDCW. The code value for MDMF message type is of binary 10000000 (decimal 128). No channel seizure precedes the data. The CID information is sent serially at a rate of 1200 bps using continuous-phase binary FSK for modulation. The two frequencies used to represent the binary states are 1200 Hz for the mark (logic-1) and 2200 Hz for the space (logic-0) as shown in Table 8.1. The transmission characteristics are similar to on-hook FSK caller ID signals.

Mark signal difference between CID and CIDCW: The CID and CIDCW differ in channel seizure signal and mark signal. In CID, 300 bits of channel seizure and 180 bits of mark signal are used. In CIDCW, only 80 bits of mark signal are used and the channel seizure signal is not applicable. Call wait ID still uses the checksum similar to CID.

Talk-off and timeout: CIDCW is applicable during active voice conversation. Talk-off is one of the problems from the voice conversation experienced with CIDCW. Talk-off occurs when the CPE falsely detects a CAS. The TE may interpret the speech signal as CAS arrived from the central office. This happens as a rare event. When talk-off occurs, both parties can hear a brief interruption in the conversation and the party on the far end can hear the ACK signal that is sent to the central office by the TE. When the handset is muted by a talk-off operation, the TE may timeout and unmute the handset after waiting a set period for an FSK signal. A TE detector behaves differently with different people because of the variations in each person's speech patterns. Speech can also cause a disturbance to CAS signal detection. This is called talk-down [URL (pic)]. Talk-off occurs when CAS is falsely detected by the TE. Talk-down occurs when TE is not detecting the valid CAS tone.

8.6.3   Call Wait ID Functioning in VoIP

This section considers a VoIP system playing the call wait ID on an FXS interface and phone. The call wait ID specific to the FXO interface is given in the next section. CID and CIDCW generations are the basic supplementary features supported in VoIP deployments. In VoIP deployment, the VoIP adapter or end-interfacing infrastructure is responsible for framing caller ID message format using MDMF, FSK generation, tone generation, DTMF, SAS, CAS generation, and ACK detection to support supplementary features such as caller ID, call wait tones, and call wait ID. The signaling protocol in the VoIP adapter controls the call flow, and the processor resides on a gateway to generate various signals required for CID display. The call wait ID and caller ID feature can be controlled for activation and deactivation in the VoIP adapter itself. The caller can restrict the caller ID supplementary service for a particular call or for a set of calls by sending the CLIR message to the proxy server in SIP.

If End User-A has subscribed to the call wait ID/caller ID service and caller ID service is restricted from VoIP adapter-A, then adapter-A sends the CLIR message in a signaling packet in negotiation to VoIP adapter-B. The adapter-B can restrict the caller ID message by keeping the display number as "P" or "O" in message framing similar to on-hook caller ID. In this case, User-B gets optional or private on phone display. If A has enabled the call wait ID service, VoIP adapter-A enables the CLIP message in a signaling packet along with the display name of the calling party. The VoIP adapter decodes the caller ID number from negotiation fields, and caller ID data are generated in the VoIP adapter to display on phone-B.

8.6.4   Implementation Care in Call Wait Caller ID in VoIP

This section describes the care in handling of CIDCW in VoIP calls. Basic caller ID sends the voice band signal to the phone. In CIDCW, real-time acknowledgments are present and these acknowledgments are time sensitive. The following points have to be taken care while handling CIDCW. Actual implementations may consider several microlevel details beyond these listed points.

- It is necessary to minimize the total interruption of the voice path during the CIDCW transmission. The adapter should not block the voice transmission for more than 1.2 s [GR-30-CORE (1998)].
- FSK data transmission should not begin if the processor is still receiving any part of the ACK signal to prevent collisions. It is achieved by avoiding any false detection of ACK or by ensuring the ACK is received for maximum detection time.
- The voice packets may be decoded during call wait ID, but they should not be played on near-end port.

- Application should not transmit the CID data during or after flash-hook.

The following requirements describe how the system should handle flash or call disconnect during the calling number display on call wait.

- False detection of flash-hook may cause interrupt in CIDCW display.
- The adapter should stop FSK data transmission as soon as it detects a flash or disconnect on the customers line.
- It is possible that the user can disconnect the current on-going call while receiving the calling number on call wait. In such cases, a power ring has to applied to the called party line after disconnect and the processor shall continue data transmission until the end of the release period.

DTMF ACK comes from phone to the VoIP system. While handling of the DTMF receiver during call wait ID, the DTMF module has to be enabled to detect the ACK from the phone. It may be possible that the gateway can send this ACK signal either in in-band or out-of-band packets to the other end that may create disturbance in voice call. So it is suggested in requirements of call wait ID, gateway should suppress and ignore the ACK digit during the call wait caller ID phase. It is also recommended that the DTMF receiver should be disabled from the line within the interval following the detection of the ACK and before reestablishment of the voice path in order to avoid any false detection of digits during the data transmission.

8.7 CALLER ID ON FXO INTERFACES

In VoIP, gateways or adapters use FXS and FXO interfaces. Some of these aspects are given in Chapter 2. Products with FXO interfaces are usually called residential gateways (RGs) or integrated access devices (IADs). As shown in Fig. 8.8(a), gateways use FXS for connecting to the telephone. In the previous sections, caller ID and call wait ID generation are given with an FXS interface. The VoIP gateway sends the required call progress tones, caller ID, and other signals to the FXS phone. On an FXS interface, the processor inside the VoIP system generates caller ID, call wait ID, and other call progress tone signals. A VoIP box uses an FXO interface for connecting to PSTN, digital subscriber line (DSL), or private branch exchange (PBX) system. As shown in Fig. 8.8(a), the FXO interface is similar to an electronic telephone and consists of a codec and data access arrangement (DAA). The codec has the same functionality as the FXS in converting analog speech to digital and vise versa. In a normal telephone, call progress tones are observed by the listener and take required actions based on call progress tones. The PSTN central office sends call progress tones and caller ID information to the FXO interface. The FXO interface

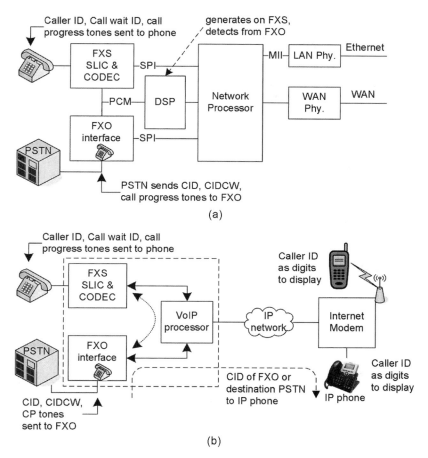

Figure 8.8. VoIP gateway with FXO caller ID. (a) Caller ID functions on FXS and FXO. (b) FXO-to-FXS caller ID as pass-through.

has to understand the way the phone understands. On an FXO interface, call progress tones are observed by the listener/phone inside the interface. On an FXO electrical interface, these call progress signals and indications are converted to the samples and commands known internally to the processor of the VoIP system. The VoIP application will initiate proper preprogrammed actions based on the state of call progress tones and caller ID signals. At a high level, a VoIP system-interfacing FXO has to analyze the caller ID signal, call wait beep, call wait ID, and call progress tones. Hence, the detection of all these signals such as call progress tones, caller ID, call wait, and call wait ID is essential within the VoIP system to connect to the FXS interface of the VoIP gateway, to any other VoIP gateway FXS interfaces, or to the IP phone on the VoIP network from a local PSTN network. In Fig. 8.8(a) DSP performs a generation for the FXS interface and a detection for the FXO interface.

As marked in Fig. 8.8(b), the VoIP gateway creates a call to terminate the PSTN call through the FXO interface on the IP network. The VoIP gateway receives calls through the FXO interface and forwards them to the IP network. It requires an account with VoIP service providers for the FXO interface that provides FXO call termination on the IP network. Some implementations use a separate E.164 number [ITU-T-E.164 (2005)] for the FXO interface to register with the VoIP service provider to originate/receive VoIP calls, and some implementations may use an existing FXS channel to route PSTN calls on the IP network.

When the FXO interface receives a caller ID from the PSTN service, the FXO analyzes the caller ID signal, and this information is used as the caller ID to display on the destination IP phone. Sometimes this caller ID information is used to analyze for any pin code checking or is sent directly to the destinations similar to FXS, IP phone, and WiFi phone. In the absence of caller ID through a PSTN interface, the FXO may decide to send its own E.164 number that was registered with the service provider as a second option, but the preference is to send the original caller ID received from the PSTN.

8.7.1 FXO FSK Detections

For incoming PSTN calls on FXO, the VoIP gateway has to detect the caller ID data on FXO and regenerate it on FXS ports to display the caller ID. The caller ID can precede any of the Type-1 service during the on-hook case. The caller ID data can be either FSK data or DTMF based on country configuration and can precede with any of the alerting signal DT-AS, RP-AS, or line reversal. During the off-hook phase, caller ID data precede with call wait and CAS tone on FXO. On detecting the CAS signal by CPTD, the PSTN state machine has to generate the ACK (digit D) on FXO to receive the caller ID data. The term "call progress tone detection" or "CPTD" refers to the detection of tones used to indicate the progress of the telephone call. In CPTD, amplitude, frequency, and cadence (timing) are analyzed. Call progress tone detection allows the VoIP gateway to monitor the progress of the resulting call, and hence, different states with respective call progress tones such as ringing tone, DT-AS, ring-back, and busy can be determined. A method used in a call progress detector analyzes the audio signal on the FXO line to detect a repeating pattern of sound and silence, such as the pattern produced by a ringing, dial tone, ring-back, CAS, DT-AS, or a busy signal. There are various types of call progress tones. These tones and country-specific deviations and techniques used for tone detections at different center frequency and frequency combinations are given in Chapter 17.

In general, FXO interface signals must be detected by the VoIP processor for Type-1 Caller ID FSK modulations and DTMF caller ID. During call wait ID, various tones such as CAS and DTMF must be detected in addition to FSK modulations. In the voice and fax chain modules, V.21 detectors are present. A similar detection scheme can be used for FSK demodulation. The baud rate of FSK is different and is 1200 baud as mentioned in Table 8.1. Caller ID FSK data can be either V.23 or BEL202 for mark and space signals. Hence, the requirements of caller ID FSK have to be taken care of in the design. Some demodulation techniques are given in Chapter 14 and can be used for FSK caller ID detections. The demodulated FSK bits are framed to extract the caller ID message format and the number. The decoded caller ID is regenerated on the FXS interface as is or using any of the type-1 caller ID services based on the country configuration of the VoIP adapter.

In an outgoing PSTN call on FXO from gateway FXS, the FXS on the gateway goes off-hook and listens for the VoIP dial tone and dials the outgoing number. If the number is identified as a PSTN number, the PSTN state machine collects all the digits and off-hooks the FXO to generate the DTMF digits on the PSTN line. The digits from the FXO interface are detected at the nearest central office. The call routing and progress is created through a central office, and the generated digits connect the subscriber to get the ring-back tone. In this situation, the central office will be generating the caller ID data on a called PSTN phone. On detecting the ring-back tone at the VoIP gateway, the FXS-to-FXO loop-back call is created and voice flow starts.

8.7.2 Caller ID Pass-Through in the FXO-to-FXS Call

Figure 8.8(b) is shown for caller ID pass-through. It is the simplest and works in several conditions without any difficulties. In some deployments, FXO is used as a lifeline as well as for local PSTN connectivity. With this option, the call combinations with FXO are FXS-to-FXO and FXO-to-FXS calls, and FXO may not be used for VoIP calls. Caller ID and call wait ID are applicable in the FXO-to-FXS call. The FXO-to-FXS call is treated as an incoming PSTN call on the FXO interface. During an incoming PSTN call on the FXO line, the PSTN sends an initial ring alert/line reversal/dual-tone alerting signal followed by caller ID data in the on-hook state of the FXO interface. The caller ID detector module on the FXO channel continuously monitors the alert signal on an FXO interface. An alert signal on the FXO has to be transferred to the FXS interface to establish an FXO-to-FXS call. At the end of the alert signal, the VoIP processor can loop back voice band samples between FXO and FXS. The voice processor has to work as a switch between FXO and FXS, and this scheme can work for multiple countries. It is widely adapted if the product feature with FXO reaches the FXS interface phone through a VoIP gateway. In FXO-to-VoIP calls, the gateway has to analyze all the incoming caller ID signal and tones and retransmit to the packet network.

8.7.3 Caller ID on WiFi and IP Phones

Figure 8.8(b) is shown with extensions of the IP and WiFi phones. These terminals are stand-alone VoIP terminals that collect the caller ID as digits and bits and send it to the liquid crystal display (LCD) directly. No hook conditions, line reversals, and FSK with such stand-alone VoIP terminals are used. Using VoIP gateway and analog phones, acknowledgments from the phone during call wait ID are provided. In IP phones, acknowledgments are not used for CIDCW as the IP phone delivers caller ID as message and digits information.

8.8 SUMMARY AND DISCUSSIONS

Caller ID, call wait ID, and call wait indication are the basic call features supported in VoIP deployments. On direct terminals of IP and WiFi phones, it is easy to implement call features compared with the VoIP gateway with analog phones. Caller ID features mainly deal with the hook status of telephone lines, signal encoding, and decoding for various modulations and tones. Several options exist, and deviations in implementations widely vary across multiple countries. Caller ID timing and hook status are more sensitive in dealing with implementations. The implementations tolerate for deviations in types of FSK modulations, signal level, impedance of subscribed line interface circuit (SLIC) and COder/DECoder (CODEC) interfaces, and so on.

In this chapter, some of the most popular approaches are given. Refer to country-specific PSTN standards for complete details on the specifications. Some country-specific deviations are also given in this chapter. These caller ID and call wait ID generation and detection features consume very little processing. When the product or software development has to cater to multiple countries, it is suggested to build a basic framework that caters to different methods in FSK, DTMF, signal levels, timing programming, and different line conditions. The telephone does not distinguish whether a call is from VoIP or PSTN; hence, these features have to emulate functions similar to a telephone interfacing with a PSTN network.

9

WIDEBAND VOICE MODULES OPERATION

To exceed public switched telephone network (PSTN) voice quality, wideband VoIP on internet protocol (IP) network is one of the main options. Traditional PSTN and VoIP services limit the analog speech bandwidth. Wideband voice systems cater to analog signals from 50 to 7000 Hz sampled at 16 kHz. Narrowband voice is from 300 to 3400 Hz, sampled at 8 kHz. Wideband transmission acoustics characteristics are given in references [TIA/EIA-470C (2003), TIA/EIA-920 (2002), Alexander (2006), ITU-T-P.311 (2005)]. The main differences of wideband voice with reference to narrowband are the telephone acoustics, wideband electrical interfaces, higher processing for the voice chain, and user experience. VoIP wideband voice is possible if an end-to-end call is through VoIP and not routed through PSTN.

Wideband voice creates more natural-sounding conversation with the sense of presence, creating a feeling of transparent face-to-face communication that eases speaker recognition. Wideband signals are also treated under classification of audio. The history of wideband voice goes back to several years. G.722 was officially released in 1988 for wideband audio/voice [ITU-T-G.722 (1988)] and operates at 64, 56, and 48 kbps. There are many new codecs for wideband VoIP voice compression such as G.729.1 [ITU-T-G.729.1 (2006)] that operates up to 32 kbps, the adaptive multirate wideband (AMR-WB) G722.2 [ITU-T-G.722.2 (2003)] that operates at variable bit rates in range between 6.6 and 23.85 kbps, and the global IP sound proprietary internet speech audio codec (iSAC) [URL (iSAC)], and iPCM [URL (iPCM)].

VoIP Voice and Fax Signal Processing, by Sivannarayana Nagireddi
Copyright © 2008 by John Wiley & Sons, Inc.

Wideband IP phones were already in the market while writing this chapter. These IP phones are made to support mainly G.722, and a few of them are in the experimental stage of launch with G.729.1 codecs with better acoustics to deliver the wideband voice. In VoIP customer premises equipment (CPE) such as the VoIP adapter, the telephone interfaces are made to support wideband, but the availability of wideband phones is limited. Hence, wideband support with regular wideband phones may be delayed for several more years. Several convergence operations with VoIP exist such as IP Multimedia subsystems (IMS), fixed mobile converge (FMC), and Femto cells. These derivative product handsets may be used for other media of audio and video that support wideband acoustics. The wideband acoustics support in handsets favor wideband voice in new derivatives and in extended multimedia services.

Wideband voice compression codecs and voice quality aspects are given in Chapter 3. In this chapter, the major changes associated with an end-to-end wideband voice call are given. The example created in this chapter is for wideband voice with VoIP adapter operation with foreign exchange subscriber (FXS) and foreign exchange office (FXO) interfaces. Similar operations are applicable to a wideband IP phone with minor deviations.

9.1   WIDEBAND VOICE EXAMPLES

In this section, some popular and feasible wideband voice call operations are presented.

9.1.1   Wideband VoIP Calls with Computer Softphones

In the actual use of VoIP, wideband voice started gaining popularity for the last few years (while writing this chapter in 2007). Some personal computers (PCs) and laptops were using software applications like Skype that was using iSAC [URL (iSAC)] as a wideband codec for end-to-end VoIP calls. PCs and laptops had full audio-capable support on the interfaces. Headsets used with analog interface are wideband capable. This makes a reasonable good example of an end-to-end simple wideband VoIP call. Skype also supports other codecs and some VoIP calls may not use wideband codecs.

9.1.2   Wideband IP Phones

Some IP phones support wideband voice mainly with G.722 and G.729.1 compression. The main requirements for wideband voice support are in the extra processing required, four-wire analog-to-digital converter (ADC)/digital-to-analog converter (DAC) hardware interfaces and acoustics capability in handset and speakerphone-mode acoustic requirements. Once acoustics are taken care of, processing and hardware interfaces may not play a major role in the wideband IP phone.

9.1.3 WiFi Handsets

WiFi phones also known as the wireless fidelity (WiFi) handset interfaces through the family of 802.11 wireless local area networks (LANs) protocols. These are self-contained VoIP phones. WiFi phones can make independent VoIP calls through the IP network as explained in Chapter 2. WiFi phones also communicate with local VoIP adapter telephones to make use of private branch exchange (PBX) functionality.

9.1.4 Wideband Phones

Some manufacturers support TIP-RING-compatible wideband phones. Several conference speakerphones support wideband voice. Currently, wideband phones are expensive compared with normal phones, but once deployments start using wideband phones, the cost may match with regular narrowband phones. The better migration to the wideband path is through handheld digital-enhanced cordless telecommunications (DECT), WiFi, and Bluetooth phones. These self-contained phones have improved acoustics, which means a better microphone and speaker. The current trend is to support wideband voice with these handsets, which may change over time. Each family of devices interfaces differently with the VoIP infrastructure as represented in Fig. 9.1.

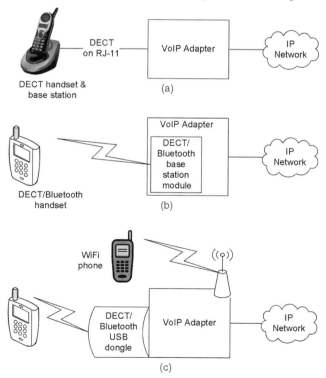

Figure 9.1. DECT, Bluetooth, and WiFi phone interface examples.

9.1.5 DECT Phones

Architecturally, the DECT phone had a handset that communicates on a wireless medium with a base station. It will work similarly to cordless phones used by several home telephone users. The base station interfaces analog signals to the processor through the TIP-RING interface. In some designs, the DECT base station is considered as part of the VoIP system. It will create direct digital communication with the processor without taking the analog signals through the TIP-RING interface, which will work much better than the analog interface while considering wideband voice.

Analog DECT phones with a TIP-RING interface use adaptive differential pulse code modulation (ADPCM) for the communication between the handset and base station that limits the quality. In recent times, some DECT handsets can communicate with the base station using G.711 and G.722 wideband codec. DECT phone base stations are also made with a LAN interface. At this stage, USB dongle-based DECT phones are getting popular in Europe. USB dongle works as a base station, and the handset communicates to the base station dongle. It will also work as a digital interface for the voice samples. This approach scales better and can be plugged into the required VoIP adapter for wideband VoIP voice calls. In general, each base station can support several handsets, allowing more users to communicate with the same base station.

9.1.6 Bluetooth Phones

Bluetooth phones interface through the Bluetooth base station as part of the VoIP adapter or through LAN interfaces such as the USB dongle. Architecturally, the interface combinations indicated for DECT phones can be used for Bluetooth with Bluetooth modules used instead of DECT modules.

9.1.7 Mobile Phones

Mobile phones are being extended for their functionality, mode of operation, and hand-over. These phones can be extended for the wideband mode of operation. Figure 9.1 is a representation for a functional usage and interface of DECT, Bluetooth, and WiFi phones. These phones can extend the wideband support without keeping dependency on the existing plain old telephone service (POTS) narrowband phones. The options indicated under wideband are time sensitive. These options may extend and change with time.

9.1.8 Wideband Calls with VoIP Adapters

A VoIP application can provide a wideband analog interface close to the user. It was not possible with PSTN voice calls. The front-end system-level integrated circuit (SLIC)–CODEC devices in VoIP are supporting wideband voice. Hence, if wideband phones are connected between two adapters, an end-to-end wideband voice call can be made. Currently available wideband-capable wire line phones are limited, and they are expensive. Mainly the handset will limit the voice bandwidth. Alternatively, wideband headsets can be used similarly to plugging the headset into the PC or laptop, which will make it a proprietary arrangement. The user has to be more knowledgeable of operational combinations.

The stand-alone direct IP calls with IP and WiFi phones are the simpler option. The other alternatives at this stage are using wideband acoustic-capable DECT and Bluetooth phones with VoIP adapters.

9.2 WIDEBAND VoIP ADAPTER

VoIP adapter software and hardware details are given in Chapter 2 for narrowband voice. The main deviations in implementation of wideband voice are marked with a sampling frequency of 8, 16, and 8/16 kHz. The common module details are given in previous chapters. In the wideband mode of operation, it is expected that phones of both narrowband and wideband are used with FXS interfaces. Even if the FXS interface supports wideband, fax machines on an FXS interface make calls in narrowband. The FXO interface is connected to the PSTN central office, and the PSTN is always of narrowband voice. Hence, several mixed operations will occur with wideband and narrowband voice combinations. The favorable part of these combinations is that the FXS interface devices support software configurations to use devices in both narrowband and wideband mode. Several VoIP systems are built with front-end interface devices made to work for both narrowband and wideband mode. Hence, the support of wideband is more of a software change on some existing VoIP systems.

VoIP adapter main functional blocks are shown in Fig. 9.2 for narrowband and wideband combinations. The explanation of modules is given under the assumption that FXS is used with wideband phone, and FXO is for narrowband PSTN. The SLIC–CODEC device for FXS is of wideband capable with the required programming or configuration. FXO devices are of narrowband. The PCM interface creates communication between FXS/FXO devices and the processor. PCM is a single shared communication interface. Several devices can communicate on the same PCM bus. On PCM interface, a required combination of wideband and narrowband slots can be used. To maintain uniformity, it is preferred to use linear 16-bit samples for both narrowband and wideband. All these samples work with the same 8-kHz frame synchronization. The frame synchronization for a mixed mode of narrowband and wideband is shown in the next section. The following major operations can be noted from Fig. 9.2. This figure is the same as Fig. 2.6 with additional marking for narrowband and wideband sampling combinations.

1. Existing modules of narrowband such as dual-tone multifrequency (DTMF) detection, fax tones, call progress tones, and FXO modules, are

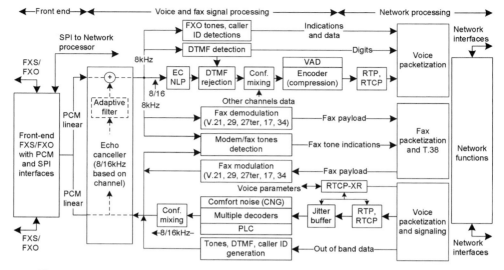

Figure 9.2. Wideband VoIP adapter representations with sampling dependency markings.

used after a band-limiting operation and downsampling to 8 kHz. These signals are of lower bandwidth.

2. Modules of echo cancellers (ECs), DTMF rejection, and voice compression have to operate directly on available sampling rate. For wideband, these modules have to operate on 16-kHz sampling.

3. For the wideband channel, the decoder chain will be providing 16-kHz samples. Decompression, voice activity detection (VAD)/comfort noise generation (CNG), packet loss concealment (PLC) and echo canceller reference have to operate at the same 16-kHz same sampling frequency.

4. DTMF, call progress tone, caller ID generation, and other call feature tones in the decoder path are of lower frequency and can be generated at 8-kHz sampling. These signals have to be upsampled before sending on the PCM interface.

The operational deviations and features of these modules for wideband operation are given in the next section.

9.2.1 Wideband and Narrowband Modules Operation in the Adapter

In this section, narrowband and wideband module operational differences and features are given.

Sampling and Front-End Devices Configuration. The front-end SLIC–CODEC devices have to work for both narrowband and wideband. SLIC is for the high-voltage drive and hybrid functions. The CODEC consists of ADC and DAC and several hardware/software functions for proper hybrid compensation. SLIC and CODEC may be part of one chip. Narrowband signals are sampled at 8 kHz; wideband signals from 50 to 7000 Hz are sampled at 16 kHz. The front-end devices have to be dynamically configurable in the middle of the call between wideband and narrowband mode. While writing this section, devices such as Si3216 from Silicon laboratories [URL (SLIC-WB)] and Legerity Le88226/246 [URL (Legerity-WB)] are found to work for both narrowband and wideband mode. These devices can be changed from one mode to another with simple programming of hardware device registers through a serial peripheral interface (SPI). The communication between CODEC and processor is through the PCM interface. PCM samples can be configurable for A-law, μ-law, and 16-bit linear. In narrowband mode, any of the three combinations can be used. Wideband mode is expected to provide better quality than PSTN. Hence, CODEC has to be configured for 16-bit linear samples. A-law and μ-law will limit the wideband voice quality even if the samples are at 16 kHz.

The existing PCM or time division multiplexing (TDM) interfaces work synchronously with 8-kHz frame synchronization. Frame synchronization (synch) is the representation of the sampling frequency. In wideband mode also, similar 8-kHz frame synchronizations is used. The narrowband and wideband samples representation is given in Fig. 9.3 for 32- and 24-slot PCM interfaces. In the figure, all PCM slots are of 16-bit linear. The top rows of boxes in Fig. 9.3(a) represent PCM narrowband slots. Narrowband slots carry one 16-bit sample per each slot marked as Ch0,Nb(0), and Ch0,Nb(1). The symbol Ch0,Nb(1) denotes channel-0 and the narrowband sample with index (1). The same 32 slots with 8-kHz frame synchronization will work as double-frame synchronization of 16 kHz when devices on a PCM interface are enabled for wideband mode [URL (SLIC-WB)]. It is shown in the middle row of PCM slots marked in Fig. 9.3(b). No physical frame synch at the 16-kHz rate exists, but wideband slots offset by 16 creates 16-kHz sampling. The third row of PCM slots in Fig. 9.3(c) indicates a mixed combination of wideband and narrowband slots. Wideband channels can keep using an offset of 16 slots, and narrowband uses slots synchronized to the regular 8-kHz frame synch pulses.

Wideband and narrowband can use multiple options. Figure 9.3(d) is for 24 slots, and in 24 slots, frame synch and bits per slot can remain the same. The PCM clock can be changed to accommodate required slots. In this mode, a combination of wideband and narrowband slots can coexist. Processors and interfacing devices can be configured to accommodate multiple configurations. It is required to configure all PCM connected devices for proper synchronization of data transfer.

DTMF Detection. DTMF tones are between 697 and 1633 Hz. Hence, lower sampling up to 4 kHz is sufficient. Most DTMF detection modules are designed to work with 8-kHz sampling. In wideband mode, the usual practice is to pass

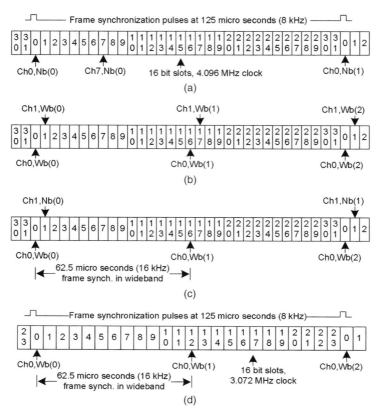

Figure 9.3. PCM interface example for narrowband and wideband combinations. (a) Narrow-band, linear 16 bit slots. (b) Wideband offset by 16 slots at 8-kHz frame synch. (c) Wideband and narrowband combination and coexistence. (d) Wideband and narrowband representation for 24 slots.

the voice samples through the band-limiting filter of a narrow bandwidth and resampling it to 8 kHz. The DTMF detection module can operate at 8-kHz mode as if signals are always coming to this module in narrowband at 8-kHz sampling. The filter and downsampler are kept in bypass mode depending on the mode of operation.

DTMF Rejection. DTMF rejection is for removing the DTMF tones from the basic signal as given in Chapter 7. With wideband input, DTMF rejection filters have to operate on wideband input. The rejection filters have to use another set of filter coefficients for wideband. The same DTMF rejection output can be used with a wideband compression codec. DTMF rejection makes use of DTMF start indication and DTMF ending information from 8-kHz-based processing modules. DTMF actual rejection has to happen on 16-kHz sampling, but RFC2833 parameters of the narrowband channel are used at destination. RFC2833 parameters can be considered as independent of the sampling selection. At the destination, depending on the sampling selection, suitably sampled DTMF tones are played based on the out-of-band RFC2833 packet information.

In the narrowband in-band mode, DTMF digits will be going to destination exactly like speech. In case of wideband, these tones have to go through the G.722 and G.729 family of wideband codecs. The codec G.722 is expected to relay DTMF in-band tones with acceptable distortion caused by ADPCM coding. At this stage, characterization of DTMF tone distortions in wideband codecs is not clearly available. Based on the additional data on DTMF distortions, such in-band modes have to be considered.

DTMF, Call Progress Tones, and Caller ID Generation. Assuming SLIC–CODEC and PCM interfaces are operating to match 16-kHz sampling, DTMF generation, tones, and caller ID are generated at either 8 or 16 kHz rate. The usual approach is to use 8-kHz samples and upsample by a factor of two. The upsampler converts 8-kHz samples to 16 kHz. In actual practice, an initial call may be established in narrowband and then will switch to wideband mode based on the negotiation through signaling. Hence, the requirements of the upsampler are considered based on the call state. Once a call is established and switches to 16-kHz mode, these states have to cater to the wideband mode. For example, after establishing the call, digits for call wait ID, call-feature-specific digits and tones may operate on the wideband signal.

Echo Canceller. The echo canceller resides on the path of actual samples. In wideband mode, the echo canceller has to operate at 16 kHz. The SLIC–CODEC device, memory buffers used in the processor in the PCM path, telephone lines, and telephone characteristics decide the echo canceller span. For a total echo span of 16 ms, 128 taps (16 ms and each millisecond is of 8 taps) are required in narrowband sampling and 256 taps (16 ms and each millisecond is of 16 taps) in wideband. The usual practice is to use a 256- or 512-tap echo canceller that works for both wideband and narrowband. Properly designed echo cancellers usually will not exhibit a major difference among wideband and narrowband signals. The echo cancellers evaluated for wideband operation can take care of narrowband requirements also. Not all narrowband echo cancellers may satisfy good voice quality for wideband mode of operation hence, thorough characterization is essential for a wideband mode of operation.

Echo Canceller Control Plane. The main control plane modules of the echo canceller are double talk (DT), nonlinear processing (NLP), modem-fax tone detectors. Most designs of DT and NLP may still work without distinction between sampling frequency. If any sampling frequency-dependent operations exist, sampling dependencies have to be addressed. Modem-fax tone detectors have frequency-dependent filters and other estimators. It is required to take suitable correction in modem-fax tones or use 8-kHz subsampled signals for tone detection.

Wideband codecs have longer basic frame durations than the narrowband codec. G.729 [ITU-T-G.729 (1996)] is of a 10-ms frame, and G.729.1 is of a 20-ms frame. The longer frames increase end-to-end delay by another 10 ms. Extra delay keeps more constraints on the echo cancellation. The echo canceller rejections have to be perfect in the case of wideband echo cancellers. Wideband is expected to pass music without distortion. Music detection, background detections, pleasant music, and background pass-through are most important in wideband voice. To help the mixed mode of voice, echo cancellers have to be made independent of 8 and 16 kHz sampling rate.

FXO PSTN Functions. The FXO is for PSTN and other PBX connectivity. All FXO signals are of narrowband. The VoIP system may be simultaneously operating with some channels as narrow, some channels as wideband, and FXO as one of the narrowband channels. The FXO may be looping back on FXS in loopback modes. These loopbacks have to operate in 8-kHz sampling mode. Call progress tones, Caller ID, and call features detection are applicable to the FXO interface. These tones have to be processed at the 8-kHz rate. Some preferred approaches for these combinations of channels arrive at a suitable configuration of the PCM interface. On most devices, the PCM interface is used as narrowband and wideband, which is possible because of the basic 8-kHz frame synchronizations and 16-kHz samples residing on the offset slots. The offset slots are given under SLIC–CODEC interface of the previous section.

PLC and VAD Algorithm. The PLC and VAD modules are applicable at 16-kHz sampling. The wideband codecs are having their own version of PLC, and VAD/CNG modules at 16 kHz sampling. These modules have to operate in relation to the selected codec and operating sampling frequency of the codec. The alternation of sampling frequency is not accepted for these modules.

Real-Time Protocol (RTP) and Jitter Buffer. The wideband codec comes with its own payload and packetization details. RTP has to take care of these features in the implementation. Jitter buffer may not distinguish the sampling, but certain time-stamp information may be used in the jitter buffer. The G.722 codec can operate at smaller frame sizes. Some wideband codecs such as G.729.1 and G.722.2 operate with a minimum packet size of 20 ms. Codec iSAC [URL (iSAC)] operates at a 30-ms frame similar to the G.723.1 [ITU-T-G.723.1 (2006)] narrowband codec. Hence, jitter buffers will be made generic to deal with required packet sizes up to 80 ms. In general, no special requirements may be included with the jitter buffer for wideband. However, with any new codec, a set of parameters is made available to the jitter buffer on per-channel basis. Some codec payload types may not be available in the existing RFC3551. The available payload from 32 dynamic payload types can be allocated to the new wideband codecs. Wideband payload aspects are given in RFC4749 [Sollaud (2006)] for the G.729.1 wideband codec.

VoIP Signaling. VoIP signaling such as Session Initiation Protocol (SIP) will distinguish codec types. In the capabilities, Session Description Protocol (SDP) sessions have to exchange capabilities with wideband codec capabilities also. Based on the mutually agreed codecs, either wideband or narrowband is used. On selecting wideband, all required module configurations have to be taken care of as per the selection.

Default Modes in Call Negotiations. Sometimes the narrowband codec may be the default codec, and in the process of negotiation, wideband may be the final requirement. In the initial dialing, if the samples are generated at 8 kHz, after the capabilities exchange, this may change to 16 kHz. All modules and sample buffers as well as memories have to be taken care of properly in the process of switching between narrowband and wideband mode.

Channel-Specific Differences. Some channels in the voice gateway may use narrowband and other channels wideband. The end-to-end operation has to be performed based on channel. All channels have to work independently. Fax and modem will be working at 8-kHz sampling. The modes of T.38 fax and fax pass-through will operate with the fax data pump modules or G.711 codec at 8-kHz sampling.

Wideband Interoperability. For end-to-end calls, interoperability is very important. Wideband is supported in a limited set of VoIP systems at this stage. It is expected that most systems in the future will support wideband voice. Wherever backward compatibility is possible, codecs will be shared between narrowband and wideband. As an example, support of wideband G.729.1 can eliminate the requirement of supporting narrowband G.729A and G.729.

Wideband Call to PSTN. PSTN operates on narrowband. Any call that requires an intermediate PSTN requirement has to be negotiated for narrowband. Even if an intermediate PSTN-to-VoIP gateway supports wideband on VoIP, end-to-end voice will be of narrowband.

Conference Mode. In a wideband conference, to get the best quality, all channels have to operate in wideband. In practice, some channels may use narrowband. To maintain better interoperability, conference bridges have to send both wideband and narrowband with several combinations of codecs. Narrowband voice is upsampled to wideband and sent to the wideband-capable channels. To send on the narrowband channel, wideband voice downsamples to 8 kHz.

9.3   WIDEBAND VOICE SUMMARY

Wideband telephones are not widely available currently. Several conference phones are available in wideband, but they are expensive. It may take some time to get these phones at the right price. With VoIP expanding in scope through various convergence operations of IMS, FMC, and other extensions, wideband may gain more access to users. Deployment boxes are incorporating the wideband-capable hardware devices, and they can work with simple software configurations.

The IP phone is a single channel processing system with three-way conferencing. The processing power should be sufficient with wideband. In VoIP adapters, one to four channels are used. It is required to cater to extra voice processing in these devices. The wideband total voice chain is expected to consume about 50% to 100% more processing depending on the codec. The codecs like G.729.1, and G.722.2 take more processing power. The waveform-based G.722 takes lower processing than does the narrowband G.729A codec. In the process of an end-to-end call, if the narrowband codec is used at any intermediate stage, there is no real benefit of using wideband. An end-to-end wideband call is the way to exceed PSTN voice quality. It is expected that wideband voice coexists with several multimedia services through an IP network.

10

PACKETIZATION—RTP, RTCP, AND JITTER BUFFER

In VoIP, voice samples from telephone interfaces are compressed using compression codecs such as G.711, G.729A, G.723.1, and G.722, and are framed as payload. Voice payload size varies with the compression codec, compression rate options, and payload duration. For G.729A, 10 ms is the frame, and for G.723.1, 30 ms is the basic frame. Voice payload may use a group of compressed frames up to 80 ms. The compressed payloads are framed as Real-Time Protocol (RTP)/User Protocol Datagram (UDP)/Internet Protocol (IP) packets for sending on the IP network. At the destination, payloads from RTP/UDP/IP are passed through the jitter buffer before decompressing in the decoder. RTP Control Protocol (RTCP), RTCP-Extended Report (XR) packets are used to convey end-to-end voice packet transmission parameters and statistics. In actual implementation, voice payload, RTP, jitter buffer, RTCP, voice quality monitoring, quality of service (QoS) mechanisms, and bandwidth management parameters work in coordination for ensuring better end-to-end packet delivery. This chapter is presented for RTP, RTCP, RTCP-XR, packet impediments, and jitter buffer. More details on voice packet headers of UDP/IP and network interface headers for Ethernet and digital subscriber line (DSL) are presented in Chapter 11.

10.1   REAL-TIME PROTOCOL (RTP)

RTP is the real-time protocol used with voice, audio, video, and several other real-time media-specific applications. RTP carries media payload and real-

VoIP Voice and Fax Signal Processing, by Sivannarayana Nagireddi
Copyright © 2008 by John Wiley & Sons, Inc.

time parameters for extracting the timing details. RTP parameters are also used in jitter buffer adjustment in VoIP applications. VoIP packets are framed with compressed voice payload, RTP, UDP, and IP header combinations as shown in Fig. 10.1 and Chapter 11. RTP and UDP combinations are used for real-time voice delivery. The VoIP packet consists of an IP header of 20 bytes given in RFC0791 [URL (RFC791)] and a UDP header of 8 bytes given in RFC0768 [Postel (1980)]. Real-time aspects of transport are taken care of by a 12-byte header of RTP RFC3550 [Schulzrinne et al. (2003)]. Voice is sensitive to delays, and RTP helps proper end-to-end delivery of real-time voice traffic. RTP header compression reduces the number of bytes. RTP header compression [Casner and Jacobson (1999)] is not considered in this chapter. The RTP header format is illustrated in Fig. 10.1 [Schulzrinne et al. (2003)].

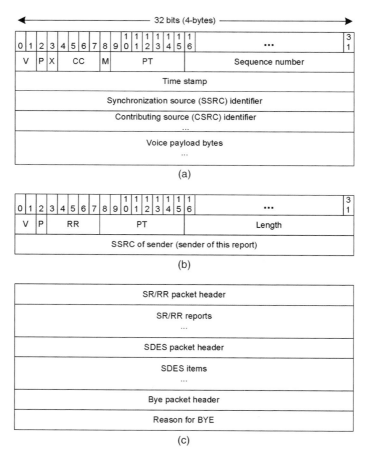

Figure 10.1. RTP and RTCP packet formats. (a) RTP packet format. (b) RTCP header. (c) RTCP packet format at a high level.

The first 12 octets up to synchronization source (SSRC) are present in every RTP packet, whereas the list of the contributing source (CSRC) identifiers is present only when inserted by a mixer. RTP header parameter details are given below.

V = RTP version (2 bits).

P = padding bit (1 bit), setting of this bit indicates more padding bytes at the end of the RTP packet.

X = header extension bit (1 bit) if set, the fixed header will be followed by one extension header. This is for the redundancy scheme.

CC = number of CSRC fields in the header (4 bits) that follows the fixed header. This caters to 0 to 15 CSRC identifiers.

M = marker bit (1 bit), it is intended to allow significant events such as frame boundaries to be marked in the packet stream.

PT = payload type (7 bits) indicates several payload types including PT for voice codecs of G.711 (PT for PCMU is 0, PT for PCMA is 8), G.729 (PT 18), G.723.1 (PT 4), G.722 (PT 9), voice activity detector (VAD) packet types [Schulzrinne et al. (2003)], and RFC2833 and Fax. Fax and RFC 2833 RTP payload for dual-tone multifrequency (DTMF) events and telephony tones uses dynamic RTP payload types from 96 to 127. For more details on RTP payload types of audio, refer to RFC3550 [Schulzrinne et al. (2003)] and RFC3551 [Schulzrinne and Casner (2003)].

A sequence number (16 bits) starts with a random value selected at the time of conversation that helps to recover sequence order of packets at the destination. Adaptive jitter buffer (AJB) and packet loss concealment (PLC) algorithms also use sequence number information to arrange the packets in the right order and to send them to the decoder.

A time stamp (32 bits) is the sampling instant of the first octet of the payload. The time stamp is incremented by the payload duration. This information is useful at the destination to derive frames from long packets and at AJB for jitter calculation.

Synchronization source (SSRC) = 32 bits; the SSRC field identifies the synchronization source. This identifier is selected randomly, with the intent that no two synchronization sources within the same RTP session will have the same SSRC identifier.

CSRC list is 0 to 15 items with 32 bits for each source. The CSRC list identifies the contributing sources for the payload contained in this packet. The number of identifiers is given by the CC field. If there are more than 15 contributing sources, only 15 can be identified. Mixers using the SSRC identifiers of contributing sources insert CSRC identifiers. In a normal end-to-end voice call, CSRC is not present.

RTP is mandatory with VoIP voice packets. RTP parameters are useful in extracting several real-time parameters and in ensuring proper playout of packets through jitter buffer.

10.2   RTP CONTROL PROTOCOL (RTCP)

RTCP is used to convey the end-to-end quality of the data stream in an RTP session. Statistics such as delay, jitter, number of packets sent, and packet loss rate allows the session participants to monitor the health of the connection. Timing parameters like network time protocol (NTP) and the parameters given in RFC1305 [Mills (1992)] establish a possible timing relation between gateways. The protocol also provides limited control functions such as BYE packets to control the status of the session. RTCP packets are transmitted on a different port than that of RTP. RTCP packets are exchanged between the end points in a periodic fashion. RTCP packets are sent at less than 5% of the RTP packets, which makes RTCP statistics to be used as steady parameter values. Packet-to-packet parameters have to be taken from RTP only. RTCP consists of five packet types:

1. Sender Report (SR)—Conveys the statistics of the active RTP sender.
2. Receiver Report (RR)—Conveys the statistics of the RTP receiver.
3. Source Description (SDES)—Description of the source regardless of whether it is a sender or a receiver.
4. BYE—Used to hang up from a session and indicates end of participation.
5. APP—Application-specific packet for experimental use.

RTCP packets are normally compound packets, which consist of a combination of several packet types from the above list. A compound packet always starts with either a sender or a receiver report, followed by the other packet types. A BYE packet if used always comes in last. A typical packet header format is given in Fig. 10.1(b). RTCP packet payload format is given in Fig. 10.1(c). In RTCP header packet, V = RTCP version (2 bits), P = padding bit (1 bit), RR = Reception report count—number of reports in this packet (5 bits), PT = RTCP packet type (8 bits), and length (16 bits) of the RTCP packet in 32-bit words minus one including the header and any padding and the synchronization source identifier (SSRC, 32bits) for the originator of this SR packet. For more detailed description of RTCP, refer to RFC3550 [Schulzrinne et al. (2003)].

10.2.1   RTCP-XR Parameters

RTCP-XR is the extension of RTCP as per RFC3611 [Friedman et al. (2003)] for monitoring voice quality parameters. VQmon [URL (Telchemy)] is one application that makes use of RTCP-XR. The VQmon and RTCP-XR framework provides a set of metrics for VoIP performance monitoring and diagnosis. It supports real-time monitoring and parameters for postanalysis. Several parameters for RTCP-XR are derived from the E-model algorithm (ITU-T

G.107- 2005) and, voice processing modules. RTCP-XR makes use of parameters from packet transmission characteristics that include packet impediments on the network, end-to-end delays, jitter buffer dynamics, and signal transmission characteristics that include signal level, noise level, gain, echo rejections, R-factor, and mean opinion score (MOS) derived from the R-model. RTCP-XR packet details are given in Chapter 20. These parameters are updated typically once in every 256 voice packets. The use of these parameters and applying feedback for improving the voice quality makes the VoIP system to deliver the highest quality under severe conditions. Jitter buffer can make use of these parameters in optimizing the performance.

10.3   VoIP PACKET IMPEDIMENTS

In VoIP, network packet drop, packet fixed delays, jitter (also called delay variation or reorder), packet errors, and fragmentation are very common. End VoIP systems and interfaces also introduce a certain amount of these impediments. The jitter buffer works on packet impediments and tries to deliver steady packets to the decoder with a minimum possible buffering delay. Depending on the end-to-end packet flow and adaptation dynamics, jitter buffer (JB) may miss the delivery of some packets. Jitter buffer and PLC will try to improve voice quality in the presence of packet impediments. At the first level, jitter buffer optimally buffers packets and tries to deliver the maximum available packets to the voice processing modules. Jitter buffer cannot replace any missing, error, fragmented, and packets arriving at unacceptable delays; it is the role of the PLC algorithm and decoder to improve voice quality beyond jitter buffer. PLC synthesizes the missing voice samples during the packet erasures without creating the noticeable artifacts. Architecturally, PLC resides on part of the decoder. Voice quality improvements in PLC vary with the compression codec, packet drop characteristics, and on the selected PLC algorithm used with the decoder.

10.3.1   Sources of Packet Impediments and Helpful Actions

This section describes sources of packet impediments. An overview on packet impediments and voice quality is given in reference [Nagireddi (2006)]. Depending on the situation, the common names used for end-to-end packet delivery issues are packet impairments, packet impediments, or network congestion. In Fig. 10.2, the sources of packet impediments are indicated at a high level. In this example, assuming VoIP customer premises equipment (CPE) has a built-in router and DSL/cable interface. Phone-A and VoIP CPE-1 are transmitting voice, and phone-B and VoIP CPE-2 are receiving voice. PLC is applicable in CPE-2, which is shown as part of the decoder block. In relation to Fig. 10.2, the sources of packet impediments are given below.

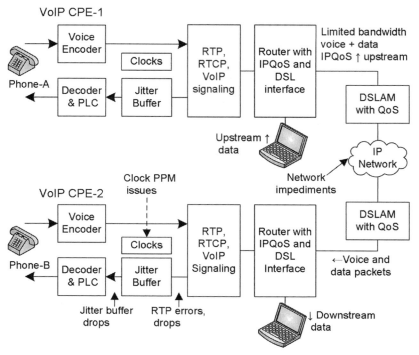

Figure 10.2. VoIP voice flow and packet drop situations for voice flow from phone-A to B. [Courtesy: printed with the permissions from Electronic Products http://www.electronicproducts. com.]

1. **Sending side jitter:** At the sending end, the router has to deal with large data packets and shorter voice packets. If voice packets are scheduled behind data packets, then it will reflect as jitter to the voice packets. Hence, voice packets are given always the highest priority more so than data packets. IPQoS and fragmentation algorithms inside the router block will take care of this issue to a certain extent. IPQoS and delay variations are presented in Chapter 18. Sending side jitter is taken care of in the receiving end jitter buffer of CPE-2. Ideally, this jitter should be maintained to zero or very less compared with the voice packet frame interval.

2. **Sending side jitter from operating system (OS) and interprocessor communication (IPC) mechanisms:** Voice payloads are created in signal processing blocks. This chapter is written with the understanding that the basic voice and fax chain is running on a digital signal processor (DSP) whereas required signaling as well as networking functions are running on another network processor. These payloads are transmitted to a network processor through an IPC mechanism. Some architectural

aspects on this mechanism are given in Chapter 19. Several applications will run on network processors. These applications may create a delay variation of a few milliseconds to voice packets.

3. **Bandwidth limitation:** The physical layer network will have a restriction for both upstream and downstream bandwidth. Usually downstream bandwidth is always more than upstream bandwidth. IPQoS gives priority to voice and releases the remaining bandwidth for data and other applications. If the router cannot track available upstream bandwidth, the IPQoS algorithm may not perform and reflect as loss to both voice and data packets in the upstream. Data packets will be retransmitted, but voice packets will be dropped. For better delivery, upstream rates have to be monitored accurately and IPQoS should work for the varying bandwidth and traffic conditions.

4. **Network IP impediments:** In IP networks, network impediments such as delay, jitter, packet errors, and packet fragmentation will be introduced at different stages of end-to-end transmission. All these impediments can result in packet drop at the input of the voice decoder. To overcome this issue, end-to-end bandwidth has to be more and QoS functions have to work in coordination.

5. **Downstream bandwidth:** Receiving end voice packets joins with downstream data and other applications. If the bandwidth from all applications exceed available downstream bandwidth on a wide area network (WAN) interface, packet drop can happen. In the case of data packet drop, Transmission Control Protocol (TCP)/IP will take care of retransmission. For voice packets, no retransmission will occur. To reduce downstream drops, it is required to restrict the demand on the overall download data from various applications. Bandwidth and packet statistics monitoring can also help in providing feedback to the applications to adjust the demanded downstream traffic.

6. Jitter buffer will try to keep most packets for proper delivery. In the jitter buffer, varying dynamics create a certain packet drop in the jitter transient conditions.

7. Clock drifts from end to end can appear as excess packets or as a shortage of packets. If CPE-1 has a negative parts per million (PPM) and CPE-2 has a positive PPM, then CPE-1 sends packets slowly and CPE-2 plays quickly. In this case, CPE-2 starves for packets, which leads to an under-run condition.

With reference to [Nagireddi (2006)], several packet impediments are marked in Fig. 10.2 in the packet and voice flow between two CPEs. As a result of all these packet impediments, jitter buffer will increase end-to-end delay and drop packets. PLC will try to enhance voice quality during packet drop. To get improved voice quality, packet impediments, delay, and jitter have to be as low as possible.

10.4 JITTER BUFFER

In public switched telephone network (PSTN), voice samples are delivered synchronously with reference clocks. In VoIP, voice samples are compressed and framed as packets. Packets will go through several impediments while traversing from the source to the actual destination.

As shown in Fig. 10.3(a), jitter buffer regulates the flow between incoming packets and the voice decoder. The input to the jitter buffer is at irregular

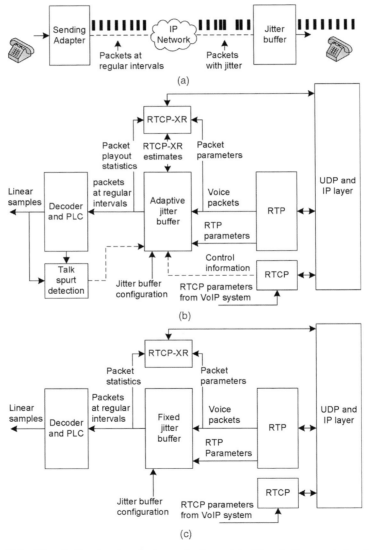

Figure 10.3. Jitter buffer input–output parameters. (a) Jitter buffer packet adjustment. (b) Packet adaptive jitter buffer. (c) Fixed jitter buffer.

intervals, but jitter buffer output is read at regular codec frame intervals. Jitter buffer removes the jitter in the arrival of the packets by holding them in the buffer for several milliseconds or several voice frame intervals. There is a trade-off between the end-to-end delay caused by the jitter buffer and the packet loss. For a comfortable voice conversation, end-to-end mouth-to-ear delay including jitter buffer should be as low as possible or at least in the range of 150 to 250 ms. The worst-case delay considered is 400 ms [ITU-T-G.1020 (2006)], and beyond this delay, voice calls are treated as not suitable for inter-active conversation. As per the E-model estimation given in Chapter 20, main-taining end-to-end delay to less than 177.3 ms helps to maintain better voice quality from delay considerations.

The size of the jitter buffer depends on the dynamic conditions of the network, trade-offs in delay, and allowed packet loss. In achieving the delay goals, jitter buffer may drop packets. A large jitter buffer size causes increase in the delay, but it reduces packet loss. In the early VoIP deployments, packet loss goals used to be of less than 5% for good quality speech. In recent times, this limit is set to 1% of packet drop [TIA-116A (2006)] as a recommendation. If the end-to-end packet delivery is free of impediments, the receiving end decoder can play packets as soon as they arrive, and this helps better voice quality. In practice, even in good conditions, some minimum buffering to the extent of one or two voice payload frames is used.

Jitter buffers are broadly classified as fixed jitter buffer (FJB) and AJB. AJB makes use of all possible conditions to maintain lowest packet drop at the lowest possible buffering delay. Fax and modem applications will not tolerate adaptive adjustments of AJB. FJB keeps a fixed buffer size that meets the requirements. FJB is sufficient in good network conditions. VoIP fax and modem calls use FJB to avoid packet adjustments. The main actions of AJB and FJB for various input impediment conditions are listed in Table 10.1.

In Fig. 10.3(b) and (c), AJB and FJB functional packet flow and parameters are given. In Fig. 10.3(b), AJB is shown to make use of a minimum number of inputs from RTP and configuration parameters. The output is at regular inter-vals to the decoder and PLC. The inputs shown to the AJB with solid lines are mandatory. The inputs marked in dotted lines are optional parameters that help to improve the AJB performance. AJB can make use of RTCP parame-ters that give round-trip delays, NTP time stamps, and other packet statistics. RTCP-XR packets will carry several inputs specific to packets flow. The param-eters from RTCP and RTCP-XR can be used to improve the AJB performance for the varying conditions. The playout adjustment algorithms can also make use of talk-spurt detection for identifying the silence zones, and this helps to improve voice quality.

FJB shown in Fig. 10.3(c) makes use of only limited parameters. Most FJB algorithms work with mandatory RTP parameters, payload, and user configu-ration. By design, FJB may not make use of RTCP, RTCP-XR, and talk-spurt information. However, FJB implementation parameters can be derived to report statistics to RTCP-XR module.

Table 10.1. AJB and FJB Actions for Various Packet Impediments

Condition	FJB	AJB
Fixed delay	No action.	Usually no action. If fixed delay is known; when fixed delay and buffer size together approaches 150 to 250 ms, AJB tries to minimize delay.
Adaptation control	No adaptation, fixed buffer size, usually preconfigured.	Adapts to dynamic conditions.
Packet jitter or reorder	Jitter will translate to fixed delay. Packets with jitter exceeding FJB thresholds will be dropped.	Jitter buffer adjusts the buffer size to accommodate the varying jitter conditions. It may drop a packet or create an extra silence.
Buffer size trade-offs	Usually higher to accommodate maximum deviations of jitter.	Keeps lowest possible buffer size while maintaining acceptable packet drop. Drops more packets if buffer size exceeds 250 ms.
Packet drop	FJB usually does not create extra packet adjustments. Input packet drop of FJB is translated to output of FJB.	AJB may drop or adjust more packets in the process of delay optimization. Total packet drop should not exceed 5% for acceptable voice quality.
Duplicate packets	Usually duplicate packets are handled in FJB. Some FJB implementations may not be having this capability.	Always takes care of duplicate packets.
Applications— distinction	Dedicated VoIP deployment when available bandwidth is more than the required. Used in fax and modem pass-through modes.	In all VoIP applications with several dynamics in the deployments. In some implementations, the same AJB is made to work like FJB for fax and modem by disabling the adaptation control.

10.5 ADAPTIVE JITTER BUFFER

Adaptive jitter buffers can take several inputs and can arrive at the best possible packet delivery while maintaining least possible buffering delay. Several algorithms for AJB exist [Ramjee et al. (1994), Pinto and Christensen (1999), Tseng et al. (2004), Moon et al. (1998)]. In this section, some popular AJB concepts are given. Each algorithm has several microlevel details and extra proprietary operational steps. The appropriate algorithm for playout delay adjustment should be chosen based on the requirements of the deployment conditions, quality goals, and available parameters from RTCP, RTCP-XR, and QoS. In a wireless environment, the network and the end-terminal conditions change frequently, and hence, a faster adaptive algorithm based on per packet interval has to be chosen. The playout adjustment algorithms are of two types based on packet adjustments.

1. Talk-spurt based—adjusts adaptively during silence periods.
2. Non-talk-spurt based—adjusts on a per-packet basis or at regular time intervals.

10.5.1 Talk-Spurt-Based Adjustments

Talk-spurt is a significant speech zone. A talk-spurt is defined as a continuous section of speech at least 300 ms in duration, containing no silent period longer than 200 ms [ITU-T-P.862 (2001)]. Talk-spurt is referred to as an utterance in voice quality measurement such as P.862. As shown in Fig. 10.4(a) with tile diagram markings, talk-spurts or utterances in solid boxes are separated by silence regions. In VAD/comfort noise generator (CNG) enabled mode, the compression codecs detect the silences as VAD silence frames or nonspeech frames. By retaining full content of talk-spurt and adjusting silence-zones, voice quality will not degrade. Adjustment of the silence zone is to increase or decrease the duration of the silence zone between the talk-spurts. Several initial articles [Ramjee et al. (1994), Pinto and Christensen (1999)] are based on a talk-spurt scheme for fixed and variable packet drops. In relation to Fig. 10.4(a), an initial jitter buffer operation will start with set initializations. The silence zones are adjusted based on the derived parameters such as packet drop and optimum playout delay from the previous talk-spurt. After adjusting the silence zone, optimum playout delay is applied on the next talk-spurt. In the regions of identified talk-spurts, jitter buffer will not allow any adjustments.

The main advantage of the talk-spurt-based scheme is improved voice quality. Speech MOS measuring instruments and human ear are not sensitive to small adjustments of silence periods. The main disadvantage of this scheme is the difficulty in detecting talk-spurt. No set algorithms are governed by standards and recommendations. Some standardized techniques used in speech quality measurements operate on several talk-spurts of data to decide on the best boundaries of talk-spurt and silence zones. When background activity is more or contains several continuous test signals, talk-spurt may not be detected. Hence, support of a non-talk-spurt scheme is also essential for working independently or for supplementing the operations of a talk-spurt-based scheme. Spike is a sudden long variation of delay that has to be treated as long bursts.

Spike can happen at any region of packets of silence or talk-spurt. Spike characteristics are presented in the later part of this chapter.

10.5.2   Non-Talk-Spurt-Based Adjustments

In a non-talk-spurt-based scheme, the parameters are extracted for every packet and jitter estimates are made continuously. In theory, the jitter buffer adjustments can happen for every new packet. In practice, these parameters are smoothed over several packet intervals, and adjustments are applied at regular intervals of a few hundred milliseconds. In Fig. 10.4(b), the adjustments are shown at selected intervals, and these intervals may change with the jitter characteristics. In the case of spike detection, the parameter updates can change at the transitions of spike and normal modes to adapt to the spiky conditions. After the end of the spike, the adjustments will be performed in normal mode. The parameter updates and jitter estimates will work based on the underlying algorithms, timing, logic, configurations, and set quality goals.

In Fig. 10.4, the spike occurrence is indicated in Fig. 10.4(b). Spike detection operation is applicable to both talk-spurt and non-talk-spurt mode. In general, it is simpler to use talk-spurt-based spike detection of Fig. 10.4(a). In Fig. 10.4(a), spike operations are not marked.

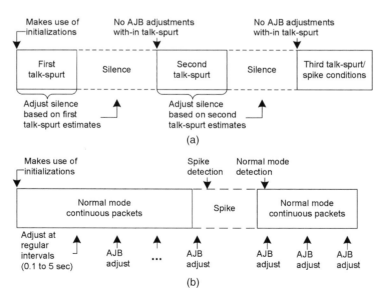

Figure 10.4. AJB adjustments. (a) Talk-spurt-based silence adjustments principle. (b) Non-talk-spurt-based adjustments.

10.5.3 Voice Flow and Delay Variations Mapping

The packet flow and impediments are indicated in Section 10.3. In this section, voice and packet flow are directly mapped to the delay and jitter (delay variations). Figure 10.5 is a simplification of Fig. 10.3 and is represented with required blocks that relate to delay and jitter. The voice and packet flow is shown from phone-A to phone-B.

Voice transmitting through phone-A is compressed in the encoder. The encoder creates compressed payload. The payload as frames of data is packetized in RTP. From phone-A to the RTP input, the delays are of a fixed nature. Depending on the processor architecture and communication mechanisms, small delay variations could occur at a sub-millisecond level. This delay is negligible compared with other IP packet impediments. The block shown with networking creates complete packets for transmitting on the network. From RTP input to launching of the packet on the IP network, the packet goes through fixed delay and a variable delay of few milliseconds. The variable delay is maintained to be of the order of less than 5 ms. More details on this jitter at network interface are given in Chapter 18.

In the IP network, several routers and switches can create network impediments to the packet flow. IP network and network interfaces are the causes of major impediments in most situations. At the destination, the packet goes

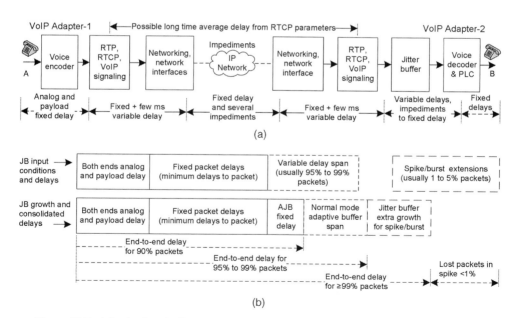

Figure 10.5. Adaptive jitter buffer end-to-end influences. (a) Delay and delay variation contributors mapped to VoIP voice call. (b) AJB buffer delays and end-to-end delay representation.

through networking and RTP blocks that introduce a few milliseconds of fixed and variable delay. Jitter buffer that interfaces RTP output has to create buffering to the variable delay packets. Jitter buffer converts all end-to-end variable delays to fixed buffer delay. Jitter buffer output is read from the decoder in a synchronous way at voice decompression frame intervals. Jitter buffer cannot take care of any of the voice encoder, decoder, analog front end, and loop length delay variations. Jitter buffer, when set properly, creates an end-to-end steady call to behave like a fixed long delay between phone-A and B.

In Fig. 10.5(b), delays are represented with tile boxes with text. The horizontal length represents the time, delay duration, or delay variation. The analog front end and both end payload creation delays are shown as fixed. All variable part of delay are grouped and shown in a dotted box. When variable delays are close to zero, jitter buffer will operate with minimum delay, which is usually of the order of 20 ms. When jitter exceeds the AJB minimum threshold, AJB keeps growing to accommodate all possible packets. The effort would be to transfer at least 99% of the available packets through jitter buffer.

The IP network characteristics or end-to-end jitter may change. Some adverse influences would be through a spike. A spike happens suddenly and behaves like huge jitter. Jitter buffer will try to grow, but it may limit the growth to reduce delay. In the process of delay optimization, a certain amount of packet discard (drop by jitter buffer) may be accepted. In most situations, packet drop in the jitter buffer adjustments is maintained to less than 1%. In some conditions, jitter buffer input may have several packet drops. Jitter buffer will try to deliver the best part of the available packets.

10.6 ADAPTING TO DELAY VARIATIONS

Figure 10.6 is an expanded version of Fig. 10.5(b) with a few more example steps of various jitter buffer algorithms and their role in arriving at optimal jitter buffer settings. Jitter buffer algorithms at a high level are expanded in the next section.

1. In Fig. 10.6(a), jitter is shown as very minimal and AJB will just stay with the minimum set delay.
2. In Fig. 10.6(b), large jitter is present that is more than the AJB minimum threshold. AJB will start increasing its threshold to accommodate almost all available packets. The jitter is marked as normal—which means jitter is not a sudden spike. In this situation, algorithms that estimate mean variation will serve the purpose of setting the playout or AJB buffer size. The buffer size usually keeps changing with time in the process of optimization of end-to-end delay. A small variation can be observed in the buffer growth as marked in Fig. 10.6(b).
3. In Fig. 10.6(c), a gap-based algorithm explained in Section 10.7.6 is marked. The algorithms based on mean value, variation and catering to

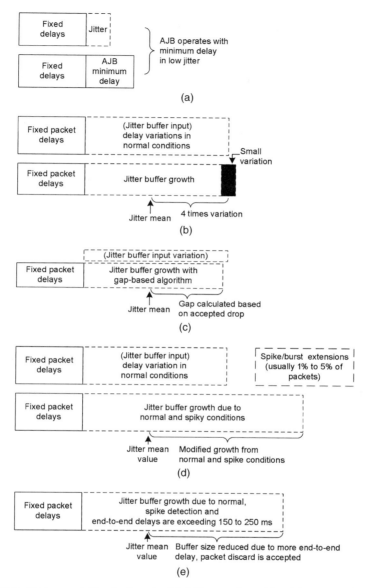

Figure 10.6. Basics on AJB adjustments. (a) AJB operating with minimum delay even with small jitter. (b) Jitter buffer adapting based on mean and variation—normal jitter will be taken care of. (c) Gap-based adjustment in spike. (d) Jitter buffer growth based on modified variation. (e) Applying trade-offs in delay and AJB buffer size.

spike conditions are not provisioned for monitoring what percentile of packets are played out in the AJB adjustments. The gap-based scheme works for the desired goals of packet drop. To meet the packet drop goals, jitter buffer may also increase or decrease significantly.

4. In Fig. 10.6(d), a large span of jitter is present that includes spike in certain intervals lasting for several hundred milliseconds. Jitter buffer algorithms will have spike detection, adapt to the sudden change, and recover back to normal conditions. The adaptation algorithm will have several conditions to deal with in the spike mode of operation.

5. In Fig. 10.6(e), jitter buffer growth is marked as limited to a certain limit. Jitter buffers keep growing at a penalty of increased end-to-end delay. End-to-end delay has to be reduced to meet other quality aspects. The importance of end-to-end delay goals are presented in Chapter 20. To reduce delay, a certain percentage of packet drop is also accepted, which results in reduced delay at the penalty of a tolerated packet drop.

10.7   AJB ALGORITHMS OVERVIEW

In the literature [Ramjee et al. (1994), Pinto and Christensen (1999)], several algorithms are explained based on the assumption—both transmitter and receiver are tracking the absolute timing information. Hence, both transmitter and receiver are mapped on the same time-scale reference. One such representation in relation to reference [Ramjee et al. (1994)] is given in Fig. 10.7. In this figure, the same symbols of [Ramjee et al. (1994)] are retained.

In Fig. 10.7, it is assumed that the sending and receiving sides maintain the same absolute timing. For a packet sent at time reference of t_i, it will arrive at a_i, stay in the jitter buffer, and be played out at time p_i. The packets will take a minimum fixed delay of D_{prop}. In normal situations, packets will arrive in the arrival limits of a_{imin} to a_{imax}. In theory, jitter buffer has to keep the span to the extent of $a_{imax} - a_{imin}$. In Fig. 10.7, playout p_i is shown to coincide with a_{imax}. When packets arrive at a delay of more than p_i, initially packets may be dropped and then jitter buffer keeps adapting to grow the jitter buffer. When the arrival time for a significant group of packets jumps (bursty delay), this can result in spiky conditions.

Jitter buffer size increases for spiky conditions, and this will result in an increase of end-to-end delay. When end-to-end delays are growing, the trade-offs would be to reduce delay at the cost of accepted packet drop. In practice, the drop and delay trade-offs have to be set in advance for algorithms to arrive at the best options. The playout delay $d_i = a_i - t_i + p_i - a_i = p_i - t_i$, should be in the range of $0 \leq d_i \leq 400$ ms. Hence, the playout time should be in the range to adjust the delay in adaptive jitter buffer algorithms $0 \leq p_i \leq 400 + t_i$.

As marked in Fig. 10.7, jitter buffer input may have a certain amount of packet loss or drop. While adapting to various conditions, jitter buffer can also discard some packets. Discard is also called a packet drop or jitter buffer drop.

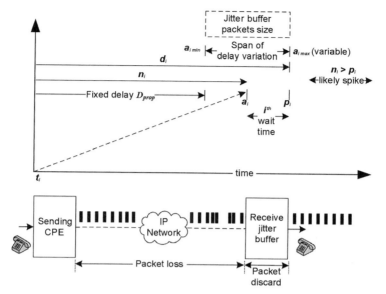

Figure 10.7. Packet delay variations and mapping to jitter buffer.

t_i: the time at which packet "i" is generated at the sending host. Sender time stamp of packet "i" as per sender's clock.

D_{prop}: the propagation delay from the sender to the receiver, which is assumed to be constant throughout the lifetime of an audio connection.

a_i: the time at which packet "i" is received at the receiving host (receiver time stamp of packet "i" as per receiver's clock).

p_i: the time at which packet "i" is played out at the output of jitter buffer.

d_i: the amount of time from the packet generated by the source until it is played out at the destination host, $d_i = p_i - t_i$. This will be referred to as the "playout delay" of the packet.

n_i: the "total delay" introduced by the network, $n_i = a_i - t_i$.

The use of discard is more correct in relation to the jitter buffer. In general, packet loss, drop, and discard are used interchangeably.

10.7.1 Playout Based on Known Timing Reference and Talk-Spurts

The network delay is the difference between the receiver (a_i) and the transmitter (t_i) time stamp. These time stamps are assumed to be derived with the same absolute timing reference such as time synchronized by NTP as given in [Schulzrinne et al. (2003), Mills (1992)]. The main inputs for the same timing reference calculations are t_i and a_i. The network delay to the packet n_i is calculated as $n_i = a_i - t_i$. In the estimation of playout delay, the mean delay estimate $\hat{d}_i$ for a packet i is given as $\hat{d}_i = \alpha \hat{d}_{i-1} + (1 - \alpha)n_i$ and the variation in the network delay is estimated as $\hat{v}_i = \alpha \hat{v}_{i-1} + (1 - \alpha)|\hat{d}_i - n_i|$. The playout time for packet "i" is defined as $p_i = t_i + \hat{d}_i + \gamma\hat{v}_i$ [Ramjee et al. (1994)]. In the talk-spurt-based scheme, assuming the packet marked "i" is the first packet of a talk-spurt, the playout time for the subsequent packets in a talk-spurt is computed as an offset from the point in time when the first packet in that talk-spurt was played out. The subsequent packet "j" in that talk-spurt can be computed as $p_j = p_i + t_j - t_i$, which implies that p_j is incremented at regular time intervals. The p_j is incremented until the end of the talk-spurt. During talk-spurt packets, the playout estimates are continuously made for every new packet, but packets are played out at regular intervals without any adjustments to the jitter buffer, which is indicated in Fig. 10.4(a).

In the above formulations, linear recursive filters are characterized by a weighting factor "α." The value for "α" used is 0.998002 [Ramjee et al. (1994)]. As the "α" reaches toward unity, the smoothing is better. The "α" value close to unity takes several seconds to settle playout. In practical implementations, "α" values up to 0.875 are used. The algorithms developed in [Ramjee et al. (1994)] differ in the calculation of $\hat{d}_i$, but $\hat{v}_i$ and p_i estimations are common across several methods.

The term $\gamma \, \hat{v}_i$ is used to set the playout time to be long enough beyond the delay estimate, so that only a small fraction of the arriving packets should be lost because of late arrival. The recommended value for γ is four that controls the synchronization and latency trade-off between sender and receiver.

If no interpacket jitter occurs, the variation part $\hat{v}_i$ of the above playout equation becomes zero and the value of mean playout delay $\hat{d}_i$ becomes the network delay (i.e., n_i), which means that as soon as the packet is arrived at the jitter buffer after experiencing the network delay, it will be played out immediately without buffering the packet. Low jitter conditions can use minimum jitter buffer. Alternatively, some minimum fixed jitter is applied even in the absence of any packet impediments as shown in Fig. 10.6(a).

10.7.2 Playout During Spike

A spike is a sudden, large increase in end-to-end network delay variation often less than one round-trip time, followed by a series of packets arriving almost simultaneously as burst, which leads to the completion of the spike.

In the playout algorithm description of previous sections, it can be observed that the playout delay algorithm changes slowly. When the spike suddenly increases the delay, the mean delay variation (MDV) algorithm reacts too slowly and fails to adapt to the new situation. The direct consequence is that the packet in this spike may be lost completely. The effect on listening is that the voice gets interrupted during the spike; hence, detection of a spike is very important to minimize degradation in voice quality. Detection of the spike has to check whether the new delay variation of the packets is large enough to be called a spike [Ramjee et al. (1994)], and the algorithm switches to spike mode. If the spike mode is reached, the $\hat{d}_i$ calculation is modified to $\hat{d}_i = \hat{d}_{i-1} + (n_i - n_{i-1})$.

With this formula under the spike condition, the jitter buffer can react to the network quickly. The formula to calculate variation and playout is the same. Therefore, the additional operation here is to decide the threshold. After entering the spike mode, the algorithm should know when the spike stops and quit to the normal mode. An experimental formula determines the end of a spike [Ramjee et al. (1994)].

$$var = var/2 + abs[2n_i - n_{i-1} - n_{i-2}]/8$$

The "var" tracks the slope of the spike. From practical observations, if the value of "var" is less than or equal to 63 ms [Ramjee et al. (1994)], then the algorithm returns to normal mode. For completeness of this topic, the summary steps of the spike detection algorithm from Ramjee et al. (1994) are given in below.

```
if (mode == NORMAL)
{
        if (abs(n_i - n_{i-1}) > abs(v̂_{i-1})×2 + 800)
        {
                var = 0; /* Detected beginning of spike */
                mode = SPIKE;
        }
}
else
{
        var = var/2 + abs[2n_i - n_{i-1} - n_{i-2}]/8
        if (var ≤ 63)
                mode = NORMAL
}
if mode = NORMAL
        d̂_i = α d̂_{i-1} + (1 - α) n_i
else
        d̂_i = d̂_{i-1} + (n_i - n_{i-1})
v̂_i = α v̂_{i-1} + (1 - α) abs | n_i - d̂_i |
n_{i-2} = n_{i-1},  n_{i-1} = n_i
```

This adaptive algorithm explicitly adjusts to the sharp spike-like increase in packet delay at a lower rate of packet drop. In the reference [Ramjee et al. (1994)], this delay is 800 ms. These thresholds have to be decided based on the parameters of the deployment. This threshold may be set to a lower value to cater to long bursts and short duration spikes.

10.7.3 Non-Talk-Spurt-Based Jitter Calculations

The interarrival jitter between two packets can be calculated independent of synchronization of sender and receiver clocks [Schulzrinne et al. (2003)]. The interarrival jitter can be computed from a sender/receiver report of RTCP

[Schulzrinne et al. (2003)]. The statistical variation of the RTP data packet interarrival time is measured in time-stamp units expressed as an unsigned integer. The time stamp reflects the sampling instant of the first octet in the RTP data packet. The interarrival jitter is defined as the mean deviation (smoothed absolute value) of the difference in packet spacing at the receiver compared with the sender for a successive pair of packets. As shown in the equation below, "D" is an instantaneous or interarrival jitter. Interarrival jitter is defined as the difference in the "relative transit time" for the two packets.

The relative transit time is the difference between a packet's RTP time stamp and the receiver's clock at the time of arrival, which are measured in the same units. Assuming S_i is the RTP time stamp from packet "i", and R_i is the time of arrival in RTP time-stamp units for packet "i", the relative transit time for the ith packet is $R_i - S_i$.

Instantaneous jitter is calculated based on the time stamps of two recent packets. For the arrived packets "$i - 1$" and "i", the instantaneous jitter is expressed as $D(i) = (R_i - R_{i-1}) - (S_i - S_{i-1}) = (R_i - S_i) - (R_{i-1} - S_{i-1})$. The instantaneous jitter does not have a mean value. The mean value denoted as $\hat{d}(i)$ is zero with this formulation of instantaneous jitter. Hence, the absolute value of $D(i)$ represented as $|D(i)|$ has to be used in the mean estimate. At packet "i," the mean estimate $\hat{d}_i = \alpha \hat{d}_{i-1} + (1 - \alpha)|D(i)|$ and the variation in the network delay are estimated as $\hat{v}_i = \alpha \hat{v}_{i-1} + (1 - \alpha)|(\hat{d}_i - |D(i)|)|$. The playout time for packet "i" without reference of transmit time is given as $p_i = \hat{d}_i + 4\hat{v}_i$. The interarrival jitter is calculated continuously as each data packet "i" is received from the same source. Before estimating the $D(i)$ value, duplicate packets are eliminated to avoid multiple updates. The algorithm uses the difference "D" for packet "i" and the previous packet "$i - 1$". The packets are considered in the order of arrival, and this is not necessarily in sequence. If no jitter occurs, the mean value $\hat{d}_i$ becomes zero as these values are computed relatively between two packets. In this method, no absolute network delay value n_i occurs. For characterizing the jitter buffer performance, it is essential to know on end-to-end delay.

10.7.4 Alternate Way of Estimating Network Delay

In literature [Ramjee et al. (1994)], symbol "n_i" is used for delay difference between sending and receiving. It is a combination of fixed and variable (jitter) delay. Variable delay is related to $D(i)$ in the previous calculations. Fixed delays cannot be recovered from RTP time stamps unless both timing references are known. A long-term average value of n_i is in RTCP or RTCP-XR packets. RTCP or RTCP-XR packets are sent at about 5% of the RTP packets count. The parameters average is usually over 256 RTP packets and lasts for few (typically 2.56) seconds. This duration is too long a duration for instantaneous jitter buffer adjustments and spreads into several talk-spurts. In practical implementation, n_i can be extracted from one-way delay available in RTCP and parameter $D(i)$. Jitter buffers can work at a near-perfection level even without n_i. What really maters to the jitter buffer is the jitter or delay variation. Fixed end-to-end delays are not required to keep jitter buffer length. The n_i plays an important role when the trade-offs are imposed between percentage of packet drop and end-to-end delay limits. If end-to-end delays are approaching close to 250 ms, it is suggested to consider trade-offs between small packet drops instead of increasing delay beyond 250 ms. It can be interpreted clearly in an R-model-based MOS calculation. The delay n_i is calculated from a "one-way delay" parameter from RTCP packets and the instantaneous jitter. Current arrival packet delay n_i = one-way delay + relative delay variation $D(i)$.

RTCP provides several timing parameters and packet delivery statistics. NTP is also used by RTP to derive timing relations. NTP gives time relative to 0-hours of coordinated universal time (UTC) on 1 January 1900 [Schulzrinne et al. (2003)]. The full-resolution NTP time stamp is a 64-bit unsigned fixed-point number with an integer part in the first 32 bits that represents relative time in seconds, and the last 32 bits are for the fractional part (fraction of a second). Use this time to arrive at the actual network delay.

10.7.5 Playout Time and Jitter Buffer Size

The playout time is the packet delivered at the output of the jitter buffer. It does not directly convey the jitter buffer size. The playout time remains constant during each talk-spurt. In the absence of talk-spurt detection, the playout time will remain the same for the duration of the selected update time window. When the playout time is steady and input packets arrive at variable delays, the following basic observations can be made.

1. Packet arriving at lowest delay of D_{prop} will stay for a long time in the jitter buffer.
2. Packet arriving just before the playout time will stay for the least amount of time in the jitter buffer.
3. Packet arriving anytime after playout delay will be lost, usually discarded by the jitter buffer. This information may be used in arriving at the next talk-spurt playout time.
4. For the playout time, the maximum jitter buffer size can be calculated as $JB\_max = p_i - a_{i\_min}$. Here, $a_{i\_min} = D_{prop}$. Users will refer to this jitter buffer size for a particular talk-spurt or block of packets in the non-talk-spurt-based scheme.

From the above points, the jitter buffer size varies based on the arrived packet. In the literature, playout time is denoted with the symbol p_i. To avoid confusion, some clarifications are given here. The timing parameter p_i is not representative of the size of the jitter buffer duration or packets. If the travel time of packets from source to jitter buffer input is 40 ms and stays in jitter buffer for 20 ms, the playout time is 60 ms. In practical implementation, the user will be concerned with how many packets are required to be kept in a jitter buffer while giving the voice packets to the decoder. Fixed delays may vary based on several parameters beyond the control of the receiver VoIP terminal and jitter buffer.

In some recent deployments, end-to-end jitter is low and packet drops are less than 1%. When jitter values are from 10 to 40 ms, the simplest approach would be using FJB of 20 to 40 ms or AJB set with maximum deviations of $D(i)$ jitter. Some guidelines specific to playout delay are given below.

- If jitter is lower (approximately 20 to 40 ms), in most situations, going by simple maximum jitter as buffer size will be sufficient. In general, some smoothing and exponential averaging techniques are used in arriving at some thresholds given in RFC0793 [Postel (1981)].
- In case of higher jitters, the adaptive algorithm has to take care of optimal buffer sizes, spiky detections, spiky recovery, and trade-offs in delay and packet drop through monitoring of the overall statistics. In Ramjee et al. (1994) and other literature, delay and variation are used for arriving at the playout buffer, which takes care of most situations. Monitoring packet drop and linking to delay and buffer size will be helpful for better voice quality.

10.7.6 Gap-Based Playout Estimation

In talk-spurt-based algorithms, the playout time during normal mode and spike situations is given in the previous sections. The mean deviation and variation adaptive methods [Jacobson and Karels (1988), Ramjee et al. (1994)] were designed to target the packet loss rate under 5% while maintaining a minimum end-to-end delay. These algorithms were not monitoring the different desired levels of voice quality by setting different packet loss tolerances. In normal mode, the gap-based scheme is popular for use in practical implementations. This algorithm works as closed-loop operation and makes use of the packet performance of the previous normal mode of talk-spurts. During spike mode, the gap-based algorithm will not update its playout parameters. The gap-based scheme helps in providing the feedback to achieve the desired performance [Pinto and Christensen (1999)].

The gap-based algorithm computes the difference between the packet's playout time and its arrival time, and it is called gap. Gap for packet "i" in talk-spurt "k" is denoted as $\text{Gap}_i^k = p_i^k - a_i^k$. Gap is positive for correctly played packet $a_i^k \le p_i^k$ and negative for discarded packet $a_i^k > p_i^k$. Assuming "n_k" packets are present in talk-spurt marked as k, the gap algorithm decides on what is the gap in the playout threshold to achieve accepted packet drop. This gap is adjusted in the next talk-spurt. The gap values are calculated for all packets of the most recently finished talk-spurt to adjust the playout time for the next talk-spurt.

Figure 10.8 is given for the gap-based algorithm. In Fig. 10.8(a), the playout time is low and more packets are lost than accepted. The playout time, accepted levels, and new playout time are marked with text. On a new talk-spurt, the adjusted playout time will be applied. This playout time increase introduces extra silence before starting the next talk-spurt. In Fig. 10.8(b), the playout time is more, and all the packets arrived at the input of the jitter buffer are played out. This type of operation is helpful, but end-to-end delay will increase. The playout time is reduced to the extent of acceptable packet loss level. Figure 10.8(c) is shown with gap-based adjustments and relating this to end-to-end delay. Packet loss is a more severe degradation in VoIP. End-to-end delays of 150 to 250 ms are usually acceptable in deployments. When one-way end-to-end delays are lower than the delay goals, effort should be on playing all the available packets. In practice, users will impose a list of precomputed

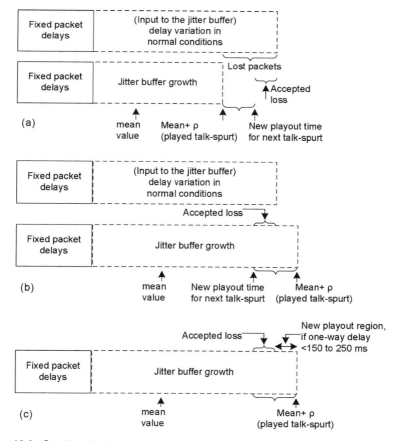

Figure 10.8. Gap-based playout time. (a) The playout time is low and more packets are lost than accepted. (b) The playout time is more, and more delay is introduced than required. (c) Making delay dependency in addition to the gap-based adjustments.

tables that give values for delay and packet loss acceptance. As a practical suggestion, if the playout time is changing by a few milliseconds and one-way delays are in an acceptable range, it is suggested to keep a slightly higher threshold and allow it to stay there instead of adjusting rapidly.

Gap-based algorithm steps are given in [Pinto and Christensen (1999)]. The gap-based adjustment replaces the variation part of the playout calculation of previous methods. Given a set of ordered network delays for talk-spurt $k - 1$, the optimum gap with required packet loss tolerance is computed as

$$\text{Opt}_{\text{gap}} = \text{Gap}\left(\sum_{i=0}^{(1-\text{toler})(n_{k-1})} \text{freq}_i \right)$$

where toler is the tolerance to packet loss specified by the user application given in the range $[0, 1]$, with 0, meaning no losses are tolerable and 1 meaning 100% losses are tolerable. The total number of packets that have to be played in the talk-spurt to maintain the desired packet loss rate is $(1 - \text{toler})(n_{k-1})$. For example, considering 5% loss for 100 packets of talk-spurt, the above value becomes 95 and needs to consider these packets for the delay of the packets to calculate the gap. Gap( ) is the operator that retrieves the gap corresponding to the entry "i" of the set. The freq_i is the number of packets that experience the same network delay and thus the same gap.

As given in Pinto and Christensen (1999), the optimum theoretical playout delay is a per-talk-spurt quantity defined as the minimum amount of delay that would have been required to add to the production time of each packet belonging to a finished talk-spurt to playout the talk-spurt at exactly the desired packet loss rate. This quantity is not known until the talk-spurt is finished and all of the networks delays and gaps are calculated.

If each packet of the talk-spurt experiences different network delay, the freq_i is equal to one and such network delays are ordered for those many numbers of packets in the talk-spurt to extract the gap. The equations below summarize the playout delay adjustment performed during "normal" operation. This optimum gap is used in updating gap parameter "ρ" that is used in updating the playout. The parameter "ρ" is updated for every talk-spurt of block of frames. The other formulations' mean estimation, playout time, and continuous playout time for the first and subsequent packets within the talk-spurt will be similar to previous algorithms [Pinto and Christensen (1999)].

$$\rho = \rho + \text{Opt}_{\text{gap}}^{k-1}$$
$$p_1^k = t_1^k + \hat{d}_i^k + \rho$$
$$p_i^k = p_1^k + (t_i^k - t_1^k)$$

Designers may use several variants of gap-based implementation when statistics are available from RTCP and RTCP-XR. The gap-based algorithm can be a simple playout time increase and decrease at a certain adjustment rate. In all these algorithms, packet drop and packet discard have to be identified clearly. As an example, when the input of the jitter buffer is with 2% packet drop, it is not possible to reach a goal of 1% drop to calculate on gap. It is required to consider available packets at the input of the jitter buffer for estimating the gap. While imposing the trade-offs in buffer size, one-way delay and packet drop are useful parameters to consider drop and discard combined influences.

In non-talk-spurt-based gap methods, packet estimates can be made at regular intervals on a block of packets. The estimate blocks may be once in 0.5 to 5 seconds based on the jitter characteristics. In the literature, it is reported [McNeill et al. (2006)] that wireless applications will need quicker updates than wire-line applications.

10.8 ADAPTIVE JITTER BUFFER IMPLEMENTATION GUIDELINES

The following summary lists the most popular conditions and guidelines followed in designing AJB. This section is an expansion of Table 10.1. In a well-behaved end-to-end packet transmission, several design points will be relaxed in AJB and will behave similiarly to a fixed buffer.

- Jitter buffer has to remove duplicate packets with minimal processing effort.
- A dropped packet before reaching jitter buffer input cannot be recovered inside a jitter buffer. In case of redundancy, a forward error correction (FEC), or duplicate packets availability, algorithm has to maintain a best effort to substitute dropped packets. Catering to such substitutions may create extra end-to-end delay. Jitter buffer giving a dropped packet (meaning null packet to the decoder) can only be enhanced through PLC algorithm that is present as part of a voice chain decoder.
- Jitter buffer on its own should minimize any additional packet drops.
- Jitter buffer has to maintain a stationary pattern of speech and drop packets. If the drop duration is for 20 ms at the input of the jitter buffer, output should also maintain the same drop duration. It should not remove or insert extra silence/null packets to get the best out of PLC.
- Most algorithms of AJB adjust packets on an as-needed basis. Any adjustment made in the middle of speech is harmful to the speech quality. As an improvement, jitter buffer should also look at the significance of speech conditions before adjustments. AJB should maintain a best effort to adjust packets in between the talk-spurts for introducing or removing silence durations. Talk-spurt consists of a bunch of speech frames with very short breaks in speech. Between two talk-spurts, several 100-ms silence zones will occur. AJB should try to adjust packets during the silence zones, which calls for talk-spurt and silence zone detection as part of voice chain modules.

- If VAD is used as part of voice compression, talk-spurt detection algorithms can also make use of VAD packet boundaries.

- In the absence of silence-sensing-based adjustments, it can readjust at a slower rate of 1 in 100 seconds. Burst adjustments should not be attempted. An intentionally created drop rate has to be maintained to less than 1%. For a 1% drop a packet drop or excess silence is created once in a second (1 silence or drop out of 100 packets). Once again talk-spurt silence is preferred to dropping a speech packet or creating silence between the talk-spurts.

- In the case of an under-run condition, the playout time should not continue in silence when receiving a new valid packet for playout. This helps in maintaining the stationary pattern for packet drop. If minimum playout depth has to be maintained after under-run, this has to be built at a slower rate.

- In the case of overflow, old packets, preferably silence durations, have to be removed to minimize the end-to-end delay.

- In a voice call between two adapters, an excess packet or packet starvation can happen depending on the relative PPM. It happens very slowly. A worst-case PPM of 100 is possible in hardware-based VoIP adapters. One hundred PPM demands a packet readjustment at the rate of 1 in 10,000, (i.e., one packet of 10 ms in 100 seconds).

- Jitter buffer and RTP events have to be timed at a resolution comparable with 1 ms. A basic frame-based time stamp of 10 ms could be used as a worst case. A 10-ms time resolution can cause an additional 10 ms of jitter and result in a 10-ms end-to-end delay increase.

- Jitter buffer output works synchronously with voice chain processing. Jitter buffer input works in coordination with RTP and the received packet.

- Jitter buffer possesses spike detection. If the jitter buffer lower threshold is lower than the spike duration, the gap-based scheme can also help monitor the packets and provide additional correction.

- Spike increases the delay. The playout delay should decrease in an appropriate rate after the spike.

- While keeping the trade-offs, packet drop is to be minimized even at the cost of increase in end-to-end delay. If buffer size grows and makes end-to-end delays to exceed 150 to 250 ms, packet drop up to 5% is considered while allowing end-to-end delays up to 400 ms. The trade-offs in packet drop and delay are dependent on a selected compression codec. These estimations can be arrived through E-model delay and effective impairments given in Chapter 20.

- VAD packets should not be dropped inside a jitter buffer at any time; however, duplicate packets handling can be used with VAD packets.

- Jitter buffer input should be able to handle multiple packetization intervals up to 80 ms and has to cater to various codec basic frame durations.
- For modem and fax pass-through, AJB should be made to work like FJB by disabling adaptation logic.

10.9 FIXED JITTER BUFFER IMPLEMENTATION GUIDELINES

FJB has to work very similarly to AJB with the main difference of lack of provision to track network conditions. In actual implementation, AJB is made to work like FJB with disabling of adapting tracking of playout time.

- Jitter buffer to remove duplicate packets with minimal processing.
- The jitter buffer lower threshold is usually from 20 to 70 ms, and the upper limit is 400 ms for voice calls. The upper threshold has to be greater than or equal to the expected spike.
- The jitter buffer lower threshold may be increased for fax and modem pass-through calls.
- Jitter buffer has to maintain a stationary pattern for packet drops and VAD silences.
- Jitter buffer should be prevented from complete under-run. Before under-run occurs, possible silence zones need to be created at a rate of 1 in 10,000 packets (meaning one 10-ms packet for every 100 seconds). Here also preference has to be given based on talk-spurt detection.
- Jitter buffer should be prevented from over-run. Before reaching the over-run, old packet removal needs to be created at a rate of 1 in 10,000 packets (meaning one 10-ms packet for every 100 seconds). Here also preference has to be given to talk-spurt detection.

11

VoIP VOICE—NETWORK BIT RATE CALCULATIONS

In VoIP deployment, the VoIP network bit rate, which is also referred as to the bandwidth requirement, is higher than the actual compression rate of the codec. It is important to know the actual network bit rate to cater to end-to-end planning and to arrive at recommendations for the right Internet service. This chapter describes the bit rate requirements for VoIP voice services with G.711, G.729A, and G.723.1 compression codecs [ITU-T-G.711 (1988), ITU-T-G.729A (1996), ITU-T-G.723.1 (2006), Kondoz (1999), Goldberg and Riek (2000), Hersent et al. (2005)] on Ethernet, digital subscriber line (DSL), and cable-based Internet services. In addition to bit rate requirements, service providers will consider several quality and interoperation aspects to take care of wider end-to-end deployment requirements.

VoIP is considered a network-bit-rate-saving voice communication compared with the public switched telephone network's (PSTN's) 64-kilobits per second (kbps) rate, which is not always true. The most popular G.729A compresses voice to 8 kbps, but it takes 70.4 kbps on the Ethernet and 84.8 kbps on the DSL interface with 10-ms frames, which is more than PSTN's 64 kbps. Internet Protocol (IP) packet headers and extra bytes from IP network interfaces are causing the 8-kbps G.729A to increase to a much higher rate of 84.8 kbps. In this chapter, bit rate calculations with interpretation of headers, examples with codec payloads, calculations, and summary tables for Ethernet, DSL, and cable interfaces are given.

VoIP Voice and Fax Signal Processing, by Sivannarayana Nagireddi
Copyright © 2008 by John Wiley & Sons, Inc.

11.1   VOICE COMPRESSION AND BIT RATE OVERVIEW

In VoIP, voice samples are collected from a telephone interface at an 8-kHz sampling rate. These samples are interfaced to the processor using a pulse code modulation (PCM) time division multiplexing (TDM) serial interface. Every sample on the PCM interface of the VoIP system uses either 16-bit linear or 8-bit A/μ-law. Voice modules operate on linear samples, and logarithmic A-law or μ-law 8-bit samples are converted to linear 16-bit samples. Modules such as echo cancellation, G.711, G.729AB, and G.723.1 compression process linear samples. For every linear input sample of 16 bits, G.711 gives an 8-bit μ-law or A-law sample. G.729A operates on a block of 80 sample (10-ms duration) frames and gives 10 bytes of compressed payload at an 8-kbps compression rate. G.723.1 operates on a block of 240 sample (30-ms duration) frames and delivers 24 bytes (6.3 kbps) and 20 bytes (5.3 kbps) depending on the selected rate.

A compressed voice frame is created into a packet with required headers to ensure end-to-end delivery on an IP network. As an example, G.729A takes 80 samples of input and gives 10 bytes of compressed frame in 10 ms. On the Ethernet interface, 10 bytes of basic payload is created as a total of 88 bytes. On the DSL interface with point-to-point protocol over Ethernet (PPPoE), the same 10 bytes of G.729A will use a total of 106 bytes to make the payload utilization 9 (10/106 = 9) percent. The headers on top of compressed payload are much bigger than the actual payload in many combinations of VoIP voice packets delivery on physical interfaces. To get better utilization, more compressed voice frames are grouped together into one packet, which improves the bit rate utilization but increases the end-to-end delay. It is also common practice to use bandwidth in place of bit rate. Data path members use both bandwidth and bit rate to present bits per second. Signal processing members may first place map bandwidth as frequency-specific information in Hz. This chapter is on the network-specific bit rate and conveys bits per second and not frequency in Hz.

In VoIP, packet intervals up to 80 ms (10, 20, 30, 40, 50, 60, 70, and 80 ms) are considered, but packet sizes of 10, 20, and 40 ms are most common in the actual deployment. For lower delay between end-to-end operations, 10- or 20-ms packetization is used as the preferred packetization. G.711 can operate on a sample basis. For every compressed sample of 1 byte, the Ethernet and VoIP header of 78 bytes gives a payload utilization of 1.27 (1/79 = 1.27%) percent on the Ethernet interface, which is not efficient. As a trade-off on bit rate and end-to-end delays, some deployments catering to 5-ms packetization with G.711, but 20 ms is most commonly used as per RFC3550 [Schulzrinne et al. (2003)].

The minimum packet interval in G.723.1 is 30 ms. Codec G.723.1 may also use packetization of 60 ms, but for maintaining better quality through lower end-to-end delay, 30 ms is preferred with G.723.1. For a summary on codecs, packetization, and bit rate combinations without going through a lot of detail, refer to the last section "Summary of VoIP bit rate for G.711, 729A, and 723.1 codecs." The detailed bit rate calculations with header level information are given in the next sections.

11.2 VOICE PAYLOAD AND HEADERS

A compressed voice frame is required to be packetized with Real-Time Transport Protocol (RTP), User Datagram Protocol (UDP), and IP headers and then encapsulated with network interface headers [Schulzrinne et al. (2003), Postel. (1980), URL (RFC791)].

The RTP header is 12 bytes. Voice is sensitive to delays. RTP helps proper end-to-end delivery of real-time voice traffic. RTP header compression reduces the number of bytes, but header compression is not considered in this book. Details on RTP are given in Chapter 10 and are available in RFC3550 [Schulzrinne et al. (2003)].

The UDP header is of 8 bytes as given in RFC0768 [Postel (1980)]. UDP offers a basic transport service of a 2-byte source port, 2-byte destination port, 2-byte length, and 2-byte checksum. The real-time aspects of transport are taken care of by RTP.

The IP header is 20 bytes in IP version 4 (IPv4) as given in RFC0791 [URL (RFC791)]. In IP version 6 (IPv6), the IP header is 40 bytes.

Compressed payload, RTP, UDP, and IP header combinations are called VoIP packets. A basic small block of samples used in voice compression is called a frame. In G.729A, 10 ms (80 samples) is the basic frame, and in G.723.1, 30 ms (240 samples) is the basic frame. Voice payload may use a group of compressed frames up to 80 ms. The group of frames is called VoIP voice raw payload or VoIP packet payload. An IP packet consisting of IP, UDP, RTP, and compressed voice payload is illustrated in Fig. 11.1. These payload and VoIP headers are encapsulated with many other headers for end-to-end delivery through physical interfaces such as Ethernet, DSL, and cable. Simple arithmetic to represent bit rate calculations is given below.

$$\text{VoIP header} = (\text{IP} + \text{UDP} + \text{RTP}) = 40 \text{ bytes in IPv4 and 60 bytes in} \\ \text{IP version 6 (IPv6)}$$

$$\text{VoIP packet} = (\text{VoIP header} + \text{voice payload})$$

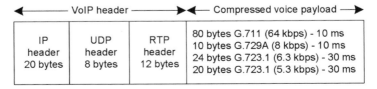

Figure 11.1. VoIP packet format with voice payload and headers.

$$\text{Physical network VoIP packet} = \text{Network interface headers}$$
$$+ (\text{VoIP header} + \text{voice payload})$$

Voice payload varies with compression codec, payload duration, and compression rate options.

In some books, bit rate (usually with name bandwidth) is calculated based on the VoIP packet without including the network interfaces. As an example, for G.711 of 80 bytes in a 10-ms frame, bit rate is considered as 120 bytes of 100 packets per second, (i.e., $100 \times 120 \times 8 = 96\,\text{kbps}$). The actual bit rate requirements are more than 96 kbps on physical interfaces.

11.3   ETHERNET, DSL, AND CABLE INTERFACES FOR VoIP

In this chapter, Ethernet, DSL, and cable interface options are considered for VoIP bit rate calculations. Figure 11.2(a) shows the VoIP adapter or customer premises equipment (CPE) with an Ethernet interface. This type of combination is referred to as a VoIP CPE on a local area network (LAN) interface. This Ethernet interface is terminated on the distributed LAN in an office or to a service provider's modem such as DSL or cable modem. VoIP CPE will send Ethernet packets, and the DSL modem will perform disassembly of the Ethernet packet and frames into asynchronous transfer mode (ATM) cells (as one of the options considered in this book) for sending on the DSL interface.

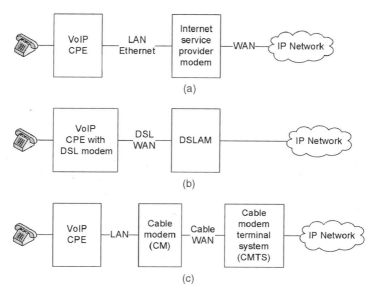

Figure 11.2. VoIP on physical interfaces. (a) VoIP on an Ethernet interface. (b) VoIP on a DSL interface. (c) VoIP on a cable interface.

In Fig. 11.2(b), VoIP CPE has a built-in DSL modem. The VoIP application gives a basic payload, and the DSL interface will send voice packets as ATM cells usually with PPPoE. In some deployments, PPP over ATM (PPPoA) [Gross et al. (1998), Heinanen (1993)] and IP over ATM (IPoA) [Laubach and Halpern (1998)] may be used. PPPoA and IPoA consume a lower bit rate than PPPoE. In this chapter, four different interface combinations are considered for bit rate calculations with G.711, G.729A, and G.723.1 codecs.

1. VoIP voice packets on an Ethernet interface
2. VoIP voice packets on an Ethernet interface with VLAN tagging [URL (IEEE-802.1Q), McPherson and Dyxes (2001)]
3. VoIP voice packets on a DSL interface with PPPoE [Mamakos et al. (1999)]
4. VoIP voice packets on a cable interface

11.3.1 VoIP Voice Packets on an Ethernet Interface

In Fig. 11.2(a), VoIP CPE is used with a telephone and an Ethernet interface. An Ethernet interface joins with an office LAN, Internet service provider's modem such as DSL, and cable in residential applications as shown in Fig. 11.2(a). Voice samples are compressed and created as raw payload. RTP, UDP, and IP encapsulate this raw payload with multiple parameters.

An Ethernet interface adds a total of 38 bytes—namely 8 bytes of Ethernet preamble, 14 bytes of Ethernet header, 4 bytes of cyclic redundancy check (CRC) and 12 bytes of transmission gap. The Ethernet packet high-level details are given below [URL (OSU-Eth), URL (National-Eth)]:

Preamble: 7 bytes of alternating ones and zeros for synchronization, followed by 8-bits of "10101011."

Ethernet header: 6 bytes of destination (dest) address, 6 bytes of source (src) address, and 2 bytes of length/type of payload.

VLAN tag: VLAN is a virtual LAN that makes use of 4 bytes of header also referred to as 802.1Q tagging.

Actual voice/data payload: For VoIP, this includes IP, UDP, RTP, and compressed voice payload as marked in Fig. 11.1.

CRC: 4 bytes of CRC is used for checksum.

Transmission gap: 12 bytes of transmission silence gap between each Ethernet frame.

In Fig. 11.3(a), an example is given for the G.711 codec at a 10-ms frame with Ethernet network interface headers of 38 bytes.

Network VoIP packet total bytes = 38 + (40 + 80) = 158 bytes.

Network bit rate = (8) (total bytes) (1000)/(packet interval in ms). This can also be expressed as

Ethernet preamble	Ethernet header	IP/UDP/ RTP header	G.711-10 ms voice payload	CRC	Transmission gap
8 bytes	14 bytes	40 bytes	80 bytes	4 bytes	12 bytes

Total bytes in G.711-10 ms packet = 158 bytes
Total bit rate from 100 packets per second = 126.4 kbps
Percentage of payload utilization = (100 × 80)/158 = 51

(a)

Ethernet preamble	Ethernet header	VLAN header	IP/UDP/ RTP header	G.711-10 ms voice payload	CRC	Transmission gap
8 bytes	14 bytes	4 bytes	40 bytes	80 bytes	4 bytes	12 bytes

Total bytes in G.711-10 ms packet = 162 bytes
Total bit rate from 100 packets per second = 129.6 kbps
Percentage of payload utilization = (100 × 80)/162 = 49

(b)

Figure 11.3. Voice packet format examples for the G.711 codec on an Ethernet interface. (a) Without VLAN header. (b) With VLAN tagging.

$$\text{Network bit rate} = (8)(\text{total bytes})(\text{Packets per second})$$
$$= (8)(158)(1000)/10 = 126,400 = 126.4 \, \text{kbps}$$

Percentage of payload utilization = 100 (voice payload 80 bytes)/(total 158 bytes) = 51%

Table 11.1 is listed for three codecs for different packet intervals. The results are noted as total bit rate in kbps and percentage of payload utilization. In Table 11.1, the G.711 (10 ms) column is the same as in Fig. 11.3(a) with some headers collapsed into a single header. For any other codec, it is required to calculate compressed total payload and number of packets per second. Total bit rate and percentage payload utilization can be calculated with simple arithmetic as per the details marked in Fig. 11.3(a).

11.3.2   VoIP Voice Packets on Ethernet with VLAN

As shown in Fig. 11.2(a), VoIP CPE with an Ethernet interface may use VLAN tagging. VLAN tagging helps prioritize the voice packets. VLAN tagging incorporates an additional 4 bytes in headers as shown in Fig 11.3(b). It is also called 802.1Q support. Engineers usually combine this support with IPQoS implementation and may not specify it explicitly. Table 11.2 is given for three codecs and different packet intervals. The results are listed as percentage payload utilization and the total bit rate in kbps.

Table 11.1. VoIP Voice Packet Bit Rate on an Ethernet Interface

Description	G.711				G.729A				G.723.1	
Compression rate	64 kbps				8 kbps				6.3 kbps	5.3 kbps
Packet interval in ms	10	20	30	40	10	20	30	40	30	30
Voice payload bytes	80	160	240	320	10	20	30	40	24	20
VoIP header	40	40	40	40	40	40	40	40	40	40
Ethernet interface header	38	38	38	38	38	38	38	38	38	38
Total bytes per voice packet	158	238	318	398	88	98	108	118	102	98
Total packets/ second	100	50	33.3	25	100	50	33.3	25	33.3	33.3
Total bits/ second (kbps)	126.4	95.2	84.8	79.6	70.4	39.2	28.8	23.6	27.2	26.1
Percentage payload utilization	51	67	75	80	11	20	28	34	24	20

Table 11.2. VoIP Voice Packet Bit Rate on an Ethernet Interface with VLAN Tagging

Description	G.711				G.729A				G.723.1	
Compression rate	64 kbps				8 kbps				6.3 kbps	5.3 kbps
Packet interval in ms	10	20	30	40	10	20	30	40	30	30
Payload bytes	80	160	240	320	10	20	30	40	24	20
VoIP header	40	40	40	40	40	40	40	40	40	40
Ethernet interface header with VLAN	42	42	42	42	42	42	42	42	42	42
Total bytes per packet	162	242	322	402	92	102	112	122	106	102
Packet per second	100	50	33.3	25	100	50	33.3	25	33.3	33.3
Total bits/ second (kbps)	129.6	96.8	85.9	80.4	73.6	40.8	29.9	24.4	28.3	27.2
Percentage payload utilization	50.6	66	74.5	79.6	10.9	19.6	26.8	32.8	22.6	19.6

11.4 VoIP VOICE PACKETS ON A DSL INTERFACE

In DSL interface-based VoIP, IP packets are sent as ATM cells as one of the popular options. The ATM cell size is 53 bytes with 48 bytes of payload and 5 bytes of ATM header error control (HEC) as given in reference [URL (Protocols-ATM)]. In the process of creating ATM cells, ATM adaptation layer 5 (AAL5) protocol data unit (PDU), segmentation and reassembly (SAR), and 8 bytes of AAL5 trailer are used. AAL5 will use headers depending on the supported protocols such as PPPoE, PPPoA, IPoA, and 8-byte trailer. The packet is then zero padded such that the packet size consists of integer multiples of 48 bytes. This process is required to split the packet into an integer number of ATM cells.

In this chapter, calculations for PPPoE are given, as PPPoE is popular in actual deployments. It is most popularly used with the DSL interface. Voice packets sent on PPPoE on a DSL interface will have more overhead bytes compared with PPPoA and IPoA. In general, many other supporting standards and RFCs exist for work on DSL and supported protocols. In this chapter, only header sizes are considered without going through the details of bit fields. Refer to the RFCs—RFC2516 [Mamakos et al. (1999)], RFC2364 [Gross et al. (1998)], RFC2225 [Laubach and Halpern (1998)] for additional details on the protocols and headers.

11.4.1 VoIP on a DSL Interface with PPPoE

In G.711 of a 10-ms packet interval, the VoIP packet (compressed voice payload + RTP/UDP/IP) size is 120 bytes. As shown in Fig. 11.4(a), the PPPoE header is 8 bytes, Ethernet header is 14 bytes, Ethernet over ATM header (RFC1483) is 10 bytes, and trailer is 8 bytes, which are added to the 120 bytes to make it a total of 160 bytes. These bytes are split into four cells by making use of $4 \times 48 - 160 = 32$ zero padding. Each ATM cell uses an extra 5 bytes of header on top of 48 bytes to make it a total of $4 \times 53 = 212$ bytes. In one second, 100 such basic packets (or 400 ATM cells in this example) are created. The total bit rate in G.711 = 212 bytes $\times$ 100 packets/s $\times$ 8 bits/byte = 169.6 kbps. In general, sending voice IP packets over ATM with PPPoE requires more overhead compared with VoIP on Ethernet. Table 11.3 is given for three codecs and different packet intervals. The results are presented as percentage payload utilization and the total bit rate in kbps.

11.5 VoIP VOICE PACKETS ON A CABLE INTERFACE

As shown in Fig. 11.2(c), VoIP CPE is used with a cable interface. An Ethernet interface joins with a cable modem in residential applications. Both VoIP CPE and a cable modem can be on the same box. The data-over-cable system has two streams: cable modem (CM) to CM terminal system (CMTS) as the

Ethernet over ATM	Ethernet header	PPPoE	VoIP header	G.711 voice payload	Zero padding	Trailer bytes
10 bytes	14 bytes	8 bytes	40 bytes	80 bytes	32 bytes	8 bytes

Zero padded with 32 bytes for 4 cells of 48 bytes ATM cells = 4, total bytes in ATM cells = 4×53 = 212 Bit rate = 100×212×8 = 169.6 kbps Percentage of payload utilization = (100 × 80)/212 = 38

(a)

FGPS	MAC OH	BPI	HCS	Ethernet header	VoIP header (IP,UDP,RTP)	G.711-10 ms voice payload	CRC
14-34 bytes	4 bytes	5 bytes	2 bytes	14 bytes	40 bytes	80 bytes	4 bytes

Total bytes on Upstream from CM = 183 (with 34 bytes FGPS) Bit rate = 100×183×8 = 146.4 kbps Percentage of payload utilization = (100 × 80)/159 = 43

(b)

←——Eth header——→

MAC OH	BPI	HCS	DA	SA	T	VoIP header	G.711 voice payload	CRC
4 bytes	5 bytes	2 bytes	6 bytes	6 bytes	2 bytes	40 bytes	80 bytes	4 bytes

(c)

Figure 11.4. Packet format examples with G.711-10 ms on DSL and cable interfaces. (a) DSL with PPPoE. (b) Cable upstream is with 34 bytes of FGPS calculation. (c) Cable downstream.

upstream direction, and CMTS to CM through the cable network as the downstream direction (traffic toward the CPE). In the data-over-cable system, the services used for VoIP are Unsolicited Grant Service or Unsolicited Grant Service with activity detection to support real-time service flows that generate fixed-size data packets (like voice) on a periodic basis as given in the DOCSIS 1.1 document [DOCSIS 1.1 (2005)]. Figure 11.4(c) shows a normal RTP packet carried on a downstream channel. It is similar to an upstream channel packet except for the physical layer overhead.

Figure 11.4(b) shows a normal RTP packet carried on an upstream channel. In this section, an overview with example headers is given. More details are available in the DOCSIS 1.1 document. The header level description of upstream and downstream packets is given here:

FGPS: Physical layer overhead F-FEC, G-Guard Time, P-Preamble, S-Stuffing bytes. The physical layer overhead varies from 14 to 34 bytes and is included in the upstream direction from CM to CMTS. In this chapter, 34 bytes are considered in the example calculations.

Table 11.3. VoIP Voice Packet Bit Rate on a DSL Interface with PPPoE

Description	G.711				G.729A				G.723.1	
Voice compression rate	64 kbps				8 kbps				6.3 kbps	5.3 kbps
Packet duration in ms	10	20	30	40	10	20	30	40	30	30
Payload bytes	80	160	240	320	10	20	30	40	24	20
VoIP header	40	40	40	40	40	40	40	40	40	40
Ethernet over ATM	10	10	10	10	10	10	10	10	10	10
Ethernet header	14	14	14	14	14	14	14	14	14	14
PPPoE	8	8	8	8	8	8	8	8	8	8
Zero padding	32	0	16	32	6	44	34	24	40	44
ATM trailer	8	8	8	8	8	8	8	8	8	8
AAL5 cells	4	5	7	9	2	3	3	3	3	3
Total bytes per packet	212	265	371	477	106	159	159	159	159	159
Packets per second	100	50	33.3	25	100	50	33.3	25	33.3	33.3
Total bits/ second (kbps)	170	106	98.9	95.4	84.8	63.6	42.4	31.8	42.4	42.4
Percentage payload utilization	38	60	65	67	9	13	19	25	15	13

MAC OH: MAC layer overhead of 6 bytes including a MAC header of 4 bytes and MAC header checksum (HCS) of 2 bytes.

BPI: 5 bytes of Baseline Privacy Interface—Extended header for MAC to provide cable modem users with data privacy across the cable network. It encrypts traffic flows between the CM and the CMTS.

Ethernet header: 6 bytes of destination address (DA), 6 bytes of source address (SA), and 2 bytes of length/type (T) of payload.

VoIP header: (IP + UDP + RTP) headers = 40 bytes in IPv4 and 60 bytes in IPv6.

Voice payload: 10 bytes to 160 bytes depending on the speech codec used.

CRC: 4 bytes of CRC for checksum.

In Fig. 11.4(b), an example is given for the G.711 codec at a 10-ms frame with cable interface headers totaling 63 bytes in the upstream direction assuming the physical layer overhead occupies 34 bytes.

$$\text{Network VoIP packet total bytes} = 63 + (40 + 80) = 183 \text{ bytes}$$

$$\text{Network bit rate} = (8)(\text{total bytes})(1000)/(\text{packet interval in ms})$$

or

$$\text{Network bit rate} = (8)(\text{total bytes})(\text{packet per second})$$
$$= (8)(183)(1000)/10 = 146,400 = 146.4\,\text{kbps}$$

Percentage of payload utilization = 100 (voice payload 80 bytes)/(total 183 bytes) = 43%

The bit rate will be reduced if payload header suppression is enabled as marked in reference [DOCSIS 1.1 (2005)]. Table 11.4 is listed for three codecs for different packet intervals. The results are noted as total bit rate in kbps and percentage of payload utilization.

Note on Profiles and Bit Rate. An upstream modulation profile will define how the information is transmitted between CM and CMTS. Several parameters are defined in a modulation profile such as guard time of the burst, preamble, type of modulation (quadrature phase shift keying—QPSK or 16-quadrature amplitude modulation—QAM) and forward error correction (FEC) protection, which are changed for a maximum throughput of a cable bit rate. The modulation profile changes depending on the application used. The CM vendors will predefine a set of profiles for various applications to avoid confusion. The bit rate calculations in Table 11.4 are considered for a

Table 11.4. VoIP Voice Packet Bit Rate on a Cable Upstream Interface

Description	G.711				G.729A				G.723.1	
Compression rate	64 kbps				8 kbps				6.3 kbps	5.3 kbps
Packet interval in ms	10	20	30	40	10	20	30	40	30	30
Voice payload bytes	80	160	240	320	10	20	30	40	24	20
VoIP header	40	40	40	40	40	40	40	40	40	40
Cable interface header	63	63	63	63	63	63	63	63	63	63
Total bytes per voice packet	183	263	343	423	113	123	133	143	127	123
Total packets/ second	100	50	33.3	25	100	50	33.3	25	33.3	33.3
Total bits/ second (kbps)	146.4	105.2	91.4	84.6	90.4	49.2	35.4	28.6	33.8	32.8
Percentage of payload utilization	43	60	70	75	8	16	22	28	18	16

maximum of 34 FGPS bytes. The FGPS ranges from 14 to 34 bytes. The FGPS will depend on the upstream modulation profile and the application used. In the case of VoIP, it also varies depending on the speech codec and multiple frame packets used. One complete example is given in the Cisco cable profile calculation for G.711-20 ms. Refer to [URL (Cisco-cable)] for various details on profiles, headers, and bit rate calculations. The payload header suppression (PHS), suppresses the UDP, IP, and Ethernet header as given in [DOCSIS 1.1 (2005)]. The PHS is not considered for calculations. PHS reduces by 40 bytes from the VoIP header and the cable-interface-specific headers. PHS can reduce the bit rate significantly compared with the values mentioned in Table 11.4.

11.6   BIT RATE CALCULATION FOR DIFFERENT CODECS

VoIP deployment may use multiple codecs in the deployment. Some codecs such as AMR, G.729.1, and G.722 have multiple compression rates. It is required to consider proper mapping of compression rate, payload size, and packets per second. Some codecs may use scalable payload, meaning dropping some of the compressed bits based on the available bit rate. For all these combinations, the variables are payload size, packets per second, and the interface overhead per packet.

Generic VoIP bit rate calculators use voice payload with RTP/UDP/IP. Other interface specific headers are not considered. It is useful to interpret the bit rate calculation up to the network interface. It is also essential to consider many other quality trade-offs and deployment-specific considerations before selecting codec and packet intervals.

11.7   BIT RATE WITH VAD/CNG

Voice activity detection (VAD) and comfort noise generation (CNG) are used mainly to reduce the bit rate on the Internet. VAD reduces the packets during inactive regions of speech. CNG is comfort noise creation at the receiver based on the received VAD packets. Usually speech conversation is half-duplex, meaning one at a time speaks, and during the speech, there are silence zones. VAD/CNG exploits these nonspeech zones and eliminates the need for sending voice packets to the network, which will give about a 40% to 60% Internet bit rate savings. In general, the use of VAD/CNG has to be eliminated based on the availability of sufficient network bit rate for VoIP voice. No exact bit rate calculation for VAD/CNG exists because of several variations in speech conditions [URL (Newport-BW)]. Practically, a significant bit rate savings with VAD exists, with the penalty of slight degradation in voice quality. Deployments have to cater to the full bit rate without considering a VAD/CNG operation.

11.8 BIT RATE WITH RTCP, RTCP-XR, AND SIGNALING

RTCP is the RTP control protocol [Schulzrinne et al. (2003)]. RTCP packets are used to send the packet reports at regular intervals. The bit rate from RTCP is controlled to less than 5% of the RTP bit rate listed in the previous sections. In practical systems, users are controlling RTCP bit rate to 1% to 3%. RTCP-XR is the RTCP extended reports [Friedman et al. (2003)]. It is used mainly for sending voice quality monitoring parameters. These packets are generated usually once in 256 basic frames of RTP. The bit rate contribution of RTCP-XR is very low, less than 0.5%, which can be accounted for as part of the RTCP 5% allocation. Signaling packets are generated while establishing the calls, tearing down the calls, and establishing the call features. In the initial and final stages of the calls, no voice packets occur in the call. Hence, signaling packets can reuse part of the voice packet bit rate. In steady state, signaling activity reduces except for any new call feature activity. As an approximation, accounting for about 10% of voice packets bit rate can take care of RTCP, RTCP-XR, and signaling bit rate requirements.

11.9 SUMMARY ON VoIP BIT RATE

A summary on bit rates is given in Table 11.5 for G.711, G.729A, and G.723.1 for different packet sizes, over Ethernet, DSL, and cable interfaces. VoIP voice packets will demand less bit rate when compared with an available Ethernet interface bit rate of 10 to 100 Mbps. The bit rate from the cable modem closely matches with DSL at small packetization. On a DSL interface, the upstream rate is lower than the downstream rate. VoIP on a DSL interface is limited by the upstream bit rate of the DSL interface. The Internet service provider also limits the bit rate on a DSL interface. G.711 is supported in most VoIP deployments. Hence, the minimum bit rate available on DSL should support the requirements of the G.711 channel bit rate. Most deployments use 20 ms as default packetization. G.711-20 ms with PPPoE on a DSL interface consumes 106 kbps. In G.711, using higher packetization such as 20 ms (106 kbps), 30 ms (98.9 kbps), and 40 ms (95.4 kbps) gives a significant savings in bit rate compared with 5-ms (254 kbps) and 10-ms (170 kbps) packetization. G.711 with 5 ms is not popular because of higher bit rate requirements. In Asia and Europe, some deployments consider this as an advantage for lower end-to-end delay and improved voice quality. For G.711 with 5-ms packetization, a calculated bit rate value is listed in summary Table 11.5, and detailed break-up is not shown in the previous tables. In the case of G.729A on a DSL wide area network (WAN) interface, the bit rate consumed comes down with packetization and 40-ms packetization may be a better trade-off on the DSL interface for bit rate savings. In VoIP, packetization up to 80 ms is considered. Most packet loss concealment algorithms work better up to 20 to 40 ms. Hence, it is suggested to limit the single packetization interval to 10 to 40 ms. VAD/CNG saves on bit rate requirements by about 40% to 60%, but it varies with

Table 11.5. Summary on Bit Rate for Ethernet, DSL, and cable Interface (values in cable interface are with FGPS 34 bytes)

Codec-packetization	Ethernet kbps	Ethernet-VLAN kbps	DSL PPPoE kbps	Cable Upstream kbps	Remarks
G.711-5ms	188.8	195.2	254.4	228.8	In Asia, this rate is considered for lower end-to-end delay
G.711-10ms	126.4	129.6	170.0	146.4	Commonly used for low delay and fax pass-through
G.711-20ms	95.2	96.8	106.0	105.2	Default suggested in RFCs, significant savings in bit rate
G.711-30ms	84.8	85.9	98.9	91.4	
G.711-40ms	79.6	80.4	95.4	84.6	Some deployments use this
G.711-50ms	76.5	77.1	84.8	80.4	Not common in deployments
G.729A-10ms	70.4	73.6	84.8	90.4	Reduces delay
G.729A-20ms	39.2	40.8	63.6	49.2	Default suggested in RFCs
G.729A-30ms	28.8	29.9	42.4	35.4	Matching rate as G.723.1-30ms
G.729A-40ms	23.6	24.4	31.8	28.6	40ms seems to be optimum from bit rate utilization
G.729A-50ms	20.5	21.1	25.4	24.4	Not common in deployments
G.723.1-30ms (6.3kbps)	27.2	28.3	42.4	33.8	G.729A-30ms gives better quality than G.723.1
G.723.1-30ms (5.3kbps)	26.1	27.2	42.4	32.8	G.729A gives better quality than G.723.1

conversation conditions. The detailed break-up for VAD/CNG is not given in this chapter. RTCP sends call statistics specific to voice packets and consumes less than 5% of the voice packets bit rate. RTCP, RTCP-XR, and VoIP signaling is expected to take less than 10% on top of voice packet bit rate consumption.

Redundancy as per RFC2198 [Perkins et al. (1997)] for the G.711 codec is presented in Chapter 16 along with fax pass-through mode. Redundancy increases the network bit rate compared with basic G.711. A T.38 fax makes use of redundancy. Redundancy aspects and bit rate calculations are given in Chapter 16. Some remarks included in Table 11.5 are the general recommendations and subjective in nature, which may not have any direct reference to standards/recommendation.

11.9.1 Packet Size Choice

The selection of codec for meeting the voice quality requirements are given in Chapter 3 and in Chapter 20. Once codec selection is made, frame selection depends on the available minimum bit rate from the Internet service provider, interoperability with the supporting infrastructure, and end-to-end delay goals.

Smaller packets reduce end-to-end delay. Reduction of delay is an improvement in voice quality. Smaller packets lost in end-to-end delivery can be recovered using packet loss concealment. In networks with no packet drop, higher packet intervals may be considered. Many deployments support packet sizes from 10 ms and multiples of 10 ms. Hence, 10 and 20 ms will be typically preferred in the deployments. The 20-ms packet gives significant savings in bit rate. The fax pass-through presented in Chapter 14 makes use of G.711 with 10-ms packetization.

11.9.2 Delay Increase Example for Large Voice Packets

The delay increase is explained with examples of G.729A packets with 10, 20, and 40 ms on the Ethernet interface. In a 10-ms packet-based delivery, one packet is delivered on the interface once every 10 ms. In 20-ms-based delivery, one packet is delivered once every 20 ms. G.729A gives compressed payload once every 10 ms. To create a 20-ms payload, at RTP-stage, two compressed 10-ms frames are used, which will delay sending the compressed voice by an extra 10 ms. Similarly, for sending a 40-ms payload, the first frame is held for an extra 30 ms compared with 10-ms-based packets. In the decoder, all packets are expected to undergo the same delay. Hence, the decoder will not add any delay. End-to-end delay with pictorial representation is given in Chapter 20. As an example summary, considering 10-ms-based packets take a 60-ms end-to-end delay, 20-ms packets can take a 70-ms end-to-end delay and 40-ms based packets can take a 90-ms delay at the increased risk of packet loss performance with large packets.

12

CLOCK SOURCES FOR
VoIP APPLICATIONS

Clock precision is one of the important aspects of VoIP for maintaining voice and fax quality. This chapter is presented for some popular clock sources used in VoIP adapters, public switched telephone network (PSTN)-to-VoIP gateways, and the key influencing clock selection parameters that help to maintain voice and fax quality. In PSTN, central offices (COs) and digital loop carriers (DLCs) are made to work with precision stratum clock references. In VoIP, every user's VoIP adapter hosts part of the services of PSTN CO and DLC. The VoIP adapter can be treated as a central office interfacing with the user. Hence, clock precision requirements will propagate to the VoIP adapter or closest interfacing VoIP system. From a clock perspective, maintaining VoIP voice interface clock accuracies matching the PSTN system clock is one of the helpful parameters to maintain better voice and fax quality in VoIP deployments that gives benefits of exceeding PSTN quality through wideband. The VoIP system designer may attempt certain trade-offs in clock selection to optimize for cost, space, and power consumption. It is helpful to know the performance trade-offs with reference to the optimization in the selection process. Providing the right clock is usually a hardware member's responsibility. In the absence of right clocks to voice interfaces, many problems will translate to degradation of voice quality and it takes a lot of effort to debug those directly influenced by clock issues. It is essential to analyze clock requirements from voice algorithms and quality requirements.

VoIP Voice and Fax Signal Processing, by Sivannarayana Nagireddi
Copyright © 2008 by John Wiley & Sons, Inc.

In several voice and fax recommendations, including G.711 [ITU-T-G.711 (1988)] are listed for clock precision of 50 parts per million (PPM). In RFCs [Demichelis and Chiniento (2002), Paxon et al. (1998), Almes et al. (1999)], it is recommended to contain a VoIP local clock source to ±50 PPM for VoIP. In the VoIP adapter, the voice hardware's interfacing clock has to be maintained at 32 to 50 PPM and the clock timing root-mean-square (RMS) jitter at less than 3 nanoseconds (ns). The RMS jitter of 3 ns corresponds to ±5.2 ns, assuming uniform distribution for jitter. A ±50 PPM means a frequency drift of 50 Hz for every 1 MHz (or 1 million) of center frequency, and this can be either side of the center frequency. For easy representation, ±50 PPM is given as 50 PPM in this chapter. Most VoIP adapters will be using an analog telephone interface with subscriber line interface circuit (SLIC)–subscriber line access circuit (SLAC or CODEC). The signals are sampled in CODEC consists of [analog-to-digital converter (ADC), digital-to-analog converter (DAC)] interface at 8- or 16-kHz sampling. The interface used between processor and CODEC is called pulse code modulation (PCM) or time-division-multiplexing (TDM). CODEC makes use of the PCM interface clock to generate samples at 8 or 16 kHz. In summary, the PCM clock has to be maintained at 32 to 50 PPM to ensure required precision on 8- or 16-kHz sampling. The same PCM clock precision is propagated on the packet intervals. As an example, 80 sample payloads are treated as one 10-ms frame. Any timing or clock frequency error in PCM clock can create deviation sampling timing that will propagate to a 10-ms Internet Protocol (IP) packet.

VoIP systems use multiple processors and interfaces. These interfaces will need different clock frequencies. As an example, Ethernet will use a 25-MHz clock, and USB will use a 24- or 48-MHz clock. Synchronous dynamic random access memory (SDRAM) or double data rate (DDR) SDRAM memories will use 125 or 133 MHz; PCI will use 33.33 or 66.66 MHz, and the main processor may operate usually at a higher frequency of 400 to 750 MHz. To transfer samples at the 8-kHz rate and each sample of 8 bits (assuming μ-law and A-law in this example), the interface will need 64 kHz. Multiple voice channels will be used on the same PCM interface, demanding the PCM interface clock to be multiples of 64 kHz. The usual clocks of 0.256, 0.512, 0.768, 1.024, 1.536, 2.048, 4.096, and 8.192 MHz are accepted by the most of the interfacing SLIC-CODECs used for telephone interface. In general, a clock of 1.536 or 2.048 MHz is used as a default for 24/32 channels communication. Some PCM interfaces are configured for linear 16-bit samples as explained in Chapter 9. For 16 bit, the PCM clock is made as integer multiples of 128 kHz instead of 64 kHz. It is to be noted that the PCM interface is required to be working at multiples of 64- or 128-kHz clock frequency. A voice processing central processing unit (CPU) can operate at any other frequency, and this need not be multiples of PCM clock frequency. In most VoIP systems, the PCM clock is derived from the digital signal processor (DSP) or host processor used for voice applications. In such systems, a processor CPU clock is made as multiples of 2.048 or 1.536 MHz with the required characteristics on precision. When the processor is given a PCM clock, ensuring the processor clock precision is important as that can impact PCM samples generation and finally IP packets duration.

12.1 PSTN SYSTEMS AND CLOCKS

PSTN systems and central offices maintain various levels of a stratum (plural is strata) clock. Depending on the clock precision, stratum suffix numbers 1, 2, 3, and 4 are included [URL (Symmetricom)]. A summary on strata clocks is listed in Table 12.1. Many VoIP boxes located at customer premises may not derive synchronization from PSTN, and it is essential to maintain required stand-alone clock stability within the VoIP system.

12.2 VoIP SYSTEM CLOCK OPTIONS

VoIP systems may use several clock options. In this section, some popular clock options are given as listed below:

1. Using precision crystal for clock generation
2. External clock oscillator at multiples of 2.048/1.536-MHz frequency
3. Deriving clock from PSTN T1/E1 family interfaces
4. Deriving clock from network timing reference (NTR)
5. Deriving clock from network time protocol (NTP)

These five options are marked in Fig. 12.1 circled with the serial number of the above list. The recommendations made in this section are broad options. In the figure, all clocks are shown as joined to the same phase locked loop (PLL) with dotted lines. It is implied that only one of these options is used in

Table 12.1. Stratum Clocks used in PSTN and Relevance with VoIP Systems

Clock classification	Stability in PPM	PSTN and VoIP applications
Stratum-1	10^{-5}	This is the primary reference for many applications. This is derived from global positioning system (GPS) in many systems.
Stratum-2	0.016	Many national PSTN systems and local telephone exchanges use this.
Stratum-3	4.6	DLCs use this. Many high-channel density VoIP gateways try to derive or implement this clock in stand-alone mode.
Stratum-4	32	DLC worst-case free running precision, most of the stand-alone VoIP CPEs, PSTN, and IP PBX systems use this.

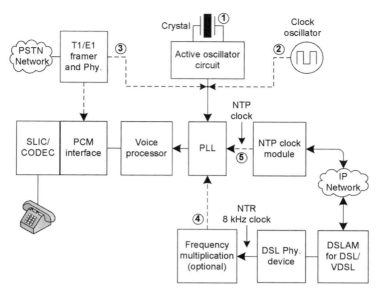

Figure 12.1. Functional Representations of Clock Sources in VoIP Applications.

the system. The user may find a different scheme in the process of reusing the existing hardware resources and clocks available as part of other subsystems and the total system. Hence, the scope of clocks is not limited to the given combinations of Fig. 12.1.

12.2.1   Using Precision Crystal to Work with Processors

Crystal in association with an active circuit is the most popular technique to building the clock oscillator. The clock is generated using two terminal passive crystals along with the active electronic oscillator circuit. The active circuit is usually kept along with the processor's PLL, and two terminals are provided on the processor to connect crystal and compensating capacitors. Engineers call the two terminals on the processor as crystal pads. The crystal pads used with a passive two-pin crystal of selected frequency and associated passive components can work as a clock oscillator. The external crystal operating at a frequency in the range of 5 to 30 MHz is used for most applications. Note that crystal always comes for one frequency with specified characteristics. A voice processor's CPU will use a higher frequency than 5- to 30-MHz. Most voice processors will have internal or external PLL, frequency multiplier, and divider chain to increase the clock frequency to higher MHz based on the processor capability and system requirements. The major precautions while using the external crystal for clock generation are given below.

Selection of Crystal. For VoIP applications, crystal has to be selected to operate with integer multiples of 128 kHz frequency and with less than 50 PPM for VoIP applications. To take care of passive components influence and parasitic effects, it may be preferred to select a crystal frequency of lower than 30 PPM. The crystal center frequency should match with the oscillator active circuit capability. A crystal and PLL combination has to take care of the required processor's maximum operating frequency.

External Passive Components Selection. Most crystal-based oscillators use two shunt capacitors connected with each pin of crystal. Passive component values vary with crystal manufacturer and influences on the center frequency. It is required to consider using passive components of recommended tolerance (usually of 1% to 5%) and the precautions suggested by the crystal manufacturer for printed circuit board (PCB) layout. Parasitic capacitors also influence and increase the equivalent value of the capacitors. In general, the selected capacitors should help in reducing the PPM. A wrong selection of passive components (mainly capacitors) can drift the crystal frequency even if the selected crystal is of good PPM.

12.2.2   External Clock Generator/Oscillator

An external digital clock is a small four-pin hybrid chip. These oscillators are made for a variety of packages, environmental conditions, voltages, logic levels, and with several extra control pins. It is an active hybrid device that integrates a passive crystal and oscillator circuit in one chip. These devices are made with metallic casing to minimize radiation influence. In most applications, clock oscillators are used from 10 to 40 MHz. For providing higher clock frequencies to the voice processor, PLL present as part of the processor chip is used. Required PPM is mainly controlled in the manufacturing process of the clock oscillator. The main PPM influencing factors for these hybrid chips are power supply voltages, ground noise, aging, and environmental influences such as temperature variations, shock, and vibration. The clock oscillator manufacturer provides enough immunity to meet the stated PPM stability under wider environmental conditions. In a passive crystal and processor-based scheme, it is possible to drift (or tune) the center frequency to about 100 PPM by changing external capacitor values, but it is not possible to change characteristics of basic clock oscillators.

Several types of clock oscillators exist, and some clock oscillators are made with an extra voltage controlled pin. Analog voltage is applied on the control pins to change the center frequency. Some devices are made with a 4- to 5-bit digital interface. A digital value is set on the pins to modify the central frequency. Temperature-controlled, oven-controlled, and a combination of features are also provided. In most VoIP customer premises equipment (CPE) such as adapters, IP phones, and wireless fidelity (WiFi) phones, simple clock generators with 30 to 50 PPM is selected from cost consideration, with some consideration for power consumption and space.

12.2.3 Deriving Clock from PSTN

The purpose of a higher end voice gateway referred to as the VoIP–PSTN gateway in Chapter 2 and given reference [Donohue et al. (2006)] is to interface PSTN-to-VoIP and VoIP-to-PSTN calls. These systems will be working in a VoIP service provider's deployment to bridge calls between VoIP and PSTN. Voice gateways use multiple interfaces of T1/E1 and T3/E3 to connect to PSTN. The interfacing devices on a VoIP gateway for T1/E1 and T3/E3 can also derive the precision clock synchronized to the PSTN exchange. Typically, the stratum-3 clock is derived from the PSTN interface. The physical layer device and framer combination interface on the T1/E1 family of devices extracts the synchronous clock.

In practical systems, the internal oscillator will be running at better than 32 PPM and it may be locked to the PSTN interface clock derived from the framer devices. As a worst case, even if locking is lost for some reason, the stand-alone clock will continue to make the system operate with acceptable degradation. Simple VoIP adapters will not have any T1/E1 support. Hence, VoIP adapters may operate from stand-alone crystal or clock oscillators of better than 50 PPM.

12.2.4 Network Timing Reference (NTR)

In some deployments, the digital subscriber line (DSL) modem and VoIP adapter are built in the same box [URL (Sagem-RG)]. These boxes are usually called residential gateways (RGs) or integrated access devices (IADs) instead of the VoIP adapter. VoIP adapters will use a simple Ethernet local area network (LAN) interface, and RG/IAD will use a direct wide area network (WAN) interface as DSL and very high-speed DSL (VDSL). NTR is popularly derived from DSL and VDSL physical layer interfaces. NTR is usually derived from the DSL central office and communicated indirectly on the physical interface of DSL and VDSL. It helps in deriving a PSTN reference clock or precision reference maintained by the DSL central office, which is similar to providing the CO clock reference extended through CPE. Usually NTR represents a PSTN clock translated to 8 kHz. It is also required to incorporate stable PLL inside the processor or external to the voice processors to up-convert this 8-kHz clock to 2.048/1.536 MHz or higher multiples [URL (IDT)] for voice processing as shown in Fig. 12.1. VoIP systems can derive stratum-3 or stratum-4 as precision while using an NTR clock reference of 8 kHz.

12.2.5 NTP for Timing and Clock Generation

NTP is used on most VoIP systems to acquire an accurate time to set the VoIP box or any system with an Internet network interface for local or coordinated universal time (UTC). NTP uses 64-bit timing information providing a 200 picoseconds timing resolution [Mills (1992), Mills (1996), URL (EECIS-NTP)].

The precisions are influenced by several factors based on the location of the servers. Servers on the WAN interface can provide 10-ms accuracy. Servers on the LAN interface can provide sub-millisecond accuracy. Servers directly mounted on the same box can provide microsecond to picosecond resolution [URL (Quartz)]. NTP packets also carry several other bytes that help in deriving a precision clock.

NTP servers are resident on the IP network. Most servers directly interface with global positioning systems (GPS) to provide stratum-1 clock precision. VoIP gateways communicating to the stratum-1 NTP server can derive the stratum-2 precision. VoIP end systems maintain the NTP client that communicates with multiple NTP servers. Most systems use NTP during their boot-up time and continue to use the same time for several days or get NTP updates on regular basis.

Many NTP servers are available for Internet public access, and user systems will be getting NTP packets from the shared NTP servers. It will delay the response, and round-trip delays may contribute to some errors. These errors are acceptable for timing on a seconds scale, but they have significant influence in deriving the secondary clock. The service provider of VoIP usually keeps their own NTP server to improve on the accuracy. NTP provides several details in the IP packets. Clock synchronization is also established with NTP packets information. To derive the clock from the NTP, dedicated hardware modules are required. It is one of the possible options in high-density voice gateways. The end boxes can derive up to stratum-2 by making use of the NTP reference. At this stage, the reference clock for VoIP adapters is not derived from NTP because of cost, space, and power consumption considerations.

12.3 CLOCK TIMING DEVIATIONS RELATING TO VoIP PACKETS

This section describes the clock drift and jitter requirements for VoIP voice adapters with foreign exchange subscriber (FXS) interfaces. The summary of this section is that the PCM clock used for interfacing hardware CODEC (ADC, DAC) sampling has to be maintained at better than ±50 PPM and that the clock timing RMS jitter has to be maintained at less than 3-ns RMS. Control of clock drift and jitter on the PCM interface clock can contribute to the improving voice quality, support several fax pages in both pass-through and T.38, and sustain modem calls for several hours. In this chapter, jitter name is used interchangeably for PCM clock jitter and jitter in adaptive and fixed jitter buffers. It is required to interpret it based on the described background text.

FXS is the main interface on most VoIP adapter and residential gateways. Telephone, fax, and modem are connected with this FXS interface. The FXS interface will have hardware devices such as SLIC and CODEC to interface with the processor. CODEC samples the signals at 8 or 16 kHz. The PCM or TDM interface communicates between voice processor and CODEC. The

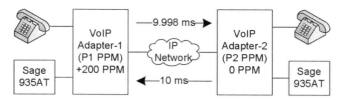

Figure 12.2. Typical setup to represent the clock drift issue at a high level.

Table 12.2. Clocks and PPM

Adapter-1 PPM	Adapter-2 PPM	Relative PPM, Adapter-1 to 2	Relative PPM, Adapter-2 to 1
+50	+40	10	−10
+50	−40	90	−90
−50	−40	−10	10
−50	+40	−90	90

voice processor compresses voice and sends it as raw payload to Real-Time Transport Protocol (RTP). On the receive path, voice payload is taken from jitter buffers and played on hardware CODEC after processing. The PCM clock influences end-to-end packet delivery timing errors. As an example, 80 PCM samples can create a 10-ms duration payload. With references to Fig. 12.2, if the clock is drifted by +200 PPM in Adapter-1, the 10-ms compressed packet on a voice processor is created for every 9.998 ms. Assuming Adapter-2 does not have any drift, it will expect a packet for every 10 ms but receives it once every 9.998 ms. As a result, there could be overflow of one packet for every 5000 packets. Similarly, packets received at Adapter-1 will be in the under-run state once every 5000 packets. In the subsequent sections, clock PPM and timing jitter interpretations are given. The analysis of Fig. 12.2 with PPM symbols marked as P1 and P2 in the figure are revisited once again in later sections of this chapter.

PPM examples between two adapters are given in Table 12.2 to make relative PPM clear. Considering an example with 50 PPM, Adapter-1 can have +50 PPM and Adapter-2 can have −50 PPM. Relatively each adapter will see the same 100 PPM. If both adapters are of the same polarity of 50 PPM, then relative PPM is zero. In general, each adapter runs with its own clock mechanism. Hence, each Adapter has to be limited to ±50 PPM. Table 12.2 lists PPM values that are different in each adapter to convey the relative PPM in a better way.

12.3.1 Interpreting Clock Drifts from the Distortion Goal of the Voice Signal

In voice front end, the maximum undistorted sine wave signal power is of 3.17 dBm. Idle channel noise (noise when phones are in mute) is required to be better than −68 dBm (20 dBrnc), which is about 71 dB below the highest signal power. Recent front-end devices can achieve idle channel noise of 10 dBrnc (−78 dBm). Typical longitudinal balances are of 63 to 68 dB. Samples in wideband codec use a 16-bit linear format that will use a 78-dB (from 13 bits of ±8159) dynamic range. Intermodulations to the voice frequencies have to be maintained to better than 40 to 50 dB as per TR-57. From the above parameter guidelines, it is preferred to consider a 71- to 80-dB signal-to-noise ratio (SNR) as the preferred distortion goal for the calculations of various clock parameters.

SNR and Jitter Relation. As per reference [Brannon (2004)], clock jitter influences on SNR in the process of quantization in the ADC of the CODEC is given as

$$SNR = -20 \log_{10}(2\pi f t_j) \tag{12.1}$$

In the equation, f is the input frequency and t_j is the RMS jitter in seconds. By substituting f as 7000 Hz for wideband voice, and SNR as 78 dB for 13-bit linear (6 dB per bit), $t_j = 3$ ns. The time jitter t_j can be rewritten as follows for direct calculation:

$$t_j = \frac{10^{\frac{-SNR}{20}}}{2\pi f} \tag{12.2}$$

As per the equations and interpretation, the PCM clock derived from the voice processor has to be maintained at an RMS jitter of 3 ns. It is treated as a close approximation, and such requirements can be easily met in any digital system. Clock jitter is given in picoseconds and is much lower than a nanosecond in practical systems. PLLs used for voice band sampling can meet this requirement of 3-ns jitter. From a typical oscillator specification, jitter maximums are noted as ±250 ps, which gives a lot of margin for voice sampling. In the datasheets of front-end SLIC-CODEC devices, higher jitter shown on the PCM interface may be listed for the hardware functioning point of view. Allowing higher clock jitter can degrade voice quality and mainly results in SNR degradation.

Interpreting Clock PPM with Phase and Amplitude Hits. As per TR-57, amplitude and phase hits allowed are 15 per 15 minutes or 1 hit in 60 seconds. With a 50 PPM clock and relative worst-case PPM of 100, clock PPM can contribute to one hit per every 100 seconds. As an example, when a VoIP call is made between two adapters; each adapter can have a worst case of 50 PPM. If Adapter-1 has −50 PPM and Adapter-2 has +50 PPM, Adapter-1 sends packets slowly and Adapter-2 plays quickly. Using 10-ms packetization, an under-run of 1 packet for every 10,000 packets (100 packets in 1 million packets) can be noticed, which will amount to one packet starvation at Adapter-2 for every 100 seconds. Similarly, in the packet flow from Adapter-2 to Adapter-1, an excess packet is created for every 100 seconds at Adapter-1. The excess of packets can create a phase hit, and a starvation of packets can create an amplitude hit. From a clock PPM basis, the achieved hits with 50 PPM will be better than maximum allowed, but the hits count has to cater to several other impediments like telephone ring noise and adjacent channel noise as explained in Chapter 1.

12.4 MEASURING CLOCK PPM

Hardware designers and manufacturers should ensure clock PPM. The simplest option is to measure the PCM interface clock or system clock used in deriving the PCM clock under planned environmental conditions.

Example. Expected PCM clock frequency = 2.048 MHz. Measured PCM clock frequency = 2.0475 MHz. Clock drift in PPM = (2.0475 − 2.048) (1 million) / 2.048 = −244 PPM. If measured frequency is more than 2.048 MHz, it will give positive PPM. A 50 PPM on a 2.048-MHz PCM clock is 102.4-Hz drift. Frequency counters of 1-Hz resolution are required to get a measurement of 1-PPM resolution, and this translates to instruments with a minimum of 7 or 6.5 digit counters for PCM clock.

12.4.1 External Estimate from Frequency Transmission Measurements

Assuming adapter boxes are available in closed form and clock terminals are not available for direct measurement, frequency drift between transmit and receive telephones can be used for estimating approximate clock drifts or PPM. If Adapter-1 is sending frequency f1s, and adapter-2 is receiving frequency f2r, usually both frequencies f1s and f2r will be the same with a few Hz or a fraction of Hz difference. If the frequency measurement can be made up to 0.01-Hz resolution, then f1s, f2r, f2s, and f1r can be used to estimate PPM. PPM from adapter 1 to 2 = (f2r − f1s)/f1s; PPM from adapter 2 to 1 = (f1r − f2s)/f2s after multiplication by 1 million. From the two PPM values, each unit PPM can be estimated. To improve the measurement accuracy, the measurements can be conducted on two or more sets of frequencies. The higher the input frequency within the voice band, the higher will be the drift for the same PPM. Hence, the highest possible frequencies close to 3400 Hz, but not harmonically related to 8000-Hz sampling, may be used instead of 1004 Hz for PPM measurements. A deviation of 0.1 Hz at 3333-Hz frequency is 30 PPM. Telephone interfacing equipment may not provide a frequency resolution of 0.01 Hz. It is required to use at least a 0.1-Hz resolution supported source and measuring devices. A suitable signal generator and counter may be coupled on the telephone interface for these measurements. Satisfying the TR-57 measurements presented in Chapter 1 between two VoIP boxes will reveal any major first-level mismatches of clock issues. It is suggested to conduct measurements in one direction at a time. The procedure given in this section is for first-level approximation of PPM to show the possibility, but this is a difficult option.

12.4.2   External Measurements from Packet Hits

The suggested method in this section is for a coarse estimation. Clock drift first-level approximations can be estimated based on IP packets. In a single VoIP gateway, all channels will use the same PCM clock. Hence, calls made within the same VoIP box will behave relatively like the "0" PPM system. To conduct some measurements and estimates, it is required to use two boxes. It is suggested to identify one of the boxes with known clock characteristics at room temperature.

As shown in Fig. 12.2, let Adapter-1 have P1 as PPM, and Adapter-2 have P2 as PPM. P1 and P2 are independent of each other and can be positive or negative. In all external measurements listed here, it is required to disable packet loss concealment (PLC) to make packet adjustments visible to the measuring instruments as phase and amplitude hits. VoIP packetization may be 20 ms.

Measuring Hits through Sage 935AT Communication Set. In Chapter 1, a description of transmission hits was given. The meaning of the hits is discontinuity to the expected tone. An amplitude hit is recorded when the amplitude discontinuity or silence is received. This silence will be noticed when some packet is missing or jitter buffer under-runs because of lack of packets. Phase hits are applicable to phase discontinuity and can be observed in an over-run situation. Figure 12.2 is taken as an example. Adapter-2 is fed with a 1004-Hz tone at the −16-dBm power level. Adapter-1 measures the tones with the hits test option using the Sage instrument. At regular intervals, Sage 935AT [URL (Sage935)] will keep accumulating the measured hits. Details on some test equipments are given in Chapter 13. As a default, hit tests are conducted in a span of 15 minutes (i.e., 900 seconds). Consider that an example between two Adapters, hits observed on Adapter-1 is H1 and Adapter-2 is H2 in 900 seconds.

Adapter-1 will see under-run when the Adapter-1 clock is faster than the Adapter-2 clock.

Adapter-1 will see over-run when it is relatively slower than the Adapter-2 clock.

For example, H1 of 10 and H2 of 6 at 10-ms packetization reveals relative PPM observed at Adapter-1 = $(10,000 \times H1)/900 = 10,000 \times 10/900 = 111$ PPM. Relative PPM observed at Adapter-2 = $(10,000 \times H2)/900 = 10,000 \times 6/900 = 66$ PPM. These calculations assume hits are not present because of any other impairment. Based on the relative PPM of one Adapter and type of hits, it is possible to extract the relative PPM of another Adapter.

It is suggested to use Adapter-2 as a reference with a known gateway and PPM through a frequency measurement. From known parameters of one Adapter, unknown parameters can be estimated accurately. The procedure indicated here takes at least 15 minutes. The oscillator of the reference adapter has to have long-term stability for at least 15 minutes.

12.5   CLOCK DRIFT INFLUENCE ON VOICE AND FAX CALLS

In voice applications, packet under-run and over-run can happen with clock drifts. This packet mismatch over time cannot be avoided in practical deployments. Based on the long-term jitter buffer growth direction, packet adjustments can be attempted without allowing for complete under-run or over-run. One of the best options of packet adjustment is to correct silence zones in the middle of utterances or talk-spurts. The jitter buffer adjustments are presented in Chapter 10.

In fax pass-through, a silence zone will not be clearly available like voice. Any attempt of removing or creating extra silence zones will create amplitude or phase hits that cause the fax to work at a lower rate. To minimize these issues, jitter buffer has to adjust smaller duration frames at a rate of slower than once every 1 minute to match the acceptance of the hits test. Fixed jitter buffer with higher initial thresholds is also used to avoid getting into requirements of packet adjustments for several pages. In some VoIP boxes, low-bit-rate V.21 modulations are continuously extracting low-speed data. In the transition of low-speed to high-speed data, a small gap of about 75 ms is available. These gaps can be used in adjusting silence zones.

In T.38 [ITU-T-T.38 (2005)] fax, PPM issues may not be dominant because of redundancy, forward error correction, and silence zones available in the middle of data packets. As an overall summary, it is preferred to maintain the clock close to stratum-4 (32 PPM) or ±50 PPM as a worst case in VoIP simple CPEs in the practical environment of use. Clock jitter has to be lower than 3 nanoseconds.

13

VoIP VOICE TESTING

VoIP voice solutions have to work at multiple customer locations in multiple deployments to meet the expectations of the users. To ensure quality, VoIP systems are subjected to several voice quality tests and interoperability conditions. In actual deployment, users usually provide subjective feedback with a possible description based on perception. To cater to several conditions, the solutions have to be tested for several simulated conditions along with deployment-specific infrastructure. The outcome has to result in objective results. Some popular tests for VoIP voice are given in Table 13.1. In this chapter, a high-level overview on the tests and some instruments used for performing these tests are presented. Even for basic VoIP voice tests, several instruments are involved as indicated in Fig. 13.1. This test setup assumes the basic working functionality of the VoIP boxes. The actual tests may include several other boxes to create a reasonably close match to the deployment. The instruments listed in this chapter are for creating an example representation. In general, several instruments are on the market for conducting similar or extended tests.

13.1 BASIC TEST SETUP

Figure 13.1 shows the basic voice quality measurement setup with two VoIP gateways. Each gateway is shown with multiple foreign exchange subscriber

VoIP Voice and Fax Signal Processing, by Sivannarayana Nagireddi
Copyright © 2008 by John Wiley & Sons, Inc.

Table 13.1. Classification of VoIP Tests

Classification of test	Suggested instruments
VoIP manual tests	This is to make several first-level observations based on the listening tests and IP packet analysis
End-to-end voice transmission similar to TR-57 tests of North America	Sage-935AT, Sage-960, PCM4
Telephone line testing and monitoring	Advent instrument AI-5120
MOS, PESQ, delay, and signal impediment	Malden Electronics MultiDSLA system
Bulk call tests for stress	Spirent bulk call generator (BCG), Ameritec BCG, Sagem 960B, Malden MultiDSLA
IP impediment tests	Spirent IPwave to create network conditions
VoIP packet analysis	Ethernet packet analysis (Wireshark, Hammer call analyzer)
VoIP compliance tests	With multiple test instruments and certification agency
VoIP interoperability (interop)	Interop with multiple VoIP boxes, soft switches in interop events, and actual deployment infrastructure
Actual deployment testing	Total deployment setup with several instruments and users
Voice quality certification	Internal tests, certification by third parties and through deployment feedback
Voice along with data traffic conditions	Smart bits
PSTN and FXO interfaces	PSTN switch like Teltone or looping back calls from PSTN service
Fax signaling and interoperability	FaxLab, fax machines, and VoIP gateways. These tests are given in Chapter 15

Abbreviations: FXO, foreign exchange office; MOS, mean opinion score; PESQ, Perceptual evaluation of speech quality; PSTN, public switched telephone network.

(FXS) interfaces. Based on the testing, one gateway could be of a known third-party reference. Voice calls are made within each gateway and between two different gateways. Several comparisons can be derived with these combinations. At the first level, several manual calls and observations can be made. To test the TR-57 transmission requirements given in Chapter 1, in the test setup, Sage communication test instruments are connected to each gateway on the FXS interface. In addition to TR-57 tests, Sage instruments are used to extract several telephone parameters such as singing return loss (SRL), echo return loss (ERL) and idle channel noise. In Fig. 13.1, the AI-5120 telephone line tester is shown in series with multiple phones. The instrument AI-5120 measures signal, timing, tones, and call features analysis by monitoring and storing

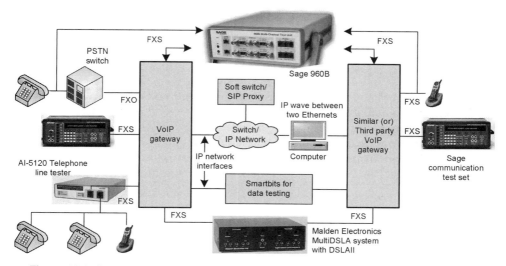

Figure 13.1. Basic voice quality measurement setup with FXS and FXO voice interfaces. [Courtesy: (Picture of "Digital Speech Level Analyser (DSLAII) as part of MultiDSLA System" is printed with permission from Malden Electronics Ltd., UK, www.malden.co.uk/multidsla.htm); (Picture of "AI-5120 Telephone Line Monitor is printed with permission from Advent Instruments Inc., Canada, www.adventinst.com); (Pictures of "Sage 935AT Communication Test Set, and Sage 960B Multi-Channel Test Unit" are printed with permission from Sage Instruments, www.sageinst.com)].

several waveforms based on the events and call progress. Multiple parallel phones on the same interface helps in evaluating the ringer equivalence number (REN) characteristics.

Digital speech level analyzer (DSLAII) is the analog interface to the VOIP gateways in the MultiDSLA test system and is used for PESQ–MOS measurements. This instrument measures PESQ–MOS. In addition, it gives signal analysis, delay, and jitter measurements. DSLA can also perform stress testing on two telephone interfaces per instrument. With multiple telephone lines, stress tests are created through BCG. These instruments create several thousand calls per hour. BCGs also provide PESQ–MOS and other signal analysis. A PSTN switch is required for testing the FXO interfaces, and some BCGs also support direct FXO interfaces. Based on the availability of the interface, tests are conducted on FXS or FXO interfaces. A PSTN switch is used with FXO to create PSTN connectivity or to convert the FXO-to-FXS bridging interface.

An IP wave creates impediments in the networking path. Along with voice and signaling tests, data tests are also important. Smart bits is a popular instrument for creating data traffic generation and measurement. A soft switch controls VoIP calls, and soft switches are also referred to as gatekeepers [Donohue et al. (2006)]. Based on the protocol, a simpler version of the soft switch may be used. In practical testing, several computers can be associated with the setup to control and configure instruments and to log various measurements.

13.1.1 Extending Basic Test SetUp

In the setup of Fig. 13.1, two gateways with FXS and FXO interfaces are considered. In the actual setup, usually there will be several other VoIP systems such as interoperability terminals, high-density PSTN–VoIP gateways, message servers, conference systems, billing and accounting, management of the deployment, and so on. Instruments are available with several capabilities with multiple interfaces. The scope of testing depends on the required level of testing.

13.2 FIRST-LEVEL VoIP MANUAL TESTS

At the first level of VoIP voice testing, it is essential to ensure all the required network and telephone interfaces, basic functions of VoIP signaling, voice, and call features are working with multiple phones and third-party VoIP gateway boxes. In manual tests between two VoIP gateways, several telephones with call features support as well as fax machines are used as the analog interface for the FXS port of the VoIP gateway. A VoIP soft switch is used for call establishing between gateways . A third-party good-quality reference box with a minimum of two telephone interfaces can be used for interoperability testing. It is suggested to conduct the following manual tests with phones and fax machines before conducting tests with several instruments:

1. Basic VoIP voice calls within the same gateway, also called hairpin voice calls, assuming the gateway has multiple telephone interfaces. If the gateway has only one voice interface, voice calls have to be conducted between two gateways.
2. Basic voice calls between similar gateways (gateways of the same manufacturer) and voice calls between gateway and another third-party gateway. The third-party gateway has to be of a good-quality VoIP box used as a known reference.
3. Manual calls for the call features of the product, such as caller ID, call waiting, call wait ID, call forward, call transfer, three-way conference call, and message waiting, have to be conducted. Some call features testing will need soft switch support based on the supported VoIP signaling.
4. Voice calls with codec combinations of the product features—G.711, G.729AB, etc.—have to be tested.
5. Direct voice calls without soft switch and voice calls with soft switch have to be conducted. Depending on the VoIP protocol used, a simpler version of the soft switch may be used for laboratory testing; for example,

the SIP Proxy is used with Session Initiation Protocol (SIP) VoIP signaling.

6. Dialing and redialing with multiple phones, for dual-tone multifrequency (DTMF) and pulse dialing have to be validated.

7. Voice quality perception with different country phones is also useful. Usually quality with multiple country phones is not ensured because of a different loudness rating as explained in Chapter 6.

8. REN evaluation—in summary, how many parallel phones are ringing together and are able to communicate without distinguishable quality degradation. PSTN supports REN of 4 to 5. VoIP gateways cater to 2 to 3 REN.

9. Perception delays in the call establishment and call features establishment.

10. Initial delays in getting dial tone, dial tone tripping, ring delays, voice initial "hello" delays, any voice ticks in the beginning and at the end of the call, RFC2833 DTMF rejection, and any annoying ticks.

11. Any annoying voice, ticks, howling sounds, and feel of more background noise in mute ON and mute OFF conditions.

12. Testing various configurations, command line interface (CLI) web pages, able to enable disable modules, setting various parameters, and observing voice and fax calls.

13. Echo canceller, voice activity detection (VAD)/comfortnoise generation (CNG) listening tests with multiple backgrounds, coupling of speech, and music.

14. Ability to make a fax pass-through call and send 10 to 50 pages successfully with the G.711 codec is a first-level indication of good quality in a gateway; however, a fax call cannot reveal issues with the echo canceller and many other voice call features.

15. Repeating fax calls between fax machines, PCs, and laptops as fax machines.

16. Switching from voice to fax mode of operation with fax calls on fax machines.

17. The scope of manual calls can be easily extended with Ethernet captures and packet analysis. It is observed that many engineers always keep Ethernet packet analysis active to supplement with voice call observations and measurements.

In the above tests, clear expectations are not given for all listed tests. The best comparison may be with a PSTN call or with calls made with PSTN boxes such as the Teltone PSTN switch. Some overall quality expectations and issues with VoIP voice quality are given Chapter 20 and at reference [URL (Netiq)].

13.3 ANALOG FRONT-END VOICE TRANSMISSION TESTS

Connecting fax machines and sending several fax pages in fax pass-through (as explained in Chapter 15) at a high speed confirms the possibility of good transmission quality in the VoIP system. FXS interfaces on VoIP gateways are similar to PSTN central office interfaces provided to the residential users. To make VoIP voice quality match PSTN, it is essential to ensure end-to-end signal transmission and switching delays should be in compliance with TR-57 requirements. TR-57 [TR-NWT-000057 (1993)] is the technical reference for North American PSTN digital loop carrier (DLC). TR-57 measurements are for narrowband voice and are classified into two categories—namely signaling or switching characteristics that analyzes various delays in call progress and transmission characteristics that play a major role in steady-state voice quality.

For TR-57 tests, a gateway has to use a G.711 call that gives the maximum possible quality. These calls are made within the same platform or are made on the local area network (LAN) interface to eliminate IP impediments. Jitter buffer is used with a slightly higher threshold or is configured to operate as fixed jitter buffer to avoid packet adjustments during the transmission tests.

TR-57 is the Telcordia/Bellcore reference for qualifying the PSTN digital loop carriers. Even though the TR-57 standard does not talk about VoIP, meeting its specifications for the FXS/FXO interface is expected to result in better speech quality under good network conditions. Several transmission requirements are given in country-specific PSTN documents [URL (NTT-E), FT ITS-1 (2007)].

TR-57 transmission tests and extensions of IEEE-743 tests with interpretation are given in Chapter 1. Some of the popular instruments with telephone-compatible interfaces are Sage 930, Sage 935AT, and Sage 960. HP voice band spectrum analyzers are used with telephone-adapting interfaces. PCM4 is one of the popular early instruments, but this is not in production. These instruments send a particular tone or combination of tones, and measurements are made on the receive port. Different benchmarks for the required results are given in TR-57 and Sage instrument application notes [URL (Sage935)]. Certain deviations in the PSTN specifications can occur based on the country of deployment.

13.4 TELEPHONE LINE MONITOR FOR TONES AND TIMING CHARACTERISTICS

VoIP gateway may be used with multiple country phones. Multiple country phones and interfaces will have different impediments. A telephone line monitor monitors the line without loading and helps in measuring various parameters such as hook-status, battery voltages, ring, tones, loading effects with multiple parallel phones, DTMF detection and analysis, signal levels, different country frequency-shift keying (FSK), and other call feature signal monitoring. This instrument in addition to the TR-57 measurements can provide several details from the whole call, including call features. Line monitor AI-5120 shown in Fig. 13.1 is from Advent Instruments [www.adventinstruments.com]. Some more details on this instrument are available from the manufacturer webpage.

AI-5120 is a compact device designed for monitoring and analyzing signals present on the telephone line. By sensing the voltage present on the telephone line, it detects and measures ringing, DTMF, pulse dialing, FSK of Bell-202 and V.23 signals, line polarity reversals, and open switching interval (OSI). Working in conjunction with the TRsSim software on a PC, the AI-5120 becomes a useful tool in analyzing and debugging caller ID and short message service (SMS) data transmission. Like an oscilloscope, it can capture and display waveforms showing various signals present on the telephone line. This instrument is used for the applications of

- Testing and troubleshooting fixed-line SMS delivery and submission for short message entity (SME) devices
- Testing caller ID signal generation from central office (CO), private branch exchanges (PBXs), and line simulation equipment
- Measuring general telephony signals such as DTMF, ringing, line flash duration, and line voltage
- Monitoring telephone line signals—tone detection and analysis.

13.5 MOS—PSQM, PAMS, AND PESQ MEASUREMENTS

Telephones on a gateway are connected on the FXS interface. This is similar to the PSTN central office interface provided to the user. Voice quality monitoring instruments are used for testing voice quality parameters such as MOS and other derived parameters. The popular measurement techniques for objective MOS quality are based on perceptual speech quality measure (PSQM), perceptual analysis measurement system (PAMS), and PESQ algorithms. PESQ is the latest in the family based on the P.862 series recommendations. Voice quality and PESQ are given in Chapter 20. The Malden Electronics Digital Speech Level Analyzer shown in Fig 13.2, which is popularly known as DSLAII, is a popular compact version of the equipment for measuring PESQ-based MOS quality as part of the MultiDSLA system.

The current active measurement is PESQ algorithm based. In the process of performing measurements, instruments provide several other parameters that help in identifying degradations. In general, there will be several common features among multiple instruments. The following list details some features of the MultiDSLA system. More features on this equipment are discussed at [www.malden.co.uk]:

Figure 13.2. Digital Speech Level Analyser-II for voice quality measurements. [Courtesy: (Picture of "Digital Speech Level Analyser (DSLAII) as part of MultiDSLA System" is printed with permission from Malden Electronics Ltd., UK, www.malden.co.uk/multidsla.htm)].

- Measures the listening quality of the speech transmission on the MOS scale of one to five.
- Narrowband PESQ, PESQ-LQ, ITU-T Rec. P.862.1, and wideband P.862.2 scores with several mappings are supported.
- Generates artificial speech test signals of both genders (male and female) in different languages, makes use of user data files, and can generate complex signals, (e.g., speech with noise using a wide range of program-mable sources).
- Makes manually or simultaneously scheduled test calls through analog, time-division multiplexing (TDM), and VoIP interfaces and stores the signals captured from the network element under test. Several audio interfaces are supported to use with multiple VoIP gateways.
- Measures key performance indicators, including mean active speech level (ITU-T Rec. P.56 Method B), peak level, noise level, activity factor, impair-ment value dual activity, delay, delay variation, echo level, VAD operation and comfort noise level in presence of speech, as well as DTMF analysis.
- Measures automatic gain control (AGC), nonlinear elements in the network, and mean delay; range of delay variation is calculated and displayed on an utterance-by-utterance or frame-by-frame time offset basis.
- Speech quality testers also emulate another VoIP virtual node to set up calls over SIP or H.323 with optional G.729A, G.729B, G.723.1, G.722, and iLBC vocoding for monitoring SIP messages, packet analysis and voice quality.

13.6 BULK CALLS FOR STRESS TESTING

VoIP voice solutions need to be stress tested to observe how they are perform-ing over time. The voice call stress tests are conducted with or without the presence of data traffic and other applications running on the gateway. Multiple voice calls are established within the gateway or from a third-party reference gateway. To make multiple voice calls, BCGs are used. Bulk call generators apart from making multiple voice calls also measure the voice quality in the call using PSQM, PESQ, and other statistics like signal-to-noise ration (SNR), one-way delay, and round-trip delay. These stress tests with bulk call generators are performed for different types of supported codecs such as G.711, G.729, and G.723.1. A BCG will not know which codec is used. Different packetization periods are also used for stress tests with BCGs. The most popularly used BCGs are from Spirent, Ameritec, Teltone, and Sage Instruments. The details on these instruments are available at the respective manufacturer's website. Spirent offers bulk call generators through its product series Abacus. Abacus II is the most current version. This equipment comes with at least two telephone interfaces up to several T3/E3 span interfaces. These instruments also give several call details and various statistics of the voice quality. Ameritec BCGs are also popularly used for stress testing of VoIP voice calls. Ameritec BCGs also give precise impairment measurement, and they calculate MOS values. All functions are performed simultaneously on every channel in the call generator without influencing the performance of the test equipment. Ameritec BCGs are also available for a wide range of physical interfaces—Analog, T1/E1 CAS, ISDN-PRI, SS7, and DS3—and provide interworking between different interfaces [URL (Ameritec)].

13.7   NETWORK IMPEDIMENTS CREATION

In a typical VoIP deployment, a VoIP voice call encounters several network impediments. The impediments are classified as delay, jitter, packet reorders, packet drop, duplicate packets, packet errors, and packet fragmentation. Before launching the VoIP voice solution into the real network, the voice solutions have to be tested for various packet impediments and expected network conditions. Several free and proprietary software programs are available that are built to cater to incorporating IP impediments into several thousand calls. An IP wave is one of the network impediment simulators supplied by Spirent [URL (Spirent-IP)]. An IP impediment simulator in a laboratory testing is shown in Fig. 13.1. An IP wave separates the network and introduces impediments as per the set of configurations. The graphical user interface (GUI) on the PC displays several details of the impediments created and monitors progress.

As shown in Chapter 2, VoIP products will be closely associated with several LAN ports such as Ethernet, universal serial bus (USB), and wireless LAN (WLAN). These interfaces are used for both voice and data. In the measurements, voice quality from the VoIP terminals operating on LAN interfaces along with data is also important. Some voice quality aspects with data are given in Chapter 18. To validate these interfaces for voice, VoIP terminals have to be used on LAN interfaces with simultaneous data and IP impediments.

13.8 VoIP PACKETS ANALYSIS

In the previous sections, voice testing is presented with manual calls, transmission and telephone line testers, DSLA voice quality tester, bulk call generators, and IP impediments. In addition to these tests, several tests are conducted for signaling and packet analysis. The most common practice of analyzing the signaling and voice packets is through Ethernet packet capture, but this will be time consuming to decode in some situations. Several protocol analyzers are available for VoIP packets analysis.

13.9 COMPLIANCE TESTS

Several manufacturers developed tools with dedicated hardware as well as PC-based tools for compliance tests. In the compliance tests, several protocol test cases are emulated and presented to the VoIP system. Based on the responses and dynamic interactions, the quality of compliance may be judged. Compliance is supposed to create a certification process, but no clear certification process is available for these tests. The users have to decide on the coverage and declare what they support and how it is tested. In some form, a compliance test is mapped as protocol testing in the absence of a certification process. On completing the compliance tests, interoperability success will be improved. The Spirent protocol tester [URL (Spirent-Protocol)] is one analysis tool for this purpose. The key operations in compliance tests are as follows:

- Checks for all possible messages and states.
- Modifies call flows based on new standards.
- Adds, removes, and modifies messages.
- Adds, removes, and modifies message headers.
- Modifies mandatory headers and messages for negative testing that helps to deal with wrong operations.

13.10 VoIP INTEROPERABILITY

In VoIP standards and request for comments (RFCs), several options and states are given for completing end-to-end calls. Compliance tests ensure signaling level test cases. Many implementations may not go through complete compliance tests. VoIP interoperability tests help with interworking in a deployment. Inter-ops tests are conducted as follows:

1. Participating in VoIP inter-op events, where several VoIP implementers will be participating in the organized inter-op events. These events usually last for few days. It may be difficult to fix all the issues in the inter-op event. The fixes may be validated in the subsequent events.

2. Major issues and results can be gathered by conducting tests with several adapters, IP phones, and gateways in the laboratory and deployment setup. This is one of the useful exercises, and most of the implementers will be taking up this exercise in their own facility. In some form, this inter-op testing also takes care of compliance testing.

3. Tests are also conducted with the certification agency. Some certification agencies conduct a minimal version of inter-op with a fixed setup. The setup can take care of part of the inter-op and compliance tests. The time available for such certifications may last for a few weeks.

4. Tests with the service provider and user experience may be the last step. In general, this testing will be conducted after successfully passing several tests as given in the previous sections. Service provider setup will be used mainly to address any pending minor inter-op issues and user experiences. Usually, deployment setup may not cater to full voice testing and interoperability.

In practice, inter-op involves coordinating with several third parties, inter-op events, and on-site fixing of the issues.

13.11   DEPLOYMENT TESTS

In actual deployment, several inter-op and voice quality tests are conducted, but the scope of testing is limited. Required additional tests are supplemented with laboratory test setup. Several voice ports from the deployment are also looped back to the laboratory test setup, and deployment performance is evaluated with several laboratory tests. It is also common practice to use several listening tests with various call combinations. Some popular options for voice quality tests are as follows:

- Multiple listeners in noisy and noise-free environment.
- Echo canceller tests through listening and measurements with Sage 960B.
- Testing using instruments like Opticom P.563 [URL (Opticom-P563)], which is based on a single-ended voice quality measure [www.opticom. de] that measures one-way voice without calling for reference speech signal.
- Voice quality monitoring using VQmon, Real-Time Transport Control Protocol-Extended Report (RTCP-XR), and E-model, which conduct

voice quality monitoring as a remote logging operation from the servers.

- PESQ–MOS measurements by looping the calls to the laboratory or using distributed test nodes with instruments like Malden Electronics MultiDSLA.

13.12   VOICE QUALITY CERTIFICATIONS

Voice quality certification is a very important action for promoting the solutions. Several focused categories for voice quality certification exist. The main are given here:

- Agencies like American Telephone and Telegraph (ATT) and British Telecom certify for echo canceller and overall listening quality. This is mainly for voice aspects and not for the complete VoIP product.
- The Telecommunication Technology Association (TTA) [www.tta.or.kr] tests [URL (TTA)] and certifies for voice and several networking functions. Usually the tests are conducted on 50 different signaling and voice operations and their parameters.
- The Tolly Group [URL (Tolly)] also conducts a certification program called the "VoIP capable infrastructure (Quality of Service)," which verifies the device under test's quality-of-service (QoS) mechanisms. It can adequately support latency-sensitive applications such as voice in a congested environment by providing sufficiently low latency as well as voice quality scores that are coincident with those deemed toll quality [www.tolly.com].
- There are some test agencies for fax over IP testing such as Commetrex Corporation [www.commetrex.com].
- CT Labs [www.ct-labs.com] also conducts several stress and voice quality tests.

The test agencies and testing support may change over time based on the time-sensitive requirements. Therefore, refer to the latest information on test agencies.

13.13   VoIP SPEECH QUALITY TESTS BY THE ETSI

The European Telecommunications Standards Institute (ETSI) conducts VoIP speech quality test events (SQTEs) in different regions. Vendors participate with their VoIP equipment to test speech quality during scheduled events [URL (ETSI-news)]. The goal of taking the products through a Plugtests event is interoperability and voice quality. Plugtests increase the probability of achieving interoperability by debugging the standard and the companies' implementations at an early stage.

All conversational aspects like speech quality, echo measurements, double talk performance, and transmission quality in the presence of background noise are considered for the speech quality test. The coder test under different packet loss rates and jitter conditions, detailed analysis of packet loss concealment (PLC), and jitter buffer implementations are also investigated in the SQTE. These test events provide important hints for the optimization of speech quality and allow for a benchmark comparison of the products. Moreover, the results of the ETSI VoIP SQTEs have become an important marketing tool for the participating manufacturers because the test methods used for the SQTE have practically developed into a standard for speech quality evaluation of VoIP equipment. The ETSI also provides the Plugtests service, which is a professional unit of ETSI specialized in running interoperability test events for a wide range of telecommunications, Internet, broadcasting, and multimedia converging standards.

13.14   USER OPERATIONAL CONSIDERATIONS

VoIP products have to deliver very good voice and fax quality, and the user has to be satisfied with the product usage. Most service providers may preconfigure customer premise equipment products, and the user can simply be using these products with Internet services and interfaces. In this case, some provisions are provided to the user for configuration, the product has to be user friendly, and the configurations have to be simple and observable. The configurations also have to check for any abnormal user inputs and should use an allowed range of correct inputs without destabilizing the basic operations. Several of these operations depend on the product features and level of support planned with the products.

VoIP products also support RTCP-XR voice quality monitoring and GR-909 telephone diagnostics. A frequent Voice Quality check, and building of, the dynamic feedback mechanisms to configure remotely also helps in improving the voice service. VoIP voice may be part of many services. Educating the user and supporting the required details and manuals can be used to help with the service.

14

FAX OPERATION ON PSTN, MODULATIONS, AND FAX MESSAGES

Fax is an abbreviation for facsimile. To access the public switches telephone network (PSTN), an analog two-wire TIP-RING interface is used for voice and fax communication. The PSTN may use analog, or a combination of analog and digital operations, for end-to-end transmission of voice and fax signals. The fax data that are in bits and the digital numbers that form inside the fax machine are modulated and created as digital samples. These samples are converted to an analog voice band of a 300- to 3400-Hz signal before delivering on the TIP-RING interface. TIP-RING interfaces are connected to the PSTN lines. High-level representation of a PSTN-based fax call is represented in Fig. 14.1. In PSTN calls, fax and voice share the same telephone lines. At the digital hierarchy of PSTN, fax or voice samples are sent as G.711 compression bit streams and take 64 kilobits per second (kbps) per call. The PSTN central office (CO) or digital loop carrier (DLC) converts an analog signal to a digital bit stream.

Fax data are divided into four groups depending on the image formatting and handling method used inside the fax machine. Each group of fax machines will take different time durations to send the same A4 size fax page or document.

Group-1 (G1) became a fax recommendation in 1968 for analog facsimile devices to communicate over analog telephone lines. Group-1 is based on the ITU-T-T.2 recommendation. It takes six minutes to transmit a standard A4 size document with G1. This transmission used two-tone combinations—one

VoIP Voice and Fax Signal Processing, by Sivannarayana Nagireddi
Copyright © 2008 by John Wiley & Sons, Inc.

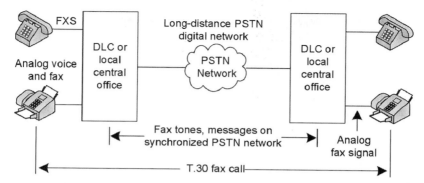

Figure 14.1. Functional PSTN fax call as a voice call. [Courtesy: used with the permission of CED Copyright © 2007. All rights reserved.]

for black and another for white. Group-1 fax machines are not in use now, and T.2 is a withdrawn recommendation.

Group-2 (G2) became a recommendation in 1976 for analog facsimile devices to communicate over analog telephone lines. Group-2 is based on the ITU-T-T.3 recommendation, and T.3 is a withdrawn recommendation. It takes three minutes for transmission of a standard A4 size document. From group-2 onward, operation of fax devices conforms to the T.30 [ITU-T-T.30 (2005)] recommendation for a fax control signaling handshake. Group-2 did use both frequency modulation (FM) and amplitude modulation (AM) depending on the geographical region. Group-2 fax machines are not in use now.

Group-3 (G3) became a recommendation in 1980 for digital facsimile devices to communicate over analog telephone lines. G3 machines are also referred to by the name G3FE, which stands for group-3 facsimile equipment. Group-3 is based on the T.4 [ITU-T-T.4 (2003)] recommendation for compression and is most popular in use. G3 devices take 1 minute for transmission of a standard A4 size document. The characteristics and the operation of G3 fax devices conform to the T.30 [ITU-T-T.30 (2005)] recommendation for a fax control signaling handshake. From group-3 onward, it is a digital facsimile operation and it makes use of the V.27ter, V.29, and V.17 [ITU-T-V.27ter (1988), ITU-T-V.29 (1988), ITU-T-V.17 (1991)] recommendations for modulation and demodulation of fax digital data. V.21 is used for transmission of control messages at 300 bps [ITU-T-V.21 (1988)]. Supergroup-3 (SG3) V.34 [ITU-T-V.34 (1998)] introduced in 1996 supports high-speed fax up to 33.6 kbps in addition to G3 features. SG3 fax is based on the T.4 [ITU-T-T.4 (2003)] and T.6 [ITU-T-T.6 (1988)] recommendations for compression. SG3 is about three times faster than G3. Group-3C is used in digital networks like an integrated services digital network (ISDN) fax and is described in T.30 Annex C [ITU-T-T.30 (2005)(2005)]. The use of error correction mode (ECM) is mandatory for all facsimile devices using a V.34 half-duplex and full-duplex fax system. The modulation and demodulation (modem) modules of V.27ter, V.29, V.17,

V.34, and V.21 are together referred to as a fax modem or as a fax data pump or simply as a data pump.

Group-4 (G4) became a recommendation in 1984 for digital facsimile devices to communicate over digital telephone lines. Group-4 is based on the T.6 [ITU-T-T.6 (1988)] recommendation and takes approximately 10 seconds to transmit an A4-sized page. The G4 fax is mainly designed for an ISDN fax at 64 kbps, and T.6 supports line width as well as the line length two-dimensional image compression. T.6 also supports black and white, grayscale, as well as color images [ITU-T-T.30 (2005), McConnell et al. (1999)]. The characteristics and operation of the G4 facsimile devices conform to the ITU-T recommendations T.6, T.62, T.70, and T.503 [ITU-T-T.62 (1993), ITU-T-T.70 (1993)(1993), ITU-T-T.503 (1991)(1991)]. The facsimile group-4 modified modified read (MMR) coding or T.6 recommendation was incorporated as an option in the ITU-T group-3 recommendation in 1992.

Group-3 or supergroup-3 facsimile devices are the most common standard fax machines in use, and they are expected to stay in use for several years to come. Group-3 devices use the T.30 protocol [ITU-T-T.30 (2005), ITU-T-F.185 (1998)] describing the formatting of non-page data such as messages that are used for capabilities negotiation and communication. Image data, coding, and format are handled by T.4 and optional T.6 [ITU-T-T.4 (2003), ITU-T-T.6 (1988)] recommendations. In some fax documents, the CCITT recommendation is referenced. The CCITT is the International Telegraph and Telephone Consultative Committee currently known as the ITU. This committee sets the recommendations for all fax equipment, thereby allowing different manufactures and faxes in different countries to communicate with each other.

This fax chapter and Chapter 15 have several timing parameters that have relevance with ITU-T-T and V.series as well as with country-specific documents. For presenting the chapter content, basic parameter values are used and at some places with parameter tolerances in the running text. There are several parameter options, tolerances, and interdependencies with other parameters and fax call states. For completeness of information, refer to the relevant recommendations and dependencies.

14.1   FAX MACHINE OVERVIEW

At a high level, the facsimile terminal device (fax) consists of a paper input device (scanner) paper output device (printer); accessories like telephone keypad, display, handset, and controller [McConnell et al. (1999), URL (Maine), URL (ATIS)]; and a fax modem as shown in Fig. 14.2. The fax modem is connected to the phone line of the PSTN. Once a call is established between two fax machines on the PSTN line, the PSTN infrastructure is transparent without keeping any distinction between voice or fax calls. A functional representation of a PSTN-based fax call is shown in Fig. 14.2.

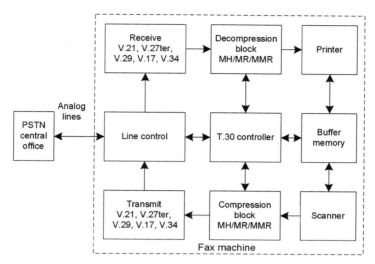

Figure 14.2. Functional representations of G3 fax equipment (G3FE) internal modules.

A fax machine scans a page and electrically breaks up a document into picture elements or pixels or pels. The most common type of scanning method used in fax machines is the flat bed type, and in this method, a document is moved across an optical unit one line at a time in the vertical direction and then the optical device scans the line in the horizontal direction. The size of each picture element or pixel is different based on the resolution of the image being scanned. A smaller pixel size generates a higher resolution of the document, but it increases the amount of picture data sent over the phone lines. These pixels generate a binary bit pattern. The scanned picture elements are coded using modified Huffman (MH), modified read (MR), or modified MR (MMR) coding methods. The compression, also called the coder block, represented in Fig. 14.2 generates code words containing the picture information in a compressed format. The compressed picture elements are modulated using amplitude, phase, and frequency variations depending on the selected modulation. These modulations appearing as samples are converted into analog signals that can be sent on a TIP-RING interface and on telephone lines. The group-3 facsimile devices use V.21, V.27ter, V.29, and V.17 modems for modulation and demodulation of fax digital data/bits. SG3 facsimile devices use a V.34 modem that supports up to a 33.6-kbps rate. End fax machines distinguish between various modes of operation and rates. The users of fax machines may not be distinguishing the fax machines by G3 and SG3, but the speed of operation in sending pages can give some idea.

The receiving fax machine assembles the picture elements together as it receives them, until a copy of the original is made. During reception, the receiver modem demodulates the received analog signal and regenerates the digital signal. The decoder or decompression block expands the facsimile digital data to picture elements. The picture elements as electrical signals are sent to a printing device that converts the bit stream into a copy of the original page. Typically, the printing device would be a thermal print head, bubble jet print head, or laser printer. The operation of the fax machine is specified by the T.30 [ITU-T-T.30 (2005)] recommendations. All group-3 facsimile devices operate in a half-duplex mode that can either send or receive at one time.

When a page is transmitted with ECM, the compressed coded data are buffered and embedded into high-level data link control (HDLC) frames with error correction before sending it for modulation block. Data of up to 256 HDLC frames constitute one partial page. More details on ECM functionality are given in Section 14.4.3. The overhead of the HDLC frame structure for each frame that includes flags, address, control, and frame check sequences will create an extra time of 2 to 3 percentage (%) as given in Annex A of T.30 [ITU-T-T.30 (1996)]. At the receiver fax machine, the receiver modem decodes the received analog HDLC frames and buffers the data. The raw payload is extracted from the buffered HDLC frame and expanded as pixel information for printing.

14.2 FAX IMAGE CODING SCHEMES

An optical scanner scans the document and generates a series of electrical signals corresponding to the picture elements on the scan line. On a standard size (8.5″ × 11″) document, the scanner gives 1728 bits of data or pixels in the horizontal direction and 1145 lines of information in the vertical direction (while using the lowest vertical resolution that is 3.85 lines per mm). This procedure produces (1728) × (1145) = 1,978,560 bits (approximately 2 million bits per page) of information from one page. Without any compression and coding techniques, it would take minimum of 207 seconds to transmit the entire one page data at 9600 bps rate [URL (Maine)].

The method that G3 fax machines use to reduce data is called coding. In voice, voice compression or codecs are used for the compression operation. In fax, the coding name is used for compression. Fax coding performs data compression, and voice codecs perform signal compression. The ITU recommendations T.4 and T.6 are used for coding the fax data. T.4 implementation comprises image information, encoding, and decoding operations that are required for a facsimile device. It also considers image size and resolution. T.4 and T.30 complement each other in a G3 facsimile device. The main parts of T.4 are the run-length coding/decoding schemes, the bilevel compression schemes, and the color/grayscale modes. Bilevel scan lines are composed of white and black areas, and groups of black and white pixels are placed in every line of picture data. These pixel groups are referred to as black run lengths and white run lengths of variable size. Based on the run-length size, identifying code words are assigned to the different size run lengths and only these codes are sent to save the bit rate for transmission. The popular coding scheme used with G3 fax machines are the 1-D MH scheme and the 2-D MR scheme

[McConnell et al. (1999), URL (Fax-theory)]. MH and MR are used in many G3 fax machines. The most recent G3 fax machines use an optional MMR scheme [ITU-T-T.6 (1988)]. A G3 color fax machine uses joint photographic expert group (JPEG) and joint bilevel image experts group (JBIG) color-coding schemes for coding the color pages. Group-4 devices follow the T.6 recommendation for image compression and coding. G4 uses MMR, trellis, and JBIG coding schemes. These coding schemes of MMR, JPEG, and JBIG color-coding are incorporated into recent versions of G3 fax machines, mainly in SG3 facsimile devices. Each one of these coding methods reduces the amount of data needed to be sent over the phone lines, and each improvement in coding yields an improvement in data transmission speed. Typical compression achieved with fax coding techniques is given here [URL (Multitech)]: Modified Huffman 5:1, Modified read 7.5:1, and Modified modified read 10:1, JBIG and JPEG 10:1 to 20:1.

14.2.1   Modified Huffman 1-D Coding

Modified Huffman is a 1-D coding scheme, which is supported as mandatory in all G3 facsimile devices. An overview on MH is given in this section. MH is the most common black-and-white encoding scheme, in which one line of data is scanned and coded for transmission. An optional ECM can also be used to transmit the total coded scan line in an MH coding scheme. This encoding scheme compresses a typical facsimile black-and-white image six to ten times. MH coding uses two different look-up tables of codes called terminating codes and make-up codes. An MH coded line always starts with a white run length. If an actual scan line begins with a black run length, a white run length of zero is sent. The run length can be either a termination or a make-up code word. If it is a terminating code word, then the next run length will be black. If it is a make-up code, then the next run length will be white. Each coded line is terminated with an end-of-line (EOL) code. The synchronized EOL code word consists of 11 zeros followed by one bit of "1." The end of the facsimile document page is marked with six consecutive EOL codes. The pattern of six consecutive EOLs is called return to control (RTC). The longer the run lengths of black and white, the compression becomes better. Each picture element is represented by either one terminating code word or one make-up code word followed by a terminating code word. Run lengths from 0 to 63 are coded using the appropriate terminating code words. Run lengths in the range 64 to 1728 are first coded using the appropriate make-up code word followed by the terminating code word, which represents the difference between the run length and the make-up code word. Different code words are available for both black and white run lengths. A list of code words for black and white run lengths is given in T.4.

> **EOL:** Each coded line of data is followed by an EOL. The EOL format is of 12 bits with the binary value 000000000001 (11 zeros and a one). The

beginning of page transmission is indicated by EOL. This unique code word is not found in a valid line of data. Hence, resynchronization is possible even in the presence of error burst.

Fill: The fill bits are used for creating pause (silence) in the message flow. Fill bits are of a variable-length string of zeros. Fill can be inserted between a line of data and an EOL but never within a line of data. A fill is inserted to ensure that the transmission time of data, EOL, and fill is greater than the minimum transmission time of the total coded scan line but less than 5 seconds [ITU-T-T.4 (2003)].

RTC: The format for RTC is EOL repeated for six times (EOL EOL EOL EOL EOL EOL). These six EOLs indicate end-of-document transmission. After the return to control signal, the transmitter will send postmessage commands in a framed format as well as the data-signaling rate of the controlled signals [ITU-T-T.4 (2003)].

14.2.2 Modified Read (MR) 2-D Scheme

The MR coding scheme is also known as two-dimensional coding that works very similar to the modified Huffman scheme. This coding/decoding algorithm is explained in the T.4 recommendation [ITU-T-T.4 (2003)] and is popularly used in G3 machines. MR gives higher compression than MH. Similar to the MH coding scheme, MR uses an optional ECM for transmitting the coded scan line. The synchronization code consists of EOL (11 zeros followed by a bit of "1") plus a tag bit. The tag bit indicates whether the following line is coded as one dimension or two dimensions. Tag bit "1" indicates one-dimensional coding, and "0" indicates two-dimensional coding. The pattern of six consecutive synchronization codes (EOL plus tag bit) is called RTC in the MR coding scheme. In MR, each line of data is compared with the previous line for making code changes. After a line has been coded, it becomes the reference line for the next coding line. This coding scheme is likely to create errors in the picture data. A limit exists on how many lines of data are compared with a reference line. If the reference line has an error, then the same error is reproduced on any line that was compared with it. To avoid transmission errors at the end of the page, a one-dimensional code line is sent for every second or fourth line depending on the scanning density.

Transmission Time Per Total Coded Scan Line. In group-1 and group-2 modes, the minimum scan line time was fixed. Because of coding in G3 and G4 modes, the amount of compressed data varies from line to line. Therefore, the minimum scan line time (MSLT) supported by fax machines must indicate to each other before actual communication takes place. MSLT is the minimum time required by the receiving device for printing or buffering an image line. During the initial handshake, the receiving fax machine sets the digital identification signal (DIS) command for minimum scan line time. This information is communicated to the sending device to ensure that image data are not transmitted faster than the receiving device can process it. The sending device will pad the image line with zeros to increase the transmission time to match the receiving device requirements. The total coded scan line is defined as the sum of data bits, fills bits, and EOL bits. For the optional two-dimensional coding schemes, the total coded scan line is given as a sum of data bits, required fill bits, and EOL bits plus a tag bit. The minimum transmission time of the total coded scan line is defined as the total time taken for transmission of the total coded scan line that is 20 ms [ITU-T-T.4 (2003)]. These values vary with image data in the page and resolution. Based on printing methods, there are several optimal minimum transmission times of total coded scan line in addition to the standard 20 ms.

14.2.3 Modified Modified Read (MMR) Scheme

MMR is the basic facsimile coding scheme used in G4 ISDN facsimile devices. For G4, MMR is used for better compression than the MR scheme. The MMR coding is also known as the extended two-dimensional coding scheme that is described in T.6 [ITU-T-T.6 (1988)]. This coding scheme became a recommendation in 1992 and is used in G3 fax machines. The MMR coding scheme works similarly to the modified read scheme, but only two-dimensional lines are transmitted without fill bits and without EOL codes for synchronization. The coding scheme uses a two-dimensional line-by-line coding. The first reference line is assumed as a white line. The line to be coded is called the coding line, and it is coded with reference to the position of the elements of the reference line. After coding, the coding line becomes the reference line for the next line of data. An error-free communication link in ISDN makes it possible to transmit pages without errors. This coding scheme has to be used with the ECM option.

14.2.4 JPEG Image Coding

Continuous-tone color and grayscale modes are optional features of G3, which enable transmission of color or grayscale images. Color and grayscale modes were incorporated into G3 in 1994 [ITU-T-T.42 (1996)] by introducing JPEG image compression. This compression is explored in the T.81 recommendation [ITU-T-T.81 (1993)]. JPEG provides a compression method that is capable of compressing continuous-tone image data with a pixel depth of 6 to 24 bits with reasonable speed and efficiency. JPEG is popularly used in still pictures compression. The JPEG specifies two classes of coding processes: loss and lossless. The loss compression processes are based on the discrete cosine transform (DCT), and the lossless are based on predictive techniques. Four modes of operation exist under which the various processes are defined, namely the sequential DCT, progressive DCT, sequential lossless, and the hierarchical mode. JPEG was not an efficient solution to transmit the text or line art. JPEG can compress the text and line art, but to get good compression efficiency, image quality suffers. JBIG is more suitable for fax images that code efficiently. JBIG gained more popularity than JPEG for fax images.

14.2.5 JBIG Coding

The JBIG group is a recommendations committee that had its origins within the International Standards Organization (ISO) and is a method for compressing efficiently bilevel, two-color, multilevel gray, and color images data. The main features of JBIG are given here:

- Lossless compression of one-bit-per-pixel image data
- Ability to encode individual bit planes of multiple bit pixels, catering to gray-level and color images
- Progressive or sequential encoding of image data
- More efficient and scalable for different resolutions

Most information transmitted through a facsimile device contains only black-and-white bilevel information. Most infrastructure of the transmission and the printing system are binary, and the output quality of this type is acceptable. JBIG addresses efficient coding of both bilevel and multilevel gray and color images as defined in T.82 [ITU-T-T.82 (1993)]. JBIG is intended to replace the less efficient MR and MMR compression algorithms used by the G3 and G4 fax devices.

14.3 FAX MODULATION RATES

In group-3 facsimile equipment, messages and data are sent as modulations in the voice band. The messages are sent as V.21 frequency-shift keying (FSK) binary modulations. The picture or scanned lines information is modulated using one of V.29, V.27ter, V.17, and V.34 modulations as given in Table 14.1. The modulation scheme is negotiated in the beginning of the fax call or in the middle of the call for any fall back to lower rate requirements based on the interoperability, line conditions, and errors.

Baud Rate and Bit Rate Differences. In this note, an overview on bit rate and baud rate is given. Bits per second (bps) or bit rate is the number of bits transmitted or received in a one-second time interval. For representing higher modem bit rates, a unit of kilobit per second (kbps) is used. A fax rate of 14,400 bps is represented as 14.4 kbps. Baud is the number of symbols per second [URL (Linux-docs)]. Baud is named after Emile Baudot, the inventor of the asynchronous telegraph printer [URL (ATIS)]. In the early development of modem and fax, the modulations used were FSK, which also used to

Table 14.1. Fax Modulations for Messages and Data

ITU-T Recommendation name	Modulation data rates in bits per second (bps)	Modulation method
V.27ter	2,400 and 4,800	Phase-shift keying (PSK)
V.29	4,800	Quadrature amplitude modulation (QAM), similar to V.27ter without amplitude change; V.27-4,800 is used in place of this modulation
	7,200, 9,600	QAM
V.17	7,200, 9,600, 12,000, 14,400	Trellis-coded modulation (TCM)
V.34	2,400 to 33,600	TCM of total 14 rates in 2,400 bps steps
V.21 for fax messages	300	Frequency-shift keying (FSK)

be called two tones, and PSK. In FSK and PSK, one bit is used per symbol. Hence, the baud rate and bit rate names are used interchangeably. In V.21, baud is 300, which means symbols are 300 and the number of bits sent on FSK is 300. In FSK, the baud and bit rate in bps are the same and each symbol sends just one bit. In advanced modulation methods, each symbol sends more bits. In general, the baud rate and symbol rate convey the same meaning. To convey the correct meaning, "symbol rate" is used in recent literature instead of "baud rate." For PSK-, QAM-, and TCM-based modulations, one symbol can take a group of bits to modulate, which gives a higher bit rate per second. The symbol rate name or symbols per second is also used for modulations with 1 bit per symbol of FSK and PSK.

14.4 PSTN FAX CALL PHASES

The PSTN fax call is divided into five phases of A, B, C1, C2, D, and E [ITU-T-T.30 (2005), McConnell et al (1999), URL (Audiocodes-fax), URL (SPRA073), URL (Sage-fax)] as shown in Fig. 14.3. Fax call messages for multiple pages of transmission are illustrated in Fig. 14.4. In this section, an overview of fax call flow messages is given. Complete details, several messages, various options, and errors are given in the T.30 recommendation Appendix A.

Phase A is the call establishment, and this is achieved either manually or automatically.

Phase B is the premessage procedure for call control and capabilities exchange. The major steps of this phase are as follows:

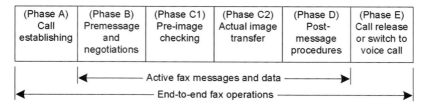

(Phase A) Call establishing	(Phase B) Premessage and negotiations	(Phase C1) Pre-image checking	(Phase C2) Actual image transfer	(Phase D) Post-message procedures	(Phase E) Call release or switch to voice call

Active fax messages and data

End-to-end fax operations

Figure 14.3. PSTN fax call phases.

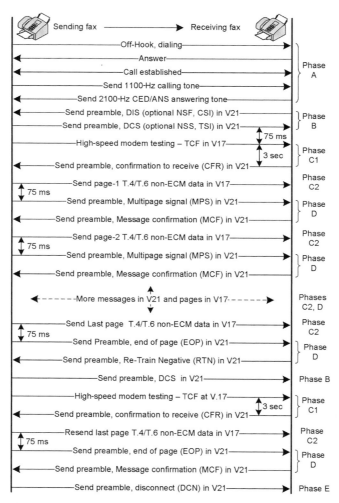

Figure 14.4. Basic fax call phases in relation to messages, pages, and multiple page extensions.

- The calling and answering fax devices identify themselves and exchange capabilities
- Transmission parameters selection

Phase C1 is the in-message procedure for message transmission and controls, and it completes signaling for in-message procedures, namely in-message synchronization, error detection, correction, and line supervision:

- The calling unit sends a test pattern to determine the maximum data rate
- The answering unit either accepts the data rate or requests a lower rate

Phase C2 is the T.4/T.6 image transmission at an accepted data rate.
Phase D is the postmessage procedure consisting of

- End-of-page (EOP) message is sent by the originating fax machine to indicate that all pages are transmitted. For a properly received page, the session will go to phase-E, and for a page with errors, the receiving fax machine may go to phase B waiting for the digital command signal (DCS) and training check.
- A multipage signal (MPS) is sent by calling the fax machine to indicate the end of the fax page and more pages to follow and returns to phase C2 to transmit the next page.
- End of message (EOM) indicates the end of the page, and both fax machines return beginning at phase B when digital identification signal (DIS) and DCS are exchanged again.
- Partial page signal (PPS) message is used in ECM to indicate the end of a partial block.
- Message confirmation (MCF) validates that the image is received and ready for more pages or to disconnect.

Phase E is for call release.

Phase-A Call Establishment. The fax call is established either through a manual process, in which the user dials a call and puts the machine into fax mode, or by automatic procedures, in which no human interaction is required. In both cases, the originating fax machine generates a 1100-Hz calling tone (CNG) and the terminating machine responds with a 2100-Hz called terminal identification tone (CED) or answering tone (ANS) at the beginning of the fax call. The SG3 V.34 fax machine generates an answer tone with an amplitude and 180° phase reversal modulation (/ANSam) tone instead of the ANS tone. These tones are used to recognize the presence of a fax call. At the end of phase-A, with a delay of 75 ± 20 ms, the called fax machine generates a mandatory V.21 signal for transmitting the fax call preamble and messages. In the case, ANS is not available or is not properly interpreted, and V.21 preamble flags are used for fax call detection. Details on CNG, ANS family of tones, and V.21 preamble are given in the fax tones detection section. In the literature, the CED and ANS names are used interchangeably. Both names convey the same meaning. In this book, both names are used.

Phase-B Premessage Procedure Capability Negotiations. During phase B, the fax machines negotiate the set of supported capabilities and agree to common fax capabilities and rates. The called fax machine transmits the mandatory DIS describing its capabilities to the calling fax. The calling fax then determines the common denominator for both machines and responds with a DCS to inform the called fax of the selected settings. The settings include

- Data rates supported
- Vertical resolution
- Image encoding
- Page width capabilities
- Maximum page length capability
- Handshake speed
- Error correcting mode

Optionally, both calling and called fax machines exchanges manufacturer-specific proprietary features through a nonstandard facility (NSF), from the called fax machine, and nonstandard facilities setup (NSS) from the calling machine. International phone numbers similar to caller ID are enchanged through called subscriber identification (CSI) and transmitting subscriber idenfication (TSI) from the calling fax machine. The modulation used for capability negotiation is defined in the V.21 modem.

The framing of V.21 messages is created by HDLC [ITU-T-T.30 (2005), URL (HDLC)]. Every message starts with a preamble of flag characters 01111110 in binary bits or 0x7E in hexadecimal number format. The pattern of 0x7E is repeated for 1 second ± 15%. The initiating fax machine maintains a timeout of 3 seconds ± 15% for responses. When timeout occurs, the initiating fax machine retransmits the frame, and after three unsuccessful attempts, it disconnects. After exchanging its capabilities to determine common modulation scheme and data rates, the calling fax machine sends a training check function (TCF) pattern. The TCF test pattern contains continuous zeros for 1.5 seconds ± 10% to test end-to-end transmission. The modulation and data rate for TCF data is V.17, V.29, and V.27ter as specified in the DCS message.

The answering unit responds to the TCF with confirmation to receive (CFR) for the correct received TCF pattern. Failure to train (FTT) is issued for the TCF test pattern identified as bad. If the calling unit receives an FTT, the calling fax machine may fall back to the next lower rate and then it sends a new DCS message followed by a TCF test pattern at the new data rate. This process is continued until a TCF test pattern is received OK (correctly) or all data rates have been attempted. Depending on the manufacturer, few fax machines may retrain at the same data rate when they receive the FTT message. This process is continued for a maximum of three times, and then the call gets disconnected. The CFR or FTT messages are sent using V.21 modulations.

Phase-C Image Transfer. In some documents, phase-C is divided into C1 and C2. After testing the phone line and agreeing to a common data rate, the calling fax machine starts sending the page data to the receiving facsimile device. The exact rate and modem used depends on the negotiations and training check. Image data characteristics are data compression, resolution, paper size, and so on, as specified in DCS and any one of the fax rates from V.27ter (2400/4800 bps), V.29 (7200/9600 bps), V.17 (7200/9600/12,000/14,400 bps), and V.34 (14 rates from 2400 to 33,600 bps).

Phase-D Postmessage Procedures. Phase-D goes through several conditional operations and options. Immediately after the image, the calling fax terminal sends an EOP, MPS, or EOM handshake message to indicate final image, more images to come, or a request for polling or to change parameters like a different resolution or paper size. The answering unit responds to the EOP, MPS, or EOM messages with an MCF if the image was received correctly, the re-train positive (RTP) image was received OK but with some errors and the phone line must be retested before receiving additional images, or the re-train negative (RTN) is applicable for unacceptable image. If the calling unit receives an MCF message, then it sends pending additional images or it sends a disconnect command (DCN) and disconnects. If the calling unit receives an RTP, then it resends a DCS/TCF sequence to retest the phone line at the same data rate before sending any more images. If the calling unit receives an RTP in response to an EOP command, then it sends the disconnect command and disconnects the call. If the calling unit receives an RTN, then it falls back to the next lower data rate, sends a new DCS/TCF sequence to retest the phone line, and retransmits the page for which the RTN was received.

Phase-E Call Release. When the fax transmission is complete, the call is disconnected with a DCN command. Several modes of phase-E exist. Some popular disconnect modes are as follows:

- When a response to a specific command is not received in time, the fax machine retransmits the signal and after three unsuccessful attempts or until timeout occurs and then it enters into phase-E by sending the DCN message. Any terminal can send this DCN message.
- If a transmitter receives, receive not ready (RNR) continuously for a period of 60 ± 5 seconds in ECM mode, it may transmit DCN and enter phase-E. The digital response RNR is used only in the T.4 error correction mode.
- Procedure interrupt disconnect (PID) is the digital response indicating that a message has been received but that additional transmissions are

not possible. After correction of all outstanding pages or partial pages, the transmitter shall enter phase-E.
- DCN is used to switch back to voice mode in fax over IP. More detail on fax-to-voice call switching is given in Chapter 15.

14.4.1   Multiple Pages and Fax Call Phases

The call phases for multiple pages are shown in Fig. 14.4. The call phases B, C1, and C2 are repeated depending on the postmessage commands. Immediately after the first page, the calling facsimile device sends an MPS to indicate that more pages will follow. At the last page transmission, the calling facsimile device sends an EOP to indicate that this is the final page. The answering unit may respond with MCF, RTP, or RTN. A few example message exchanges of multipage transmission are given below.

Example Situation-1. If the answering terminal responds with an MCF as a response to an image followed by an EOP, then the originating fax machine enters into phase-E. In this example, the call flow goes through the normal phases of phase-A, B, C1, C2, D, and E. This situation occurs only a simple page transfer without any errors. Upon receiving the MCF message as a response to an image followed by an MPS command, the originating fax machine enters into phase-C2 to transmit the next page as shown in Fig. 14.4. The call flow phases for multiple pages are phase-A, B, C1, C2, D, C2, D, ..., C2, D, and E. This call flow indicates that all pages are transmitted completely without any errors. If any one of the pages is received with errors and it requires retransmission, then the call phases may change for multiple pages. These cases are explained in examples 2 and 3.

Example Situation-2. If the answering unit responds with an RTP as a response to the image followed by an MPS, then the originating fax machine enters into phase-B. The additional messages follow after retransmission of a DCS and training at the same or lower rate. The rate may fall back if fallback support is enabled in the DIS and DCS commands. In this case, the call flow phases for multiple pages are phase-A, B, C1, C2, D, C2, D, ..., C2, D, B from DCS onward (DCS, training), C1 and C2 (page retransmission), D, and E. If the answering unit responds with an RTP as a response to the image followed by an EOP, then the originating fax machine will go to phase-E.

Example Situation-3. If the receiving terminal responds with an RTN as a response to the image followed by an EOP/MPS, the originating fax machine enters into phase-B. The additional messages will follow the retransmission of a DCS and training at a lower rate. In this example, the call flow phases are the same as in example-2 [i.e., phase-A, B, C1, C2, D, B from DCS onward

(DCS, training signal at a lower rate), C1 and C2 (page retransmission), D, and E]. The difference between the response to RTP and RTN is that on receiving an RTP, the originating terminal enters phase-E if it has completed the transmission of all pages. If transmission is not complete, it will enter phase-B and subsequently transmit the next page. On the contrary, with RTN, the originating terminal enters phase-B and then retransmits the current page for which RTN was received as shown in Fig 14.4.

Example Situation-4. In a few situations, the originating fax machine may change the mode or rate for the subsequent pages by sending an EOM instead of an EOP/MPS. In such cases, both fax machines will go to the beginning of phase-B. The receiving fax machine responds with an MCF in response to the EOM and enters into phase-B. Upon receiving the MCF in response to the EOM, the originating fax machine enters into phase-B. In this example, DIS and DCS are exchanged again after the MCF. The call flow phases are phase-A, B, C1, C2, D, B (DIS, DCS, Training, CFR), C1 and C2 (page retransmission), D, and E.

14.4.2   Fax Call Timeouts

Fax communication takes place through complex and noisy networks where a packet can be corrupted or lost. To compensate for these impairments, the T.30 protocol specifies the retransmission handshaking messages. If a response to the message is not received within a specified time usually of within 3 seconds, then the fax machine resends the messages again. The messages are repeated up to three times or until an acknowledgment is received. The unacknowledged DIS messages are repeated every 3 seconds until the end of the 35-second timeout. The delay between the transmissions of the DIS signals shall be 4.5 seconds ± 15% for a manual receiving terminal and shall be 3 seconds ± 15% for automatic fax terminals. The delay between two consecutive messages is 75 ± 20 ms [ITU-T-T.30 (2005)].

14.4.3   Fax Call with ECM

ECM is the error correction mode. The switched telephone network is prone to error during the transmission of fax data. These errors are created because of several impediments, such as telephone line noise from long-distance analog lines and from fax machine interaction. When such a transmission error occurs in a PSTN, page data may be corrupted. Two error control options are defined by the ITU. Before the optional ECM was approved in 1988 by the ITU [McConnell et al. (1999)], error concealment techniques were used to minimize the effect of transmission errors in the PSTN. The error concealment technique is possible when EOL code is transmitted between two scan lines. This technique was considered for MH and MR coding techniques used by G3 fax machines. In this technique, if the receiver does not detect the EOL code at the expected location in the line data stream, it indicates that an error has likely occurred on the line after the recent correctly received EOL code. This option limits the effect of errors without correcting them.

The second option employs true error correction technology called ECM that is an optional transmission mode built into G3 fax machines. The ECM is mandatory when SG3 fax machines are used with a V.34 fax modem. The MMR line encoding requires the use of ECM as it does not use any EOL between scan lines. In ECM, errors are detected at the receiving end, but they are not corrected automatically; the receiving end reports these errors to the transmitting end. The transmitting fax machine will retransmit only those frames that were reported to be received with errors. Eventually, the errors are corrected by repetitive retransmissions of such frames. The T.30 Annex-A defines the error control mechanism to cope with transmission errors over PSTN for T.4 image data. The ECM is based on a half-duplex page-selective automatic repeat request (ARQ) technique.

An HDLC frame structure is used for all binary-coded facsimile message procedures. More details on framing are given in the section "V.21 HDLC Framing and Deframing." Specific to HDLC framing in ECM mode, the facsimile information field (FIF) is a length of 257 or 65 bytes and is divided into two parts: the frame number (1-byte) and the facsimile coded data (FCD) field (256 or 64 bytes).

In error correction mode, each page data is divided into 64 or 256 byte frames. These frames are transmitted as payload inside HDLC frames with a unique frame number and frame check sequence (FCS) that allows the receiver to detect any errors in the transmitted frame. Several frames constitute a block or partial page. Each partial page contains a maximum of 256 frames. The exact number and frame size is negotiated during phase-B using DIS and DCS messages. The receiving terminal must be able to receive 256 or 64 byte frames up to a maximum block size of 256 frames.

The transmitting terminal may send the block whose size is less than 256 frames at the end of each partial page if page data contain more than 256 frames or for simple page data. This block is called a short block. The frame size should not be changed during a transmission of one page. The T.30 fax call flow with ECM for multiple pages is shown in Fig. 14.5. After sending the partial page data, the delineation is obtained by the transmission of the return to control for partial page (RCP) frame. This transmission tells the T.4 modulation system to drop off the line and be replaced by binary coded modulation with a delay of 75 ± 20 ms for postcontrol message command exchange. The receiving fax machine initiates the postmessage command reception if it detects at least one RCP frame. At the end of each block or partial page, the originating fax machine sends the T.30 PPS message and waits for MCF from the receiving fax machine. The originating fax machine can send any of the following messages after each partial page. The partial page signal-NULL (PPS-NULL) is transmitted between partial pages or a block of page data. The partial page signal-multipage signal (PPS-MPS) is transmitted between pages.

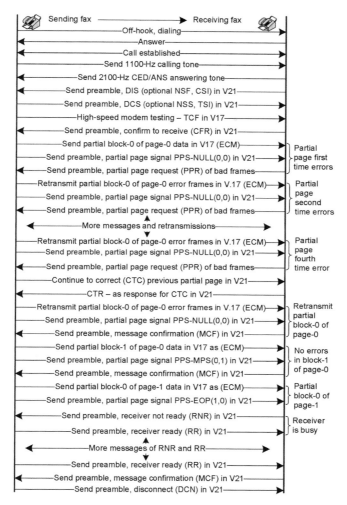

Figure 14.5. Fax call flow with ECM for multiple pages.

The partial page signal-end of procedure (PPS-EOP) is transmitted at the end of the last partial page of the last page data. The partial page signal-end of message (PPS-EOM) is transmitted at the page boundary to change the modulation speed and frame size.

The receiving fax machine unit responds to these postmessage commands with either MCF if the partial page was received correctly, or the partial page request (PPR) if the partial page is not received correctly, or the RNR if the receiver is not ready to receive more data. When the partial page is not received satisfactorily, the receiver requests the retransmission of bad frames by frame number. The receiver respond with a PPR to seek retransmission of the frames received with errors. When a PPR is received, the transmitting terminal retransmits the requested frames specified in the PPR information field. Multiple PPRs are possible. When a PPR is received four times for the same block because of errors, either an end of retransmission (EOR) command is transmitted or a continue to correct (CTC) command is sent for continuous retransmission. The calling fax terminal decides to terminate the retransmission of error frames in response to the fourth PPR received; it sends the EOR command. The calling unit transmits the next partial block upon receiving the ERR command in response to the EOR command. As indicated in Fig. 14.5, the command CTC comes from calling (transmitting) the fax terminal in response to the fourth PPR received, and it indicates that the transmitting terminal will send the frames requested in the PPR information field. When the transmitter receives PPR four times, the modem speed may either fall back or continue at the same speed in accordance with the decision of the transmitting terminal using a CTC command. The receiving terminal can accept the CTC message and can respond back with a CTR response to continue to correct. The fax call with two pages of data with ECM is shown in Fig. 14.5. The first page is divided into two partial blocks (0,0) and (0,1), and the second page has only one partial page block (1,0). As illustrated in Fig. 14.5, the receiver can respond to a postmessage sequence with RNR indicating that the receiver is not ready to receive any more data. The calling terminal then queries the receiving fax terminal with a receiver ready (RR) message. If the receiver unit is still busy (e.g., busy with printing), then the receiving terminal repeats the RNR message. The sequence RNR/RR, RNR/RR, RNR/RR can be repeated up to 60 ± 5 seconds as defined in the T.30 recommendation.

14.5 FAX AND MODEM TONES BASICS

Group-3 fax machines work up to 14400 bps. A fax call starts with a 1100-Hz calling (CNG) tone from the originator and a 2100-Hz ANS acknowledgment tone from the responder. These tones are used to detect the fax call in VoIP. The V.21 preamble is generated immediately after the ANS tone from the responding fax machine. The V.21 preamble indication is also used in addition to the CNG/ANS tones for detecting the fax call. In Fig. 14.4, the tones are indicated in the beginning of the call. SG3 fax machines use a V.34 modem for high speed that generates an answering tone consisting of amplitude and $180°$ phase reversal modulation (/ANSam). In this case, the fax detector has to support the detection of an /ANSam detection indication to switch to a fax call. Some fax machines send modem tones resulting from various options and violations in implementations. To cater to these impairments, it is necessary to analyze the combinations of these tones, timings, and V.21 preamble detection to arrive at a final decision on fax and modem. Many deployments do not use a modem in VoIP. It is suggested to favor the decision to fax when any fuzzy conditions occurs between fax and modem detections.

14.5.1 CNG Tone

CNG is the calling tone generated as a first indication of a fax call from the originating fax machine. The CNG tone is a pure tone with a frequency of 1100 ± 38 Hz (tone frequency tolerance is ±38 Hz) with 0.5 seconds ON and 3 seconds OFF. The timing tolerance on 0.5 and 3.0 seconds is ±15%. The CNG tone is generated with a periodicity of 3.5 seconds until receiving the CED tone as a response from the destination fax machine or the CNG timeout of 60 ± 5 seconds [ITU-T-T.30 (2005)]. The silence period of 3 seconds is meant for the CED tone. The CNG sends for 0.5 seconds and expects CED in silence period of 3 seconds. On recognizing the CED, the CNG will be terminated. When the CED is not received from destination, the CNG will be retransmitted. From the observations, it is noted that most G3 fax machines generate CNG tones at the −12-dBm to −24-dBm power level. CNG detections are not sufficient conditions for fax call detection. CNG helps, but a final decision is made based on CED and V.21 detections. In case a fax machine is trying to call a normal telephone, it will not get a CED tone as a response from a telephone. In the absence of CED, the CNG will be sent for 60 ± 5 seconds and the fax call times out [ITU-T-F.185 (1998)] as expected. Manual calling fax machines manufactured prior to 1993 may not transmit this CNG tone. To make it work, fax calls will progress even without CNG tones. A destination fax machine responds with a CED for ring alerts even in the absence of a CNG tone. The V.21 preamble flag FSK detection is also helpful for fax call detection.

14.5.2 CED (or ANS) Tone

The CED is the called tone, which is also called the answering tone (ANS). The tone is a continuous tone signal of 2100 ± 15 Hz for a duration of 2.6 to 4 seconds [URL (Audiocodes-fax), URL (Sage-fax), ITU-T-G.168 (2004)]. On receiving a CNG tone, the answering fax machine responds in 1.8 to 2.5 seconds after recognizing the CNG. The CED is used as a voice and fax call discriminator. The power levels of the CED are from 0 to −30 dBm. In VoIP, detection of CED is also used to control the echo canceller operation. With CED detection, echo canceller nonlinear processing (NLP) is disabled and basic echo adaptation is kept in the enabled mode. In general, fax signals are continuous (without silence zones) with a minimum of a −30-dBm power level, and this minimum level ensures that NLP is disabled even if it is not configured to disable on fax detection.

14.5.3 Modem or /ANS Tone

This tone is generated by the modem. It is similar to fax CED/ANS with added phase reversal of the tone at regular intervals. The symbols slash "/" preceding the ANS in "/ANS" is read as "ANS bar," which means phase reversals of ANS. It is similar to binary phase-shift keying (PSK). The same notation is used for

/ANSam for V.34 modems. The modem answering (/ANS) tone is a sine wave signal at 2100 ± 25 Hz with phase reversals for every 450 ± 25 ms [ITU-T-V.25 (1996)]. The phase shifts of $180°$ with a tolerance of $25°$ is detected as valid, and those in the range of $0 \pm 110°$ will not be detected. The usual power levels of the tone are from 0 to -30 dBm. To detect this tone it is required to observe a minimum of two-phase reversals occurring at 450 ± 25-ms intervals.

In the field, it is observed that some fax machines are sending this modem tone in place of fax ANS. Validating ANS or /ANS with additional CNG detection is one of the useful techniques for fax and modem discrimination based on ANS or /ANS. With modem tone detection, the echo canceller is fully disabled or bypassed.

14.5.4   ANSam Tone

This tone is generated by the modems. The modified answer tone is ANSam. The suffix or lowercase "am" after ANS indicates "amplitude modulation." The ANS tone is maintained at the tight tolerance of 2100 ± 1 Hz without phase reversals. Amplitude modulation waveform is of a sine wave at 15 ± 0.1 Hz. AM is at 15 Hz; hence, a frequency tolerance of 2100 Hz is only ± 1 Hz to discriminate 2100- and 15-Hz modulations [Schulzrinne and Petrack (2000)]. The depth of amplitude modulation is $20 \pm 1\%$. After detecting ANSam, echo cancellers are fully disabled or bypassed.

14.5.5   /ANSam Tone

Supergroup-3 fax machines and few modems send /ASNam [ITU-T.V.8 (2000)] tones. G3 fax machines support up to a 14,400-bps rate using V.17, and SG3 supports up to a 33,600-bps rate using V.34. The modified answer tone (ANSam) with phase reversals (/ANSam) is a sine wave signal at 2100 ± 1 Hz with phase reversals at intervals of 450 ± 25 ms, amplitude-modulated by a sine wave at 15 ± 0.1 Hz. The modulated envelope is of $20 \pm 1\%$ on tone average amplitude. The average transmitted power of 0 to -30 dBm is in accordance with recommendation V.2. In G3 fax up to 14,400 bps, echo enable and NLP disable is suggested. In SG3, it is suggested to disable the echo canceller unless it is thoroughly validated for SG3 and a combination to be working.

14.5.6   V.21 Preamble Sequence

After completion of CED from the answering fax machine, the preamble flag starts with a delay of 75 ± 20 ms [ITU-T-T.30 (2005)(2005)]. If CED is not detected, V.21 preamble flag detection is used as a fax answer indication. Two channels in V.21 are based on the center frequency. Channel-2 is used for G3 fax calls. In channel-2, FSK modulations are at frequencies of 1750 ± 100 Hz. Power levels of greater than -43 dBm is considered a valid V.21 signal. V.21 bits are framed with a preamble flag sequence followed by a handshake message of CSI and DIS. All handshake messages start with a preamble of flag character bits of 01111110 (0x7E hexadecimal number) repeated for 1 second ± 15% (0.85 to 1.15 seconds). FSK demodulation is explained separately in the V.21 section in this chapter.

14.5.7 Modems Call Setup Tones

On dialing, a modem call starts with a call indication (CI) signal that is the V.8 alternative to the call tone (CT) signal defined in V.25 [ITU-T-V.25 (1996)]. Modems not starting with the V.8 CI signal will use this CT. After detecting the CI signal, the receiving modem responds with the answer tone (/ANS) as an acknowledgment tone for a duration of 2.6 to 4.0 seconds. If the data modem chosen is in V.34 mode, it responds with a modified answering tone (ANSam), and some V.34-based modems also send the /ANSam tone. The CI or CT signal from the originating modem and the /ANS or ANSam tones from the receiving modem are used to detect the modem call in VoIP.

The CT signal is a 1300-Hz tone with an ON-period duration between 0.5 and 0.7 s and an OFF-period duration in between 1.5 and 2.0 s. The CI signal is an FSK-modulated signal with V.21 (L) at 300 bps and ON /OFF cadence [ITU-T-V.25 (1996)]. The CI signal ON period is not less than three CI sequences and not greater than 2 s in duration [ITU-T-V.8 (2000)]. The CI signal OFF periods is between 0.4 and 2 s in duration. The CI sequence consists of 10 ones followed by 10 synchronization bits and call function octet. These 10 ones and 10 synchronization bits precede with each V.8 signal sequence, and it is called a preamble sequence for the CI sequence. Synchronization bits will vary for different V.8 sequences. The call function octet is preceded by a start bit zero and followed by a stop bit one. The CI signal uses V.21 channel-1 frequencies for modulation with a center carrier at 1080 Hz, bit-0 at 1180 Hz, and bit-1 as 980 Hz.

The modem call will not be detected entirely based on the answering tone as some data modems may respond with /ANSam. So it is required to analyze both the CI signal from the originator and the answering tone from the responder modem before switching to modem pass-through or MoIP. The CI indicates whether the calling terminal wants to function as a PSTN multimedia terminal (H.324), text phone (V.18), fax send/receive, or data modem. In practice, it is also reported that few data modems generate CNG tones in place of CI for a non-fax call; such modems would generate a false trigger for fax. Validating CI with /ANS or ANSam detection is one of the useful techniques for modem or fax discrimination.

14.6 TONES DETECTION

In the detection of tones for fax and modem, three different families of tones are considered.

- CNG pure fax tone at 1100 Hz
- ANS family ANS, /ANS, ANSam, and /ANSam with 2100-Hz phase and amplitude modulations
- V.21 FSK modulations

A summary on the fax and modem tones are given in Table 14.2. For complete specifications, refer to the ITU recommendations [ITU-T-V.8 (2000), ITU-T-V.25 (1996), ITU-T-G.168 (2004)] and RFC2833 [Schulzrinne and Petrack (2000)].

Table 14.2. Summary on Fax and Modem Tones and Parameters

Signal name	Frequency and ± tolerances	Phase, amplitude, frequency modulation	Power levels	Timing/ duration	Fax- or modem-specific information
CNG tone	1100 ± 38 Hz	No phase, No AM	Typical −12 dBm, usually >−40 dBm	0.5 ON, 3 s OFF repeats	sending side fax machine tone
CED or ANS tone	2100 ± 15 Hz	No phase, No AM	0 to −30 dBm	2.6 to 4.0 s duration	terminating fax tone
/ANS tone	2100 ± 25 Hz	Phase reversal 180 ± 25°, rejects 0 to 110°, no AM	0 to −30 dBm	minimum of two phase reversals for 450 ± 25 ms	mainly modems and some fax machines as anomaly
ANSam	2100 ± 1 Hz	No phase reversal, AM 15 ± 0.1 Hz	0 to −30 dBm	2.6 to 4.0 s	modems
/ANSam	2100 ± 1 Hz	Phase reversal 180 ± 25°, rejects 0 to 110°, AM 15 ± 0.1 Hz	0 to −30 dBm	With phase reversal and duration of 2.6 to 4.0 s	V.34-capable fax and some modems
V.21 FSK, DIS, etc.	1750 ± 100 Hz FSK	Binary FSK, channel-2 frequencies	greater than −43 dBm	1 s ± 15%	after answering fax ANS tones with a delay of 75 ± 20 ms

14.6.1 CNG Fax Tone Detection

The CNG detection involves passing the samples through a band-pass filter (BPF) centered at 1100 Hz. A typical bandwidth of BPF is between 64 and 100 Hz. An output of BPF is envelope detected and checked for proper power levels, duration, and repetition. Power levels have to be calibrated usually with a μ-law sine wave treating an 8159 sine wave amplitude as +3.17-dBm power.

The accepted power level for CNG is wider in range. Based on power level, several signals could pass through a 1100-Hz centered BPF and appear as CNG detection. To avoid false detections, it is required to monitor out-of-band energy. To take care of this requirement, it is required to pass the input through band elimination or a notch filter centered at 1100 Hz. The power levels observed in the notch filter path act as a reference for out-of-band energy. The fax machine generates CNG; hence, there is no extra voice or background acoustic pick-up in the fax call. Thus, a threshold of about 24 dB is allowed, which means out-of-band energy is at least 24 dB below the in-band (1100 Hz BPF path) power level for the duration of the tone. This 24 dB depends on the notch filter, notch bandwidth, and role off. The notch filter has to be wider than the BPF bandwidth. The notch filter may use a notch bandwidth of the order of 200 Hz. Broadly this approach with BPF and notch filters avoids the wrong detection of CNG with voice conversation. The logic can be perfected in many ways based on filter characteristics, various states, and timing of the voice and fax call.

Interpretation on Notch Filter Requirements. Assuming test input is at 1020 Hz at a power level of −10 dBm. A BPF at 1100 Hz may provide a response of 12 dB down at 1020 Hz with reference to a peak response at 1100 Hz. A 1020-Hz tone will appear in BPF output as −22-dBm power, because of 12-dB attenuation on top of −10-dBm input level. The signal at −22 dBm may be detected in the CNG detection process. With a notch filter channel, the notch filter will estimate a power of −10 dBm and the BPF path will estimate a power of −22 dBm, which will cause detection to fail. This sort of BPF and notch filter channel processing ensures proper first-level protection from multiple strong disturbances. A similar interpretation can be extended to speech, tones, and background disturbances.

14.6.2 ANS Family Fax and Modem Detections

ANS family tone detections are given in Fig. 14.6. ANS family tone detection has to take care of basic 2100-Hz tone validation, phase, and amplitude modulations. The principles followed in the CNG detection with two filters are essential for ANS 2100-Hz detection. Several holding and guard band characteristics are defined in the G.168 recommendations [ITU-T-G.168 (2004)] for 2100-Hz tones. These conditions will translate in the BPF, notch filter

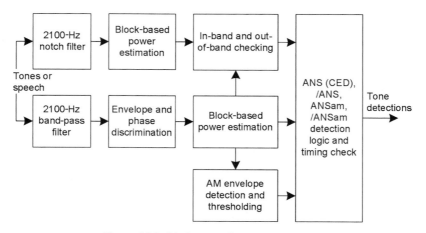

Figure 14.6. Modem and fax tone detections.

bandwidth, power decision thresholds, and timing on the BPF output envelope.

Phase Demodulation. Phase demodulation of $180 \pm 25°$ happens once in 450 ± 25 ms on /ANS and /ANSam signals. The BPF filter output for a $180°$ phase shift and a $110°$ phase shift is shown in Figs. 14.7 and 14.8, respectively. The waveforms are shown with phase modulation. BPF is a narrowband filter. Passing PSK signals through a narrowband gives amplitude change. It also gives very interesting observations. Phase discontinuity makes the filter change the waveform from one phase to another phase. This process takes the BPF output amplitude through deep nulls, and these nulls are required to be detected for the decision. If phase is $110°$, nulls amplitude is low. To distinguish between $180 \pm 25°$ and 0 to $110°$, it is required to look at the nulls depth.

In the case of CNG, 2100-Hz ANS and ANSam tones, a block-based power envelope is sufficient for comparison and decision making. In the case of /ANS and /ANSam, it is required to estimate the exact envelope on a sample basis to identify phase discontinuity. For this estimation, a simple single-pole infinite impulse response (IIR) filter given in [URL (SPRA073), URL (SPRA576)] can be used. If BPF filter output is $x(n)$, this sample-based envelope output $y(n)$ is given as

$$y(n) = y(n-1) + \frac{|x(n)| - y(n-1)}{16} \tag{14.1}$$

In the equation, $|x(n)|$ is the absolute value of $x(n)$. Notch filter output can be used as a block-based power estimation for threshold comparison.

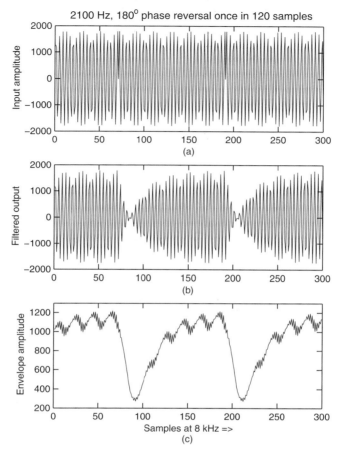

Figure 14.7. BPF output for 2100-Hz tone with 180-degree phase reversal once in 15 ms (15 ms used in place of 450 ms just to show waveform clearly). (a) Tone at 2100 Hz with phase reversals. (b) Band-pass filter output. (c) Amplitude envelope.

14.6.3 Detection Steps for /ANS

Initial validation of the 2100-tone is based on in-band (BPF path) absolute power in the range of 0 to −30 dBm, and this power is much greater than notch filter channel power, which ensures first-level criteria of 2100-Hz pure tone detection based on power levels and band energy. Overall, 2100-Hz tone power is from 0 to −30 dBm. With varying power, it is difficult to decide on the right threshold to distinguish between $180 \pm 25°$ and $110°$. Interestingly, these thresholds can be made adaptive. Tone average power is already known from pure tone. This power need not be calibrated on an absolute scale of dBm. It is for relative comparison. The nulls threshold can be derived as a factor of several dB down from the average power. This will track power and works

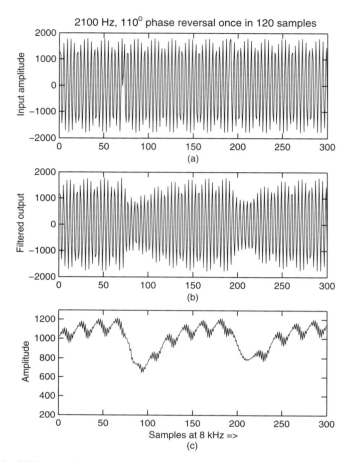

Figure 14.8. BPF output for 2100-Hz tone with 110-degree phase reversal once in 15 ms (240 samples, i.e., 15 ms used in place of 450 ms just to show waveform clearly with multiple phase changes). (a) Tone at 2100 Hz with phase reversals. (b) Band-pass filter output. (c) Amplitude envelope.

properly for a wide range of power levels. This type of adaptive threshold eliminates requirements of automatic gain control (AGC).

In general, the null width depends on the filter bandwidth. If BPF is too wide, then the null width is too small in duration. It is required to use a BPF of 64 to 100 Hz to get about a 10-to 15-sample null width at 8000-Hz sampling for 180°. The null width will keep reducing by a few samples for a 110° phase shift. This additional measure discriminates in width in addition to the null depth.

14.6.4 Amplitude Demodulation for ANSam and /ANSam

The steps of ANS and /ANS are applicable for ANSam and /ANSam for validating a pure 2100-Hz tone and for phase modulations of 180° for /ANSam. The envelope of the 2100-tone will include 15-Hz amplitude modulation. The envelope of a pure 2100-Hz tone has to be subsampled to greater than 30 Hz. A typical sampling of 100 Hz may be used as it matches with 10-ms VoIP packet intervals. AM envelope detection requires passing the envelope through 15-Hz BPF. The bandwidth could be about 2 to 5 Hz centered on 15 Hz. After ensuring proper filter gain calibration, the envelope can be validated with a mean envelope for the depth of the modulation of 20%. There could be a few logic steps and additional threshold windows to see that modulations are within a span of the 15% to 25% range. In this summary note, the basic principle for AM detection is given. Many variants could be based on the overall algorithm's design and reuse of any other developed modules.

14.6.5 Summary on Fax and Modem Detections

CNG and ANS detection can be simple detections with pure tones extraction with BPF and notch filter output power level comparison and absolute power thresholds. /ANS, ANSam, and /ANSam require several processing steps to get phase and amplitude modulations. The basic steps in processing for tone detections are given in this chapter. Using similar steps with an adaptive threshold scheme, a robust tone detector can be designed. V.21 is additionally used for discrimination of a fax/modem call, as V.21 is always present irrespective of the CED tone. In general, most tones are generated from fax machines at −10 to −20 dBm. Some tones are sensitive up to −43 dBm as per recommendations. In practical implementation, it may be sufficient to cater to power levels up to −20 to −24 dBm as a default with the configuration to work up to the lowest levels as per the recommendations. A V.21 detection scheme is given in the next section.

14.7 FAX MODULATIONS AND DEMODULATIONS

In the fax machine, extracted bits from the scanned images are modulated into the voice band of 300 to 3400 Hz. Several modulation techniques are mainly decided by the data rate requirements and support for interoperation. The popular group-3 modulations are ITU-T-V.21, V.27ter, V.29, and V.17 supporting up to 14,400 bps. Recently, V.34-based fax machines have also been launched for 33,600-bps fax transmission. The fax modem modules are together known by the name "fax data pump," "fax modem data pump," or simply "data pump." This section describes the basic modulation techniques and the functional blocks of facsimile modem modules of V.21, V.27ter, V.29, V.17, and V.34 [ITU-T-V.21 (1988), ITU-T-V.27ter (1988), ITU-T-V.29 (1988), ITU-T-V.17 (1991), ITU-T-V.34 (1998)]. Refer to Berg (2000), and documents [URL (SPRA080), URL (SPRA073), URL (ADI Vol-2)] for more details on fax modules, coding, formulation on modulation and demodulation.

14.7.1 Modulation

Modulation plays a key role in any communication system. It accepts a bit stream as input and converts it to an electrical waveform for transmission.

It can be used effectively to minimize the effects of channel noise, match frequency spectrum of the transmitted signal with channel characteristics, multiplex many signals, and overcome some equipment limitations. The most important parameters of the modulators are amplitude, frequency, phase, and bandwidth used. The modulator minimizes the effects of noise by the use of large signal power and bandwidth and by the use of waveforms that last for longer durations.

Modulation consists of superimposing the user information signal on a carrier signal, which is more adequate to the transmission medium. The carrier can also be used in a voice band of 300 to 3400 Hz for modem and fax applications. Typically, three basic types of modulation are used in fax equipment, namely amplitude modulation (AM), frequency modulation (FM), and phase modulation (PM). Quadrature amplitude modulation (QAM) is also used for higher rates of data. In QAM, amplitude and phase modulations are used simultaneously. The QAM system is a very efficient modulation technique in terms of bandwidth usage.

14.7.2 Demodulation

Demodulation is the inverse process of modulation. The modulated signal is transmitted to a receiver at the receiving station. The modulation information is extracted to bits in demodulation. Various modulation schemes used in fax processing are listed below:

- FSK—FSK is used with V.21 at a 300 baud rate. Each baud is of 1 bit.
- PSK—4 and 8 PSK are used in V.27ter demodulations.
- QAM is used in V.29.
- QAM in relation to TCM coding is used in V.17.
- QAM with TCM is also used in V.34. It is more complex than modulations in V.17.

Various modulation schemes and their principles are explained in the next section.

Baud Rate and Sampling Rate Consideration. In the telephone analog front end, sampling CODEC hardware consists of an analog-to-digital converter (ADC) and digital-to-analog converter (DAC) that sample the signal at 8000 Hz. Narrowband voice processing modules directly operate at 8000 Hz. To process for fax channels, ADC path samples of 8000 Hz are rate converted to arrive at integer multiples of a baud rate that keeps the required analog fax signal bandwidth. In the modulation DAC path, the sample rate converter modifies the samples back to 8000 Hz and delivers on the pulse code modulation (PCM) interface. Depending on the fax module in operation, the fax signal sampling rate has to cater to sampling at multiples of the baud rate (i.e., 300, 1200, 1600, and 2400 baud rates). The original samples at 8000 Hz can work directly for the 1600 baud rate. Catering to the sampling rate with an integer multiples baud rate helps in better alignment at symbol or bit boundaries. This type of sampling rate conversion is commonly used in multirate signal processing [Vidyanathan (1992)].

14.8 V.21 FAX MODEM

V.21 is the ITU-T recommendation for sending the data and messages at a 300 baud rate. Each baud is of 1 bit, which is the slowest rate used by the fax machines and data modems. All G3FE devices must incorporate this standard for exchanging the initial capabilities of two fax machines. V.21 uses a modulation type called FSK with 1 bit for modulation. V.21 and a summary of FSK specifications are given below. One frequency is used for logic "1," and another frequency is used for logic "0." Modems that use full-duplex communication use two channels with four different frequencies. Channel-1 is used for the modem that originally made the call, whereas channel-2 is used by the modem receiving the call. Group-3 fax machines communicate in half-duplex mode (i.e., one direction at a time) using channel-2 for both directions. Both G3 fax machines use the same frequencies with bit value "0" at 1850 Hz and bit value "1" at 1650 Hz. The key parameters of the V.21 recommendation are listed here:

Baud rate and modulation: 300 and FSK with capability for full duplex

Channel-1 frequencies: Center carrier 1080 Hz, bit-0 is 1180 Hz, bit-1 is 980 Hz

Channel-2 frequencies: Center carrier 1750 Hz, bit-0 is 1850 Hz, bit-1 is 1650 Hz (carrier is the mean of bit-0, bit-1 frequencies)

Power levels: Power output should not exceed 1 mW, and nominal level −13 dBm as per V.2 [ITU-T-V.2 (1988)]

Signal detection: Bit or signal is detected if signal level is of >−43 dBm

In full-duplex mode, the originating modem transmits a V.21 signal in low band and receives in high band and the receive modem transmits a V.21 signal in high band and receives in low band. This section is presented for V.21 basic modulation and demodulation techniques and the functional blocks of the V.21 fax data pump module. All G3 fax machines use V.21 for exchanging the initial capabilities of two fax machines. After the optimal capabilities of the two fax machines have been determined, fax machines switch to another form of modulation. V.21 is not used for sending the actual fax image data.

14.8.1 V.21 Implementation Aspects

The primary advantage of low-bit-rate FSK systems is that no requirement of carrier phase recovery exists, which reduces the processing complexity. FSK is a constant envelope signal that is robust against amplitude nonlinearities and modulations. The signal is preserved even while passing through many compression codecs like G.729A and G.726. The main disadvantage of the FSK system is its low spectral efficiency. On a 300- to 3400-Hz voice bandwidth, a 300-bps data rate is a very little utilization of bandwidth. In this section, V.21 implementation aspects are given with trade-offs.

14.8.2 V.21 Demodulation

FSK demodulation is required in V.21 demodulation and caller ID detections. V.21 demodulation is shown in Fig. 14.9 for the input samples at 8000 Hz.

In the rate conversion, 8000-Hz sampled signals are rate converted to the fax signal-sampling frequency. The rate-converted samples are band-pass filtered with bandwidth sufficient to pass FSK signals without any significant attenuation. Some processing schemes may use automatic gain control. In most situations, AGC is not required if the threshold design is made adaptive and using the adaptive threshold is a better option than AGC.

In the demodulator, the instantaneous sample is multiplied by samples that arrived 90° ($\pi/2$) phase shifted with reference to carrier frequency. In case of V.21, the channel-2 carrier is 1750 Hz. The phase shift of 90° may be implemented as a digital filtering operation. The original signal and the delayed signal are multiplied (product multiplication) and low-pass filtered. The product signal contains double the frequency component and a useful signal directly proportional to the sin(frequency deviation). The low-pass filter removes the high frequency, and the filtered low frequency represents the bits of level "0" and "1."

The demodulation process is described below, treating all the signals as analog signals. The same results are valid for digital samples also. For easy presentation, analog signal representation is used in the following equations [URL (SPRA080)]. The V.21 FSK modulated signal as input is taken as $r(t)$

$$r(t) = \cos[(\omega_c \pm \delta\omega)t + \varphi] \tag{14.2}$$

where

$$r(t) = \text{received V.21 analog signal}$$
$$\omega_c = \text{carrier frequency} = 2\pi f_c$$
$$\delta\omega = \text{frequency shift} [(2\pi)(100\,\text{Hz}) \text{ for V.21}]$$
$$t = \text{time}$$
$$\varphi = \text{random phase shift}$$

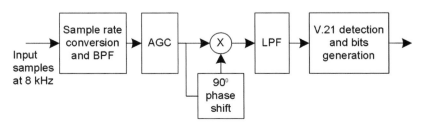

Figure 14.9. V.21 demodulation functional block diagram.

In a binary FSK system, the frequency of the signal is either $\omega_c + \delta\omega$ or $\omega_c - \delta\omega$, depending on whether a "0" or a "1" value is in the bit. The received signal $r(t)$ is multiplied by a delayed version by delay of τ.

$$r(t-\tau) = \cos[(\omega_c \pm \delta\omega)(t-\tau) + \varphi] \qquad (14.3)$$

The product of the received signal and the delayed received signal is

$$= 2\cos[(\omega_c \pm \delta\omega) \times t + \varphi]\cos[(\omega_c \pm \delta\omega)(t-\tau) + \varphi] \qquad (14.4)$$

Using the product formula for trigonometric identities

$$\cos(u)\cos(v) = \frac{1}{2}[\cos(u+v) + \cos(u-v)] \qquad (14.5)$$

Equation (14.4) can be reorganized as

$$= \cos[2(\omega_c \pm \delta\omega)t - (\omega_c \pm \delta\omega)\tau + 2\varphi] + \cos[(\omega_c \pm \delta\omega)\tau] \qquad (14.6)$$

If $\omega_c \tau$ is set to $\pi/2$, and the signal in Equation (14.6) is low-pass filtered to remove the double frequency component, the resulting signal is

$$\cos(\pi/2 \mp \delta\omega\,\tau) = \sin(\pm\delta\omega\,\tau) \approx \pm\delta\omega\,\tau \qquad (14.7)$$

For small angles, $\sin(\theta) = \theta$, where "θ" is in radians. If $\delta\omega\,\tau$ is greater than 0 (for FSK signals considered in V.21, this is valid), then the sign of the low-pass filter will be negative or positive depending on whether $\omega_c + \delta\omega$ or $\omega_c - \delta\omega$ is transmitted originally. The amplitude of this bipolar signal depends on $\delta\omega\,\tau$. In V.21, deviation from the center is 100 Hz, and $\tau = 142.86$ microseconds (i.e., one-quarter period of 1750-Hz carrier). $\delta\omega\,\tau = 2\pi(100) / (142.86$ microseconds$) = 0.0898$. In this analysis, amplitude is taken as unity and any other amplitude can be superimposed on the same formulation. The amplitude, sign change, and timing information of a 300 baud rate is used to extract the bits from V.21 demodulations. The above description is a theoretical formulation, and in actual implementation, engineers will impose several conditions, amplitude scaling, and timing analysis.

14.9   V.27ter FAX MODEM

V.27ter is a half-duplex voice band modem used in fax applications. The suffix "ter" or "Ter" along with V.27 conveys that it is the third revision. Some ITU recommendations were marked with the suffix "bis" and "ter." In French, *bis* means second and *ter* means third. When recommendations with "bis" and "ter" suffix are available, it is required to use "ter" as the latest. V.27ter is supported as a mandatory modem in fax. For simplification in naming, V.27ter is referred to as V.27 or V27 in representation. A summary on V.27 features are given below:

Baud rate and modulation: 1200 symbols at 2 bits per symbol (4-DPSK), making a total bit rate of 2400 bps as per V.26 Alternative A [ITU-T-V.26 (1988)] and 1600 symbols at 3 bits per symbol (8-DPSK), making a total bit a rate of 4800 bps

Center frequency: 1800 Hz and no separate pilot tones are used

Power levels: Power output should not exceed 1 mW, and nominal level −13 dBm as per V.2

3-dB bandwidth frequencies: 1000 and 2600 Hz for 4800 bps; 1200 and 2400 Hz for 2400 bps

Signal detection: Bit or signal is detected if signal level is greater than −43 dBm, and no signal detection if signal level of <−48 dBm. When transmission conditions are good, the signal detectors can be made less sensitive.

All amplitudes for 4-phase or 8-phase are the same. V.27 is supported in most fax machines. Because of a low rate and slow symbol rate, V.27 tolerates more distortions and it takes more time to send a fax page in V.27.

14.9.1 V.27 Modulator

The V.27 modulator and demodulator works similarly to most quadrature phase-shift keying (QPSK) based digital communication systems. The function of modulator or transmitter is to use bits as input and delivers modulated samples as output. A generic V.27 modulator is illustrated in Fig. 14.10(a). Fax image bits from a scanner are T.4 coded, and bits are sampled with a 4800-Hz or higher clock. These bits are mapped and modulated on the carrier. The carrier is usually sampled at multiples of 1200, and 8000-Hz sampling may be directly used for a 1600 baud rate because of dual baud rate support in V.27. These samples are rate converted to 8000-Hz sampling to deliver synchronously on the PCM interface. The modulator blocks are given in Fig. 14.10(a).

Scrambler: The input serial bit stream coded with T.4 is first scrambled by a self-synchronizing (requires no clock signal) scrambler. Scrambling takes the input serial bit stream and randomizes the data sequence that helps with whitening of the spectrum of the transmitted data. Employing a scrambler improves on carrier locking, uses the bandwidth of the channel more efficiently, and makes adaptive equalization and echo cancellation possible. A V.27 scrambler uses a $1 + x^{-6} + x^{-7}$ generating polynomial function with additional guards against repeating patterns of 1, 2, 3, 4, 6, 8, 9, and 12 bits. In implementation, a scrambler is built with a shift register and Exclusive OR (XOR) operations on binary bits.

Encoders: The scrambled bit stream is divided into groups of 3 bits (tribits) for the 4800-bps data rate and 2 bits (dibits) for the 2400-bps data rate.

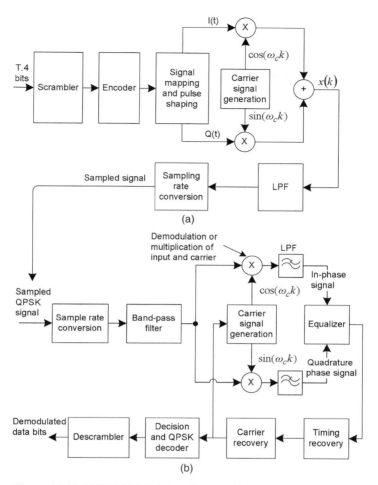

Figure 14.10. V.27. (a) Modulation diagram. (b) Demodulation diagram.

These combinations are given in Table 14.3. These bits are differentially encoded. Each tribit is encoded in steps of 45 degrees as a phase change relative to the phase of the preceding signal element. The left-hand bit of the tribit data stream enters the modulator portion of the modem first after the scrambler. In 2400-bps mode, each dibit is encoded in steps of 90 degrees as a phase change relative to the phase of the immediately preceding signal element.

Signal mapping: The 3- or 2-bit symbols are mapped into the signal space as per the V.27 recommendation for modulation. The signal space mapping produces two coordinates, one for the real part of the eight-phase or four-phase PSK modulator and another for the imaginary part. In V.27, all magnitudes are the same and take different phase shifts based

Table 14.3. Summary on V.27 Modulation Parameters

2400 bps		4800 bps		Normalized I-Q	
Dibits	Relative phase shift degrees	Tribits	Relative phase shift degrees	$I(t) = m_I$	$Q(t) = m_Q$
00	0	001	0	1.0	0
		000	45	0.707	−0.707
01	90	010	90	0	−1.0
		011	135	−0.707	−0.707
11	180	111	180	−1.0	0
		110	225	−0.707	0.707
10	270	100	270	0	1.0
		101	315	0.707	0.707

on the dibits or tribits value. The normalized in-phase is denoted with m_I and quadrature phase with m_Q. In QAM, amplitudes will vary in addition to phase shifts.

Pulse shape filters: The pulse shape filter is based on the impulse response of a raised cosine function. These filters attenuate frequencies above the Nyquist frequency that are generated in the signal mapping process. The filters are designed to cancel intersymbol interference. In 4800 bps, 3-dB bandwidth points are maintained at 1000 and 2600 Hz with a linear phase between 1100 and 2500 Hz. In 2400 bps, 3-dB bandwidth points are at 1200 and 2400 Hz with a linear phase between 1300 and 2300 Hz. In general, maintaining 3-dB points at 1000 and 2600 Hz will take care of both 4800- and 2400-bps rates.

Modulation: In V.27 modulations, in-phase and quadrature phase carriers are used to modulate the complex mapped symbols. The carrier frequency is 1800 Hz, and the modulation rate is 1200 symbols/second for the 2400-bps data signaling rate and 1600 symbols/second for the 4800-bps data signaling rate. These modulations will happen at multiples of the symbol rate (i.e., multiples of 1200 Hz and 1600 Hz in V.27 implementations). Sampling considerations for fax modulations are explained in the V.21 modulation section. Cosine and sine wave carrier are modulated with phase shift (i.e., in 45 degrees ($2\pi/8$ or $\pi/4$) in 8-PSK). In QPSK, a phase shift of 90 degrees ($2\pi/4$ or $\pi/2$) is used, and in binary phase-shift keying (BPSK), a phase shift of 180 degrees (π) is used. The following equations are given for representing the I-Q modulation output for 8-PSK:

$$s_I(t) = m_I(k)\left[\cos\left(2\pi f_c t + k\frac{\pi}{4}\right)\right] \qquad (14.8)$$

$$s_Q(t) = m_Q(k)\left[\sin\left(2\pi f_c t + k\frac{\pi}{4}\right)\right] \qquad (14.9)$$

where k is the symbol-running index with 0, 1, 2, etc. at the symbol rate.

Low-pass Filter (LPF): The QPSK-modulated signals are bandlimited with LPF. It can be implemented either at the fax sampling rate or after rate conversion to 8000 Hz.

Sampling rate conversion: Sampling rate conversion is required to interface with the PCM interface at 8000 Hz. This process is also explained in basic demodulation.

14.9.2   V.27 Demodulator

The V.27 demodulator takes PCM or sampling rate converted samples and delivers a demodulated bit stream for V.27. A generic V.27 receiver block diagram is given in Fig. 14.10(b). The description of blocks is given here.

Input filter: The received signals at 8000 samples per second are rate converted for required fax signal sampling. The sampled input is filtered with a band-pass filter. The band-pass-filtered output is fed to the V.27 receiver module. In general, sampling rate conversion is reused by several fax modules.

Demodulator: The band-pass-filtered output is now multiplied with a local carrier of cos and sin components and low-pass filtered to extract the baseband signal. The extracted baseband signal containing both in-phase and quadrature-phase components is then multiplexed with the final baseband signal.

Equalizer: The samples received by the modem may suffer from intersymbol interference (ISI). The ISI results from the linear amplitude and phase dispersion in the channel, which broadens the transmitted signals and causes them to interfere with one another. To increase the transmission performance, a filter is required to estimate the inverse transfer function of the channel and to perform channel equalization. Typically, a linear adaptive transversal filter is used along with the least mean square (LMS) algorithm that adjusts the coefficients.

Timing recovery: The purpose of timing recovery is to recover a clock at the symbol rate or a multiple of the symbol rate from the modulated waveform. This clock converts the received continuous-time signal into a discrete-time sequence of data symbols.

Carrier recovery: The carrier frequency is generated by a timing reference in the transmitter. Coherent demodulation of a signal requires exactly the same carrier frequency and phase. When a receiver first starts receiving data, it should first derive timing, then estimate the carrier phase, and finally adapt the equalizer.

Decision unit and differential decoder: Once the amplitude and phase of the signal point received is known, the corresponding symbol must be back mapped to decode the encoded bit.

Descrambler: The descrambler is implemented using a delay line, which is similar to the scrambler. The descrambler is the last functional block that the data passes through in the receiver. The data that are input to the descrambler are in effect multiplied by the appropriate generating poly-nomial. This multiplication performs the inverse operation of the scrambler.

Training the modem: The modem V.27ter and other modems such as V.29 and V.17 will have training patterns. On arriving at capabilities and on arriving at mutually supported parameters, training ensures the capabil-ity of the channels to send an actual fax at that rate. The features of the V.27 training patterns are listed below:

• Ensures data can be sent using the selected rate through training.
• Channel and equalizers can adapt to work at required rates.
• Falls back to lower rates and retrains if the selected rate is producing several errors.
• Training is conducted in four to five segments, with each one lasting for several symbols or tens of milliseconds. The longest training for V.27 is 1.158 seconds.
• Training segment combinations and duration varies with selected rate and in echo canceller protection mode.

14.10   V.29 MODEM

The V.29 modem operates in duplex and half-duplex mode. It can operate at data rates of 9600, 7200, and 4800 bits per second (bps). It uses 16-QAM for data rates of 9600 bps, 8-QAM for 7200 bps, and 4-PSK for 4800 bps. V.27ter has 4800-bps PSK support. To avoid duplicity, the 4800-bps mode of V.29 is supported from V.27ter. Data transmissions with V.29 and signal details are listed below [ITU-T-V.29 (1988), URL (SPRA073)]:

Baud rate and modulation: 2400 symbols, 16-QAM for 9600 bps, 8-QAM for 7200 bps, and 4-PSK for 4800 bps

Center frequency: 1700 ± 1 Hz, and no separate pilot tones are used

Power levels: Power output should not exceed 1 mW, and nominal level −13 dBm as per V.2

Signal detection: Signal is detected if signal level is greater than −26 dBm

QAM at 9600 bps. To get a 9600 fax rate with 2400 symbols per second, it is required to send 4 bits on each symbol as one group. A group of four consecu-tive data bits is called quadbits. The first bit (Q1) in time of each quadbit is used to determine the amplitude of the carrier to be transmitted. The second (Q2), third (Q3), and fourth (Q4) bits are used to encode the phase shift. The

Table 14.4. Phase and Relative Amplitude at 9600 bps

Q1 bit-0 magnitude (I-Q amplitudes m_I, m_Q)	Q1 bit-1 magnitude (I-Q amplitudes m_I, m_Q)	Q2	Q3	Q4	Phase change in degrees
3 (3.0, 0)	5 (5.0, 0)	0	0	1	0
$\sqrt{2}$ (1.0, −1.0)	$3\sqrt{2}$ (3.0, −3.0)	0	0	0	45
3 (0, −3.0)	5 (0, −5.0)	0	1	0	90
$\sqrt{2}$ (−1.0, −1.0)	$3\sqrt{2}$ (−3.0, −3.0)	0	1	1	135
3 (−3.0, 0)	5 (−5.0, 0)	1	1	1	180
$\sqrt{2}$ (−1.0, 1.0)	$3\sqrt{2}$ (−3.0, 3.0)	1	1	0	225
3 (0, 3.0)	5 (0, 5.0)	1	0	0	270
$\sqrt{2}$ (1.0, 1.0)	$3\sqrt{2}$ (3.0, 3.0)	1	0	1	315

Table 14.5. Phase and Relative Amplitude at 7200 bps

Relative fixed magnitude and (I-Q components m_I, m_Q)	Q2	Q3	Q4	Phase change in degrees
3 (3.0, 0)	0	0	1	0
$\sqrt{2}$ (1.0, −1.0)	0	0	0	45
3 (0, −3.0)	0	1	0	90
$\sqrt{2}$ (−1.0, −1.0)	0	1	1	135
3 (−3.0, 0)	1	1	1	180
$\sqrt{2}$ (−1.0, 1.0)	1	1	0	225
3 (0, 3.0)	1	0	0	270
$\sqrt{2}$ (1.0, 1.0)	1	0	1	315

relative amplitude of the transmitted signal element is determined by the first bit (Q1) of the quadbit and the absolute phase of the signal element as shown in Table 14.4.

QAM at 7200 bps. At the fallback rate of 7200 bits per second, the scrambled data stream to be transmitted is divided into consecutive three bits called tribit groups, which works similar to a 9600 fax rate except for amplitude. The relative amplitude is fixed at different phase shifts and takes only 3 and $\sqrt{2}$ for the phase combinations shown in Table 14.5.

QAM at 4800 bps. At the fallback rate of 4800 bits per second, the scrambled data stream to be transmitted is divided into groups of two consecutive data bits called dibits. The first data bit in time determines the Q2 of the modulator quadbit, and the second data bit determines the Q3 of the modulator quadbit. The Q1 of the modulator quadbit is a data ZERO for each signal element, meaning fixed amplitude. The Q4 is determined by inverting the modulo-2 of Q2 + Q3 using an Exclusive NOR (XNOR) operation. A QAM at 4800 is not used.

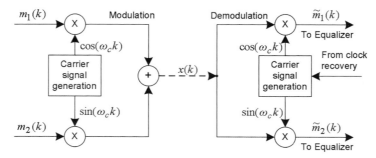

Figure 14.11. QAM modulator and demodulator.

QAM Formulation. The QAM is a very efficient modulation technique in terms of bandwidth usage. In QAM, two quadrature (90° phase-shifted) carriers, $\cos(\omega_c k)$ and $\sin(\omega_c k)$ are amplitude-modulated by two separate information-bearing signals as shown in Fig. 14.11. For input information-bearing sequences $m_1(k)$ and $m_2(k)$, the synthesized digital sequence can be expressed as

$$x(k) = m_1(k)\cos(\omega_c k) + m_2(k)\sin(\omega_c k) \tag{14.10}$$

In the above equation, m_1 is used here in place of I-component m_I; m_2 is used in place of Q-component m_Q. The spectrum components of the information-bearing signals overlap here. However, the quadrature phase relationship in the carrier components $\cos(\omega_c k)$ and $\sin(\omega_c k)$ allows the receiving end of the V.29 system to separate the two signals. The demodulation is performed as shown in Fig. 14.11. A digital phase locked loop is used to obtain the carrier component $\cos(\omega_c k)$ and to generate $\sin(\omega_c k)$. At the receiver, the received sequence is multiplied by the two quadrature carriers. This multiplication results in two signal sequences:

$$x(k)\cos(\omega_c k) = \frac{1}{2}m_1(k) + \frac{1}{2}m_1(k)\cos(2\omega_c k) + \frac{1}{2}m_2(k)\sin(2\omega_c k) \tag{14.11}$$

$$x(k)\sin(\omega_c k) = \frac{1}{2}m_2(k) + \frac{1}{2}m_2(k)\cos(2\omega_c k) + \frac{1}{2}m_1(k)\sin(2\omega_c k) \tag{14.12}$$

The information-bearing signal components $\tilde{m}_1(k)$ and $\tilde{m}_2(k)$ can be recovered by passing each sequence through a filter that rejects the double-frequency terms centered at $2\omega_c$. In the particular V.29 implementation, the carrier frequency (f_c) is 1700 Hz and the symbol rate is 2400 Hz.

V.29 Modulation and Demodulation. The block diagram of the V.29 transmitter and receiver will be similar to the V.27 presented in the previous section.

There will be some deviations in the blocks and their functions. Some main deviations are that the scrambler polynomial that operates on the data is different and the QAM mapping for different rates is similar to PSK mapping with different amplitudes. Filter characteristics, bandwidth, and training patterns are also different in V.29.

14.11 V.17 MODEM

The V.17 modem operates in full-duplex and half-duplex mode with a continuous or controlled carrier. V.17 with a half-duplex modem is used for high-speed facsimile applications that operate at data rates of 14,400, 12,000, 9600, and 7200 bps. The transmitter and receiver of V.17 modems make use of Trellis coded modulations (TCMs) for intermediate bits coding and QAM for final modulations. The V.17 receiver implements the receive side of a V.17 modem, which demodulates the modulated signal and produces the bit stream. The V.17 transmitter implements the transmit side of a V.17 modem, which takes the coded data and produces a modulated signal. V.17 uses 2400 symbols per second and bits varying in each symbol. After considering one bit per symbol for Trellis coding, this gives usable bit rates of 7200, 9600, 12,000, and 14,400 bps. Refer to [ITU-T-V.17 (1991), Berg (2000), URL (SPRA073), URL (ADI Vol-2)] for more details on V.17. V.17 data transmission and signal details are listed below:

Baud rate and modulation: 2400 symbols, TCM coding followed by QAM modulations. Inclusion of data scramblers, adaptive equalizer, and eight-state trellis coding

Two sequences for training and synchronization: long train and resynchronization

Data rates: 14,400, 12,000, 9600, and 7200 bit/s synchronous

Center frequency: $1800 \pm 1\,Hz$ and no separate pilot tones are used

Power levels: Power output should not exceed $1\,mW$, and nominal level $-13\,dBm$ as per V.2

Signal detection: Signal is detected if signals level is greater than $-43\,dBm$

Signal Element Coding for 14,400 bps. The basic steps for 14,400 rate modulation are listed below:

- Scrambling the input data bits.
- Making them as blocks of 6 bits for 14,400 rate (reduced number of bits for other rates).
- First 2 bits $Q1_n$ $Q2_n$ are differentially coded to generate $Y1_n$, $Y2_n$. The bits $Y1_n$ $Y2_n$ are used to generate $Y0_n$ using convolution coding.
- Seven bits $Y0_n$, $Y1_n$, $Y2_n$, $Q3_n$, $Q4_n$, $Q5_n$, $Q6_n$ are mapped to QAM for 128 levels.

Table 14.6. Convolution of Bit Pairs and Previous States for TCM

Input		Previous output		Outputs derived from $Q1_n$ $Q2_n$	
$Q1_n$	$Q2_n$	$Y1_{n-1}$	$Y2_{n-1}$	$Y1_n$	$Y2_n$
0	0	0	0	0	0
0	0	0	1	0	1
0	0	1	0	1	0
0	0	1	1	1	1
0	1	0	0	0	1
0	1	0	1	0	0
0	1	1	0	1	1
0	1	1	1	1	0
1	0	0	0	1	0
1	0	0	1	1	1
1	0	1	0	0	0
1	0	1	1	0	1
1	1	0	0	1	1
1	1	0	1	1	0
1	1	1	0	0	0
1	1	1	1	0	1

The scrambled data stream to be transmitted is divided into groups of six consecutive data bits $Q1_n$ to $Q6_n$, which are ordered according to their time of occurrence. The first two bits in each group, $Q1_n$ and $Q2_n$ (where n designates the sequence number of the group), are first differentially encoded into $Y1_n$ and $Y2_n$ according to Table 14.6. This follows the relation $Y1_n = Q1_n$ (XOR) $Y1_{n-1}$ and $Y2_n = Q2_n$ (XOR) $Y2_{n-1}$.

The two differentially encoded bits $Y1_n$ and $Y2_n$ are used as inputs to a systematic convolution encoder that generates a redundant bit $Y0_n$. This redundant bit and the six information-carrying bits ($Y0_n$, $Y1_n$, $Y2_n$, $Q3_n$, $Q4_n$, $Q5_n$, and $Q6_n$) are then mapped into the coordinates of the QAM modulation. For 6 bits of original input, it is required to map 7 bits as 128 QAM. Six bits at 2400 symbols per second is 14,400 bps.

Signal Element Coding for 12,000, 9600, and 7200 bps. In the 12,000-bps rate, the scrambled data stream to be transmitted is divided into groups of five consecutive data bits $Q1_n$, $Q2_n$, $Q3_n$, $Q4_n$, $Q5_n$. As shown in Table 14.7, $Y0_n$, $Y1_n$, $Y2_n$, $Q3_n$, $Q4_n$, $Q5_n$ bits are generated that are modulated in QAM-64. The bits marked in {Q} and {Y} are to show the correspondence between Q and Y bits in the process of convolution.

In the 9600-bps rate, the scrambled data stream to be transmitted is divided into groups of four consecutive data bits $Q1_n$, $Q2_n$, $Q3_n$, $Q4_n$. As shown in Table 14.7, $Y0_n$, $Y1_n$, $Y2_n$, $Q3_n$, $Q4_n$ bits are generated that are modulated in QAM-32.

Table 14.7. Bit Mapping in V.17 14,400, 12,000, 9600, and 7200 bps

Rate, modulation description at 2400 symbol rate	Input bits	Output bits for TCM/QAM
14,400 bps, 6-bits input, TCM/ QAM-128 with 7-bits	$\{Q1_n\ Q2_n\}\ Q3_n,\ Q4_n,\ Q5_n,$ $Q6_n$	$\{Y0_n\ Y1_n\ Y2_n\}\ Q3_n,\ Q4_n,\ Q5_n,$ $Q6_n$
12,000 bps, 5-bits input, TCM/ QAM-64 with 6-bits	$\{Q1_n\ Q2_n\}\ Q3_n,\ Q4_n,\ Q5_n$	$\{Y0_n\ Y1_n\ Y2_n\}\ Q3_n,\ Q4_n,\ Q5_n$
9600 bps, 4-bits input, TCM/ QAM-32 with 5-bits	$\{Q1_n\ Q2_n\}\ Q3_n,\ Q4_n$	$\{Y0_n\ Y1_n\ Y2_n\}\ Q3_n,\ Q4_n$
7200 bps, 3-bits input, TCM/ QAM-16 with 4-bits	$\{Q1_n\ Q2_n\}\ Q3_n$	$\{Y0_n\ Y1_n\ Y2_n\}\ Q3_n$

In the 7200-bps rate, the scrambled data stream to be transmitted is divided into groups of three consecutive data bits $Q1_n$, $Q2_n$, $Q3_n$. As shown in Table 14.7, $Y0_n$, $Y1_n$, $Y2_n$, $Q3_n$ bits are generated that are modulated in QAM-16.

Training and Synchronizing Sequences. Two separate sequences of training and synchronizing signals are used in V.17. The training sequences are separated as four segments, and each one has a specific function in training. The long train sequence that lasts 1393 ms is for initial establishment of a connection or for retraining when needed. The resynchronization (re-sync) sequence of 142 ms is for resynchronization after a successful long train. The details on various training patterns for different rates are given in the V.17 recommendation.

14.11.1 V.17 Modulator

The generic block diagram of the modulator or transmitter is illustrated in Fig. 14.12(a). V.17 uses TCM, and the data to be transmitted is scrambled. The least significant 2 bits of each symbol are differentially encoded using a simple look-up approach or exclusive OR operation. The resulting 2 bits are convolution encoded producing 3 bits. The extra bit is the redundant bit of the Trellis code. The other bits of the symbol are unchanged. The resulting bits define the constellation point to be transmitted for the symbol. The redundant bit doubles the size of the constellation, and so it increases the error rate for detecting individual symbols at the receiver. However, when a number of successive symbols is processed at the receiver, the redundancy actually provides several dB of improved error performance.

The standard method of producing a QAM-modulated signal is to use a sampling rate that is a multiple of the baud rate. The raw signal is used as a series of complex pulses, each lasting for an integer number of samples. These pulses can be shaped, using a suitable complex filter, and multiplied by a complex carrier signal to produce the final QAM signal for transmission. The encoder consists of a shell mapper, differential encoder, precoder and Trellis

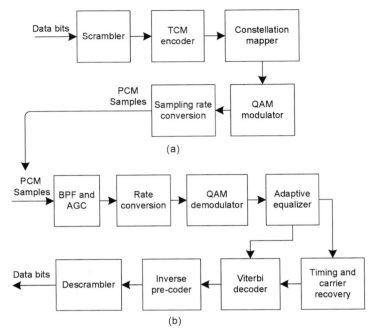

Figure 14.12. Trellis coded modulation. (a) modulation. (b) demodulation.

encoder, as well as a nonlinear encoder. Three new techniques have been employed in the V.17 encoder: rotationally invariant TCM (RI-TCM), shell mapping, and precoding equalization. The concept behind the RI-TCM is that the coding and modulation are combined so that the modem can benefit from high reliability with a limited transmission frequency bandwidth. The standard recommends three convolution encoder structures for the transmitter. The three Trellis codes, 16-state rate-2/3, 32-state rate-3/4, and 64-state rate-4/5, all with four-dimensional QAM constellation, can achieve asymptotic coding gains of about 5 dB. Shell mapping reduces the average signal energy and thus improves the modem performance without changing the signal-to-noise ratio (SNR). The precoding technique, which is a nonlinear equalization scheme, can reduce the effects of noise enhancement in the equalizer.

14.11.2   V.17 Demodulator

The generic V.17 receiver block diagram is shown in Fig. 14.12(b). The received analog signal from an analog front end (AFE) at 8000 samples per second is rate converted to fax signal sampling. The sampled input is filtered with a BPF. The band-pass-filtered output is fed to the receiver modules. The AGC operates over a wide dynamic range and maintains the output signal at required levels based on receiver design. This is necessary for the proper operation of the modem receiver algorithms. In the receiver, several modules use amplitude thresholds to make their decisions. During every sample interval, the AGC is executed. However, the gain update is disabled until the valid V.17 signal is detected. AGC module output is passed through the complex Hilbert filter, which gives both the real and the corresponding quadrature component of the signal. The complex Hilbert output is demodulated by the local receive carrier.

The samples received by the modem suffer from ISI. The ISI results from the amplitude and phase dispersion in the channel, which broadens the transmitted signals and causes them to interfere with one another. To increase the transmission performance, the channel equalizer performs the required inverse operation of the channel and distortions. The timing and carrier recovery unit recovers the clock at the symbol rate, or a multiple of the symbol rate, from the modulated waveform. This clock converts the received continuous-time signal into a discrete-time sequence of data symbols. The carrier frequency is generated by a timing reference in the transmitter.

The output of the QAM demodulator and the equalizer is a sequence of noise-corrupted points. The Viterbi decoder attempts to map these points to their ideal constellation points; simply, the decoder determines the most likely sequence of transmitted points. Viterbi decoding is made possible from the usage of the Trellis encoder in the transmitter to create a Trellis sequence. The output of the Viterbi decoder corresponds to the estimated Trellis coding of the signal at the transmitter. Because the signal is precoded at the transmitter before Trellis coding, the Trellis sequence at the transmitter is not the same as the one selected by the mapping unit at the transmitter. For this reason, inverse precoding is applied at the receiver.

The V.17 modem uses the self-synchronized descrambler having the polynomial $1 + x^{-18} + x^{-23}$. At the receiver, the received polynomial, of which the received data sequence represents the coefficients in descending order, is multiplied by the scrambler-generating polynomial to recover the message sequence.

14.12 V.34 FAX MODEM

V.34 gained popularity because it allowed dial-up modems to get higher speed Internet connectivity. The popular mode of operation is at 28,800 bps with the option of supporting up to 33,600 bps. This rate is three times faster than V.29 and twice as fast as the previous V.17 modems. V.34 [ITU-T-V.34 (1998)] supports 14 rates starting from 2400 to 33,600 bps in steps of 2400 bps [URL (Canata)]. Fax machines used to use the popular speeds of 9600 and 14,400 bps. Recently, fax machines were upgraded to 33,600 bps, which has increased fax transmission by two times compared with previous 14,400-bps machines. Up to 14,400 bps, the machines are called group-3 supported machines. V.34-supported machines are SG3. V.34 also makes use of TCM and QAM. In general, V.34 is a more involved processing and memory-taking module compared with V.17. To upgrade an existing solution, it is required to account for this difference. For new design, catering to V.34 may be the requirement.

Some counterarguments on the use of V.34 fax over IP exist. In many countries of Asia (Japan and Korea) and Europe, broadband data connections are provided to the users through asymmetric digital subscriber line (ADSL2), very high-speed DSL (VDSL), VDSL2, and fiber. This broadband with proper implementation of quality of service (QoS) are making end-to-end packets delivery with end-to-end delays comparable with 50 ms and absolutely no packet drops. Thus, a justification can be made for using the fax pass-through mode. The front-ends and pass-through modes are made perfect. Hence, future higher speed fax may be just fax pass-through.

Some benefits of higher speed fax are given in [ITU-T-V.34 (1998)]. In VoIP, cost and duration of call may not be an issue. With the current improvements in various technologies, a user will be more comfortable in sending a fax quickly, which creates a better perception (time response) to the user with V.34 support. V.34-based fax solutions have created few additional requirements in the detection process.

- ANSam, /ANSam answering tones are generated in SG3 fax calls.
- Additional signaling messages are conveyed in V.21 as per V.8. Incorporation of V.8 became a requirement in SG3.
- On the module level, the G3 mode was using an echo canceller with the option of disabling it. In SG3, it has to be disabled.
- The pass-through mode has to disable echo cancellers for V.34.
- G3 modules of V.17/V.29 are not compatible with scaled rates of V34. To interoperate between G3 and SG3 fax machines, the SG3 machine cannot use V.34 in fax over IP. It may use V.17 or V.29 modes in fallback modes.

14.13 V.21 HDLC FRAMING AND DEFRAMING

The main purpose of the V.21 FSK modem is to exchange control information between fax modems. This exchange of control information between two fax modems is established using the HDLC protocol. The HDLC protocol specifies a packetization standard for serial links. The basic HDLC structure consists of several frames, each of which is subdivided into many fields. It provides frame labeling, error checking, and confirmation of correctly received information. A basic HDLC frame structure is given below [ITU-T-T.30 (2005), URL (HDLC)].

Flag field: The 8-bit HDLC flag sequence is used to denote the beginning and end of the frame. For the facsimile procedure, the flag sequence is

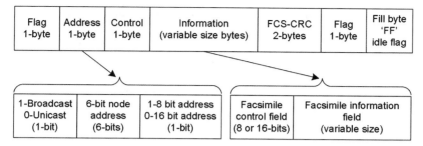

Figure 14.13. HDLC packet format.

used to establish bit and frame synchronization. The trailing flag of one frame may be the leading flag of the next frame. Continued transmission of the flag sequence may be used to signal to the distant terminal that the terminal remains on line but is not currently prepared to proceed with the facsimile procedure. The 8-bit flag sequence is represented in the format "0111 1110" in binary or "0x7E" in hexadecimal format. In the idle state or in between HDLC frames, fill flags may be transmitted with the flag identifier as "11111111 – byte value 0xFF" or 0x7E. The fill bytes are optional and vary with the implementation.

Address field: The length of the address field is of 0, 8, or 16 bits depending on the data link layer protocol. The 8-bit HDLC address field is intended to provide identification of specific terminals in a multipoint arrangement. In the case of transmission on the general switched telephone network, this field is of one byte. Most of the fax implementations use format "11111111" in binary or "0xFF" in hexadecimal format.

Control field: The 8-bit HDLC control field provides the capability of encoding the commands and responses unique to the facsimile control procedures. The control field byte is in format "1100 X000" in binary. Where a bit designed as "X" selects two options. X = 0 for nonfinal frames within the procedure and X = 1 for final frames within the procedure. A final frame is defined as the last frame transmitted prior to an expected response from the distant terminal.

Information field: The HDLC information field is of variable length and contains specific information for the control and message interchange between two facsimile terminals. In this recommendation, it is divided into two parts, the facsimile control field (FCF) and the FIF.

Facsimile control field (FCF): The facsimile control field is defined as the first 8 or 16 bits of the HDLC information field. This field contains complete information regarding the type of information being exchanged and the position in the overall sequence. Most fax messages use 8 bits as FCF. The commands that come under the FCF category are DIS, NSF, CSI, TSI, DCS, CFR, TCF, MPS, MCF, EOP, and DCN. An FCF of 16 bits is used for the optional T.4 error correction mode. The commands that come under the 16-bit FCF category are PPS-NULL, PPS-EOP, PPS-MPS, and PPS-EOM. The 16-bit FCF is represented as FCF-1 (first 8 bits) and FCF-2 (upper byte). FCF-1 indicates the signal for error correction such as PPS and EOR. FCF-2 indicates postmessage commands in the case of ECM mode. The commands that come under FCF-2 in are NULL, EOP, MPS, and EOM. The FCF-2 field is represented as part of the FIF field in some standards. The bit assignments within the FCF are given in Table 14.8 for some FCF categories as an example. The first bit of FCF is set to "1" by the originating terminal when it receives a valid DIS signal and is set to "0" by the called terminal when it receives a valid response to the DIS signal.

Facsimile information field (FIF): The facsimile information field is of variable length containing specific information related to facsimile control messages. In many cases, the FIF is followed by the transmission of additional 8-bit octets to clarify on the facsimile procedure. This information for the basic binary-coded system would consist of the definition of the information in the DIS, DCS, digital transmit command (DTC), calling subscriber identification (CSI), TSI, nonstandard facilities command (NSC), nonstandard facilities (NSF), nonstandard setup (NSS), password (PWD) for polling, selective polling (SEP), Subaddress (SUB), file diagnostic message (FDM), CTC, PPS, and PPR signals. Each of these messages will consider several options accommodated in 8-bit combinations. These messages and the details of the FCF and FIF fields are given in the T.30 recommendations [ITU-T-T.30 (2005)]. As shown in Table 14.8, the FIF of different classes/types of data is present. Some possible combinations of the information field are listed below:

- Only the FCF value, no FIF, and the commands/messages that come under this category are CFR, FTT, MCF, and DCN.
- 16-bit FCF (FCF-1 and FCF-2) and 3 bytes of FIF information, example: PPS-EOP, PPS-MPS, EOR-MPS, and EOR-EOP.
- FCF and several bytes of FIF to help FCF and data, example: DIS, DCS, and DIS.
- FCF and FIF as only data, example: optional T.4 error correction mode indicated in Fig. 14.14(b).

FCS field: Frame check sequence, 16-bit cyclic redundant check–committee consultative international telegraph and telephone (CRC–CCITT) polynomial is used to calculate the CRC. By checking the FCS, the receiver can discover bad data. If the data are correct, it sends an "acknowledge" packet back to the sender. The sender can then send the next frame. If the receiver sends a "negative acknowledge" or simply drops the bad frame, the sender either receives the negative acknowledge or runs into

Table 14.8. FCF and FIF Values for Some Example Messages

Message	FCF value	FIF field details
Facsimile message from the called to the calling terminal		
NSF	0x04	1-byte country code, and several bytes of nonstandard capabilities of receiving fax
CSI	0x02	phone number of the receiving fax
DIS	0x01	Several bytes indicating capabilities of called fax
Facsimile messages from the calling to the called terminal		
NSS	0xC4/0x44	country code (1 byte) and several bytes of nonstandard capabilities of calling fax device
TSI	0xC2/0x42	phone number of the calling fax
DCS	0xC1/0x41	capabilities of originating fax terminal
Pre-message responses		
CFR	0x21	no FIF
FTT	0x22	no FIF
Post-message commands		
MPS	0xF2/0x72	no FIF
EOP	0xF4/0x74	no FIF
EOM	0xF1/0x71	no FIF
Post-message command responses		
MCF	0x31	no FIF
RTP	0x32	no FIF
RTN	0x33	no FIF
DCN	0x5F/0xDF	no FIF
Post-message commands in ECM		
PPS-NULL	FCF-1: 0xFD/0x7D (PPS)	3 bytes of facsimile information fields
	FCF-2: 0x00 (NULL)	FIF field 1 for page count
		FIF field 2 indicates partial page block count
		FIF filed 3 indicates the number of frames transmitted inpartial page or block (up to 255 frames)
PPS-EOP	FCF-1: 0xFD/0x7D (PPS)	FIF bytes same as in PPS-NULL
	FCF-2: 0xF4/74 (EOP)	
PPS-MPS	FCF-1: 0xFD/0x7D (PPS)	FIF bytes same as in PPS-NULL
	FCF-2: 0xF2/72 (MPS)	
PPS-EOM	FCF-1: 0xFD/0x7D (PPS)	FIF bytes same as in PPS-NULL
	FCF-2: 0xF1/71 (EOM)	
RR	0xF6/0x76	no FIF
CTC	0xC8/0x48	two bytes of FIF, same as 1 to 16 bits of DCS
Post-message command responses in ECM		
PPR	0x3D/0xBD	FIF of 0 to 255 bits (32 bytes) with each bit indicating the HDLC FCD frame delivery status. Bit set as 0 for correct, and bit as 1 for wrong delivery
RNR	0x37/0xB7	no FIF, With errors same as PPR
CTR	0x23/0xA3	no FIF

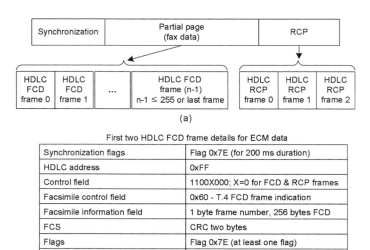

(a)

First two HDLC FCD frame details for ECM data

Synchronization flags	Flag 0x7E (for 200 ms duration)
HDLC address	0xFF
Control field	1100X000; X=0 for FCD & RCP frames
Facsimile control field	0x60 - T.4 FCD frame indication
Facsimile information field	1 byte frame number, 256 bytes FCD
FCS	CRC two bytes
Flags	Flag 0x7E (at least one flag)
HDLC address	0xFF
Control field	1100 0000 (0xC0)
Facsimile control field	0x60 – T.4 FCD frame indication
Facsimile information field	1 byte frame number, 256 bytes FCD
FCS	CRC two bytes
Flags	Flag 0x7E (at least one flag)
... (Third to n-2 HDLC FCD frames)	(Third to n-2 FCD frames, continued ...)

(n-1) or last HDLC FCD frame and HDLC RCP frame (0) details

HDLC address	0xFF
Control field	1100 0000 (0xC0)
Facsimile control field	0x60 - T.4 FCD frame
Facsimile information field	1 byte frame number, 256 bytes FCD
FCS	CRC two bytes
Flags	Flag 0x7E (at least one flag)
HDLC address	0xFF
Control field	1100 0000 (0xC0)
Facsimile control field	0x61 – RCP frame indication (no FIF)
FCS	CRC two bytes
... (RCP frames 1 & 2 details)	(RCP frames 1 & 2 continued ...)

(b)

Figure 14.14. HDLC format for ECM frames. (a) Representation for FCD and RCP frames. (b) Details on FCD and RCP.

its time limit while waiting for the acknowledgment. It then retransmits the failed frame. The 16-bit CRC–CCITT polynomial is given as $1 + x^5 + x^{12} + x^{16}$. Some HDLC frames may use 32 bits for CRC to improve the reliability given in [ITU-T-V.42 (2002)] as $1 + x + x^2 + x^4 + x^5 + x^7 + x^8 + x^{10} + x^{11} + x^{12} + x^{16} + x^{22} + x^{23} + x^{26} + x^{32}$.

Bit stuffing: Actual binary data may have a sequence of bits that is the same as the flag sequence. Therefore, the data bits sequence must be transmitted in such a manner that it does not appear to be a frame delimiter/flag sequence. It is achieved using bit stuffing. The bit stuffing is done as follows:

- Transmitter inserts extra "0" after each consecutive five "1"s inside the frame
- Receiver checks for five consecutive "1", i.e., "11111" as a bit pattern
  - If the next bit = 0, it is removed.
  - If the next two bits are 10, then the flag is detected.
  - If the next two bits are 11, then the frame is identified as an error.

14.14 HDLC MESSAGES IN ECM

The ECM aspects are given in Section 14.4.3. In this section, the ECM payload-specific overview is given. The G3 facsimile machine uses a basic HDLC frame structure, including the flag sequences, address field, control field, information field, and frame check sequence for transmitting the image data in ECM mode. The HDLC structure for a partial page is shown in Fig. 14.14(a). The details on HDLC FCD and RCP frames are listed in Fig. 14.14(b).

In ECM mode, every new partial page block starts with a series of synchronization flags for a nominal duration of 200 ms. The synchronization flags are used prior to the first HDLC frame. These flags are transmitted for about 1 second ±20% for the V.21 facsimile messages procedures. Similar to the V.21 HDLC frame, an 8-bit flag sequence is used to denote the beginning and end of an HDLC frame. The flag sequence is also used to establish bit and frame synchronization.

Synchronization sequence flag: This byte is part of the synchronization sequence. The synchronization training sequence is a series of flag sequences for nominal 200 ± 100 ms.

Address field: The address field is the same for both V.21 facsimile messages and T.4 ECM data of value: 0xFF.

Control field: The 8-bit HDLC control field provides the capability of encoding the command unique to the facsimile message procedure. Format: 1100 X000, where the bit X is set to 0 for the T.4 FCD frame and return to control for the partial page (RCP) frame.

Information field: The HDLC information field is of variable length and contains specific information of T.4 data. The information field is divided into two parts, the FCF, and the FIF. The FCF is defined as the first 8 bits HDLC information field.

FCF for T.4 FCD frame: 0110 0000 (0x60)
FCF for RCP frame: 0110 0001 (0x61)

The facsimile information field for the T.4 ECM HDLC frame is again divided into two parts: frame number of 1 byte and T.4 data of length of 256 or 64 bytes. It does not include bit stuffing for ECM T.4 data. RCP is the final frame of a partial page that does not contain the FIF.

Fax machines will send multiple RCP frames up to three at the end of each partial page. In the above sections of V.21 and HDLC, focus is given on the big picture for continuity on the topic. Several details and messages are coupled with V.21, HDLC format, and ECM-specific details with HDLC. Several key-words used in this section are not explained to the details. Refer to [ITU-T-T.30 (2005), ITU-T-T.4 (2003)] for more details on these topics.

14.15 SUMMARY AND DISCUSSIONS ON FAX

In this chapter, the basics on fax machine, coding techniques for fax data compression, fax call messages, timing, fax data modulations, HDLC framing, and deframing are given. The PSTN fax will be successful in most situations. Because of interoperability of fax machines and analog telephone lines, fax calls in PSTN may fail. Several mechanisms are built into fax coding, messages, timing, and supporting error correction modes for data that helps smooth fax transmission. In this book, the main interest is VoIP fax or fax over IP. Fax over IP systems mainly deals with fax modulations, messages, and timing. Certain additional techniques are incorporated into fax over IP systems specific to packet network impediments that are presented in Chapter 15. End-to-end fax over IP systems are made to work like PSTN, but some of these IP network techniques and features are different from the PSTN system.

15

FAX OVER IP AND MODEM OVER IP

The fax over IP (FoIP) application enables standard fax machines to work with VoIP infrastructure. The voice interfaces on VoIP customer premises equipment (CPE)—also called the VoIP adapter or gateway—can support fax calls much the same way public switched telephone network (PSTN) telephone interfaces support them. The two most popular methods to send fax over IP networks are store-and-forward fax service and real-time fax over IP. The main difference between the real time and store-and-forward approach is the delivery and method of reception confirmation. This chapter describes the VoIP fax, Session Initiation Protocol (SIP) based T.38 fax call, fax pass-through, fax over IP testing, and interoperability issues in fax over IP. The modem over IP (MoIP) is presented at a high level as an introduction to this topic.

15.1 FAX OVER IP OVERVIEW

The classification of FoIP is given in Fig. 15.1. Store-and-forward fax is based on the T.37 recommendation [ITU-T-T.37 (1998)]. It is used for off-line fax transmission. Store-and-forward fax is similar to e-mails, but at both the end terminations, fax machines and computers are used to interface for fax pages. In the store-and-forward mode, the caller sends the fax messages stored on the simple mail transfer protocol (SMTP) server to another SMTP server, and

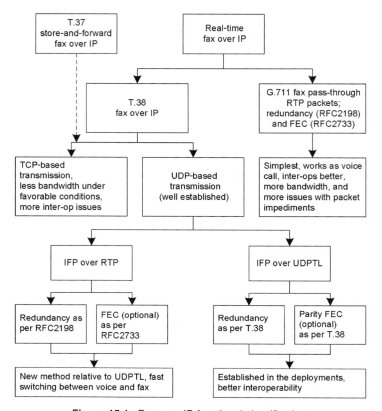

Figure 15.1. Fax over IP functional classification.

finally reaches the destination VoIP fax interface. At the destination, a fax can be delivered as an e-mail message either with attachment or as a fax to a standard PSTN-based fax machine. The caller is notified on the fax delivery status. Store-and-forward fax is used with both computers and fax machines. Store-and-forward fax does not give users some fundamental advantages of fax technology. It lacks guaranteed delivery because store-and-forward fax services treat fax transmissions like e-mail without any guarantee of delivery. Psychological satisfaction will be missing, because of off-line operation. Store-and-forward fax allows retrieval of fax later similar to e-mails offering this as one of the service advantage. Store-and-forward fax details are not given in this book.

Real-time fax over IP works similar to a regular fax call. Fax machines synchronize and send data over the IP link between the two connections. When the fax is busy, the caller gets a busy signal and the user has the option to retry sending fax later. Internet-aware fax (IAF) devices may have the option to revert to store-and-forward mode on identifying the busy conditions. The simple analogy can be given in relation to Internet Protocol (IP) phones. IP phones are having built-in phone and VoIP functions. IAF is a complete fax machine with built-in VoIP functions. The popular real-time modes of sending real-time fax on VoIP are fax pass-through and T.38 fax relay. Fax pass-through is similar to a G.711-based VoIP voice call [ITU-T-G.711 (1988)] with additional care in modifying certain functionality of echo cancellation (EC), voice activity detection/comfort noise generation (VAD/CNG), packet loss concealment (PLC), dual-tone multifrequency (DTMF) rejection, signal gain–loss settings, and jitter buffer. A fax call based on T.38 [ITU-T-T.38 (2005)] is a true real-time FoIP call. T.37 is a store-and-forward equivalent of T.38. FoIP with T.38 or T.37 uses fax modules of ITU-T-V.21, V.27ter, V.29, V.17, and V.34 [ITU-T-V.21 (1988), ITU-T-V.27ter (1988), ITU-T-V.29 (1988), ITU-T-V.17 (1991), ITU-T-V.34 (1998)]. These modules are used to extract bits for delivering payload on FoIP using Internet facsimile protocol (IFP) packets. These V.series modules are also known by the name "data pump." The management of a fax call is done through VoIP signaling protocols such as SIP, T.38, and T.30. A minimal part of T.30 [ITU-T-T.30 (2005)] is usually built inside the T.38 to handle T.30 timeouts and to simulate spoofing techniques, and hence, T.30 is usually not mentioned separately in VoIP fax calls. In the absence of T.38 and fax data pump support in the VoIP gateway or if interoperation of T.38 and data pump cannot progress with the call, a fax call is established in G.711 fax pass-through mode. Fax pass-through is used when there are no major network impediments and available bandwidth is more. It will be closely matching to a PSTN-based pulse code modulation (PCM) μ-law (PCMU) PCM A-law (PCMA) voice call under best end-to-end packet and signal transmission characteristics. Pass-through is closely related to the VoIP voice call. Refer to Chapter 2 for a VoIP voice modules overview. The T.38 standard is used for facsimile transmission in real time over the IP network, in which the T.30 fax from the PSTN is demodulated at the sending gateway. The demodulated bits of fax content are encapsulated into IFP packets. The IFP packets are sent over the network and the T.38 relay on the receiving gateway de-packetizes, performs the remodulation, and it sends the T.30 fax through the telephone interface to the answering fax device. At both fax machines, analog fax signals will appear very similar to a fax that is coming from PSTN lines.

As shown in Fig. 15.1 of fax calls classification, the T.38-based FoIP call makes use of user datagram protocol (UDP)-or transmission control protocol (TCP)-based transport protocols to deliver IFP packets on the IP network. TCP is a session-based, confirmed delivery service that does not need error control techniques such as redundancy and forward error correction (FEC) [Perkins et al. (1997), Rosenberg and Schulzrinne (1999)]. In TCP transport, the IFP payload is encapsulated in a transport protocol data unit packet (TPKT). The TPKT header as defined in RFC1006 [Rose and Cass (1987)] precede the IFP packet in TCP implementation. TCP implementation is more effective when the bandwidth for facsimile communication is limited and packet impediments are minimal. The usage of UDP with UDPTL or real-time transport protocol (RTP) may be more effective when the bandwidth for facsimile communication is sufficient.

In UDP transport, IFP packets are encapsulated using either RTP or UDP transport layer protocol (UDPTL). Sometimes RTP [Schulzrinne et al. (2003)] is shown as a separate method. RTP makes use of UDP for voice and fax transport and is shown under UDP in main classification. The T.38 fax over UDP with UDPTL for transport is the most popular and well-established method used in the current deployments. UDPTL makes use of redundancy and FEC techniques for error correction as per ITU-T-T.38 recommendation. Parity FEC algorithms takes care of burst packet losses based on the number of FEC messages encoded. The IFP with RTP and G.711 fax pass-through encapsulation uses the redundancy and FEC techniques as defined in RFC2198 and RFC2733, respectively, for error correction. The format for RTP headers is different for redundancy and FEC schemes. The error recovery techniques with UDPTL make use of procedures given in the ITU-T T.38 recommendation. The payload and packet formats are given in Chapter 16.

15.2 FAX OVER IP BENEFITS

A major part of PSTN voice service is migrating to the VoIP service. Hence, providing VoIP fax services is essential and useful for the migration path of fax services. Some main advantages of fax over IP are listed below:

- When VoIP service is used for fax, the fax machine does not need a separate PSTN line.
- Existing fax machines can be used, which is similar to existing phones used for voice in VoIP.
- The fax machine is connected within a few feet of the VoIP gateway; hence, there will not be any line distortions or noise issues. PSTN lines reach several miles from the PSTN digital loop carrier.
- Multiple levels of redundancy can be created without a major increase in bandwidth in T.38-and T.37-based fax calls.
- All the benefits of VoIP cost savings, able to receive and send faxes without location dependency is applicable to fax over IP.
- Features like fax store and forward can cater to storage of the fax communication, and allows later retrievals of fax similar to e-mails.
- Fax over IP also offers many other benefits such as secure, guaranteed delivery and fax broadcast. The management will be easier by combining e-mail, voice mail, and fax into a single universal platform.
- Internet faxing with IAF also offers many other benefits of several modes of VoIP fax services for real time, and store and forward.

The main issues with fax over IP are interoperation failures because of packet impediments such as an increase in end-to-end delays, link quality, bandwidth variations, packet loss, timing issues, and packet formats. The other issues are of voice to fax call switching, end-to-end transmission characteristics, clock drifts, various configurations, interactions between voice and fax modules, and not handling all the optional implementations that lead to failures in handling certain messages and errors in the middle of the fax call. With a well-designed system and deployment infrastructure, proper T.38 software implementations and FoIP calls can be improved for interoperability. The interoperability improvements are addressed in a separate section of this chapter.

15.3 FAX BASIC FUNCTIONALITY AND DETECTING FAX CALL

In this section, a summary on fax functionality mapped to PSTN and VoIP are presented. This section has an extended summary of Chapter 14 for creating continuity to map a PSTN fax call to VoIP. A fax scans a page and generates binary bits of "ones and zeros," which are modulated using amplitude, phase, and frequency variations depending on the selected modulation. These modulations are within the voice band frequencies of 300 to 3400 Hz. A PSTN network works as a transparent line for the voice band signals between two fax machines. At destination, the fax machine demodulates the analog signal to extract the bit pattern. These bits are finally composed as an image for printing.

A group-3 (G3) fax machine generates a 1100-Hz calling (CNG) tone and a 2100-Hz called terminal identification tone (CED) or answer tone (ANS) at the beginning of the fax call. The CNG name in voice is comfort noise generation of voice processing. In fax, the CNG name is for the calling tone. Some V.34 super group-3 (SG3) fax machines generate a call indication (CI) signal in place of the CNG tone. CI is a V.8 signal used as an alternative to call tone (CT), which is usually generated in the modem call setup phase. On detecting the CNG tone, the receiving SG3 fax terminal responds with an /ANSam tone. Signal /ANSam is a 2100-Hz answering tone with an amplitude-and phase-modulated tone.

The CNG tone is generated from the originating fax machine, and the CED or /ANSam is the acknowledgment tone generated from the terminating/destination fax machine. CNG and CED tones are sine wave tones lasting for a few seconds. The CNG and or CED tones are used to detect the fax call, which is also known as voice/fax call discrimination. In some VoIP systems, the V.21 preamble indication is used in addition to CNG, CED, or ANS, and/or the /ANSam tones for detecting the fax call. V.21 is a 300-bps frequency shift keying (FSK) modulation used to convey messages and acknowledgment bits in fax and modem calls. In general, a call switches from a voice to a fax call

(T.38 mode or pass-through) when it detects either ANS (CED) or /ANSam, or V.21 preamble or combinations of V.21 preamble detection and ANS or /ANSam signals. CNG is used to support the decision process, but CNG is not sufficient to make the decision. Fax call switching is made even without considering CNG.

VoIP gateways are designed to handle the several combinations of tones and indicators for initial fax call detection. More details on fax tone detections are given in the modem and fax tones detection section of Chapter 14.

- CNG tone, CED or ANS, /ANSam, and V.21 preamble tones are used as in-band packets passing through a voice call. The call setup tones generated by the fax terminal are sent as in-band (RTP) payload using preferred voice codecs such as G.711, G.729AB, G.723.1A, and G.726. The codecs G.729AB and G.723.1A compressed tones are also found to be accepted by the fax machines and tone detectors. When the CED or /ANSam or V.21 preamble indication is detected, the VoIP gateway switches to a T.38 fax call or fax pass-through call based on the gateway capability negotiations.

- In fax call CNG, CED, and V.21 preamble tones are sent as both in-band and T.38 IFP indicator packets. Assuming an initial voice call is established in-band mode, the fax tones are exchanged using in-band RTP payload with a preferred negotiated voice codec. After switching to a T.38 fax call, the gateways may also send the CNG, CED, and V.21 preamble indicator IFP packets over UDP/TCP. The peer gateway has to ignore these indicator packets when identifying tones exchanged through RTP media.

- CNG, CED, and V.21 preamble tones are sent in RFC2833 out-of-band packets in the voice call. On establishing an initial voice call in out-of band mode using RFC2833 capability exchange, the gateways may relay CNG or CED fax call setup tones in RFC2833 RTP payload format. The details of the RFC2833 RTP payload format for tones are explained in Chapter 7. In such modes, the gateway detects and suppresses the fax tones and the information of the tone event is passed in RFC2833 RTP payload. The peer gateway regenerates these tones to the fax terminal for establishing a T.38 or fax pass-through call.

- The CNG and CED tones are also sent directly as T.38 IFP indicator packets without establishing a voice call and are then switched to the T.38 fax mode. This type of implementation is mainly observed in IAF devices and VoIP fax servers. IAF devices directly start with indicator packets from the originating fax terminal starting with a CNG tone.

The fax call discrimination based on V.21 preamble flag detection at the receiving gateway is a generic method that allows regeneration of tones on the other end. After switching to T.38, the V.21 flags and additional T.30 messages are passed as IFP packets over UDPTL, RTP, or TPKT. The information of CNG,

CED, and V.21 as T.38 indicator packets are mainly implemented in fax-only Internet-aware facsimile devices.

15.4 T.38 FAX RELAY

The end-to-end functional representation of a real-time FoIP call is shown in Fig. 15.2. The calling G3 or SG3 facsimile equipment is connected to an emitting gateway, and the called G3 facsimile terminal is connected to a receiving gateway through the RJ-11 connector. The emitting gateway is connected to a receiving gateway through an IP network, which makes a VoIP call to the called gateway. Once the VoIP call is established at both ends, the two G3 facsimile terminals are linked virtually. An alternative option is available of connecting an Internet-aware fax device directly to the IP network. At the calling facsimile terminal, the optically scanned fax image data in the form of run lengths of black-and-white picture elements are coded according to the one-dimensional code book and optionally use the two-dimensional coding techniques as defined in T.4 [ITU-T-T.4 (2003)].

The coded fax data are then modulated to one of the transmission bit rates of 33,600, 28,800, 14,400, 12,000, 9600, 7200, 4800, and 2400 bps as defined in the standards ITU V.17, V.27ter, V.29, and V.34. ITU V.21 is defined as the low-speed 300 bps fax modulation used to exchange the binary procedural data. V.17, V.27ter, V.29, and V.34 are defined as the high-speed fax modulation protocol for T.4 and T.6 [ITU-T-T.6 (1988)] based fax equipment. The T.38 fax relay resides in the VoIP gateway that consists of the fax data pump, T.38, and VoIP signaling [URL (Cisco-fax1), URL (Cisco-fax2), URL (Audiocodes-fax)]. The fax data pump modules of V.21, V.27ter, V.29, V.17, and V.34 on the sending side perform the demodulation of voice band signals from an analog interface and send the decoded data to the T.38 protocol. T.38 reads the data from the fax data pump and packetizes the data as defined by the T.38 standard for transmission over the IP network. The T.38 on the receiving gateway

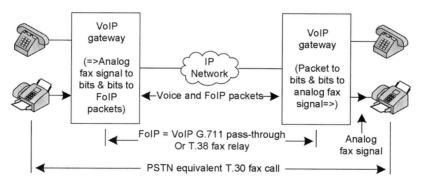

Figure 15.2. End-to-end functional FoIP call. [courtesy: used with permission of CED copyright © 2007. All rights reserved.]

depacketizes the packets obtained from the IP network. The data bytes are sent to the data pump. The data pump on the receiving relay performs the remodulation and sends the T.30 fax signals through the telephone interface to the answering fax machine. As shown in Fig. 15.2, both fax machines work as if the fax is on an analog interface and the fax is coming from the PSTN. Between VoIP gateways, a fax is sent as IP packets. The management of a fax call is done through the T.38 and VoIP signaling. Refer to the Chapter 14 to obtain PSTN fax call details and to become familiar with several keywords used in the fax call.

VoIP voice and fax calls can be established using ITU-T-H.323, media gateway control protocol (MGCP), SIP signaling protocols. In this chapter, an SIP [Rosenberg et al. (2002)] based fax call is considered as an example. In SIP-based fax calls, T.38 Annex–D procedures are used for the changeover from VoIP voice to fax mode during a call. An initial VoIP voice call is established with a compression codec such as G.711, G.729AB, G.723.1A, and G.726 as shown in Fig. 15.3. After starting the fax, the call setup tones generated by the fax equipment are passed through a compression codec with acceptable degradation. As described in Section 15.3, these tones are used to detect the fax call, which is also known as voice-to-fax call discriminator. When ANS or /ANSam or V.21 preamble indicator is detected by the receiving gateway, the receiving gateway switches to a fax call in the T.38 relay mode or in the simple pass-through mode. The signaling protocol at the receiving gateway is used to renegotiate the T.38 capabilities for a T.38-based fax call. After detecting these tones, a SIP INVITE request with T.38 capabilities in the Session Description Protocol (SDP) is sent to the emitting gateway with the same caller ID as the voice call. SIP-based T.38 fax relay call flow is shown in Fig. 15.3. Fax call events are given below:

1. The end user decides to send a fax through the IP network from the fax terminal and enters the phone number and then press the start fax button on the fax machine.

2. In FoIP, an initial voice call is established with any of the preferred voice codecs of G.711, G.726, G.729AB, or G.723.1A.

3. The receiving or terminating gateway detects a CED, which is also called a ANS tone or /ANSam or uses V.21 preamble flag sequence, and sends an SIP Re-INVITE request with T.38 negotiation parameters in the SDP field to the emitting/originating gateway or to the SIP proxy server, depending on the network topology.

4. The originating gateway receives the SIP Re-INVITE request, validates the T.38 negotiation parameters in SDP field, and sends back a 200 OK message with appropriate T.38 parameters in the SDP field. The gateway optional T.38 negotiation fields are show in Table 15.1. The fax bit rate in the negotiation is the maximum bit rate supported by the gateway, and it can be given a user-configurable parameter.

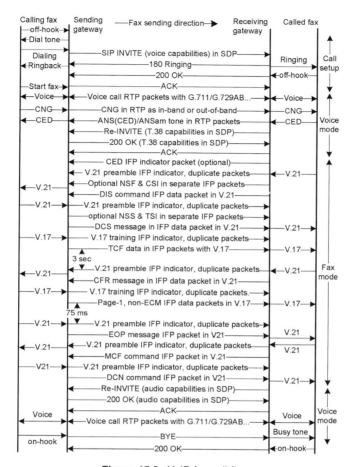

Figure 15.3. VoIP fax call flow.

5. The receiving gateway acknowledges the 200 OK messages and sends an acknowledgment (ACK) message to the originating gateway.

6. A fax call will be established in T.38 mode if the originating gateway supports the T.38 relay; otherwise, a fax call will fall back to the fax pass-through mode using the G.711 codec.

7. In T.38 mode, the receiving gateway starts sending T.38 IFP packets over UDPTL/RTP/TPKT based on the T.38 parameter negotiations between the gateways. Most gateways use UDPTL support to transmit the IFP packet on the IP network.

8. At the end of the fax transmission, the gateways can disconnect the fax call or switch to voice mode as shown in Fig 15.3. It also depends on the end fax terminal. When the end fax terminal is configured for fax-only mode, it immediately disconnects the fax call after sending the DCN

Table 15.1. Gateway Capability Support Indications

Option	Description
Version	Version number of ITU-T-T.38 and new versions shall be compatible with previous versions.
Data rate management	Method-1 is the local generation of TCF that is required for IFP over TCP. This method is optional over UDP. Method-2 is transferred to TCF used with UDP (UDPTL or RTP) and not recommended for use with TCP.
Data transport protocol	The fax-initiated gateway may indicate a preference for UDP/UDPTL, UDP/RTP, or TCP/TPKT for transport of T.38 IFP packets. The peer gateway selects the transport protocol.
Fill bit trans-coding	Indicates the capability to remove and insert fill bits in phase-C, non-ECM data to reduce bandwidth in the packet network. Optional field.
MMR trans-coding	Indicates the ability to convert between modified modified read (MMR) and line formats for increasing the compression. Optional field.
JBIG trans-coding	Indicates the ability to convert to/from joint bilevel image expert group (JBIG) to reduce bandwidth. Optional field.
Maximum buffer size	Maximum jitter buffer size (optional) used for UDP (UDPTL or RTP) modes; this option indicates the maximum number of octets that can be stored on the remote device before an overflow condition occurs.
Maximum datagram size	This is an optional field for maximum IFP packet size. This field is used to determine the IFP packetization period for high-speed data.
Error protection schemes	Error protection modes are UDP redundancy (mandatory) or UDP parity FEC (optional) for UDPTL as per T.38. RTP mechanisms use packet redundancy (RFC2198) and FEC protection (RFC2733) negotiated with RTP audio field in SDP. No error protection negotiated with TCP.
MaxBit rate	Maximum bit rate supported by the gateway; e.g., V.17 maximum bit rate field should be 14.4 kbps. Bit rates are 14,400/12,000/9600/7200/4800/2400.
T.38 vendor information	Fax machine-specific optional proprietary information.

message. In this case, the gateway sends the SIP BYE request to the other end after detecting the on-hook event from the end fax terminal. The other gateway responds with a 200 OK response after receiving the on-hook event from the fax terminal.

9. Figure 15.3 is given for the functional fax call. Several optional fields and call states are available as T.38 and T.30 while catering to the VoIP fax call. These options are not shown in Fig. 15.3.

In FoIP-to-VoIP voice call switching, the gateways can revert to the VoIP voice connection based on the following events:

- Detection of a T.30 disconnect (DCN) message.
- Detection of bidirectional silence as defined in the T.38 standard. It is recommended to detect transition back to voice mode on detection of more than 7 seconds of bidirectional silence to allow for the T.30 T2 timers within the G3 facsimile equipment to time out [ITU-T-T.38 (2005)].
- Detection of voice.
- Reception of re-INVITE request with audio media SDP descriptors.

At the end of the fax transmission, another SIP INVITE request is sent to the receiving gateway after detecting the DCN message or 7 seconds of bidirectional silence to return to VoIP voice mode. In the case of fax-to-voice mode, after receiving the DCN message, the gateway send the SIP re-INVITE request with the preferred voice codec option in the SDP field. The other gateway has the capability of switching to voice call, and it responds with 200 OK with audio media in the SDP field as shown in Fig. 15.3. Otherwise, the SIP BYE request is transmitted. In FoIP-to-VoIP switching, many other modules like echo canceller, VAD/CNG, and PLC may be enabled and voice may go through low-bit-rate codecs like G.729AB or G.723.1A based on negotiation.

A T.38 fax operation falling back to the fax pass-through mode of operation is given below. The gateways can fall back to fax pass-through mode in the following scenarios:

- Assuming the receiving gateway supports the T.38 fax relay and the originating gateway does not support the T.38 fax relay, then after detecting the CED/ANSam tone, the receiving gateway initiates the fax call in T.38 mode by sending the INVITE request and the originating gateway responds with 200 OK with audio media in SDP. Then a fax call will fall back to fax pass-through mode using the G.711 codec or G.726 at 32/40 kbps. All VoIP gateways support the G.711 codec.
- If the calling gateway receives a fail to train (FTT) message in response to the training check (TCF) test pattern generated at 14.4 kbps, the calling fax machine may fall back to the next lower rate and then it sends a new digital command signal (DCS) message followed by a TCF test pattern at the new data rate. This process is continued until a TCF test pattern is received OK or all data rates have been attempted. In some implementations, gateways are falling back to the pass-through mode by sending the SIP re-INVITE request using the G.711 codec if continuously the FTT message is received in response to the TCF test pattern more than three times.
- It is observed in the field that few gateways may establish the fax directly in pass-through mode if both or any of the fax machines connected are found to be an SG3 facsimile device based on fax tone /ANSam tone detection.

T.38 defines two types of IFP packets, namely T.30 indicator and T.30 data packets. The T.30 indicator packet defines a type of facsimile signal CED or CNG, preamble flags, and modulation indications. The T.30 data packet consists of signal type field and IFP data bytes. The IFP packet formats for indicator and data packets are given in Chapter 16. The IFP data element is a sequence of subpackets having data field type, optional size, and payload data. The data packets are sent during data demodulation of the V.21 handshake or T.4 image data transfer [ITU-T-T.4 (2003)] as shown in Fig. 15.3. T.38 packets are sent on a UDP or TCP transport protocol [Rose and Cass (1987), URL (RFC791), Postel (1980)]. RFC2327 [Handley and Jacobson (1998)] SDP provides mechanisms for describing a session for T.38. Several T.38-specific parameters have to be negotiated when establishing a T.38 media stream. The main classifications of parameters are given in Table 15.1 [ITU-T-T.38 (2005)].

T38 Version: T.38 version numbers are 0, 1, 2, and 3. IFP over TCP support is added in version 1. The abstract syntax notation.1 (ASN.1) notation is modified in version 2 with TCP support. The modified ASN.1 notation in version 2 and previous notations in version 0 or 1 cannot interoperate with each other. Version 3 supports V.34 fax. In the field, most initial deployments support version 0, and it recommends that version numbers are a mandatory field in T.38 to indicate which ASN.1 syntax is used. If no version number is provided, the default version 0 is assumed.

TCF Modes: The T.38 standard defines two methods of handling the TCF training signal. Method-1, the TCF signal is generated locally by the receiving gateway, and in method-2, the TCF signal is transferred from the originating gateway to the receiving gateway using IFP packets. TCF local generation is mandatory for IFP over TCP implementations and is optional for UDP implementations. Transferred TCF is mandatory for IFP over UDPTL/RTP. On identifying both gateways as IAF devices through digital identification signal (DIS)/DCS messages, it may use local generation of TCF. The TCF rate management is negotiated in SDP. IFP over TCP implementations should set the T.38 version to one or higher. T.38 version 0 implementations will not support the IFP over TCP.

15.4.1 HDLC Messages in PSTN and Fax over IP

A PSTN-based fax call is explained in Chapter 14. As continuity on V.21 and high-level data link control (HDLC), the fax over IP-related aspects are given in this section. In the case of fax over IP, initially a voice call will be established. Based on the CNG, CED, and V.21 preamble flag detections, a voice call will switch to fax pass-through or T.38 fax depending on the supported features and configurations. In the case of fax pass-through, all fax modulations are sent as voice using A-law or μ-law compression. In the case of T.38 support, several options in the implementation exist. In most implementations, CNG is allowed like voice. While receiving a CED, some implementations sending CED as voice and others as CED as a packet sending in out of band as explained in Section 15.3. With CED, a call will switch to T.38 without waiting for V.21. The signal in V.21 is usually sent as a HDLC packet, which implies a fax call will switch to out-of-band mode for sending V.21. Several deviations could occur here. In a PSTN call, HDLC stuffed bits as analog or digital will be delivered end-to-end. In VoIP, HDLC unstuffed information is delivered as IFP packets. At the destination, stuffing is recreated from IFP packets before delivering the V.21 data to the fax machine.

The V.21 flag sequences are sent for about 1 s whenever new transmission begins with V.21 messages in G3 fax. The flag sequences (0x7E) are indicated as a preamble indication in IFP packet over the IP network to the other end VoIP gateway. The flags (0x7E) are also generated between two HDLC frames but of shorter duration. The HDLC bytes like address, control, and information—facsimile control field (FCF) and facsimile information field (FIF) are sent in one IFP packet in fax over IP. HDLC frame data of called subscriber identification (CSI), DIS, DCS, and confirmation to receive (CFR) commands are packetized in one IFP packet as shown in Fig. 15.3. Some implementations send the HDLC frame bytes in separate IFP packets. In such cases, 1 byte of HDLC frame is packed as one IFP packet with the V.21 data type. Frame check sequence (FCS)–cyclic redundancy check (CRC) information is sent in a separate IFP packet or a single IFP along with V.21 data.

15.4.2 T.38 Fax Relay with ECM Support

Error correction mode (ECM) is designed to cope with transmission errors on the network. ECM handles error control during the image transfer. More details on the ECM technique in the PSTN network is given in Chapter 14. In ECM mode, the high-speed image T.4 bit stream is divided into 64 or 256 byte frames. The exact number of frame size is negotiated during phase-B through digital information signal (DIS) and DCS messages. The transmitting terminal divides the coded data into a number of frames and transmits them with frame number. Up to 256 frames constitute a T.4 or T.6 partial page. These frames are transmitted as payload inside HDLC frames with frame check sequence (FCS) for error detection by fax machines. The use of ECM mode is mandatory for SG3 facsimile devices, which uses a high-speed fax modem V.34 and T.6 coding scheme. The ECM is optional for the V.27ter, V.29, and V.17 fax modem that uses the modified Huffman (MH) and modified read (MR) coding schemes. The ECM mode is a mandatory option with fax machines that uses modified MR (MMR)/joint bilevel image experts group (JBIG), joint photographic expert group (JPEG) coding schemes. End fax machines using DIS/DCS fields negotiate the coding schemes. In FoIP, based on the coding method, the fax modem negotiated in DIS/DCS messages, and based on the gateway capabilities, the ECM mode can be modified in control message negotiation.

The originating gateway performs the demodulation of voice band signals from an analog interface and the decoded high-speed an HDLC frames may contain an HDLC header, data, and flag sequences. The T.38 relay on the sending gateway collects the whole HDLC frame and packetizes the HDLC frame data using IFP packets as defined in the T.38 standard and sent on the IP network. One HDLC frame can have a maximum of 260 bytes with 4 bytes of the HDLC header without FCS information and 256 bytes of facsimile data. The HDLC header is composed of address, control field, facsimile control field, frame number, and FCS. It takes about 150 ms at V.17 14.4 kbps to transmit the entire HDLC data. The ECM HDLC frame is also divided into several IFP packets based on the packetization period and is sent on the network in regular intervals. Because of HDLC packet overhead, T.38 fax transfer in ECM mode takes more time to transmit the entire page and demands more bandwidth than the non-ECM mode.

In ECM mode, a lower packetization interval of 20 or 40 ms is the preferred option. A higher packet interval like 100 ms creates additional delay in ECM mode, which may cause a failure in processing of postmessage sequences. At a 40-ms IFP packetization period, the HDLC frame at the V.17 14.4 kbps rate is fragmented into four IFP packets and transmitted on the network at regular intervals. T.38 relay has to transmit all these frames sequentially without adding delay between these IFP packets. If any one packet is dropped on the network, it may create an error and request retransmission. At the receiving gateway, during the ECM T.4 image transfer mode, the T.38 fax relay jitter buffer should be generic to handle the IP impediments such as burst packet drop within the HDLC frame. Some techniques to handle this case are to create a minimum threshold of two to three HDLC frames in jitter buffer to take care of IP impediments like jitter and high network delays. If T.38 does not receive any HDLC packets during image transfer, the data pump has to continue to generate the flag sequence or known pattern until the next frame is received, which will avoid retransmission of error frames in ECM. Retransmission of error frames will cause long delays and the cause fax call to time out.

15.5 FAX PASS-THROUGH

The fax pass-through call flow is shown in Fig. 15.4. In the absence of T.38 and data pump support, or if interoperation of T.38 and a data pump is an issue, a fax is transmitted in pass-through mode using the G.711 codec [URL (Cisco-fax3)]. Fax pass-through is the simplest technique for sending a fax over IP networks. In all VoIP deployments, G.711 is supported. Once a VoIP G.711 call is established, the fax in pass-through mode works similarly to a PSTN-based fax call. In fax pass-through mode, gateways do not distinguish a fax call from a voice call. When an initial voice call is established with low-bit-rate compression codecs like G.729AB/G.723.1A, on detection of CED or ANS or /ANSam or V.21 preamble flag by the gateway, the codec will be switched to G.711 (PCMU or PCMA) with no VAD and no echo cancellation for the duration of the fax session. If an initial voice call is established with G.711 or G.726–32 kbps, then on detection of the ANS family of tones, the gateway modifies

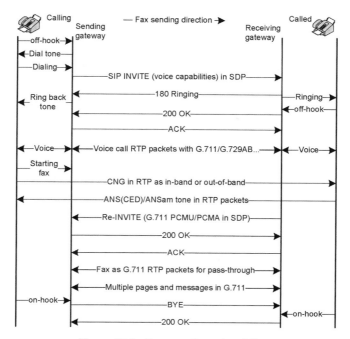

Figure 15.4. Fax pass-through call flow.

the other modules like EC and VAD for the duration of the fax session. In this scenario, a codec will not be modified. The VoIP gateway in pass-through mode performs the following functional SIP call events to establish a fax pass-through call:

- The initial call is established like a voice call with a dial plan.
- The receiving or terminating gateway detects an ANS/ANSam/V.21 flag sequence from a called fax machine, while the terminating gateway exchanges the voice codec for a G.711 codec and turns off EC and VAD. In SIP signaling, this process is done by sending a Re-INVITE request with session description parameters that contain negotiated preferred codec options of PCMU/PCMA, packetization period, and RTP/Real-Time Control Protocol (RTCP) transport address to the emitting/originating gateway or to the SIP proxy server depending on the network topology.
- During negotiation, many module configurations are modified. Echo cancellers, VAD/CNG, PLC, and DTMF rejection are disabled. Jitter buffer is set to fixed jitter buffer with an option to handle duplicate and reorder packets.
- The originating gateway receives the Re-INVITE message and responds with an 200 OK response to the receiving gateway with SDP information that contains negotiated codecs, packetization period, and RTP/RTCP transport address.

- The receiving gateway generates the ACK message to the 200 OK response and sends an ACK message directly to the originating gateway.
- The fax will go through in normal G.711 media packets.
- At the end of the fax transmission, the fax call will be disconnected as shown in Fig. 15.4 or switched to a voice call.

15.5.1   T.38 and Fax Pass-Through Trade-Offs

In some deployments, the Internet bandwidth provided to the user is very high. The bandwidth required for voice or to fax even with redundancy-1 will be much lower than the available bandwidth to the user. Quality of Service (QoS) mechanisms are undertaken to prioritize voice. In such situations, end-to-end delays are lower, packet impediments are not present, and mostly packet drop is maintained to be less than 0.01 % (100 packets drop in one million packets). Fax pass-through with these options is the simplest and works fine in most situations. The fax pass-through works similar to a PSTN call and issues of fax machine and VoIP signaling interoperability will not be present. While writing this chapter, most VoIP fax calls in Japan were being planned with the G.711 pass-through mode. T.38 takes much lower bandwidth than the G.711 pass-through, caters to huge packet drops, and incorporates higher redundancy with minimal bandwidth increase. Other derivatives of T.37 store-and-forward service can be derived with VoIP fax. The main concern from T.38 and T.37 VoIP fax is interoperability and not handling all the optional implementations that leads to failures in handling certain messages and errors in establishing and continuing the fax call.

Comparison Between a PSTN and a VoIP Fax Pass-Through Call. The PSTN is a circuit switched network that will make a call with fixed delay, whereas VoIP is a packet based operation. In VoIP, some packets may get delayed, go through jitter, duplicate, or packet drop. In the PSTN, every representation of the G.711 sample reaches the fax and modem, whereas in VoIP pass-through, it is not guaranteed. A packet drop in 1 minute duration may sustain the fax call, but a frequent miss usually referred to as amplitude and phase hits in the TR-57 [TR-NWT-000057 (1993)] standard, can force the fax machine to operate at a lower speed or to fail the fax and modem calls. In addition, the IP transmission paths should have reasonably low delays to meet the F.185 [ITU-T-F.185 (1998)] requirements for optimal performance in FoIP.

15.6   FAX OVER IP INTEROPERABILITY CHALLENGES

The primary challenge in FoIP implementation is end-to-end interoperability. By improving interoperability, FoIP calls can be comparable with PSTN-based fax calls. In most situations, PSTN fax calls are completed successfully. Nevertheless, PSTN fax calls can occasionally fail because of line conditions from the user fax machine to the PSTN digital loop carrier or central office. Fax machine anomalies, mismatched messages, and capability exchanges contribute to these failures. Even in the PSTN, intermediate long-distance routing of PSTN fax calls may use intermediate VoIP or FoIP calls.

The type of fax machines used, VoIP gateway features, fax call switching, and deviations in fax call tones can influence FoIP interoperability. IP network impediments—such as delays, bandwidth variations, jitter, packet loss, timing issues, packet formats, redundancy, error correction mode, end-to-end transmission characteristics, clock drifts, as well as various configurations and interactions between voice and fax modules—affect interoperability. However, interoperability concerns can be alleviated by ensuring that FoIP implementations are tolerant of the many anomalies that may occur [URL (CED)]. FoIP interoperability issues are given here.

15.6.1 Interoperability with Fax Machines

Several deviations exist among the available fax machine timings in delivering messages and responses. T.38 FoIP adds delays that may exceed fax-timing limits, resulting in failed fax calls in T.38 mode or forcing the fax call to the G.711 pass-through mode. The new SG3 with V.34 support creates interoperability with other low-speed G3 fax machines and computers. Personal computers, in combination with a telephone interface, are also used for faxing document files. In many situations, users may not have upgraded the bug patches or use of the computer as a fax machine is not evaluated fully with hardware and operating system combinations. The goal here is to make the VoIP gateway interoperate with several fax machine anomalies.

15.6.2 Deviations in Fax Call Tones

The ANS family of tones consists of ANS, /ANS, ANSam, and /ANSam that have several deviations. Here the suffix "/" in keywords denotes phase modulation. Some fax machines also send modem tones in place of fax tones, which requires extra validation for fax and modem that is not supported in all gateways. Validating ANS or /ANS with additional CNG or V.21 preamble detections is a useful option. Some fax machines omit the ANS and just begin with the V.21 preamble as a first handshake message. A G3 fax machine sends an ANS tone, whereas a high-speed SG3 fax machine sends the /ANSam tone during the call setup phase. These deviations create interoperability challenges. When a fax is sent between G3 and SG3 facsimile devices, end fax machines will fall back to the G3 mode. Fax call detection can be managed using V.21 preamble detection itself without having /ANSam tone detection, whereas the /ANSam tone detection is mandatory for interoperation between two SG3 fax machines and for fall back to the G3 mode. A fax call can be managed using

V.21 preamble detection between two SG3 fax machines, by suppressing the CM message at both the sending and the receiving gateways to make fax machines fall back to the G3 mode. The CM message is transmitted by the originating gateway after detecting the /ANSam signal.

The T.30 standard requires the answering fax device to send an answer tone of 2100 Hz for approximately 3 seconds before sending the first handshake message. Some fax machines send a 1650- or 1850-Hz tone instead of a 2100-Hz tone. Some machines send an 1100-Hz tone, and few fax machines omit the answer tone altogether and just begin with the first handshake message.

15.6.3 Handling of Voice to T.38 Fax Call Switching

Gateways initially establish VoIP voice calls using compression codecs, such as G.729A, G.723.1, and G.711 and then switch to the fax call after analyzing for calling tone, (CNG), ANS family of fax/modem tones, and V.21 preambles. Some FoIP gateways also wait for the V.21 preamble from the originating fax machine to switch from voice to a T.38 fax call. In an in-band operation, call setup tones and V.21 preamble tone and DIS signals may be distorted with compression codecs. Out-of-band packet creation can introduce huge delays in detecting and regenerating tones and messages. In general, voice-to-fax call switching is established with CNG, ANS or /ANSam, and V.21 preamble in-band and out-of-band combinations. In all these situations, the T.38 relay should be generic to handle several combinations of fax call switching modes using in-band and out-of band combinations as explained in Section 15.3. Some fax call setup issues are list below:

- In general, the receiving gateway detects and analyzes the ANS family of fax tones—CED or ANS, /ANSam, and V.21 flags. Few VoIP adapters initiate the fax call from the originating gateway after the CNG tone, which may create an issue when the peer gateway cannot handle it.
- Few VoIP adapters and gateways initiate the fax call after detecting the V.21 preamble followed by the DCS at the originating gateway. In such a case, fax tones and preamble and DIS messages are exchanged as in band through the RTP media. The T.38 fax relay should be generic to initiate the state machine and to configure the T.38 mode. With some boxes, the initial voice call is established with a low-bit-rate codec. CSI and DIS frames are exchanged in in-band signaling through compression codecs, which may lead to failure at the initial call setup.

15.6.4 Interoperability with VoIP Adapters at Different Rates

Few VoIP adapters or gateways can support the fax data pump up to the V.29 fax modem with a maximum bit rate of 9600 bps. When G3 fax machines that support the data rate up to 14.4 kbps are used to send a fax with low speed using V.29, the fax may fail. T.38 implementation should be able to handle the rate control in phase-B to fall back to 9.6 kbps for interoperability with VoIP adapters at different rates.

1. When the DIS frame is received from the fax machine on the receiver side with V.27ter, V.29, and V.17 capabilities and when the maximum rate negotiated is 9.6 kbps, then the DIS frame with V.27ter and V.29 modems is modified before sending it on the network.
2. Similarly, when DIS is received from the remote gateway, the rate negotiated in the DIS frame is compared with the required SDP negotiated rate. If the negotiated rate in the DIS frame is above the maximum required negotiated rate through SDP, then it is required to force the DIS field with a modified data rate as per the maximum negotiated rate in SDP information.
3. The DIS frame is forced to the required SDP negotiated rate, for example, by 9.6 kbps on either the originating or the receiving side, and still the fax machine on the originating side responds with a maximum data rate of 14.4 kbps. The following cases need to be checked on the originating and the receiving T.38 relay:
   - On the originating T.38 relay, when the DCS frame is received from the fax machine, it is analyzed and compared with the required maximum negotiated fax data rate; the rate control is modified as per the maximum rate negotiated through the SDP before sending it to the network. If the fax rate in the DCS frame is above the maximum data rate negotiated through SDP information, then it is required to corrupt the high-speed TCF data until it falls back to the required negotiated data rate. Another possible scenario is when the CFR is received from the remote fax relay, the rate negotiated is compared with the required rate. On sensing that the negotiated rate is above the required configured rate, it is required to send the FTT from the originating relay to the fax machine. This process has to be continued until the confirmed negotiated rate is less than or equal to the maximum negotiated rate though the SDP.
   - On the receiving relay, when the DCS frame is received from the remote T.38 relay, it is analyzed and compared with the required maximum fax data rate. When the fax rate in the DCS frame is above the maximum data rate and the SDP information is greater than or equal to the fax rate in the DCS frame, then manipulation of high-speed TCF data is useful until it falls back to the required negotiated data rate.

15.6.5 Interoperability with VoIP Adapters and Gateways

Several VoIP gateways primarily support voice services and may not be capable of supporting T.38. Even if gateways support fax, many optional items could not be interpreted. Several revisions have been made to fax standards to add new features and optional messages as given in RFC4161 [Mimura et al. (2005)] that have to be handled in a fax call. These changes introduce gateway interoperability issues. To ensure proper function, VoIP gateways must be upgraded with the latest fax revisions.

Some fax equipment vendors have incorrectly implemented the augmented Backus–Naur form (ABNF) as defined in ITU-T-T.38 Annex D for SIP/SDP call establishment procedures for several parameters of T38faxFillBit removal, T38faxTranscodingMMR, and T38faxTranscodingJBIG. These implementers have made incorrect use of the colon (":"). Implementers should avoid this mistake and make their implementations robust by interpreting ":1" as support for the attribute, and ":0" as nonsupport for the attribute.

Interoperation with IAF Devices. The Internet-aware fax devices do not use the RTP media for initial call setup. The CNG and CED tones are exchanged as T.38 IFP indicator packets. In some implementations, the IAF device switches to fax mode at the receiving gateway by detecting the CNG tone in the receive path. For example, the VoIP adapter calls the soft fax server or IAF device sending the fax; then the initial call may be established with the preferred voice codec. The calling tone is transmitted on RTP inband or out of band using RFC2833 media to the IAF device. The IAF device detects the CNG tone on the receive path and switches to T.38 mode. In practice, most IAF devices do not send CED tone as in-band RTP packets. The CED tone is transmitted as an IFP indicator packet to another VoIP adapter. The peer gateway must regenerate the indicator packets for successful interoperability with fax machines. Some implementations do not close the voice session. In this case, it is not required to regenerate the CNG/CED indicator packets.

T.30 Timer Value with Calls Between IAFs. When both implementations are IAFs, the T.30 timer value may be extended by two or three times [ITU-T-T.38 (2005)]. Extension of the timers allows two terminals to have successful facsimile transactions when a high degree of network delay and/or loss of packets occurs or with fax machine anomalies that create timing and delivery of messages. The bit number 123 (bit-123) in the DIS/DCS message (FIF field) is the negotiation bit that indicates an IAF device.

15.6.6 Packet Payload and Format Issues

FoIP has several options in IP packet creation. T.38 IFP fax data are sent as TCP using TPKT and as UDP using UDPTL or RTP. The problems increase with combinations of fax machines, redundancy, FEC, and ECM modes. The T.38 relay should be capable of handling several combinations of packets. Some critical deviations are given below:

- In disassembly of one signal in a packet, some T.38 implementations send one T.30 signal HDLC frame in one packet and other implementations send it in multiple IFP packets. Therefore, a T.38 implementation should handle both situations and assemble the multiple packets when necessary. This principle applies to image packets as well. Some implementations place an entire HDLC frame between flags into a single packet; others may ignore the frame boundaries when inserting the data into packets.

- HDLC frames and packets have different frame boundaries when inserting the data into packets.
- Redundancy and duplicate packets are used without distinction.
- T.38 relay implementations follow different packetization intervals and redundancies for low-speed data and indicator packets.
- Most T.38 implementations follow duplicate indicator packets with the same sequence number multiple times.
- In ECM, fax image data are split into blocks and frames. Frames are sent with HDLC and CRC. Transmitting and receiving fax machines work in coordination to get complete error-free blocks through retransmission. Retransmission of lost packets can cause long delays and cause the fax call to time out.
- In dealing with a preamble packet between T.30 signals, some implementations incorrectly send a preamble packet between T.30 signal packets. A T.38 implementation that receives a sequence of this type should handle it properly. For example, the received preamble packet before "sig-end" in field-type should be regarded as flag value 0x7E.
- Some implementations limit the packet size to receive even in TCP mode. The limitation often relates to the size of one ECM packet. It is the responsibility of the sender to address this situation. One possibility is to use the same packet size regardless of whether the transport protocol is TCP or UDP and regardless of whether the remote side is an IAF or a gateway. In UDP mode, the t38faxMaxDatagram value negotiated in call setup should be used to determine the size of the packets.
- In transferred TCF or local TCF, a series of zeros as bits for 1.5 seconds is sent in one or more packets in transferred TCF, which is based on the negotiated modem speed in DIS/DCS exchange. In general, IAF devices may use local generation of TCF irrespective of the transport protocol used and negotiation support. An IAF sender must generate the TCF when it identifies that the receiving T.38 device is not an IAF device.
- Some T.38 implementation may follow a different packet interval for high-speed packets. The receiving relay should be capable of handling these situations by employing spoofing and line padding techniques.
- Some T.38 implementations pack the multiple data fields into a single IFP packet with separate field type and length header fields. For example, the T.4 data packet and sig-end packet are packetized in a single IFP packet.

15.6.7 IP Network Impediments

The IP network creates impediments of delay, jitter, packet drop, packet fragmentation, and errors in end-to-end transmission that can result in packet drop. The sending fax packets join with voice and other applications like data. If a limited bandwidth in upstream occurs and IPQoS is not performed for fax, a packet drop can happen. A fax packet can tolerate the delay, but it is very sensitive to packet drop. The redundancy and FEC techniques can take care of packet drop up to some extent in UDP mode. In IFP over TCP, TCP takes care of retransmissions. A fax machine with ECM mode is sensitive to packet drop. These fax machines may disconnect the call after several retransmissions. To overcome this issue, IPQoS has to give priority to voice and fax packets (IFP over UDP) than other applications like data.

15.6.8 Miscellaneous Topics on Fax Call Packets and Timing

In the previous section, several interoperability issues and possible high-level options are given. Several other aspects of T.38 fax calls have to be perfected. In this section, fax call packet and timing issues are presented that can influence fax interoperability.

Lost Packet Compensation. The packet lost is a more severe constraint in fax over IP depending on the type of transport protocol used to transmit the fax packets. This problem again varies with the type of fax machine used and whether ECM mode is enabled. UDP has good real-time performance, but UDP is a reduced quality of service. It is possible for packets to get lost, arrive out of order, or be duplicates in the network. These challenges in UDP transport protocol are addressed through redundancy and forward error correction schemes. UDP uses UDPTL and RTP for transport. More details on these two combinations and packet formats are given in Chapter 16. These error mechanisms are sender-based packet lost mechanisms that are implemented on the sending path. Another option is TCP-based communication at the expense of added delay and bandwidth requirements. TCP transport is the session based, confirmed delivery service, which provides full correction. In TCP mode, it is not required to use redundancy and FEC for packet drop. Packet drop is taken care of in the TCP, and IFP packets are delivered in order.

  If a receiving fax machine cannot receive an error-free page, the fax transmission may fail, and one of the fax machines may disconnect. Fax machines with ECM enabled are highly sensitive to packet drop. Some fax machines drop the call after several retransmissions, whereas others disable error correction. To take care of the network packet lost on the receive side with ECM enabled, analyze the packet loss levels and control the ECM functionality as in the disabled ECM mode or switch to pass-through with redundancy if a network has packet-loss levels greater than 2% [URL (Cisco-fax1)] because of the ECM low tolerance for packet loss.

Fax Call Delays and Spoofing. The round-trip delay is the most important factor on packet networks. In the T.30 protocol, the procedures are defined with a time-out mechanism. When response to the message is not received within a specified time of 3 ± 0.45 seconds, then the fax machine retransmits the message or disconnects the call after repeating the message three times. Under network congestion, higher end-to-end delays may be encountered. Spoofing techniques and jitter compensation allow the fax machines to sustain the fax call without disconnection [URL (Cisco-fax2)].

In a fax call, for a command to reach its destination and get back an acknowledgment, the PSTN in-band fax had round-trip delays that were usually low. Fax pass-through increases delays by 50 to 80 ms in each direction, but this is relatively lower delay compared with T.30 fax message timings. In T.38 fax relay, each gateway performs additional modulation, demodulation, and IFP packetization of demodulation bits, and depacketization creates additional higher delay. These extra delays under network congestion create delayed commands, timed out acknowledgments, and message collisions. In summary, increased delay, jitter, equivalent packet drop, or discarding conditions in collisions and fax machine anomalies demands fax call-sustaining operations, which can be achieved through spoofing techniques.

In early VoIP deployments, spoofing technique were planned to cater to higher delays of the order of 10 seconds. In the current deployments, it should be sufficient to take care of up to 3 seconds or lower. It is justified because of the use of the same networks for voice and fax. The end-to-end voice calls are maintained to achieve less than 150 ms and a worst case of 400 ms on inter-regional calls. A fax call is benefited with the improvements achieved for the voice call.

Fundamentally, in fax spoofing, a known pattern or redundant data is continued with modulations at a corresponding bit rate to sustain the fax call. In the timing part, message collisions are resolved by delaying the required messages and continually applying a suitable time slot to create a successful fax call operation and to eliminate retransmissions and additional collisions. To solve possible collisions, the T.38 fax relay should buffer the received IFP packet from the network while the data pump is engaged with demodulation of fax signals from the local fax machine. After demodulation is complete, the buffered data are sent to the data pump for remodulation and are required to drop the retried command [URL (Audiocodes-fax)].

Spoofing techniques are used with both the low-speed and the high-speed modes of a fax call. Spoofing techniques are decoder based and make use of several proprietary techniques. The implementations widely vary based on the supported product features and deployment.

15.6.9 Improving FoIP Interoperability

FoIP transmission works similar to VoIP voice calls in some aspects. However, significant key differences can be found. T.38-based FoIP calls take relatively lower bandwidth than do G.711 VoIP calls. Voice calls are interactive, and listeners can adapt to voice impediments. In fax calls, tones, messages, and page data have to operate in an automated way without human interaction [URL (CED)]. Several deviations and options have to be taken into consideration when implementing FoIP solutions. Despite these issues with T.38 fax interoperability, it is possible to achieve quality comparable with PSTN fax calls by making the system tolerant of anomalies and standards violations, while giving the solution the intelligence to handle several unexpected events. FoIP seamlessly integrates with VoIP and provides the advantages expected from an IP network. FoIP working as a G.711 pass-through, T.38, or T.37 in combination with perfecting the solutions to provide robust high-quality fax services can serve to support the migration of PSTN-based fax calls to VoIP.

15.7 MODEM BASIC FUNCTIONS ON PSTN

A modem is used with a computer for data communication, such as with Internet connectivity. Computer or data terminal generates digital data bits, which are loaded into a modem box through either RS-232, the universal serial bus (USB), or peripheral component interconnect (PCI) interfaces. Inside the modem box or modem hardware, these bits are modulated to generate the samples and finally the analog signal on the telephone TIP-RING interface. The PSTN network works like a transparent line. At the destination, the modem server, which is usually referred to as the remote access server (RAS) modem, demodulates the analog signal to extract the bit pattern. The RAS modem interface with the IP network [ITU-T-V.150 (2003), Bingham (1988), URL (dialup)] for extending the data communication.

Many functions of a modem are similar to fax except for the scanner and printer operations. The calling modem generates a V.25 CT or V.8 CI signal. The receiving modem generates 2100 Hz tones with phase reversal as an indicator. The phase reversal happens once in 450 ± 25 milliseconds. This signal is passed through the compression codec with acceptable distortion, and this tone can be detected at the receiving side. Most modems use higher data rates compared with fax transmission. The modulation and demodulation modules are referred to as the "modem data pump". The management of a modem call is done through modem high-level applications. Modem signals are made to appear in a voice band of 300 to 3400 Hz. Usually, modulations happen on digital samples, and are delivered as analog voice band signals through digital-to-analog conversion in a hardware COder/DECoder (CODEC). The main functions of hardware CODECs are analog-to-digital conversion (ADC) and digital-to-analog conversion (DAC). While receiving data from a modem, a CODEC samples the data and demodulates the samples for deriving the digital data. A modem modulator sends the bits from the user is personal computer (PC) to the outside network. The main functional blocks [Bingham (1988)] of the modem modulator are as follows:

- Asynchronous–synchronous converter to synchronize data coming from the PC or data terminal
- Error correction scheme on digital bits
- Scrambler is used to randomize the data
- Channel encoding and filters to cater to transmission effects
- Modulators and filters to create required modulation and to limit the signals to voice band
- DAC of hardware CODEC to deliver analog signals to the modem

The demodulator is positioned to receive bits from the external network to the user PC. The main functional blocks of modem demodulator are as follows:

- ADC, filters, and demodulators to convert the analog signal to bits
- Channel decoding to get the bits after channel influence
- Descrambler, error correction, and asynchronous/synchronous mechanisms to get the right format bytes to PC and data terminals

Many call progressing blocks exist in the modem, namely

- The dialer takes care of proper dialing on the PSTN
- Auto-answer includes ring and tones generation and detection
- The protocol controller manages the total operation of the modem

Integrated services digital network (ISDN) and T1/E1 lines are used for PSTN digital transmission. These lines support the 56-kbps modem rate. Analog modems connected to the regular TIP-RING telephone lines support data rates up to 33.6 kbps, and few of them support up to 56 kbps with V.90 modem support. Many messages and low-speed modes use 300 bps with FSK. Moderate speeds are achieved through phase-shift keying (PSK). A higher speed of operation is achieved through quadrature amplitude modulation (QAM). Multiple ITU-T-V series recommendations provide guidelines for various error connection schemes of bits, modulation, and demodulation. Some popular ITU-T-V.series recommendations for modulation and demodulation are listed below:

- V.21 (300 bps)
- V.23 (1200 bps)
- V.22 (600 or 1200 bps at 600 baud) and V.22bis (1200 or 2400 bps at 600 baud)
- V.32 (4800 or 9600 bps at 2400 baud), V.32bis (up to 14,400 bps), and V.32ter (19,200 bps)

- V.33 (12,000 or 14,400 bps)
- V.34 (up to 33,600 bps and falls back), V.34bis (33,600 bps)
- V.42 is used along with a modem module such as V.34 to provide error correction functionality
- V.90 and V.92 for 56 kbps

Remote Access Server (RAS). RAS is the remote access server used to host the services to the dial-up modem. RAS is under the control of an Internet service provider (ISP) for dial-up service. RAS will have multiple telephone lines and will be connected through digital PSTN lines such as T1/E1 or multiple ISDN. A dial-up modem can reach RAS by dialing into the RAS modem phone numbers. RAS allows connection of a dial-up modem after validating proper authentication like username and password. The RAS and dial-up modem exchanges training sequences and arrives at the suitable bit rate and modulation scheme for communication. RAS may also incorporate certain security features based on capabilities. Once a connection is established, one channel of the RAS is occupied for the duration of the modem dial-up connection.

Modem Connectivity Through the PSTN Interface. The dial-up modem and PSTN setup for Internet connectivity are shown in Fig. 15.5. In this setup, the dial-up modem inside the computer works as an electronic telephone. The same telephone line is used for a normal plain old telephone service (POTS) phone and fax machine. On initiation of the call from a computer (PC/laptop), a dial-up modem receives a dial tone from the PSTN central office and sends the digits. Based on the dialed digits, the PSTN exchange routes the call to the RAS modem. The RAS modem hosts the required services for the multiple dial-up modems. On receiving the call, the RAS modem and user dial-up modem exchanges modem indicator tones, training, and capability messages for establishing the call with the right options. The RAS modem will receive the data bits modulated on to the voice band and will demodulate into bits. The RAS modem interfaces with the IP network, and the dial-up modem interfaces with the computer. Depending on the line conditions and modem capabilities, the dial-up modem can send data from 2.4 to 56 kbps.

15.8   MIGRATING MODEM FUNCTIONS TO IP

A telephone is used along with a VoIP adapter and gateways for voice communication on IP. A fax may also use the same telephone interface of the adapter. A dial-up modem was used on telephone lines along with a computer for Internet connectivity. The use of a modem through the same telephone interfaces of the VoIP system is also possible. The modem is usually part of a PC or laptop or is external to the computer. The modem inside the computer

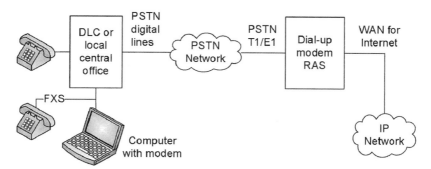

Figure 15.5. Functional PSTN setup to connect the PC modem for Internet connectivity.

will have a telephone interface. An external modem will have a computer interface for data connectivity on the one side and a telephone interface on the other side. VoIP adapters and integrated access devices (IADs) will have multiple interfaces for broadband connectivity as given in Chapter 2. Internet connectivity is extended directly through broadband interfaces instead of a modem. Dial-up modem connectivity through a telephone interface has limited use in VoIP adapters and gateways. Hence, modem connectivity through a dial-up modem did not gain much popularity. In the following sections, an overview on modem connectivity for migrating the PSTN to VoIP infrastructure is given. While writing this section, the end users were not using MoIP service.

15.8.1 Modem Simple Connectivity Through an FXO

Some VoIP IADs has digital subscriber line (DSL) for wide area network (WAN) Internet connectivity. DSL will use a PSTN line, and IAD may use a foreign exchange office (FXO) interface for PSTN lifeline support. On dial-up modem requirements, a dial-up modem can reach the FXO line directly. A VoIP IAD will simply switch the dial-up modem connected on the foreign exchange subscriber (FXS) port to FXO. Some prefix dialing digit could create FXS-to-FXO switching. This switching works similarly to the dial-up modem directly connected to the available telephone interface. As shown in Fig. 15.6, an FXO interface will reach the DSL service DSL access multiplexer (DSLAM). DSLAM will split wideband DSL signals and PSTN signals. PSTN calls are routed to the nearest or local central office. Usually RAS servers are located as local numbers and are mapped to a local telephone exchange. Supporting this dial-up modem through a PSTN interface will not be difficult, and the performance would be the same as a PSTN-based dial-up service, assuming foreign exchange subscriber (FXS), FXO front-end interfaces, and switching are working properly. Figure 15.6 is shown for the possibility of dial-up calls through IADs with a DSL interface. This type of dial-up modem scheme is given to provide an overview. When VoIP IAD is used, IAD will have several

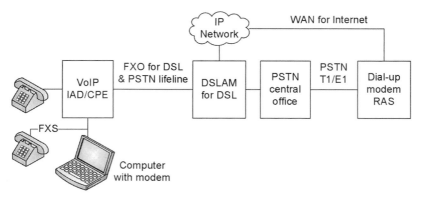

Figure 15.6. Dial-up modem connections through an IAD FXO interface.

wideband network interfaces as shown in Chapter 2. The user may not find a significant benefit out of another dial-up service that gives a low data rate service telephone interface.

15.8.2 Modem Connectivity Through a VoIP Pass-Through

In the mode of G.711 pass-through, a modem call will use a VoIP voice call similar to fax pass-through G.711 compression. For dial-up purposes, a call can be routed to the nearest voice gateway that has access to a local PSTN. The gateway and VoIP adapter will not be negotiated for a modem over an IP call. At the VoIP gateway, a call will be passed to the FXO interface from the VoIP adapter similar to a VoIP-to-FXO call. From the voice gateway, a call is routed to the nearest RAS as shown in Fig. 15.7. The dial prefixes and local number mapping in the destination voice gateway resolve routing the call to the RAS modem. Once a call is connected to RAS mode, it responds with an /ANS or ANSam modem tone. However, modem tones are detected at the gateway and VoIP adapter for arriving at the right configurations of a G.711 modem pass-through call. If a modem tone with phase reversal is detected, an echo canceller has to be disabled. An echo canceller up to 128 ms is used at the gateway in a normal FXO-to-VoIP modem pass-through call. In Fig. 15.7, a line marked with IP is also shown for joining the RAS to the IP network. The RAS is connected to the IP network for data communication.

15.8.3 Modem over IP in the VoIP Gateway

The modem relay over IP or MoIP [ITU-T-V.150 (2003), URL (Surf-com), URL (Cisco-modem), URL (Tldp-modem)] is mainly governed by V.150.1 [ITU-T-V.150.1 (2003)]. The purpose is to send modem data in IP packets and finally on the PSTN interface. In this mode of operation, the voice adapter will have a built-in MoIP function. The VoIP adapter takes an analog signal from

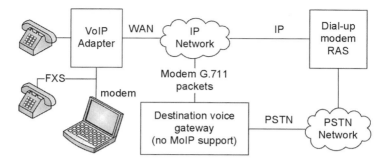

Figure 15.7. RAS modem with VoIP pass-through.

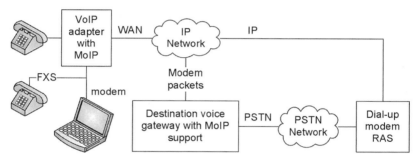

Figure 15.8. RAS built inside the voice gateway as MoIP.

the modem and performs certain functions of RAS locally within the VoIP adapter. Modem data bytes are sent as IP packets that will reach the destination gateway. The destination gateway will decode the MoIP packets and regenerate modulations and other messages on the PSTN interface. The RAS on the PSTN interface will work similar to a modem call received from the PSTN dial-up modem. The setup in Fig. 15.8 is a functional representation. Usually a RAS modem can be part of the destination voice gateway without going through another PSTN interface to the RAS, and a separate RAS modem may not be required. A failed MoIP call may revert like a G.711 pass-through call, and the destination voice gateway routes the call to a RAS modem. The functionality will be similar to Fig. 15.7, which shows the modem pass-through mode of operation.

An MoIP call may be established in multiple phases. Initially a modem call will progress as a voice call. On detecting modem tones and V.8 signals, the codec has to switch to G.711 from any other compression codec like G.729A. After exchanging messages between VoIP gateways and end modems, the call is switched to the MoIP call. An MoIP call uses packets and payloads similar to fax over IP calls.

15.9   GUIDELINES FOR FAX AND MODEM PASS-THROUGH IN VoIP

In a VoIP G.711 call, several modules such as echo canceller, VAD/CNG, PLC, and adaptive jitter buffer (AJB) will be used for voice enhancements. For a successful fax and modem pass-through call, these modules operations have to be properly controlled. The guidelines for a successful pass-through call are given below [URL (Cisco-fax4)].

Codec Switching.   A compression codec has to be switched or renegotiated for the G.711 codec with either PCMU or PCMA. In some countries, default preferences will be selected based on the local country PSTN standard. In Europe, PCMA is preferred, and in North America, PCMU is preferred.

Direct Current (DC) Blocking High Pass Filters.   In some voice processing implementations or as part of codec processing, DC blocking high-pass filters are used. In pass-through mode, these filters have to be bypassed to minimize distortions. In some designs, these filters are used after characterizing for fax and modem pass-through modes of operation.

Packetization.   For the available bandwidth, selected packetization has to be as small as possible. In pass-through mode, 10 ms packetization will be the preferred option, and more than 20 ms should not be considered. Higher packet intervals like 30 and 40 ms will create a long silence in the case of packet drop. Modem and fax will create a disconnect on such long interruptions or will fall back to slower speeds of 2400- or 4800-bps support.

Echo Cancellation.   An echo canceller (EC) is disabled or used with some restrictions during fax pass-through modes. The EC-enable mode may work up to a typical bit rate of 9600 bps in fax. Some VoIP adapters have been found to work for all G3 fax machine rates up to 14.4 kbps with the EC-enabled state, and this depends on how cleanly the signal passes through the EC. In practice, it is possible to send a fax even with EC on. It is required to evaluate the performance of EC before deciding on such enable conditions. In some designs, EC is enabled initially, but EC is kept in a hold state from the end of V.21 or the initial training data. In hold mode, all adaptive filter coefficients stay at their last update, which is referred to as EC in the freeze state. Several working options exist with EC module association of a fax pass-through call. Nonlinear processing (NLP) that is part of EC is not required during a fax call. Disabling the echo canceller NLP is recommended. In practice, fax and modem signals are continuous with strength in the range of −24 to −10 dBm. Hence, NLP may not operate even if it is not disabled. In the presence of modem tone and with SG3 fax, the echo canceller has to be kept in a disabled state.

PLC.   A PLC operation may not help fax and modem pass-through because of sensitivity of phase information present in modem and fax modulations. Hence, the PLC module working along with G.711 may be disabled.

VAD/CNG. The VAD/CNG module has to be disabled in the pass-through mode of operation. The VAD/CNG may not operate because of strong continuous signals from fax and modem. To prevent any wrong operations and any clippings, it is recommended to disable the operation of VAD/CNG modules.

End-to-End Delay and Jitter Buffers. Fixed end-to-end delays are more helpful to the pass-through call. This assistance can be achieved by making the jitter buffer work like fixed jitter buffer (FJB) with a long initial threshold— typically 50 to 150 ms. Packet overflow or an under-run situation is created based on the end-to-end relative clock parts per million (PPM) and network conditions. A packet correction of anticipated under- or over-run has to be controlled to once in 1 or 2 minutes. Allowing complete under- or over-run may disconnect a call or lower the operation rate of fax and modem. This issue is more predominant with long fax calls and multiple pages. The packet corrections have to be as short as possible, typically of 10 or 20 ms. Duplicate packet drop and reorder should be implemented in fixed jitter buffer. Some fixed jitter buffer may not have this feature. With a higher threshold in FJB, round trip increases and forces the modem to operate reliably up to 28,800 bps or lower. It is suggested to use a low-speed modem of V.34 and to operate at less than 28,800 bps to produce a sustained modem call through VoIP. In good network conditions, the jitter buffer thresholds can be lowered, which means that jitter buffer can be adaptive with minimal adjustments.

Packet Redundancy. Payload redundancy is applied on RTP packets as per RFC2198. Redundancy helps to create continued packets even in a packet drop situation. Redundancy is described in RFC2198. Packet redundancy may create an interoperability issue with some gateways. Redundancy increases the bandwidth requirements. In a bandlimited network, applying redundancy may drop the packets. Hence, a bandwidth-limited network may perform better without redundancy. Jitter buffer and RTP should be able to handle the redundancy of RFC2198 packets, when redundancy is used. Packet redundancy by one in PCMU and PCMA is useful in case of isolated packet drops. Forward error correction as per RFC2733 is also helpful for bit erasures and bit errors. FEC is an option in most implementations.

DTMF. Any wrong detections of DTMF create burst erasure with a DTMF rejection module. A DTMF rejection module is used for creating RFC2833 out-of-band packets. A DTMF rejection module has to be disabled for the duration of a fax and modem pass-through call.

Loss and Gain. End-to-end gain has to be avoided in the pass-through mode. On monitoring any deviations from nominal values in the transmit or receive path, these values have to be adjusted to the correct level at the end of fax and modem tones and before the start of fax and modem modulations. Most

PSTN services keep about 3-dB loss in each path of transmission. The gain or losses have to match with local PSTN standards. It is also required to consider country-specific compliance on losses.

Switching Back to a Voice Call. It is possible to switch back from a fax to a voice call. In practice, many users disconnect after a fax call and make a new voice call. For continuity of fax to a voice call, the compression codec has to be switched back to a required selection. FoIP-to-VoIP operations are marked in Fig. 15.3.

Analog Front-End Considerations. In addition to voice modules configuration, many other system aspects should be taken care of for better fax and modem operation. The subscriber line interface circuit (SLIC) and CODEC (ADC and DAC) used in the VoIP adapter controls the performance of the fax and modem. The main governing guidelines are given in TR-57 tests for North America in Chapter 1.

Transmission Characteristics for Fax and Modem. Fax and modem performance is mainly controlled by transmission performance. The transmission recommendations of TR-57 are mainly achieved through analog interfaces and the selection of SLIC/CODEC devices. TR-57 provides guidelines for transmission and signaling tests for telephony applicable to North America.

Hardware PCM Interfaces—Linear, μ-law, and A-law Choice. A CODEC working in linear mode works much better for T.38 fax calls and in pass-through modes. Most hardware devices support 16-bit linear format. In practice, μ-law or A-law is also acceptable for pass-through calls, but the linear format is more helpful. This selection is applicable on the hardware CODEC and on the processor interfacing the PCM samples.

15.9.1   Views on VoIP Fax and Modem Deployments

Views on VoIP Fax. Assuming an end-to-end network delivers packets properly, using good voice gateway hardware and software will help pass-through calls. Service providers with broadband service are trying to stick to this approach. After obtaining perfection in packets transmission and the availability of more Internet bandwidth, many deployments may use the fax pass-through approach. In practice, T.38-based VoIP calls are more suitable for VoIP-based fax migration, and T.38 takes care of several impediments that occur in the migration to VoIP.

In moderately controlled bandwidth deployments, such as some deployments in Europe and Asia, T.38 is incorporated in the service. In North America, VoIP service is not marketed for fax. PSTN lines are used for fax in most situations. The deployment and service options may change over time. In general, Fax over IP coexists with voice because of the benefits listed in Section 15.2.

Views on Modem Over IP. Modem over IP is not in demand in deployments. Telephones, fax, and modem work on a PSTN interface. Hence, modem support may be considered on IP. Some data terminals use a modem for their business transactions and for communicating and sharing data and records. In the absence of a supporting PSTN infrastructure, a user will need modem over IP. It is expected that a dedicated adapter may be supplied that will be used specifically for MoIP on the IP networks.

The data pump for modem modules take more memory and processing than voice and fax modules. The V.150 series of recommendations had several details on modem modules. MoIP was one of the checklist items asked by customers in the years 2003 and 2004 when different V.150.1 versions were introduced by the ITU. While writing this book, demand was not being made by end users for modem over IP commercial service.

15.10 VoIP FAX TESTS

A fax test with VoIP is more complex than testing a few fax pages between two fax machines in PSTN. The ITU has provided an umbrella series with the ITU-T-E.450 recommendations for fax quality tests, out of which ITU-T-E.453 and E.458 [ITU-T-E.453 (1994), ITU-T-E.458 (1996)] classify fax image quality and call success and E.453 [ITU-T-E.453 (1994)] addresses transmission-induced scan line errors. Instruments incorporate these ITU recommendations for fax quality measurements. The fax tests listed in this chapter are broadly classified as listed below:

- Testing with multiple fax machines that include
  - Instrument-based fax tests, QualityLogic® FaxLab® (Moorpark, CA) as a tool that emulates multiple facsimile vendors and other fax testing instruments working as multiple fax machines
  - Multiple types of physical fax machines
  - Computer used as a fax machine
- Interoperability tests
- Fax testing with data traffic
- Network impediment tests in a combination of multiple fax machines and VoIP gateways

15.10.1   Testing with Multiple Fax Machines

Several types of fax machines are available worldwide. Fax machines before 1996 had some deviations in implementation compared with the most recent revision of the standards. Most of these deviations will not create a problem on PSTN-based systems mainly because of lower end-to-end delay and synchronized end-to-end digital transmission. In VoIP-based fax, these deviations can create fax call failures caused by increased delays in end-to-end transmission, packet impediments, and huge delays in modulation and demodulation while interfacing with packetized data. It is difficult to get several fax machines and to keep experimenting with them for proper fax operation. With fax machine pass-and-fail conditions, it is difficult to identify issues in the fax call. Instruments are used that identify the various fax call states, timing, and fax signal characteristics. However, instrument-based tests have to be supplemented with multiple types of physical fax machines to cater to analog signal distortions.

Instrument-Based Fax Tests. Instruments are available for emulating multiple fax machines. QualityLogic's FaxLab [URL (Qualitylogic)] emulates 166 fax machines of nearly 50 different manufacturers. FaxLab comprises two models: FaxLab 6.1 for V.34 testing and FaxLab 4.5 for V.17 testing. Together, these modules support V.34 rates from the lowest of 2400 to the highest of 33,600 bps. This fax testing system is the most popular one used worldwide. Recently several other instruments have been introduced that include basic fax call testing as part of voice quality testing, but these instruments do not create several models of fax machines. A summary on FaxLab features is given below to provide an overview of the features considered in fax testing:

- FaxLab emulates about 166 fax machines up to V.17 14,400 rates and about 27 V.34 fax machines in FaxLab 5.1. This list will keep increasing with upgrades.
- End-to-end capabilities automate testing of communication networks and facilitate T.38 fax testing.
- Support for V.34 (super G3), MH, MR, MMR, and JBIG encoding and both color and monochrome JPEG.
- Support for 100×100, 300×600, 400×800, 600×600, 600×1200, and 1200×1200 dpi resolutions in addition to the older 200×100, 200×200, 200×400, 300×300, and 400×400 resolutions.
- The timeline view shows all program decisions as well as all events and status changes of the modem.
- Real-world fax traffic generation and multiple channel fax traffic generation.
- Automates communication network testing and reduces test time while increasing thoroughness.

In its simplest form, FaxLab can be treated as two electronic fax machines with analysis available through a computer connected to the instrument. During testing, FaxLab sends fax calls and receives calls from the device under test using data from the profiles of the selected devices. ChannelTrap is a hardware device that interfaces with the computer. ChannelTrap, along with FaxLab software running on a computer, works as an electronic fax machine. All the

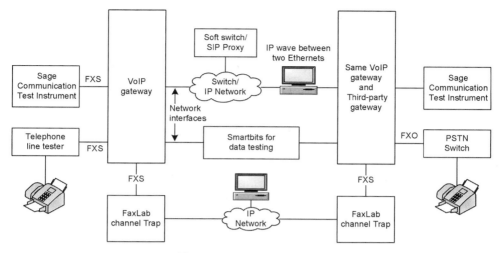

Figure 15.9. VoIP fax testing.

analysis and configurations are managed through the computer. The computer communicates to ChannelTrap through the RS-232 or Ethernet interface. Two ChannelTraps are used in the testing. A typical FaxLab usage in test setup is indicated in Fig. 15.9. The FaxLab profiles are also manipulated for the following errors to ensure that the T.38 fax relay is working in all error cases. FaxLab profiles can be manipulated for different impairments, and some of these are listed below:

Low-speed data errors:

- Affect percentage of frames
- Affect each Transmit signal ⟨n⟩ times with FCS error
- Affect each Receive signal ⟨n⟩ times with FCS error
- Affect T.30 signal until failure with FCS error
- Affect T.30 signal ⟨n⟩ times with FCS error

TCF errors:

- Affect TCF/CFR ⟨n⟩ times
- Affect TCF/CFR until failure

Message data errors:

- Affect scan lines/thousand
- Affect ⟨n⟩ contiguous scan lines
- Affect one 5-second scan lines

Partial page errors:

- Affect ⟨n⟩ partial pages continue to correct
- Affect partial pages until (CTC)
- Affect partial pages until end of retransmission (EOR)

Need for Physical Fax Machines. Instruments such as FaxLab create fax machine timing and anomalies. The front-end electronics interfaces will be fixed on FaxLab or any other instrument. In practice, fax machines will have several mismatches in the front end. In some fax machines, output amplitude modulations are observed in signals. FaxLab or other instruments will not create such analog signal impairments. FaxLab will not cater to country-specific deviations of call progress signals and matching impedances. This limitation demands the use of some physical fax machines for end-to-end fax tests.

Fax machines are supplied with several user options such as fax-only mode, telephone mode, fax and voice mode, and some auto modes. It will be simpler to test these modes and combinations with physical fax machines. In practice, fax machines meant for dedicated fax calls are kept in fax mode.

Computer Used as a Fax Machine. The summary of this section is to conduct tests by using a computer as a stand-alone fax machine. Because of the mobility advantage of small form factor computers such as laptops, the use of a soft fax machine from a computer is an interesting application. In testing, it is essential to make sure that fax over IP solutions work with a PC as a normal fax machine. Computers and laptops incorporate an RJ-11 telephone interface to use with a dial-up modem. Extra utilities are available in computers that can be used as software-based fax machines. Fax calls can be made from a computer, and fax pages are sent from data in the files. Fax pages are created in files instead of printing. The software fax machines inside the PCs usually operate at lower rates of 4800 to 9600 bps.

15.10.2 Fax Interoperability Tests

Interoperability or inter-op is required to ensure that a VoIP solution works with multiple fax machines and multiple vendor VoIP boxes. The interoperability of fax over IP includes collecting several manufacturers fax machines, VoIP boxes, and service provider boxes [URL (commetrex] to keep testing for interoperability. Many products in the market may not exactly track to all revisions of ITU standards. Once a product is released, these enhancements may not propagate. Hence, backward compatibility of previous versions of standards is also important. Interoperability tests are divided into multiple levels for the convenience of reaching the inter-op goals.

1. FaxLab creates multiple fax machines.
2. Multiple fax machines, PCs, and laptops create multiple physical fax machine situations.
3. Multiple VoIP boxes used along with FaxLab and multiple physical fax machines will enhance the inter-op with other VoIP boxes.
4. It is also helpful to conduct several tests in the presence of IP impediments.

Usually a service provider will limit the region of interest. In general, a service provider may limit the VoIP gateways to two to four manufacturer boxes, which limits the inter-op requirements to a few selected boxes and to some retail market available boxes like CISCO adapters and PSTN–VoIP gateways.

15.10.3 Fax Testing with Data Traffic

In a bandlimited service or with several data-centric applications, a QoS operation gives priority to the voice. Fax packets tolerate more delay compared with the voice packets, but fax is sensitive to the packet drops. Hence, a QoS operation for fax over IP is mainly essential to minimize packet drops. In fax, packet drop can be taken care of using error recovery techniques, but frequent packet drop causes fax pages to retransmit and calls may be disconnected. The QoS has to be applied for fax packets similar to real-time voice packets. Fax tests with multiple pages have to be performed under data traffic to verify the fax image quality. Instrument like FaxLab can be used to create several pages at different rates while conducting data and fax tests.

15.10.4 End-to-End VoIP Fax Testing with IP Impediments

IP impediments result in increased delays, jitter, and packet loss. With IP impediments, fax errors will increase or fax calls may be disconnected. IP impediment tests have to be conducted over a long duration of calls. Hence, automated tests with FaxLab-type instruments are preferred. FaxLab automates different fax machines and emulates the T.30 deviations. To incorporate IP impediments, an IP wave is shown as creating the IP impediments in Fig. 15.9. Here fax calls are established between two VoIP adapters. SIP proxy helps in establishing the VoIP call. An IP wave that is running on a separate PC separates the path and isolates the two networks. An IP wave works as a router with packet impediments created to the fax packets. An IP wave with Spirent software sits on the computer. A computer is required to keep two Ethernet ports. IP impediments are created through software running on the computer, which creates impediments on both Ethernet interfaces. Thus, impediments like delay, jitter with selected distribution, dropped packets with random, burst, and other distributions, packet errors, and fragments are produced. Several other products consisting of software, hardware, and combinations that create IP impediments are available.

FaxLab also creates several impediments. A user has to create impediments. A user typically selects the required fax machine and edits the parameters to create new fax test calls with impediments of interest. While testing with FaxLab, the system identifies the fax message degradation caused by the network transport, as well as fax protocol violations in gateways. ChannelTraps can be operated over a network or through an RS-232 connection. The system can diagnose problems at remote locations, reducing support costs. FaxLab implements the ITU figure of merit (FOM) recommendations ITU-T-E.453 and E.458 [ITU-T-E.453 (1994), ITU-T-E.458 (1996)] for classifying fax image quality and call success. Its expanded statistical report incorporates the ITU recommendations to make statistical summaries of user-selected test call groups. FaxLab can also measure post-dial delay parameters.

15.10.5   Difficulties with Fax Tests

In telephone voice testing, a user cooperates with the call. Once a voice call is established and voice packets are active on the network, a user will continue interacting with the call. Even from the customer location, or deployment, a user or testing person easily generates a description on quality. In the case of fax and modem, a machine will be interacting. Several messages and options are available even after establishing the fax and modem call. The inter-op issues with fax over IP and modem over IP will be a lot more involved compared with voice. Hence, fax testing requires several levels of testing and implementations that take care of several options and timing tolerance. Testing reveals issues, but using the correct stabilized implementation in software, hardware helps in achieving better success with fax over IP.

16

FAX OVER IP PAYLOAD FORMATS AND BIT RATE CALCULATIONS

In VoIP fax transmission, T.38 and G.711 pass-through are the two real-time fax transmission methods. A. T.38-based fax call makes use of Transport Control Protocol (TCP)- and User Datagram Protocol (UDP)-based transport protocols to deliver packets on the Internet Protocol (IP) network [ITU-T-G.711 (1988), ITU-T-T.38 (2005)]. On UDP, Internet facsimile protocol (IFP) packets are sent using either Real-Time Transport Protocol (RTP) or UDP transport Layer (UDPTL). T.38 fax over UDP with UDPTL is the most popular and well-established method in the deployments, and many VoIP gateways support UDPTL-based T.38 fax and G.711 fax pass-through. G.711 fax pass-through is similar to a VoIP voice calls with RTP packets. UDPTL makes use of redundancy and forward error correction (FEC) techniques as defined in T.38 to improve performance even with IP impediments. Network bandwidth/bit rate calculations and tables are provided in this chapter for multiple fax-machine data rates. Example calculations are presented for a packet redundancies scheme with UDPTL-based transport. The fax over IP network bit rate results are compared with G.711 fax pass-through on Ethernet and digital subscriber line (DSL) interfaces. Store-and-forward T.37 fax transmission can use similar packetization to T.38-based fax transmission, but T.37 may use the TCP method because of the non-real-time nature of fax delivery.

In this chapter, the keyword "bit rate" is used as a default to denote the fax over IP bit rate that includes fax bytes payload, IP headers, and network interfaces overhead. The keyword "bandwidth" is also used in normal usage for

VoIP Voice and Fax Signal Processing, by Sivannarayana Nagireddi
Copyright © 2008 by John Wiley & Sons, Inc.

expressing network-consumed bits per second. For distinction, fax modulation and demodulation rates of 2400, 4800, and so on, are referred to as "fax rate or fax machine bit rate."

16.1   OVERVIEW ON T.38 AND G.711 PASS-THROUGH BIT RATE

In the public switched telephone network (PSTN), fax and voice will share the same telephone interface. At the digital hierarchy of PSTN, fax or voice takes 64 kbps. In some VoIP deployments, fax is sent as a pass-through mode using G.711 [PCM μ-law (PCMU) or PCM A-law (PCMA)] or G.726 at 32/40 kbps as the compression. When network conditions are perfect, fax pass-through works similar to a PSTN-based fax. In VoIP deployments, some amount of packet impediments is unavoidable. To cater to packet impediments, RTP [Schulzrinne et al. (2003)] makes use of multiple redundant packets that are sent as per RFC2198 [Perkins et al. (1997)] and forward error correction technique as per RFC2733 [Rosenberg and Schulzrinne (1999)]. A basic G.711, 10-ms packetization-based fax pass-through call takes 126.4 kbps on an Ethernet interface, and at redundancy three (R3), it takes 328.8 kbps. In practical systems, a G.711 fax pass-through is used with 10-ms packets with redundancy one (R1—one extra payload) in reasonably acceptable IP networks. Redundancy with G.711 is limited to R1 mainly to reduce the Internet bit rate requirements and to improve on interoperability conditions. T.38-supported gateways cater to a higher level of redundancy by design. The redundancy implementation differs for RTP and UDPTL as indicated in Fig. 15.1 of Chapter 15. The redundancy scheme based on RFC2198 is applicable to voice packets, fax pass-through, and T.38 IFP over RTP packets. UDPTL-based redundancy works based on the method given in the T.38 recommendation. In practice, T.38-based redundancy up to R3 (three extra payloads) is used in deployments without a major increase in IP network bit rate.

In this chapter, bit rate calculation examples are considered with G.711 at 10-ms intervals for fax pass-through and T.38 non-ECM IFP payloads at 40-ms intervals. Packetization of 40 ms is most popular with T.38-based fax calls. Few T.38 implementations in the field also use high packetization up to 100 ms for T.4 high-speed non-ECM IFP packets. Higher packetization requires a lower bit rate, and smaller packetization reduces end-to-end delay. FEC used with voice packets, fax pass-through, and T.38 IFP over RTP packets is based on RFC2733. For T.38 UDPTL, a parity-based FEC scheme is used as described in T.38. In Fig. 15.1 of Chapter 15, these distinctions are marked for RTP, pass-through, and UDPTL-based fax transmission.

For quick reference, the T.38 bit rate and the G.711 pass-through bit rate combinations are listed in Table 16.1. In the table, redundancies are marked with the symbols "R0, R1, and R3." Redundancy R0 means no redundancy and primary payload; R1 is redundancy-1, which means one extra previous payload; and R3 is redundancy-3, which means three previous payloads. Details

Table 16.1. Summary on T.38 and G.711 Pass-Through Fax Bit Rate on Ethernet and DSL Interfaces

Fax rate in bps	Redundancy	Ethernet		DSL with PPPoE		Remarks
		T38—40ms kbps	G711—10ms kbps	T38—40ms kbps	G711—10ms kbps	
V27ter 2400	R0	17.6	126.4	21.2	169.6	T38 of 17.6 kbps is 7 times lower than G711
	R1	21.2	194.4	31.8	254.4	
	R3	28.4	328.8	31.8	381.6	
V27ter 4800	R0	20	126.4	31.8	169.6	T38 is 11 times lower
	R1	26	194.4	31.8	254.4	
	R3	38	328.8	42.4	381.6	
V29/V17 7200	R0	22.4	126.4	31.8	169.6	
	R1	30.8	194.4	42.4	254.4	
	R3	47.6	328.8	53	381.6	
V29/V17 9600	R0	24.8	126.4	31.8	169.6	
	R1	35.6	194.4	42.4	254.4	
	R3	57.2	328.8	63.6	381.6	Most popularly used and T38 is 6 times lower
V17 12,000	R0	27.2	126.4	31.8	169.6	
	R1	40.4	194.4	53	254.4	
	R3	66.8	328.8	74.2	381.6	
V17 14,400	R0	29.6	126.4	42.4	169.6	
	R1	45.2	194.4	53	254.4	
	R3	76.4	328.8	84.8	381.6	T38 is 4 times lower
V34 28,800	R0	44	126.4	53	169.6	
	R1	74	194.4	84.8	254.4	28,800 is most popularly supported in V34
	R3	134	328.8	148.4	381.6	
V34 33,600	R0	48.8	126.4	63.6	169.6 (=T38 with R3)	T38 can send 3-times redundancy compared with G711 of 169.6 kbps
	R1	83.6	194.4	95.4	254.4	33,600 rate is an optional requirement in V34
	R3	153.2	328.8	169.6 (=G711 R0)	381.6	T38 is 2-times lower than G711 or T38 with R3 the same as G711 with R0

on these calculations are given in the subsequent sections of this chapter. In the tables, the name G711 is used in place of ITU-T-G.711 and T38 is used in place of ITU-T-T.38.

At 40-ms packetization with redundancy R0, a low-speed V.27-2400 bit rate fax with T.38 takes only 17.6 kbps on an Ethernet interface. A high-speed fax at 9600 bit rate with three times redundancy takes 63.6 kbps on a DSL interface with point-to-point protocol over Ethernet (PPPoE) that is comparable with PSTN 64 kbps. In general, T.38 gives a bit rate utilization advantage between 2 and 11 times compared with G.711-based fax pass-through in VoIP. This advantage is created because of higher compression in fax modules and higher packetization used with the T.38 fax.

16.2   G.711 FAX PASS-THROUGH BIT RATE

In this section, the details on fax pass-through bit rates for redundancies zero to three are given. The G.711 codec operates at a 64-kbps rate. A loss of a G.711 pass-through packet can create impediments to the end fax delivery, and to counter this, redundancies are used. In redundancy, some immediate previous payload bytes are grouped together in a specified format. Some more details on RTP packets and redundancy format are given in Section 16.7. In G.711 pass-through, 10 or 20 ms may be used, but a 10-ms packet interval is preferred and all the calculations in this book are given for a 10-ms packetization period. For a packetization period of 10 ms, a typical G.711 payload is 80 bytes. With redundancy = one, the payload is created as present 80 bytes and previous 80 bytes grouped together as 160 bytes. In Tables 16.2 and 16.3, fax pass-through modes with redundancies zero to three on Ethernet and DSL interfaces are listed. The details on different VoIP packet headers used in Tables 16.2 and 16.3 are given in Chapter 11. From the list presented in tables, fax pass-through takes approximately 128 to 380 kbps, which is higher than the T.38 bit rate. A fax bit rate with 20-ms fax pass-through saves on bit rate, but it is still higher than the T.38 fax bit rate. Fax pass-through is independent of fax modulations. The bit rate calculations for any modulations of G3 or SG3 V.34 fax in G.711 pass-through mode are the same.

16.3   T.38 BASIC PAYLOAD BYTES FOR V.27TER, V.29, V.17, AND V.34

This section is presented on fax machine basic modulation–demodulation bit rates and basic payload byte sizes. The bit rate mentioned in this section does not include any packet headers. Fax tolerates longer round-trip delay up to 3 seconds, which allows higher packetization intervals compared with voice packets. In practical deployments, 40- to 100-ms packet intervals are used with T.38. The packet interval considered in this chapter is 40 ms (0.040 seconds).

Table 16.2. VoIP Fax Pass-Through Packets Bit Rate on an Ethernet Interface

Parameter	Details for redundancies zero to three as per RFC2198			
Redundancy	R0	R1	R2	R3
Packet interval ms	10	10	10	10
Ethernet preamble	8	8	8	8
Ethernet header	14	14	14	14
IP + UDP header	28	28	28	28
RTP header with redundancy	12	17	21	25
Fax pass-through payload bytes	80	160	240	320
CRC(4) + Transmission Gap (12)	16	16	16	16
Total bytes per voice packet	158	243	327	411
Packets per second	100	100	100	100
Total bit rate in kbps	126.4	194.4	261.6	328.8

Table 16.3. VoIP G.711 Fax Pass-Through Bit Rate with Redundancy on a DSL Interface with PPPoE

Parameter	Details for redundancies zero to three as per RFC2198			
Redundancy	R0	R1	R2	R3
Packet interval	10	10	10	10
Ethernet over ATM header	10	10	10	10
Ethernet header	14	14	14	14
PPPoE header	8	8	8	8
IP + UDP header	28	28	28	28
RTP header with redundancy	12	17	21	25
Fax pass-through payload bytes	80	160	240	320
Zero padding bytes in AAL5 cells	32	43	7	19
ATM trailer bytes	8	8	8	8
AAL5 cells each of 48 bytes	4	6	7	9
Total bytes per voice packet	212	318	371	477
Packets per second	100	100	100	100
Total bits per second (kbps)	169.6	254.4	296.8	381.6

In 0.040 seconds, a fax machine at the 9600 rate generates (9600) (0.040) = 384 bits, which is equivalent to 384/8 = 48 bytes. In Table 16.4, raw payloads for V.27ter, V.29, V.17, and V.34 modules [V27ter (1988), V29 (1988), V17 (1991), V34 (1998)] are given for various supported fax rates. V.27ter supports 2400 and 4800 rates; V.29 supports 7200 and 9600 rates; and V.17 modem supports 7200, 9600, 12,000, and 14,400 rates. The byte size values listed in Table 16.4 are reused in subsequent sections, while calculating the payload along with multiple IP headers. The name "V27" is used instead of "V.27ter" in some tables to save on space. The advanced super group 3 (SG3) fax machines support an ITU-T V.34 data modem for fax transmission at high data rates. A

Table 16.4. Fax Data Basic Payload Sizes for V.27ter, V.29, V.17, and V.34 with 40- and 100-ms Packets

Fax module and fax rate	Payload bytes in 40 ms	Payload bytes in 100 ms
V.27ter 2400	12	30
V.27ter 4800	24	60
V.29/V.17 7200	36	90
V.29/V.17 9600	48	120
V.17 12,000	60	150
V.17 14,400	72	180
V.34 28,800	144	360
V.34 33,600	168	420

V.34 fax modem operates in both half-duplex and full-duplex mode. V.34 supports up to 14 data rates from 2400 to 33,600 bps in increments of 2400 bps.

The V.34 rate of 28,800 bps is the most popular even though 33,600 is the highest rate in V.34. In Table 16.4, V.34 payloads are given for 28,800 and 33,600 rates. The bit rate 33,600 with SG3 requires a higher Internet bit rate in the G3 and SG3 famlies. In the table, the column marked with "payload bytes in 40 ms" denotes bytes generated in 40-ms frames. The column marked with "payload bytes in 100 ms" denotes bytes generated in 100-ms frames.

Number of payload bytes in a frame = (Fax rate in bps) (frame interval in seconds) / 8. For V.17 at 9600, payload bytes in 40 ms = (9600) (0.040) / 8 = 48 bytes.

16.4   OVERVIEW ON REDUNDANT AND DUPLICATE FAX PACKETS

Duplicate packets are the same packets sent multiple times with the same sequence number. Redundancy packets use an increasing sequence number and secondary payloads. Some implementations provide indicator packets as single events. Similar issues for single packets are presented in Chapter 4 while dealing with isolated voice activity device (VAD) packet transport on the network. In this situation, duplicate packets are used to emulate the benefits of redundancy.

In a redundancy-based scheme, packets are sent with present and previous payloads. The redundancy technique is given in RFC2198. Redundancy techniques are popularly used in fax pass-through and T.38-based fax data transmission. Fax signals are composed of phase and frequency modulations. In fax transmission, packet loss concealment (PLC) techniques are not useful because of the limitations in PLC implementations to recover phase and frequency modulations.

Table 16.5(a) shows the basic principle of redundancy. In this example, each frame is considered for a duration of 10 ms. In row-1 of the table, frames marked as 10, 11, 12, and 13 are transmitted for every 10 ms from the trans-

Table 16.5a. Representation of Redundancy Scheme

Payload frames for redundancy = 0 (R0)	10	11	12	13
Payload frames for redundancy = 1 (R1)	10,9	11,10	12,11	13,12
Payload frames for redundancy = 2 (R2)	10,9,8	11,10,9	12,11,10	13,12,11

Table 16.5b. Representation of Duplicate Packets Controlled Through Redundancy

Payload frames for redundancy = 0 (R0)	10			11			12			13		No signal
Payload frames for redundancy = 1 (R1)	10	10		11	11		12	12		13	13	No signal
Payload frames for redundancy = 2 (R2)	10	10	10	11	11	11	12	12	12	13	13 13	No signal

mitter side to the IP network. In row-2, packets with redundancy = one that groups a current 10-ms frame and previous 10-ms frame as one payload. In row-3, the previous two frames and the current frame is transmitted as one payload for redundancy = two. Redundancy = one takes care of one packet loss. As an example, if one packet with frames of 12, 11 is lost, a packet with frames 11, 10 and 13, 12 can make it to recover all the required frames of 10, 11, 12, and 13. When packet drop exceeds more than one packet, as an example, packets 12, 11 and 13, 12 are lost, it is not possible to recover all the frames. In this example, frame 12 is not recoverable. Redundancy = one is sufficient for one-packet drop at a time. Redundancy is advantageous for single packet drops or consecutive packet drops does not exceed redundancy. Redundancy increases the bit rate on the IP network.

Duplicate packets are required with one-time events. Some indicator packets are one-time events. As an example, called terminal identification (CED) tone detection may be reported as a one-time event. Several messages and acknowledgments are generated as one-time events. It is not possible to introduce previous valid payloads. To overcome this difficulty, many implementations adapt sending duplicate packets for the same purpose of redundancy. Table 16.5(b) is presented for redundancy R0, R1, R2, but the entries in the columns are for duplicate packets. In the example, the tenth packet is sent to the IP network two or three times with the same sequence number. Duplicate packets operate on the same sequence number. At the receiver, extra packets with the same sequence number are discarded after getting the correct packet.

In T.38 IFP over UDPTL-based transport, many implementations adapt duplicate packets only for T.30 indicator packets. No hard rules exist on duplicate packets. The simplest approach is to provide duplicate packets at regular intervals of 40 to 100 ms. This process can stop after getting another new valid packet that increments the sequence number.

Table 16.5c. Representation of Redundancy with Duplicate Packets

Payload frames for redundancy = 0 (R0)	10	10	10	11	11	11	No signal	12
Payload frames for redundancy = 1 (R1)	10,9	10,9	10,9	11,10	11,10	11,10	No signal	12,11
Payload frames for redundancy = 2 (R2)	10,9,8	10,9,8	10,9,8	11,10,9	11,10,9	11,10,9	No signal	12,11,10

The other approach is to send both duplicate and redundant packets as represented in Table 16.5(c). In general, a duplicate packet operation can be applied on a basic packet as in Table 16.5(b) or on redundant packets as in Table 16.5(c). Duplicate count can be independent of redundancy count.

16.5 T.38 IFP PACKETS

IFP or the name "IFP packets" refers to T.38 packets even though the name "IFP packet" is more correct. The IFP packet contains IFP header bytes, raw payload bytes based on the packet interval, and selected fax modulation. Viewing IP packets on a network, the IFP packet is a payload that forms IP packets using RTP, UDPTL, or Transport protocol data unit packet (TPKT) headers. In the usage, the fax payload is referred to by the name IFP packet because of the inclusion of the IFP header with raw fax payload bytes. Voice packets do not have any corresponding headers that are similar to the IFP header.

Every message of the analog fax maps to unique bytes in T.38. At the source, bytes are extracted from the fax-sampled signals. At the destination, bytes are used to regenerate the signals to a fax machine. Two broad classes of T.38 packets exist—namely, the T.30 indicator packets for tones, no signal, preamble flags, and modulation indications and the T.30 data packets for commands, messages, training, and actual fax pages.

16.5.1 T.30 Indicator Packets

The T.30 indicator type is used by the gateways to indicate the detection of signals such as calling (CNG), CED tones; high-level data link control (HDLC) preamble flags, no signal indications, and modulation training selection. The basic IFP indicator packet of two bytes is marked in Fig. 16.1(a), which contains length of packet as first byte and type of message or indicator as second byte. The indicator payload format is given below. Several indicator type names are listed here to create an overview on what information goes on this

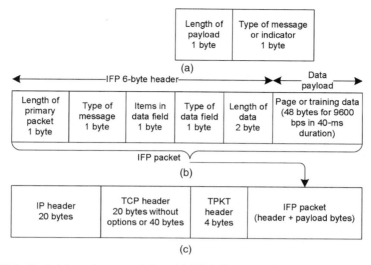

Figure 16.1. Packet format representation. (a) IFP indicator packet. (b) IFP header and data payload. (c) High-level TCP/TPKT/IFP packet structure.

byte. Refer to the T.38 recommendations for completeness of this list and updates on revised documents:

- Two bytes of primary IFP indicator packet, constructed as
  - One byte for length of payload of IFP packet
  - One byte for type of indicator consists of
    - No signal
    - CNG (1100 Hz)
    - CED (2100 Hz)
    - V.21 preamble
    - V.27 2400 modulation training
    - V.27 4800 modulation training
    - V.29 7200 modulation training
    - V.29 9600 modulation training
    - V.17 7200 modulation short training
    - V.17 7200 modulation long training
    - V.17 9600 modulation short training
    - V.17 9600 modulation long training
    - V.17 12,000 modulation short training
    - V.17 12,000 modulation long training
    - V.17 14,400 modulation short training
    - V.17 14,400 modulation long training

- V.8 ANSam signal
- V.8 signal
- V.34-cntl-channel-1200
- V.34-pri-channel
- V.34-CC-retrain
- V.33 12,000 modulation training
- V.33 14,400 modulation training

"No signal" indicator may be sent whenever no signal or silence exists. As an example, it may be used when a fax operation is changed, from V.21 to V.17 or from V.17 modem to V.21. No signal packet type is also used to adjust any buffer sizes in the middle of the T.38 call or to sustain the fax call.

16.5.2   T.30 Data Packets

The T.30 data packet is used for HDLC control data and phase-C T.4/T.6 image data. When V.34 modulation is used the data type indicates V.8 control signal data and V.34 control and primary channel data. Data packets can be classified broadly as primary and secondary. The secondary packet name is used for defining the additional or previous payload for redundancy packets. As shown in Fig. 16.1(b), the primary IFP data packet consists of 6 bytes of IFP header and actual payload data from high-speed or low-speed HDLC. This T.38 IFP packet format is common for UDPTL, RTP, and TCP transport protocols. Each T.38 IFP packet is a payload for the UDPTL/RTP/TPKT transport layer. The IFP packet over the UDPTL/RTP/TPKT structure is explained in the next few sections. The 6-bytes IFP packet header is shown in Fig. 16.1(b) along with T.38 data. The end of 6 bytes is the actual low-speed HDLC data or high-speed T.4/T.6 image data. IFP header details are listed below. More details on data fields and messages are available in references [ITU-T-T.38 (2005), ITU-T-T.30 (2005)]. Six bytes of primary IFP packet header is constructed as

- 1-byte length of primary packet
- 1-byte type of message given as
  - V.21 Channel-2
  - V.27ter 2400
  - V.27ter 4800
  - V.29 7200
  - V.29 9600
  - V.17 7200
  - V.17 9600
  - V.17 12,000
  - V.17 14,400

- V.8
- V.34-pri-rate
- V.34-CC-1200
- V.34-Pri-Ch
- V.33 12,000
- V.33 14,400
- 1-byte on number of items in data field
- 1-byte data field type given as
  - HDLC Data
  - HDLC-SIG-End
  - HDLC-FCS-OK
  - HDLC-FCS-Bad
  - HDLC-FCS-OK-Sig-End
  - HDLC-FCS-BAD-Sig-End
  - T-4 NON-ECM-DATA
  - T-4 NON-ECM
  - T.4-Non-ECM-Sig-End
  - cm-message
  - jm-message
  - ci-message
  - V.34-rate
- 2-bytes length of data field

IFP packets are created based on the transport protocol of UDP, RTP, and TCP [Schulzrinne et al. (2003), Postel (1980), URL (RFC791), Rose and Cass (1987)]. TCP operates on the same payload as given in Fig. 16.1(b) with additional TPKT and TCP headers. UDP makes use of an extra 2-byte UDPTL header for error recovery and secondary packets count. RTP also needs extra redundancy payload headers. The details on IFP packet and redundancy packet creation is given in the next section. The headers applied on data packets are also applicable to indicator packets. Indicator packets and data packets are not generated at the same time from the fax over IP system. The bit rate from the indicator packets is very low compared with data packets generated during actual page and training. In this chapter, bit rate calculations are not given for low-speed indicator or low-speed HDLC data packets, as this is much lower than high speed data.

16.6 IFP OVER TCP (TCP/IP/IFP)

TPKT/TCP is also considered for the transport of IFP T.38 messages, but it is better suited for T.37 store-and-forward fax. The TPKT header is defined in

RFC1006 and will precede the IFP packet in TCP implementations as shown in Fig. 16.1(c). Implementations using TPKT set the T.38 version to "1" or higher. Versions with "0" implementations will not use TPKT. TCP transport is a session-based, confirmed delivery service, which provides full correction. In TCP mode, it is not required to use redundancy and FEC for packet recovery. Packet drop is taken care of in TCP through retransmission. The delays can increase because of retransmissions, and the message collisions or timeouts can occur with TCP. TCP-based fax transmission requires a lower bit rate under reasonably good network conditions. The high-level TPKT/TCP/IP packet structure shown in Fig. 16.1(c) will also use extra network interface headers for actual delivery. In TCP implementation, a TCF training signal for determining the high-speed data rate is generated locally by the receiving gateway. Several gateways support UDPTL-based fax transmission and do not interoperate in the TCP mode of operation.

16.7   IFP OVER UDP

UDP is the most popular method for T.38 packet transmission. A T.38 fax can be transmitted over UDP using one of the techniques of IFP data encapsulated in the UDPTL RTP or protocol. UDP has good real-time performance, and routers do not discard the datagram during congestion. However, the downside to UDP is a reduced quality of service. It is possible for packets to get lost, to arrive out of order, or to be duplicated. To overcome these impediments, well-designed redundancy and FEC algorithms are used in UDP mode.

16.7.1   IFP over RTP

The RTP protocol is part of the UDP-based fax transmission and can be used as an alternative to IFP over UDPTL. The RTP protocol is used when both gateways negotiate this capability during call setup. Additional capabilities such as error protection schemes are also negotiated for redundancy [Perkins et al. (1997)] and FEC [Rosenberg and Schulzrinne (1999)]. For the RTP transport, the IFP packets are encapsulated in RTP as shown in Fig. 16.2(a). Within an RTP packet, an IFP packet may be combined optionally with a redundant IFP packet or with a combination of redundancy and FEC packets.

Advantage of Using RTP Protocol. RTP allows improving T.38 to permit faster switching from the VoIP voice call to the FoIP call. The switching from VoIP to FoIP could be accomplished by simply changing payload types "on the fly" rather than by renegotiating, which thus improves the interoperability and switching speed. An RTP-based fax is a recent incorporation in fax over IP calls. Many gateways in the deployments are already made with UDP–

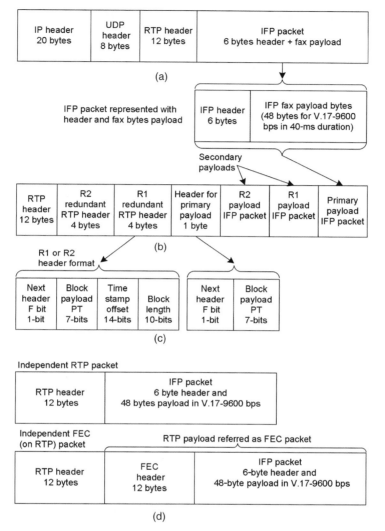

Figure 16.2. IFP over RTP packet format representation. (a) Representation of IFP/RTP/UDP/IP primary packet structure. (b) Basic format of RTP/IFP payload shown for redundancy = 2. (c) Four-byte RTP extender header format used with extra redundant payload. (d) FEC/IFP packet format for separate RTP stream.

UDPTL-based transport. Hence, the use of RTP is limited in some existing deployments.

RTP-Based Format for Primary Packets. The RTP packet starts with a fixed RTP header of 12 bytes. The next section describes the payload-specific fields of the RTP, when the RTP packet encapsulates fax IFP packets. Voice packets always use RTP. The main differences in RTP for voice and fax are in the payload definition and marker bit.

Payload type (PT): The payload type for a fax is a dynamic payload type identified by the name "t38." Marker (M) bit: The marker bit is not used for fax and must be set to zero. The marker bit has to be ignored by the receiver.

RTP-Based Format with Redundancy. An RTP-based redundancy scheme uses a different format for header combinations compared with UDPTL given in the next section. For every extended redundant payload, RTP makes use of an additional 4-byte header for redundancy and an extra 1 byte for primary payload type with termination indication. The data blocks primary and redundant are stored immediately followed by final header as shown in Fig. 16.2(b), which is different than the UDPTL-based redundancy scheme. The IFP payload format is the same in RTP- and UDPTL-based packets.

The bits in the 4-byte redundant RTP header are given in Fig. 16.2(c), and a description is given below.

F: (1-bit) The first bit in the header indicates whether another header block follows. Bit value "1" indicates that additional header blocks follow, and bit value "0" indicates the last header block.

Block PT: 7 bits of RTP PT for this block. The payload type for this fax is a dynamic payload type identified by the name "t38." If redundancy is used per RFC2198, the payload type must indicate the payload format RED (Redundancy) as per RFC2198.

Timestamp offset: A 14-bit unsigned offset of the time stamp of this block relative to the time stamp given in the RTP header. The use of an unsigned offset implies that redundant data are sent after the primary data. Hence, a time to be subtracted from the current time stamp to determine the time stamp of the data for which this block is the redundancy.

Block length: A 10-bit length in bytes of the corresponding data block excluding the header. The primary encoding block header is placed last in the packet, which does not contain a time stamp and a block length. These fields are considered from the RTP header and overall packet length. The last 1-byte header is the final header for a primary block that comprises only a "zero F-bit" and the "7 bits" of block payload type information making it a total of 8-bits.

RTP-Based Format with FEC. An IFP over an RTP scheme uses the generic forward error correction as defined in RFC2733 to compensate for packet loss in the Internet similar to G.711 pass-through. For IFP over RTP transport, the RTP packets are the combinations of an IFP packet and RTP header. The RTP payload is a single IFP packet when no redundancy and FEC schemes are used. An RTP-based FEC scheme uses a different format of header combinations compared with UDPTL given in the next section. The FEC packets are generated based on an exclusive-or (XOR) operation and sent in the same RTP stream as the RTP payload. The IFP encapsulated RTP-based FEC packets can be sent as a separate RTP stream or as an RTP redundant codec as defined in RFC2198. The FEC packet is constructed by placing an FEC header and FEC payload in the RTP payload [Schulzrinne and Petrack (2000)]. The FEC payload is one or more IFP packets generated by the use of an XOR operation. The packet format for FEC packets sent as a separate RTP stream is shown in Fig. 16.2(d).

RTP Header of FEC Packets. The RTP header for FEC packets is 12 bytes with the version field set to two. The synchronization source (SSRC) value is the same as the SSRC value of the media stream it protects. The payload type for the fax FEC packet is a dynamic payload type identified by the name "t38."

FEC Header. The FEC header is 12 bytes and consists of a sequence number (SN) base field (2-bytes), length recovery field (2-bytes), E field (1-bit), PT recovery field (7-bits), mask field (24-bits), and time-stamp (TS) recovery field (4-bytes) as defined in RFC2733. The E-bit indicates a header extension and must set this bit to zero. The length recovery field is used to determine the length of any recovered packets. The SN base field is set to the minimum sequence number of those IFP packets protected by FEC. The FEC packets can also be sent as a secondary codec in redundant RTP packets as defined in RFC2198, and the packet format is the same as the RFC2198 redundant packets as shown in Fig. 16.2(b). The redundant RTP payload (R1/R2) becomes the FEC packet generated based on the XOR operation of the IFP packets plus the 12-byte FEC header.

RTP-Based Bit Rate Considerations. The RTP header and the RTP redundancy header contribute significantly to RTP-based fax transmission. In the next section, it can be observed that UDPTL-based fax transmission has shorter headers than RTP. Bit rate calculations for RTP can be achieved based on Fig. 16.2(a) and 16.2(b). In practice, deployments will be more concerned with interoperability than with a marginal bit rate increase.

16.7.2 IFP over UDPTL—Primary and Secondary Packets

UDPTL packets comprise a sequence number, a variable length of payload, and the additional header information required for error control over UDPTL. The sequence number is essential to reorder and to extract the packets with the right order in the presence of network impediments. In voice packets, RTP takes care of a real-time sequence number, and fax packets do not maintain their own sequence number. The UDP header does not have a sequence number. The UDPTL header used in fax over IP incorporates a 2-byte sequence number similar to the sequence part of the RTP header. An IFP packet with provisioned redundancies will appear as UDPTL payload. UDPTL packets are based on the principle of framing, and each packet may contain one or more IFP packets in its payload section. The first IFP packet in a UDPTL payload is known as the "primary." Additional fields included in a payload after the primaries are called secondary IFP packets. The high-level IFP/UDPTL/UDP/IP primary packet structure is illustrated in Fig. 16.3(a).

To improve the reliability of the T.38 protocol, redundancy and parity FEC schemes are supported in UDPTL as defined in T.38. Most gateways support redundancy as a default, and FEC is used as an option. Redundancy sends

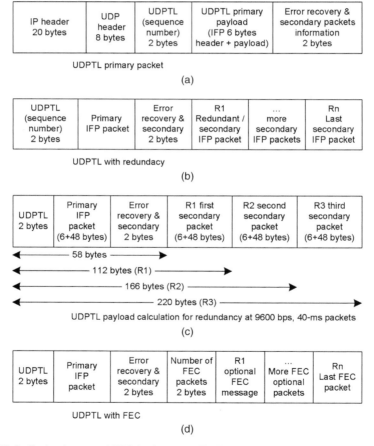

Figure 16.3. Redundancy and FEC for fax over UDPTL. (a) Representation of UDPTL/UDP/IP primary packet structure. (b) T.38 basic format of UDPTL payload section with redundancy. (c) T.38 UDP payload for primary and secondary redundancy packets. (d) T.38 basic format of UDPTL payload section with FEC.

some previous packets along with primary and FEC sends extra bytes derived with polynomial. A gateway should indicate its support of the available error protection schemes during capabilities exchange. Based on the capabilities, a choice may be made on which scheme is used for error protection. If the capability is indicated to receive both parity error correction frames and redundant frames, then either scheme may be used. If the gateway indicates a capability to receive only redundant error protection frames, then the transmitting gateway may not send parity FEC frames [Rosenberg and Schulzrinne (1999)]. The basic format of an UDPTL payload with redundancy is as shown in Fig. 16.3(b). In practical situations, even if packets are received with redundancy or FEC, a receiver can decide to use or discard the extra payload. In this chapter, the bit rate calculation for the T.38 fax over UDPTL on Ethernet and DSL interfaces is presented for the redundancy option.

The support of the parity FEC scheme is optional. A gateway providing parity FEC receiver services should also be capable of receiving redundant messages. An FEC contains a parity-encoded representation of several primaries. The number of primary IFP packets represented by an FEC field is given by the number of FEC packets usually denoted as fec-n-packets of the UDPTL packet. The basic format of the UDPTL payload with FEC is shown in Fig. 16.3(d). In comparison with redundancy, the FEC creates 2 bytes more through the number of FEC packets indication.

16.8 T.38 UDPTL-BASED BIT RATE CALCULATION WITH REDUNDANCY

UDP primary and secondary payload for redundancies from R = 1, 2, and 3 are given in Fig. 16.3(c). In Table 16.6, UDP payload for primary packet and redundancies up to R = 3 (i.e., first, second, and third secondary packets) are given. In general, this payload can be extended to higher redundancies. For data packets, redundancies of one to three are commonly used in the deployments. The bit rate calculations for UDPTL are given with a single data field item of T.4/T.6 non-ecm data. Summary points are given here.

- For R = 0, UDP payload = UDPTL header + primary IFP + 2-byte error recovery and secondary bytes = 10 bytes + raw fax bytes payload.
- For R = 1, UDP payload = UDPTL header + primary IFP + 2-byte error recovery and secondary bytes + first secondary IFP = 10 bytes + primary raw payload + 6 byte + secondary raw payload.
- For R = 2, UDP payload = UDPTL header + primary IFP + 2-byte error recovery and secondary bytes + first secondary IFP + second secondary IFP.
- For R = 3, UDP payload = UDPTL header + primary IFP + 2-byte error recovery and secondary bytes + first secondary IFP + second secondary IFP + third secondary IFP.

Table 16.6. UDP Payload Bytes for Multiple Fax Rates with Redundancies One to Three

Fax module and bit rate	UDP payload bytes for different redundancies with 40-ms packets			
	R0	R1	R2	R3
V.27ter 2400	22	40	58	76
V.27ter 4800	34	64	94	124
V.29/V.17 7200	46	88	130	172
V.29/V.17 9600	58	112	166	220
V.17 12,000	70	136	202	268
V.17 14,400	82	160	238	316
V.34 28,800	154	304	454	604
V.34 33,600	178	352	526	700

Table 16.6 summarizes the actual data and header combination as UDP payload for various fax modules and redundancy factors using IFP over UDPTL. Table 16.4 raw payload byte sizes for 40-ms packetization are mapped onto Fig. 16.3(c) representation to arrive at Table 16.6. The entries in Table 16.6 shown as UDP payload includes a 2-byte UDPTL header, 2-byte UDPTL error recovery and secondary packets information, 6-byte of IFP header, and IFP payload. Table 16.6 entries are used in calculating the bit rate for various fax rates and redundancies.

16.9 FAX UDPTL-BASED BIT RATE ON ETHERNET AND DSL INTERFACES

In Tables 16.7 and 16.8, T.38 fax bit rates are listed for basic fax rates and redundancies. In Table 16.7, bit rates are presented for no redundancy (R0) and 3-times redundancy (R3) on an Ethernet interface. The details on various headers [URL (National-Eth)] and their meaning are given in Chapter 11. For any other redundancy, the row marked "UDPTL, IFP, error recovery, redundancy" has to be changed with the corresponding values given in Table 16.6 for different redundancies.

On a DSL interface, the bit rate increases because of the additional header of PPPoE [Mamakos et al. (1999)], AAL5 trailer header, ATM cell splitting, and zero padding as shown in Table 16.8. More details on the headers are given in Chapter 11 for calculating the voice bit rate. A summary on T.38 bit rate requirements for Ethernet and DSL interfaces are given below:

- The minimum bit rate requirement from Ethernet is 17.6 kbps for the 2400 rate.
- The minimum bit rate from the DSL interface is 21.2 kbps for the 2400 rate.

Table 16.7. VoIP Fax Packet Bit Rate on an Ethernet Interface

Fax rates of 2400, 4800, 7200, 9600, 12,000, 14,400, 28,800, and 33,600 without redundancy (R0) and with three times redundancy (R3) on Ethernet interface, 40-ms packets

Fax bit rate & redundancy	2400 R0	2400 R3	4800 R0	4800 R3	7200 R0	7200 R3	9600 R0	9600 R3	12,000 R0	12,000 R3	14,400 R0	14,400 R3	28,800 R0	28,800 R3	33,600 R0	33,600 R3
Packet interval ms	40	40	40	40	40	40	40	40	40	40	40	40	40	40	40	40
Eth. preamble	8	8	8	8	8	8	8	8	8	8	8	8	8	8	8	8
Eth. header	14	14	14	14	14	14	14	14	14	14	14	14	14	14	14	14
IP header	20	20	20	20	20	20	20	20	20	20	20	20	20	20	20	20
UDP header	8	8	8	8	8	8	8	8	8	8	8	8	8	8	8	8
UDPTL, IFP, error recovery, redundancy	22	76	34	124	46	172	58	220	70	268	82	316	154	604	178	700
CRC	4	4	4	4	4	4	4	4	4	4	4	4	4	4	4	4
Transmission gap	12	12	12	12	12	12	12	12	12	12	12	12	12	12	12	12
Total bytes per fax packet	88	142	100	190	112	238	124	286	136	334	148	382	220	670	244	766
Packets/second	25	25	25	25	25	25	25	25	25	25	25	25	25	25	25	25
Total bits per second (kbps)	17.6	28.4	20	38	22.4	47.6	24.8	57.2	27.2	66.8	29.6	76.4	44	134	48.8	153.2

Table 16.8. VoIP Fax Packet Bit Rate on DSL with PPPoE

Fax bit rate & redundancy	Fax rates of 2400, 4800, 7200, 9600, 12,000, 14,400, 28,800, 33,600 without redundancy (R0) with 3-times redundancy (R3) on DSL interface with PPPoE, 40-ms packets															
	2400 R0	2400 R3	4800 R0	4800 R3	7200 R0	7200 R3	9600 R0	9600 R3	12,000 R0	12,000 R3	14,400 R0	14,400 R3	28,800 R0	28,800 R3	33,600 R0	33,600 R3
Packet interval	40	40	40	40	40	40	40	40	40	40	40	40	40	40	40	40
Eth. over ATM	10	10	10	10	10	10	10	10	10	10	10	10	10	10	10	10
Ethernet header	14	14	14	14	14	14	14	14	14	14	14	14	14	14	14	14
PPPoE	8	8	8	8	8	8	8	8	8	8	8	8	8	8	8	8
IP header	20	20	20	20	20	20	20	20	20	20	20	20	20	20	20	20
UDP header	8	8	8	8	8	8	8	8	8	8	8	8	8	8	8	8
UDPTL, IFP, error recovery, redundancy	22	76	34	124	46	172	58	220	70	268	82	316	154	604	178	700
Zero padding	6	0	42	0	30	0	18	0	6	0	42	0	18	0	42	0
AAL5 trailer	8	8	8	8	8	8	8	8	8	8	8	8	8	8	8	8
ATM cells	2	3	3	4	3	5	3	6	3	7	4	8	5	14	6	16
Total bytes per fax packet	106	159	159	212	159	265	159	318	159	371	212	424	265	742	318	848
Packets/second	25	25	25	25	25	25	25	25	25	25	25	25	25	25	25	25
Total bits per second (kbps)	21.2	31.8	31.8	42.4	31.8	53	31.8	63.6	31.8	74.2	42.4	84.8	53	148.4	63.6	169.6

- RFC2198 considers redundancy for the fax pass-through mode. For redundancy one, the RFC2198-based G.711 scheme takes about five to six times more bit rate than T.38 as shown in Table 16.1.
- V.34-33,600 is the highest rate of fax over IP call and takes more bit rate.
- On a DSL interface, FEC may require more bit rate if the two extra packet bytes split into additional ATM cell.

16.9.1   Bit Rate Change Among Redundancy and FEC

Bit rate requirements from redundancy and FEC can vary slightly on DSL-based deployment. In Table 16.8 for DSL-based calculations, one row of zero padding appears for creating ATM cells. It can be observed that zero padding reduces the bit rate utilization, but it is required for creating ATM cells. In the examples considered here, the combinations of packetization and redundancy = three make optimum utilization of the bit rate. The zero padding bytes with R3 are not present in Table 16.8. As explained in FEC, an extra 2 bytes will be present with FEC. These two an extra bytes can create an extra ATM cell with redundancy three of Table 16.8. For the same fax rate of 2400 bps, redundancy = three takes 31.8 kbps and FEC takes 42.4 kbps. In bandlimited deployments, this significant bit rate change has to be accounted for while selecting between redundancy and FEC-based methods. The bit rate variation is insignificant for the Ethernet-based calculations.

16.9.2   Bit Rate Change in Silence Zones

In the fax call of five phases (A, B, C, D, and E), the fax bit rate is calculated only for active phases of C1 and C2 where training and high-speed T.4/T.6 image data are sent as IFP packets. During premessage, postmessage procedure, and call disconnect phases, several call control and capability messages are exchanged using low-speed V.21 HDLC data packets. In the middle of a fax call, during low-speed-to-high-speed or high-speed-to-low-speed transition, no signal indicator packets are sent. During these periods, fax requires much less bit rate.

An example on bit rate savings is measured for four pages at 9600 bps using FaxLab. A test with FaxLab for V.29 at the 9600 bit rate requires approximately 99 seconds for the entire fax session to transmit four simple pages using MR coding. It is observed that a total time of 69.5 seconds is required to transmit the high-speed T.4/T.6 image data, including the training signal. The remaining 29.5 seconds is used for exchange of premessage and postmessage control information using V.21 at 300 bps and silence zones. For simple pages, it is approximately 30% savings in bit rate. The time to transmit the high speed data will vary with coding method (MH/MR/MMR) used. When complex images are used, the total fax call time for four pages is observed as 337 seconds. The high-speed data, including training, require 302 seconds to transmit four complex pages, and the low-speed data transmission and silence zones require 35 seconds. For four complex pages, it is a savings of 10%. This bit rate savings is not applicable to the G.711 pass-through mode.

Fax page transmission is half-duplex, and this is the major savings in bit rate. In G.711 pass-through, packets are transmitted in both directions.

16.10 T.38 BIT RATE RECOMMENDATIONS

Most available fax machines support up to 14,400 bps. For bit rates up to 14,400 bps, T.38 takes less bit rate compared with G.711. In Europe, 33,600 supported fax machines are also coming to the market. At the 33,600 V.34 rate, T.38 can be sent at 3-times more redundancy compared with G.711 pass-through mode in the same bit rate. This process can be observed from Table 16.8 and 16.1 for a network bit rate of 169.6 kbps. In some countries like Japan and Korea, many users receive a 50- to 100-Mbps Internet bandwidth/bit rate (bandwidth is more commonly used, even though the bit rate coveys more correct meaning). It is expected that more users will receive higher bandwidth in the future. The G.711 pass-through requirements even with redundancies are negligible compared with the available bandwidth. In this situation, G.711 pass-through should take care of VoIP fax requirements. In general, when end-to-end packet delivery is perfect and network bit rate is unlimited compared with fax pass-through requirements, G.711 should be sufficient for VoIP fax transmission. T.38 and T.37 provide bandwidth advantage, and T.37 provides the benefit of fax pages that can be retrieved off-line.

17

COUNTRY DEVIATIONS OF THE PSTN MAPPED TO VoIP

In the public switched telephone network (PSTN) system, the closest central office (CO) or digital loop carrier (DLC) drives the telephone lines. DLC is part of the PSTN system located in between the subscribers and the central office to take care of distribution of CO physical interfaces. A user connects multiple phones, fax machines, and dial-up modems on the telephone TIP-RING interface. The CO/DLC is capable of ringing and driving the required current to use three to five telephones referred to as ring equivalent numbers (RENs). As shown in Fig. 17.1(a), the DLC is capable of driving the lines to several miles, usually given in kilo feet (kft) or 1000 feet units. Multiple DLCs and COs communicate using A-law or μ-law compression schemes for digital transmission. The impedance shown in Fig. 17.1(a) at the telephone interface and central office is for North America.

In a complete PSTN system, several deviations are taken care of while dealing with solutions for multiple countries. It is required to take care of deviations and to match these to the local PSTN for better voice quality. The right combination of phones, DLCs, central offices, transmission lines, and other supporting infrastructure can help provide better quality. To map the influences to VoIP, major deviations of the PSTN for multiple countries are categorized as central office deviations, transmission lines, and telephones. VoIP supporting infrastructure focused on customer premise equipment (CPE) (adapter) will differ from the PSTN mainly in shorter loops and in lower REN drives. VoIP deployment is also required to take care of matching interface specifications of the PSTN for better voice quality.

VoIP Voice and Fax Signal Processing, by Sivannarayana Nagireddi
Copyright © 2008 by John Wiley & Sons, Inc.

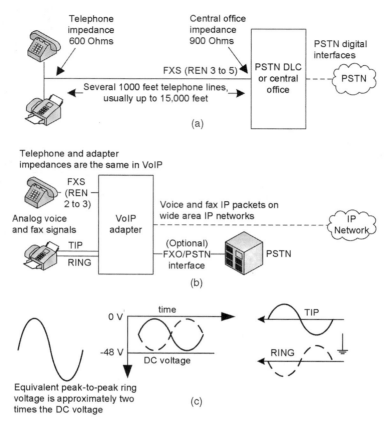

Figure 17.1. Interfaces and impedance examples for North America. (a) PSTN long-distance analog interfaces. (b) VoIP short-distance analog interface. (c) Balanced ring voltage representation (referred to in Section 17.3.4).

17.1   COUNTRY-SPECIFIC DEVIATIONS

This section is presented with the country-specific deviations mapped to the PSTN and VoIP. Catering to required deviations in VoIP can closely match the feel of PSTN-based calls.

17.1.1   Central-Office-Specific Deviations Mapped to VoIP

The VoIP adapter is like a PSTN CO for the telephone user. Hence, most functions of the CO have to be emulated through the VoIP adapter. Figure 17.1(b) is a functional representation of the VoIP adapter with foreign exchange subscriber (FXS) and foreign exchange office (FXO) interfaces. The following major parameters can influence the design and configuration of the VoIP adapter for multiple countries:

- Impedance of VoIP adapter telephone interfaces of FXS and FXO
- REN—number of simultaneous phones that can be used on the same lines
- Subscriber line interface circuit (SLIC) line voltages, battery voltages, line reversals, grounding, and various protections
- Ringing voltage, frequency, and cadence
- Diagnostic methods performed on the telephone interfaces
- Call progress tone generation and specifications
- Dual-tone multifrequency (DTMF) digit generation, detection specifications
- Caller ID, call feature tones, and timings
- Overall loudness rating (OLR), gain/loss planning in send and receive path
- Country-dependent transmission characteristics similar to TR-57 [TR-NWT-000057 (1993)] transmission characteristics
- Call switching timing, which is similar to TR-57 switching characteristics that varies with country
- Based on local PSTN A-law/μ-law compression mode, G.711 VoIP calls can use the same preferences for minimal transcoding

17.1.2 Transmission Lines

The deviations in transmission for multiple countries are as follows:

- Telephone cable length, wire gauge, and other cable characteristics between DLC or CO and the telephone equipment
- Various loading coils and matching on telephone lines
- Digital subscriber line (DSL) support on telephone lines [URL (TI-impedance)]

PSTN service makes use of several 1000-feet lines. Transmission-dependent distortion, drive strength, and losses need to be taken into consideration in PSTN-based systems. VoIP systems use short loop wires, and distortions are insignificant because of the lines. Wire gauge selection will not influence the performance because so few feet lines are used in VoIP.

17.1.3 Telephone Deviations

PSTN- and VoIP-based calls will use the same telephones. The following parameters of a telephone can influence the design for multiple countries:

- Telephone impedance
- Caller ID detection; various indicators such as caller ID display, call wait display, message waiting

- Pulse or tone (DTMF) dialing options, speed and redial timing deviations
- DTMF digits and parameters
- Loudness rating of phones
- On-hook, off-hook, flash hook key, and respective debounce settings
- Telephone connectors and adapters
- Certification of phones to regional requirements

VoIP should consider the above parameter deviations with local country standards. Each country or group of countries may follow its own set of standards in combination with other global standards from the ITU, ETSI, and IEEE. Telephones are tested for meeting electrical and acoustic specifications. A typical test list of instrument-based tests is given at [URL (Microtronix-country)] that lists the tests conducted on a Europe TBR21-complaint phone. Several baseline documents are used that are specific to the country of operation. Instruments like [URL (Microtronix-country)] perform multiple country-specific phone tests.

17.2   COUNTRY-SPECIFIC DEVIATIONS ON VoIP INTERFACES

The VoIP adapter is a CO to the telephone user connected at an FXS interface of a VoIP box. The VoIP adapter has to emulate the characteristics of a PSTN CO. The front-end SLIC of a VoIP adapter is programmed for required voltages, line conditions, REN drive, impedances, gain/loss, and diagnostic features. Some parameters of call progress tones, DTMF, and call feature generation are programmed through a processor attached to the SLIC and CODEC. Many of these parameters are country-dependent variations that have to be taken care of in the VoIP adapter, and these features and parameters run into several pages in local PSTN standards.

17.2.1   Telephone Impedance Programmed on the VoIP Adapter

As represented in Fig. 17.2, the alternating current (AC) terminating impedance can be modeled through resistors and a capacitor as an R_s, $(R_s + C_s)$ or $R_s + (R_p \parallel C_s)$ that takes care of multiple countries. The values vary for different countries. As an example, France has $270\,\Omega + (750\,\Omega \parallel 150\,\text{nF})$, France (TBR21) is $275\,\Omega + (780\,\Omega \parallel 115\,\text{nF})$, and the United States/Korea/Japan uses $600\,\Omega$, or $600\,\Omega + 1\,\mu\text{F}$ as an equivalent telephone impedance. More country-specific details are available at references [URL (Cisco-impedance), URL (Silab-DAA), URL (Si3015), URL (WinSLAC), URL (Microtronix-470C)]. Refer to the country-specific standards for the exact and complete information, as some discrepancy exists among the published literature on this subject.

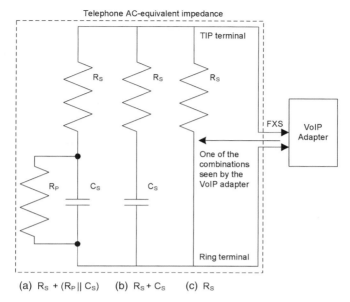

(a) $R_S + (R_P \| C_S)$ (b) $R_S + C_S$ (c) R_S

Figure 17.2. Equivalent impedance representation of telephone. (a) Complex impedance in series with parallel R_P C_S. (b) Complex impedance of R_S in series with C_S. (c) Real Impedance R_S.

The convention of representing numerical values of R_s, R_p, and C_s varies in the literature and in other documents. As an example, when R_p is not present, it will be left as blank, "-," or the symbol for infinity (∞, meaning open or not present). Refer to the description in the standards for exact details on these component values. The values of R_s, R_p, and C_s varies with the country. For reference, a few examples are given in this section.

17.2.2 Hybrid Matching for Multiple Countries

VoIP adapters drive telephones through the FXS interface. The FXS interface is built with SLIC–CODEC functionality devices. Historically, SLICs were made with several passive components, battery, and ring combining circuits. Recently, most SLICs are of active circuits. For each country, these devices may use different extra passive components. The goal is to match telephone and the SLIC–CODEC interface hybrid to get better echo return loss (ERL). In the case of the PSTN, central office impedance in combination with lines and telephone are approximately matched. Proper impedance matching also helps in reducing echo level in the PSTN and VoIP.

Recent designs have taken care of hybrid implementation in two stages. The first stage is of a fixed hybrid implemented with a fixed combination of components as part of the SLIC. This stage is shown in Fig. 17.3(a). The SLIC manufacturers will decide the required fixed hybrid passive components. The

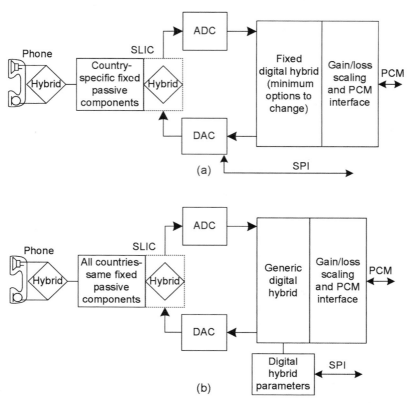

Figure 17.3. Impedance combinations. (a) Fixed impedance hardware software scheme. (b) Country selection through fixed passive components and digital hybrid parameters.

mismatched signal after passing through a fixed analog hybrid of a SLIC is sampled in hardware CODEC—[analog-to-digital converter (ADC) and digital-to-analog converter (DAC)]. These digital samples are used to form a digital hybrid. Digital hybrids are implemented as filter networks and as amplitude summing operations. These coefficients are fixed in nature for the populated passive components and country requirements. This scheme also calls for changing passive components for different country impedance combinations. Several aspects of impedance matching and measurements are given in references [URL (Seemix), URL (Microtronix-501), URL (Elect-matching), URL (Cisco-impedance)].

A country-specific programmable scheme is shown in Fig. 17.3(b). In this scheme, a SLIC will have a fixed set of analog components that are valid for multiple country requirements. The perfect hybrid functions are achieved as a combination of a fixed analog hybrid and a variable digital hybrid. Depending on the country, software will send different coefficients into a digital hybrid that makes an analog hybrid and an updated digital hybrid into the perfect combination. This process will occur usually through serial peripheral interface (SPI) communication used to interface with SLIC–CODEC devices. The generic way is to send required coefficients depending on the country and the populated passive components with a SLIC analog hybrid. In some devices like Si3015 [URL (Si3015)], the required coefficients are programmed in the CODEC device. A digital hybrid will use one of the available sets of coefficients to meet multiple country requirements. Such a country impedance selection table can be found at reference [URL (Si3015)]. Currently, this type of digital hybrid with fixed passive components (meaning a fixed analog hybrid), with a programmable digital hybrid, is most popular for taking care of multiple countries.

As an advanced option, these hybrids can be made adaptive. Even after making the country selection, hybrids can be mismatched based on phones connected or parallel phones used on the same interface. For this reason, a combination of adaptive and or country-specific selection can be used. From the ADC, DAC digital signals, a hybrid path can be calculated, which can help in deriving the correct coefficients. This path will work as an echo canceller with coarse estimation. After estimation of the coefficients, the values can be held fixed for that call. In most situations, the coefficients can be changed only on a new call. In practice, VoIP solutions use high-performance echo cancellers that are more powerful in taking care of any variation that may happen with different or multiple phones.

17.3 CALL PROGRESS TONES FOR MULTIPLE COUNTRIES

This section is presented with the electrical characteristics of various tones used in PSTN or VoIP services. In telephone service, the outcome of communication would be either visual or acoustic. Most actions in the telephone are audible acoustic tones and speech. The tones are known by name call progress tones (CPTs) as they represent the progress of the call. The tones in different countries and regions will vary. Some basic tones such as dial tone, busy tone, ring back tone, congestion tone, and call waiting tone are given in E.180 [ITU-T-E.180/Q.35 (1998)] for most countries. Many extended tones are available in local PSTN standards, for example, [URL (NTT-E)].

The most popular visual information on a phone are message waiting indicators, caller ID, call wait ID display, any light blinking associated with off-hook, ring, or call progress indication. Some phones with sufficient size display may also display dialed digits, date, time, and new call indication through selected symbols in the same screen of caller ID display.

17.3.1 Basic Call Progress Tones

Most tones mentioned in this chapter are applicable to both the PSTN and VoIP. In the case of the PSTN, a central office will deliver these tones. In VoIP, the VoIP adapter in association with call control signaling delivers the tones. These tones are given in a generic way. Specific details have to be noted from country-specific PSTN standards and E.180. The basic tone parameters also vary based on country.

Dial Tone. A dial tone is part of the originating call. When off-hook, the VoIP adapter delivers a dial tone. A dial tone is the first audible tone that indicates the proper working of the VoIP adapter and phone combination. On hearing the dial tone, digits (either pulse or DTMF dialing) can be dialed through a phone. The dial tone has to be generated in 50 ms from off-hook [TR-NWT-000057 (1993)]. Once the first digit is dialed, the dial tone generation is terminated.

Ring-Back Tone. The ring-back tone will be generated on the originating terminal while the called terminal is ringing. While a call is being established, the destination phone rings and the originator has to get a ring-back tone in relation to the destination ring tone. This tone will be terminated when an answer signal (off-hook) is received at the receiving terminal. The cadence of the called party ring may not be synchronized with the calling party ring-back tone. In some call presentation configurations, the recorded voice may also replace the ring-back tone.

Busy Tone. The busy tone is provided to the calling party when the called party is found busy or when the call fails for unknown reasons. This tone times out after a minimum duration of 60 seconds if the subscriber remains in the off-hook state. After a required timeout of a busy tone, usually an error tone will be played in place of the busy tone.

Congestion Call Tone. Several names of congestion, reorder, error, and false tone are used for the same tone. This tone is provided on the line of the calling party, when a false call situation occurs. In some services, this tone will be elapsed after a particular duration of 60 seconds, and no tone will be played on the line until making on-hook before making the next off-hook. A false call situation may occur by failure of a call (number unobtainable) or wrong operation from the subscriber (incomplete dialing).

Call Waiting Tone. This tone, which is also called an incoming identification tone, is used in the call waiting services to identify an incoming call. A call waiting service consists of presenting an incoming call on a line already busy on another call. The call waiting tone is provided to the called party for a short duration of 300 ms superimposed on the ongoing communication.

17.3.2 Other Call Progress Tones

Many supplemental tones widely vary in options in relation to country-specific requirements. Some countries may not have similar mapped tones or may use a different name and functionality in place of these call progress tones.

Busy verification/warning tone: Generates a busy verification tone, also known as a busy operator tone, which indicates to an operator that a line is engaged in an active call.

Confirmation tone: A confirmation tone indicates the success of a previous action. It corresponds to a "positive indication tone" as given in E.182 [ITU-T-E.182 (1998)].

Special dial tone or second dial tone: A second dial tone is used with register-type services like call forwarding and automatic telephone answering service or to create conditions of preventing terminations in universal call forwarding.

Message waiting indicator tone: Generates a message-waiting tone, which indicates that a message is waiting for the user on a remote voice messaging system. A message waiting tone corresponds to a "message waiting tone" as defined in Telcordia GR-506 [GR-506-CORE (1996)]. This tone is played in place of a regular dial tone when a message is waiting for the user. Once messages are cleared or read, the next call onward regular dial tone is played in place of the message waiting tone. Visual message waiting indicators are also used for message waiting.

ROC busy tone: ROC expansion is in the French language, which means completion of calls to a busy subscriber. This tone is like a special ring. The ring ON period is usually longer, and the OFF period is usually shorter than standard ring. This is provided to the calling party when the called party is busy, and the calling party is entitled to invoke the completion of call to a busy subscriber service known by the trade name "Auto-Rappel" in France Telecom services [FT ITS-3 (2005)]. This tone may be followed by voice announcements.

Special information tone: This tone is used as a "jingle" on some network failure announcements indicating that the called number cannot be reached for reasons other than subscriber busy or congestion. Some other cases of using a special information tone are when the call is connected to a recorded voice machine; the tone is then given during the silent intervals between transmissions of the announcement.

Howler tone/off-hook warning tone: This is called the receiver off-hook (ROH) tone. The network notifies a terminal by a howler tone/off-hook warning tone that an unused telephone receiver has been in off-hook for a certain time to urge that the handset be placed on-hook. ROH is given in Telcordia GR-506 [GR-506-CORE (1996)] and NTT documents [URL (NTT-E)]. The ROH tone is generated by combining four tones at frequencies of 1400, 2060, 2450, and 2600 Hz at a cadence of 0.1 second ON, 0.1 second OFF, in repetition.

Positive indication or acceptance tone: The network notifies the originating terminal that it has received the service request. This tone is used with register-type services such as call forwarding and the automatic answering service.

17.3.3 Basic Tones and Ring—Example

In Table 17.1, some popular tones and their parameter details are given for three different countries. The countries listed here are the United States [TIA-470C (2003)], France [FT ITS-3 2005], and Japan [URL (NTT-E)], and can be noted from references [ITU-T-E.180 (1998), Foster and Andreasen (2003)]. The table contains basic tones, country-specific impedance, and ring characteristics. Many parameters here will have tolerances, and it is suggested to refer to exact country-specific local PSTN standards for a complete list of parameters and tolerances.

Tone generators are designed as a combination of two tones with required frequency/phase, ON–OFF periods, and power. In the case of more than two tone

Table 17.1. Example Tones and Ring for the United States, France, and Japan (tone details of frequency, cadence, and power)

Tone Name	United States	France	Japan
Dial tone	350 + 440 Hz Continuous, −13 dBm each tone	440 Hz Continuous, −3.5 dBm	400 Hz continuous, −22-L to −19 dBm (Loss "L" is 0 to 7 dB)
Ring-back tone	440 + 480 Hz, 2 s ON, 4 s OFF; each tone −19 dBm	440 Hz, 1.5 s ON, 3.5 s OFF; −8 dBm	400 Hz modulated with 15 to 20 Hz, 1 s ON, 2 s OFF; −29-L to −4 dBm
Busy tone	480 + 620 Hz, 0.5 s ON, 0.5 s OFF; each tone −24 dBm	440 Hz, 0.5 s ON, 0.5 s OFF; −8 dBm	400 Hz, 0.5 s ON, 0.5 s OFF; −29-L to −4 dBm
Congestion/ reorder/ error/ false call tones	480 + 620 Hz, 0.25 s ON, 0.25 s OFF; each tone −24 dBm	Same as busy tone	Same as busy tone
Call wait tone or incoming identification tone	440 Hz, 0.3 s for every 8 to 10 s; −13 dBm	440-Hz tone of 0.3 s for every 10 s; −8 dBm	400-Hz amplitude modulated with 16 Hz for 0.5 s, followed by 0- to 4-s gap, 400-Hz pure tone bursts appears two times, and then burst of 400 Hz with gap. −25-L to −14 dBm
Country impedance	600 Ω	270 Ω + 750 Ω ‖ 150 nF	600 Ω + 1 μF
Ringing voltage	45 to 67 V RMS	70 to 90 V RMS	75 V RMS
Ringing frequency and cadence	20 Hz, 2.0 s ON, 4.0 s OFF	48 to 52 Hz, 1.5 s ON, 3.5 s OFF	15 to 20 Hz, 1 s ON, 2 s OFF

requirements, multiple two-tone generators are combined to generate complex melody tones. In some countries, a VoIP dial tone is used in place of a normal dial tone. It is also observed that amplitude modulations are present on tones. In practice, a generic tone generator can produce multiple frequency combinations and amplitude modulations with required power. These tones are generated digitally in the processor. The digital tone generation module is placed on the decoder path of a voice chain of VoIP. As per the received command from the signaling protocol to generate a required call progress tone and country settings, the tone generators will deliver the tone samples and send them on to the pulse code modulation (PCM) interface.

17.3.4 Ringer Equivalent Number (REN)

The two-wire telephone interface is usually referred to as a TIP-RING interface, which is applicable in the PSTN and VoIP. This interface is capable of a certain drive strength referred to as REN. REN indicates the number of phones that can be used simultaneously with the RJ11 port or FXS interfaces on VoIP. This section content is mainly prepared based on specific North American parameters as given in references [URL (TIAonline-2003), URL (TIAonline-2006)]. Deviations could exist for different countries. As marked in Fig. 17.1, PSTN interfaces cater to up to 5 REN. An analog phone without main external power is marked as ring equivalence 1.0 B, which means this phone really draws one ring equivalence current. The suffix B is for a non-bell ringer indication, and bell refers to a mechanical bell here. When a ring sound is created using a bell inside the phone, suffix A is marked as 1.0 A on North American phones. Class-A bell-based phones need a lower ring frequency of 20 to 25 Hz. Cordless phones are marked as 0.2 to 0.4 B, which means cordless phones draw less current to ring a phone. Class-B phones accept ring frequencies of 17 to 68 Hz, but 20 to 25 Hz is more commonly used. Five phones of 0.4 B connected in parallel are expected to draw the same ring current as that of two analog phones marked with REN 1.0 B. One REN is of 7000 to 8000 ohms (Ω) ring impedance at 20 Hz, which varies slightly (even for North America) depending on the standard referenced. A 5-REN impedance value is 1400 Ω (i.e., the same as 7000/5 Ω). A 5-REN load (1400 ohms at 20 Hz) will draw a loop current of 40 milliamperes (mA) peak [i.e., (40V) ($\sqrt{2}$)/1400 Ω = 40 mA peak from a sinusoidal 20-Hz ringing voltage of 40 V RMS]. The corresponding average current is 25.5 mA [(2/π) 40 mA = 25.5 mA]. Ring voltages are of 40 to 100 V RMS and vary by country. PSTN central offices operate with higher voltages and more capable ring drives. FXS interfaces on VoIP adapters drive up to 3 REN, and it is possible to make it up to 5 with the available devices to operate with lower voltages than the PSTN. For clarification, a few summary points specific to Ring are given below.

Balanced Ring. Ring is usually a sine wave at a frequency of 15 to 25 Hz. Higher ring frequencies are also used in some countries. A negative battery is used for the telephone interface and ringing. During ringing, on-hook battery voltages are high. In off-hook and actual speech conversation, battery voltages are made low by the central office or VoIP adapter. In the unbalanced ring mode, the ring is directly applied on a single RING lead [URL (Legerity-Ring)], which requires more DC line voltages and it is difficult to get the required ring strength. The usual practice is to use a balanced ring in VoIP interfaces. In balanced ring, both wires carry a sine wave as metallic (differential). And each of the lead sine waves differs by 180 degrees or in anti-phase as shown in Fig. 17.1(c). A sine wave of 48 V peak to peak can result in total of 96 V peak to peak as an equivalent sine wave ring. It is more common to use balanced ring in VoIP adapters. Compared with a single-ended ring, balanced ring is preferred in VoIP because of lower battery voltages and less adjacent channel disturbance.

Telephone Impedances. The telephone offers several mega-ohms ($>5\,M\Omega$ in North America) of DC impedance [URL (TIAonline-2003)] during on-hook conditions. The central office (or VoIP adapter) will supply a few microamperes of current during on-hook. During ring, the telephone offers about $7000\,\Omega$ impedance at ring frequency. TIP-RING DC and AC voltages are high during initial conditions and ring state. On achieving off-hook (lifting the phone) from the ring condition, a phone offers $600\,\Omega$ AC impedance or country-specific impedance. DC voltage is lowered to the order of 6 to 8 V in the off-hook state. In the off-hook state, either ring impedance is bypassed or its influence on low $600\,\Omega$ is negligible. During a voice conversation, a 100-mV voice analog signal will be riding on a few Volts of DC voltages on TIP-RING leads.

BORSHT. In SLIC functionality and telephone interfaces, the keyword BORSHT is used. It is expanded as battery, overvoltage protection, ringing, supervision, hybrid, and test functions. SLIC or SLIC–CODEC hardware devices having this classification BORSHT are more suitable for extended functionality and better diagnostics. Recent interfacing devices support BORSHT. It is good to watch for this support while selecting the interfacing devices.

17.4   CALL PROGRESS TONE DETECTORS

This section is presented for the call progress tone detectors (CPTDs) with VoIP services. The tone detection requirements vary with country. In the PSTN telephone service, the CO or DLC generates the call progressive tones, and usually no requirement is made for detecting the tones by the PSTN. The VoIP adapter has an FXO/PSTN interface as marked in Fig. 17.1(b) that allows for dialing into or receiving from the PSTN network. The FXO interface of the VoIP adapter emulates the electronic telephone and continuously monitors the progress of the call. Call progress tone detection allows a VoIP adapter that dials into or receives from the PSTN network to monitor the progress of the resulting call, and hence, different states with respective call progress tones like dial, ringing, and busy can be determined. In CPTD, amplitude, frequency, and cadence (timing) are required to be analyzed. An audio cadence is a repeating sequence of sound (tone-on) and silence (tone-off) similar to the ON–OFF pattern produced by a ring-back or a busy tone.

CPTD analyzes the signal on the line to detect a repeating pattern of signal and silence, such as the pattern produced by a dial tone, ring-back tone, or busy signal. Once a cadence has been established, it can be classified as a single ring, double ring, or busy signal that allows a PSTN application or any other signaling protocol to keep track of the progress of an incoming or outgoing PSTN call. Various types of call progress tones exist. These tones and country-specific deviations were given in the previous sections.

To detect call progress (CP) tones, many implementations of CPTD comprise filtering for required tones, power estimation, timing analysis, and protection against false detections. Generic tone detection modules are given under DTMF in Chapter 7 and fax over PSTN in Chapter 14 for higher center frequency. Some of these techniques can be used for tone detections at different center frequency and frequency combinations. Basic call progress tones are of low frequency in the range of 330 to 620 Hz.

Call progress tones and cadences may differ from country to country. Basic tones are given in E.180 [ITU-T-E.180/Q.35 (1998)] for multiple countries. Thus, one of the most difficult problems facing the user who requires analog tone recognition is the variability of these tones when calling to or from different parts of the world, thereby necessitating the implementation of required tone detection. Tolerances of detections of the transitions could occur from the ON period to the OFF period. The algorithm has to be implemented by considering the tolerances.

18

VOICE PACKETS JITTER WITH LARGE DATA PACKETS

VoIP voice calls make use of Internet service. End-to-end packet delivery on an Internet Protocol (IP) network can face impediments. Voice packets are smaller, and data packets can be bigger in byte size. As explained in Chapter 2, VoIP customer premise equipment (CPE) also has a wide area network (WAN) interface and several local area network (LAN) interfaces. The WAN interface is the main outlet on CPE for connectivity to the Internet. As shown in Fig. 18.1(a), the CPE WAN interface is terminated on another main LAN interface of an Internet service modem [e.g., digital subscriber line (DSL) or cable]. As shown in Fig. 18.1(a) and Fig. 18.1(b), the LAN interface and some applications in CPE can send large data packets along with voice. These large packets can introduce delay and delay variations (jitter) to voice packets. The effect of jitter is mainly dominant with limited bandwidth physical interfaces. For completeness of the representations, Fig. 18.1(c) is included that conveys the usage of VoIP CPE with built-in direct WAN connectivity. This direct WAN connectivity could be DSL or cable. In this chapter, a DSL physical interface is considered for creating example calculations with bandwidth limits; however, similar conditions are also applicable with other interfaces. Many residential applications use DSL service. Various options in DSL service exist, which range from 256 kbps to several 100 Mbps of bandwidth support. In the lower end basic residential service, DSL upstream bandwidth is usually limited to 256 to 768 kbps. As marked in Fig. 18.1(a), upstream is for sending packets from VoIP CPE to the external IP network. The data from the external network

VoIP Voice and Fax Signal Processing, by Sivannarayana Nagireddi
Copyright © 2008 by John Wiley & Sons, Inc.

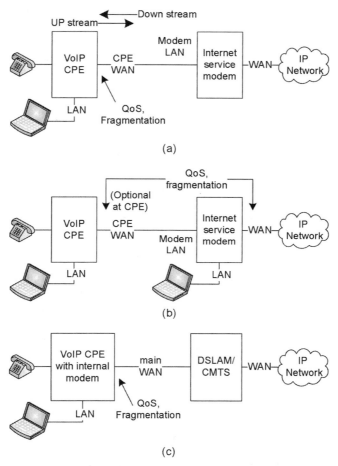

Figure 18.1. Representations of LAN and WAN connectivity. (a) VoIP CPE connected to another ISP modem. (b) VoIP CPE with ISP modem and data path shown at VoIP CPE and ISP modem. (c) CPE with built-in modem.

to the CPE is the downstream. ADSL2+, and a very high-speed DSL (VDSL) based service, gives more bandwidth than simple DSL, and the issues highlighted in this chapter are not applicable to these wideband interfaces. In many countries, residential users may get limited bandwidth because of limitations in availability of service options. Users may also opt for lower bandwidth service to reduce the expenses or because of their limited usage of the Internet. When VoIP calls are used with limited bandwidth service, VoIP calls occupy a significant part of available bandwidth. Many Internet-based applications and LAN data traffic generate large (e.g., 1514 byte) packets. IP quality of service (IPQoSs) prioritizes voice packets by holding low-priority large data packets. Once large data packets are scheduled, a scheduler may not preempt these large packets. Voice packets will join behind the large scheduled data packets. As an example, for a DSL upstream bandwidth of 256 kbps, a scheduled 1514-byte data packet can delay voice packets up to 53 ms. In the absence of large packets, this delay is not present to the voice packets. This delay varies between 0 and 53 ms, appearing as jitter that makes destination jitter buffers grow in size. In this situation, depending on the jitter buffer design, packets will be dropped or jitter buffer conditions will increase the end-to-end delay by approximately 53 ms. Both packet drop or buffer growth can degrade voice quality significantly.

In a bandwidth-constrained packet network, it is not possible to eliminate packet delay variations, but large packets can be fragmented to multiple smaller packets to limit the delay variations to less than 10 ms. Fragmentation divides a large packet into multiple smaller packets. Once large packets are fragmented, voice packets are interleaved with the fragmented small data packets, which reduce the delay variation. At the receiver, fragmented smaller packets are reassembled for creating the original large packet. In this chapter, various DSL upstream bandwidth rates, packet delay variations, and suggested fragmentation sizes to limit jitter to less than 10 ms are presented. In general, no set mandatory goals have been placed on this 10 ms. In practice, most service providers would accept a delay increase of the order of 10 ms. Some specific DSL examples are given in the subsequent part of this chapter.

18.1 ATM CELLS AND TRANSMISSION

On a DSL interface, IP packets are sent as asynchronous transfer mode (ATM) cells [Black (1999)]. ATM uses smaller byte cells to reduce delay variation while multiplexing multiple streams. Each ATM cell is of fixed 53 bytes consisting of a 5-byte mandatory header and maximum 48-byte payload. An ATM 5-byte header carries information for end-to-end ATM packet delivery. Application adaption layer 5 (AAL5) accepts transmitted data packets and maps them to ATM cells. The AAL5 process consists of protocol data unit (PDU), segmentation and reassembly (SAR), and 8 bytes of an AAL5 trailer. The trailer consists of 2 bytes for length, 4 bytes for cyclic redundancy check (CRC), 1 byte for user-to-user (UU), and 1 byte for common part indicator (CPI). To incorporate trailer bytes, the last cell of the total bytes will have a maximum of 40 bytes of payload and 8 bytes of trailer bytes. In many applications, packet sizes are much larger than 40 bytes. On an ATM network, packet sizes of 65,535 bytes can be sent [Gross et al. (1998)]. In the process of fragmentation and ATM cell creation, every original IP packet is appended with interface-specific headers, an 8-byte trailer, and zero padding to create an integer number of ATM cells. The number of ATM cells = the smallest integer of cells that satisfies the condition, AAL5 PDU bytes + zero padded bytes + 8 bytes AAL5 trailer = (smallest integer) (48).

As an example, an IP packet of size 87 bytes after adding an 8-byte trailer, 1 byte of zero padding, and two ATM cell payloads of 48 bytes each is created.

For optimum bandwidth utilization, a minimum number of cells has to be created with the least possible zero padding. Zero padding bytes can vary in size from 0 to 47 bytes, and as an example, an AAL5 PDU of 88 bytes does not need zero padding bytes and splits into two cells. An IP packet of size 89 bytes will need three cells and 47 bytes of zero padding. ATM cells consist of a 5-byte interface-specific header making each cell of 53 bytes in length. The packet sizes and cells created make a difference in ATM-based transport. Similar actions occur in cable interfaces as given in Chapter 11.

The DSL upstream bit rate of 768 kbps is considered in the examples of Table 18.1. From the table, accommodated packet bytes for integer cells and

Table 18.1. ATM Cells and Possible AAL5 PDU Bytes on 768-kbps Link

AAL5 PDU Bytes Range	ATM Cells	Total Bytes in ATM Cells	Total Bits per Packet	Time in ms on 768-kbps Link
1–40	1	53	424	0.55
41–88	2	106	848	1.10
89–136	3	159	1,272	1.66
137–184	4	212	1,696	2.21
185–232	5	265	2,120	2.76
233–280	6	318	2,544	3.31
281–328	7	371	2,968	3.86
329–376	8	424	3,392	4.42
377–424	9	477	3,816	4.97
425–472	10	530	4,240	5.52
473–520	11	583	4,664	6.07
521–568	12	636	5,088	6.63
569–616	13	689	5,512	7.18
617–664	14	742	5,936	7.73
665–712	15	795	6,360	8.28
713–760	16	848	6,784	8.83
761–808	17	901	7,208	9.39
809–856	18	954	7,632	9.94
857–904	19	1,007	8,056	10.49
905–952	20	1,060	8,480	11.04
953–1000	21	1,113	8,904	11.59
1001–1048	22	1,166	9,328	12.15
1049–1096	23	1,219	9,752	12.70
1097–1144	24	1,272	10,176	13.25
1145–1192	25	1,325	10,600	13.80
1193–1240	26	1,378	11,024	14.35
1241–1288	27	1,431	11,448	14.91
1289–1336	28	1,484	11,872	15.46
1337–1384	29	1,537	12,296	16.01
1385–1432	30	1,590	12,720	16.56
1433–1480	31	1,643	13,144	17.11
1481–1528	32	1,696	13,568	17.67

the time to transmit the packet can be observed. For large 1514-byte packets, delay is 17.67 ms. This delay increases with reduction of available network bandwidth. The delay marked in the tables appears as bursty delay variation (jitter) to voice packets.

18.2   IPQoS AND QUEUING JITTER ON AN INTERFACE

IPQoS creates improved voice quality conditions. Several methods and options are available in the implementation of the IPQoS for voice and data applications [Black (1999), Ferguson and Huston (1998), URL (Cisco-QoS), Nichols et al. (1998), Blake et al. (1998)]. The main function of IPQoS for VoIP voice packets is to assign high priority to the voice packets. IPQoS eliminates packet drop or reduces the packet drop at scheduling, which assumes available bandwidth is at least more than the voice traffic requirements. IPQoS reduces fixed delay, delay variations, and takes care of dynamic conditions of data, voice, and other priority media. IPQoS in its basic form will not take care of handling the delays created from previously scheduled large packets. To help, based on the available link rate, fragmentation and interleaving are used with large packets.

18.2.1   Fragmenting the Packets for Lower Jitter

The purpose of fragmentation of large packets is to reduce the delay variations created by the large packets [LaBarba (2000)]. In fragmentation, large packets are split into smaller packets. The fragmentation size requirements vary with the available link rate, and fragmentation may be used dynamically based on the traffic and link bandwidth conditions. Most VoIP deployments target for delay variation in the range of 5 to 10 ms. It is good to achieve 1- to 2-ms delay variations, but this is not possible unless upstream bandwidth supported is of several Mbps.

A fragmentation operation also takes care of checking the presence of real-time traffic and adapts to the required mode of operation. In the absence of real-time traffic, fragmentation may be disabled. In the products with voice and data applications, IPQoS is used at first level for ensuring end-to-end packet delivery with minimal or no packet drops. Fragmentation along with IPQoS reduces the delay variations. While fragmenting the large packets on DSL interfaces, it is preferred to split the original IP packet to create a minimum number of ATM cells possibly with no zero padding.

Table 18.2 provides different upstream bit rates, Ethernet packet sizes, time delay variations, and the suggested fragmentation sizes for creating a delay of less than or equal to 10 ms. An example interpretation of Table 18.2 is given here for a 768-kbps link and 1514-byte Ethernet packet. For a link rate of 768 kbps and packet size of 1514 bytes, the number of ATM cells required is 32 (i.e., 1514 + 8 + 10 + 4)/48). The AAL5 trailer header is 8 bytes, the Ethernet

Table 18.2. Jitter from Larger Packets and Suggested Fragmentation Sizes

Upstream Rate in kbps	Ethernet Packet Bytes	Total Bits in ATM Cells	Time to Transmit or Jitter for Voice Packets	Fragmented Size to Create Less than <10 ms Jitter. $n = 1$, $2 \ldots$ (without IP, Eth., Eth. over ATM Headers)
768	1,514	13,568	17.67	812, 812-48n
	1,024	9,328	12.15	812, 812-48n
	512	5,088	6.63	<10 ms, no fragmentation
	256	2,544	3.31	<10 ms, no fragmentation
512	1,514	13,568	26.50	524, 524-48n
	1,024	9,328	18.22	524, 524-48n
	512	5,088	9.95	<10 ms, no fragmentation
	256	2,544	4.97	<10 ms, no fragmentation
384	1,514	13,568	35.33	380, 380-48n
	1,024	9,328	24.29	380, 380-48n
	512	5,088	13.26	380, 380-48n
	256	2,544	6.62	<10 ms, no fragmentation
256	1,514	13,568	53.00	236, 236-48n
	1,024	9,328	36.43	236, 236-48n
	512	5,088	19.9	236, 236-48n
	256	2,544	9.93	<10 ms, no fragmentation

over ATM header is 10 bytes, and 4 bytes of zero padding have to be added to 1514 bytes before segmenting into ATM cells. ATM will create 32 cells of 53 bytes (a 5-byte interface specific header will be appended to each cell), which results in $32 \times 53 \times 8 = 13{,}568$ bits in each packet. The time to transmit 13,568 bits at the 768-kbps upstream link rate is $13{,}568/768{,}000 = 17.67$ ms. The same 1514-byte packet on the 256-kbps link bandwidth can create 53-ms delay variations. In Table 18.2, four different rates and four different packet sizes are given. In the last column, suggested fragment sizes are included that give delays of lower than 10 ms to the voice packets. As an example, for an 768-kbps and 1514-byte packet combination, the fragmentation column is marked as "812, 812-48n," which means basic payload bytes will fragment to 812 bytes or lower than 812 with integer multiple decrements of 48 bytes as 812, 812-48, and 812-2 × 48. This process can achieve optimum utilization of ATM cells by avoiding zero padding. Some more interpretation of delay increase is given in the next section in relation to Fig. 18.2.

Fragmentation and interleaving consumes extra bandwidth. The fragmentation size should be as large as possible to get better bandwidth utilization with the constraint to maintain delay variations to be less than 10 ms for voice applications. It is helpful to monitor the available link bandwidth and maximum packet sizes for dynamic fragmentation. Fragmentation schemes and various approaches are beyond the scope of this book.

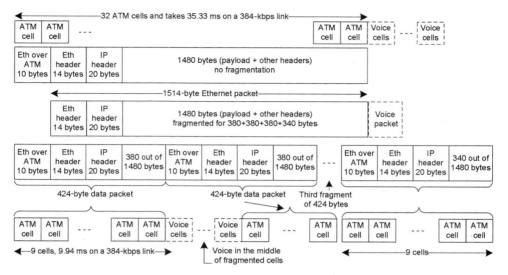

Figure 18.2. Large data packet fragmentation and voice packet interleaving (35.33-ms delay variation is reduced to 9.94 ms in fragmentation).

18.2.2   Fragmenting of 1514-Byte Packet Example

The suggested fragmentation for a 1514-byte Ethernet packet on a 384-kbps link is given to create 380 bytes or any packet sizes of lower than 380 in decrements of 48 bytes. The example given in this section is illustrated in Fig. 18.2. Considering 1514-byte Ethernet packet as 1480 bytes + 20-byte IP header +14-byte Ethernet header, 1480 bytes can split as 380 sized packets without including the IP and the Ethernet header. After adding the IP header of 20 bytes, Ethernet header of 14 bytes, and Ethernet over ATM header of 10 bytes, the fragmented packet is 424 bytes and segments into nine cells. The last 340 bytes (out of 1480 bytes) will segment into nine cells after 40 bytes of zero padding. Overall, the original 32 ATM cells from the 1514-byte packet occupy 36 cells after fragmentation. The efficiency of (32/36) 89% is better accepted with fragment boundaries that avoid zero padding in multiple fragmented packets.

In the absence of fragmentation, voice cells may be in queue behind 1514-byte packet cells for 35.33 ms for the example of 384 kbps. After fragmentation, the voice can be joined immediately after 9.94 ms with a fragmentation size of 380 bytes. Just to clarify this topic, the ATM cells of voice are not interleaved with data cells of a large packet. As shown in Fig. 18.2, voice cells are interleaved with one complete fragmented IP packet that generated nine cells.

In Fig 18.1, markings are made on QoS and fragmentation. These markings indicate the location where the implementation has to guarantee a low delay. The option indicated in Fig. 18.1(a) will be the preferred one compared with that in Fig. 18.1(b) (i.e., all the data joining at VoIP CPE). These options may widely vary based on product features, Internet service, and deployment support.

18.2.3   Voice Packet Fragmentation

VoIP voice with G.711 [ITU-T-G.711 (1988)] takes a maximum number of bytes. In most deployments, 20-ms packetization is used as the default. As discussed in Chapter 11 for DSL with Point-to-Point protocol over Ethernet (PPPoE), G.711 with 20 ms creates five ATM cells with 265 bytes, which can be treated as a smaller packet. In VoIP, voice packetization up to 80 ms is also possible. Large packetization increases end-to-end delay as explained in Chapter 3. Some VoIP boxes may have multiple voice channels. In limited bandwidth support, the large packetization will introduce more delay to the other channel operating with small packets. Hence, it is essential to fragment large voice packets. In practical deployments, packetization is limited to 40 ms for lowering end-to-end delays, and this eliminates the requirements of fragmentation. Fragmenting them increases the bandwidth requirement. The other alternative is to use a low-bit-rate codec like G.729AB that reduces payload bytes and eliminates the need for fragmentation. In practice, based on the link bandwidth, codecs, QoS implementations, and voice packetization, codec selection can be negotiated to avoid fragmentation. Overall, practical systems will not use voice packet fragmentation.

18.2.4   Summary on IPQoS and Fragmentation

IPQoS is essential for voice working along with other non-real-time and low-priority data applications. IPQoS ensures voice packets are delivered in the minimum possible time, but in a bandwidth-limited system/network, this is not guaranteed. Thus, fragmentation has to be adapted for voice applications along with IPQoS. This chapter is mainly focused on DSL-based VoIP voice applications. Similar requirements are applicable for the system with other interfaces like Ethernet and cable. Upstream bandwidth in DSL is limited, or a user may subscribe to limited bandwidth to save on subscribing expenses. VoIP voice packets will get priority with operations of IPQoS. A bandwidth of 256-kbps upstream is also common in some residential applications, which can give jitter up to 53 ms with a 1514-byte data packet. To minimize delays, it is required to fragment large data packets. It is suggested to use fragmented packet sizes that operate with "no or a minimum" zero padding for optimum ATM cells utilization. To avoid fragmentation by managing with IPQoS for a 1514-byte packet, and to contain jitter to less than 10 ms, an upstream bandwidth of 1.35 Mbps is required. Such higher upstream bandwidth may not be available to all residential customers.

19

VoIP ON DIFFERENT PROCESSORS AND ARCHITECTURES

VoIP voice and fax have gone through several parallel implementation paths while being stabilized in laboratory experiments and deployments. It is difficult to classify them, as some architectures may not be current and relevant. Some popular approaches that have been followed in VoIP processing are presented here.

1. Computers running VoIP softphones or software applications. This process is broadly treated as a software-based VoIP solution. A personal computer (PC) was used for processing voice, networking functions, and available audio interfaces for voice. Accessories such as universal serial bus (USB) handsets, built-in speakerphones, and Bluetooth headsets are used to create comfortable acoustic interfaces for the end user.

2. Computer add-on cards, mainly peripherals component interconnect (PCI) family cards, providing network and voice interfaces. Processing was shared between PCI add-on cards and the computer.

3. Digital signal processors (DSPs) executing voice chain modules, signaling, and networking functions in conjunction with data path applications on network processors. In high-channel solutions, multiple DSPs are often used with a single network processor.

4. Extended instruction set network processors performing both network- and DSP-type voice functions. The combination and count of processors

VoIP Voice and Fax Signal Processing, by Sivannarayana Nagireddi
Copyright © 2008 by John Wiley & Sons, Inc.

varies with channels, features, and other data applications provided along with voice.

5. DSP processors extended with suitable interfaces and network processing providing for a stand-alone total VoIP solution.

These architectures work in different end-user VoIP solutions such as PC, Internet Protocal (IP) phones, VoIP adapters, and high-channel gateways.

19.1 VoIP ON PERSONAL COMPUTERS

As a quick summary of a VoIP call, a VoIP voice call will provide the following main items to the user. In this section, a PC can also be a laptop computer. Figure 19.1 is presented with VoIP functionality on a laptop computer.

VoIP implementation architecture should provide the following elements:

1. User acoustic voice interfaces
2. Dialing options
3. Processing that need not be directly visible to the user
4. Network interfaces and Internet network connectivity

Audio interfaces and sound cards internal to the PC are typically used for acoustic interface. In some computers, a built-in microphone and speaker are available. These interfaces are reused for VoIP calls. These built-in devices often create severe echo problems; hence, separate external acoustic interfaces and headsets are typically used with a PC. A PC keyboard will be used to enter digits for dialing. PCs have a significant amount of processing power that can

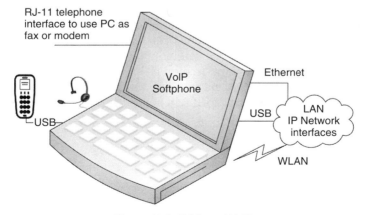

Figure 19.1. PC-based VoIP.

be used to take care of the processing needed for multiple channels of voice and associated network functions.

PCs often have several interfaces to get access to Internet connectivity. The popular network interfaces on a PC are Ethernet, USB, and wireless local area network (WLAN). A PC can also use an internal or external modem for dial-up network connectivity. These network interfaces all connect a PC to an Internet service. In practice, the PC will be connected via the LAN interfaces to digital subscriber line (DSL) and cable modems or directly to Ethernet or WLAN in residential or corporate offices.

Several PC-based applications are available, including MSN messenger, Skype, and Google talk, that provide PC-based VoIP voice functionality. These applications use the PC's processor to provide voice and network functions and to deliver voice on the available audio interfaces and extended interfacing devices.

19.1.1   PC as a Fax Machine and Internet-Aware Fax (IAF)

Figure 19.1 is a functional representation of PC based VoIP calls. As shown in the figure, the PC has a telephone RJ-11 (two-wire) foreign exchange office (FXO) interface that can be directly connected to the PSTN office [foreign exchange subscriber (FXS) interface]. This RJ-11 FXO interface is used like a dial-up modem for Internet connectivity. The PC can also work as a fax machine on the same FXO interface. The PC software will have drivers to work with the internal modem. Any printable document on a printer can be sent as a fax page by the application. A software application on a PC will provide a graphic user interface (GUI) to dial and to observe the fax status. The application detects the ring voltage on the FXO interface and answers the call to receive the fax data. Received fax pages are typically stored in tagged image file format (tiff) in specified default file folders. In summary, the PC operates as a fax machine through the fax software application. Note that when the PC RJ-11 interface is used for a dial-up modem, the same interface cannot be used for other applications.

When a PC is connected to the Internet through Ethernet, USB, or WLAN, a PC software application can also send fax pages over the network interfaces. In this situation, the PC works as a combination of fax machine and VoIP fax adapter. This functionality can be treated as an internet-aware fax (IAF). A separate VoIP adapter and fax machine are not required. Although a single PC could provide both fax and VoIP functions, it is often convenient to have a separate VoIP adapter so that the PC does not need to be running constantly to allow voice calls.

19.2   VoIP ON PC ADD-ON CARDS

VoIP is also supported by PC add-on cards. These cards typically have PCI interfaces. The VoIP voice processing is performed by the PCI cards. In most of these systems, VoIP signaling and networking functions are processed directly on the PC main processor. PC-based add-on cards may support up to several hundred telephone lines depending on the capabilities of the card. At the time of writing, PCI cards with multiple FXS, FXO, and even Quad T1/E1 interfaces were available as commercial products. Most of these add-on cards will need extra supporting software like Astrek [URL (Astrek)] or proprietary software running on a PC to complete the calls. PC-based voice applications, PCI, and compact PCI cards are discussed in references [URL (NMS-Access), URL (Sangoma)].

19.2.1 PC Add-On Cards for VoIP Instruments

PC add-on cards are extended to PC-based VoIP voice quality measuring instruments. The same telephone interfaces on the PCI cards are used for voice quality instruments. PC and measurement software measures voice quality instead of working as a VoIP gateway. PCs are modified to meet different instrument form factors and are used as complete instruments with these add-on cards. This architecture is scalable with appropriate add-on cards [URL (NMS-Access)]. In these instrument applications, the processor inside the computer will perform major computation tasks along with other user applications.

19.3 VoIP ON DEDICATED PROCESSORS

Historically, the following processor classifications have gained popularity for VoIP voice and fax applications:

a) DSPs for voice and network processor for networking software
b) Network processor performing both voice and networking
c) DSPs performing both voice and limited networking

An overview of these three families of processors and processing is given in this section. In Fig. 19.2(a), one or more DSPs will perform voice chain processing, and network processors take care of VoIP signaling, system software, and other data and networking functions. Several solutions are on the market using this type of approach. In many cases, separate DSP and network processors are used; however, modern implementations typically combine both processors into a single system on a chip (SoC) to minimize cost and simplify system implementation. Several families of devices, including the TI [URL (TI-54x)], Mind speed [URL (Mindspeed)], Verisilicon [URL (Verisilicon)], and Ikanos [URL (Ikanos-Fusiv)], use this approach. The network processors were referred to by different names such as host processor, microprocessor, control processor, reduced instruction set computer (RISC)

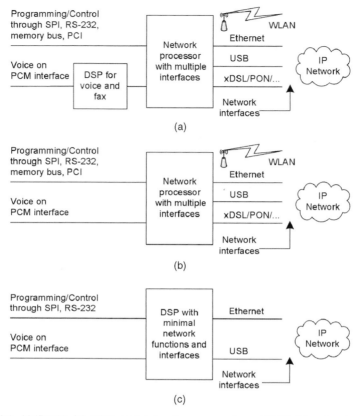

Figure 19.2. VoIP with (a) DSP- and network-processor-based VoIP, (b) network-processor-based VoIP, and (c) DSP with network extensions for VoIP.

processor, and power PC. Several processor families are used for networking functions. In early devices, the DSP operated in the range of 80 to 100 MHz and the network processor in the range of 50 to 100 MHz, supporting two to four channels of VoIP processing. Subsequently both networking and DSP speeds have increased. At the time of writing, DSPs were operating at more than 600 to 750 MHz, whereas network processors were operating at 500 to 750 MHz. This speed is about 10 times more processing power than early two- to four-channel VoIP voice applications.

Original network processors provided arithmetic logic units (ALUs) that supported only simple addition, subtraction, and multiplication operations. Many processors used in these systems now provide enhanced signal processing types of operations in their instruction set, which makes network processors suitable candidates for voice processing. Currently, one- to four-channel VoIP voice could be processed directly on network processors with DSP instruction extensions. The main advantage of this architecture is to minimize the size (and, therefore, cost) of the SoC by removing the need for a separate DSP [URL (Broadcom)]; however, care must be taken to ensure that sufficient processing time is guaranteed for VoIP operations; otherwise, voice quality may suffer. Total software may be maintained as C-code on network processors, with in-line assembly as needed to optimize time-critical operations.

DSPs have also been enhanced with network functions, operating system capability, and network physical interfaces, which makes use of a signal processor for VoIP. Analog Devices Blackfin [URL (ADI-BF536), URL (ADI-VoIP)] processors with built-in Ethernet and USB interfaces fall into this category. This approach is typically seen in a VoIP customer premises equipment (CPE) or low-channel density enterprise class dedicated to VoIP-systems. DSP-based voice is less common than the other two approaches.

19.4 OPERATING SYSTEM ASPECTS ON DIFFERENT PLATFORMS

In this section, VoIP-specific operating system aspects are considered for the following combinations:

1. Computer-based VoIP
2. DSP- and network-processor-based architectures
3. Network processor with built-in DSP instruction extensions
4. DSP with network extensions

In the following sections, the keywords "MIPS", "mips", "MCPS", and "MHz" are used. To convey clarity in information, the following clarifications are provided for these keywords.

19.4.1 Keywords MHz, MCPS, MIPS, and DMIPS Association

Several terms are used to describe processor performance. This section deals with the description of those terms. Processors typically have a master clock signal that times the execution of instructions. The frequency of this clock signal is normally measured in MHz, where 1 MHz represents a clock frequency of one million (1,000,000) cycles per second. The term "MCPS" may be used interchangeably with "MHz", but it is less common.

Modern processors used in VoIP systems typically execute one instruction per clock cycle; however, some processors require more than one clock to execute each instruction (older processors and some DSPs), and others can execute more than one instruction per clock (superscalar architectures). Thus, it is necessary to have some approach to measure processor performance independent of this clock speed. One common approach is to state the number of instructions executed per second, which is often called million instructions per second (mips or MIPS). Although this measure is useful, it does not allow for the number of instructions needed to perform a given function in each different instruction set or for the number of instructions executed in each clock cycle.

Many tests have been devised to measure the real performance of a processor for a particular function. One of the earliest tests was the Dhrystone MIPS test. This test measured the processing performance of the processor relative to a DEC VAX 11/780 processor (nominally a one Dhrystone MIPS machine) for typical processing functions at that time. Unfortunately, this test was so small and so simple that the Dhrystone MIPS results were extremely dependent on compiler optimization, cache size, cache architecture, and memory technology for a given processor—a characteristic not strongly shared by real applications. Subsequently, several benchmarks have been developed that better simulate the performance of many real-word applications. For example, the EDN Embedded Microprocessor Benchmarking Consortium (EMBCC) telecommunications benchmark is intended to model operations used in VoIP algorithms, and the EMBCC networking test is intended to model typical networking processing.

Adding to the confusion, the term "Dhrystone MIPS" is often abbreviated simply as "DMIPS" or just MIPS. For modern processors, these terms are not the same. In addition, the term "MIPS" is also used to denote of a family of processor architectures from MIPS Technologies. So a MIPS 24KEc processor core [URL (MIPS32)] running with a 625-MHz clock is rated at 900 Dhrystone MIPS.

19.4.2 Operating System (OS) Aspects on Computers

Many popular early VoIP applications were hosted on PCs. These computers used multiple operating systems such as Windows, Linux, and Mac-OS. The computer could also be running several other applications. An important characteristic of many of these operating systems was that they were originally unable to guarantee timely execution of any application, including the VoIP application. Embedded real-time operating systems (RTOS) now allow very deterministic scheduling of the various applications. A modern version of these PC operating systems supports some form of real-time scheduling, which is important as VoIP compressed packets typically have to be serviced in less than 10 ms, otherwise voice quality will suffer. With OS improvements, real-time software modules, and increasing computer performance, VoIP on computers can support real-time packet interval duration similar to dedicated hardware implementations with real-time operating systems. Early implementations of PC-based VoIP had several voice quality concerns. Some voice applications use larger jitter buffers to take care of the timing resolution and network impediments, which can increase end-to-end delays and create lower voice quality and increased echo perception. Some applications may not have echo cancellers with a PC speakerphone, thus degrading voice quality.

Some other aspects like a lack of quality-of-service (QoS) mechanisms on PCs and lack of proper packet loss concealment (PLC) techniques may also degrade voice quality. PC-based voice has several variables, and the PC platform under user control does not allow several voice quality controls. Overall, the PC as a complete VoIP gateway has not remained popular. To achieve good voice quality with a PC-based VoIP application, it is generally necessary to minimize both other network-related activities and to reduce the processing load from other applications on the computer.

19.4.3 Operating System Aspects for DSPs

Most critical algorithms for voice processing are written in assembly language to optimize MCPS and memory consumption. High-level interfaces and logic are written in C using fixed-point arithmetic. DSP manufacturers supply their own tools. Several third parties also provide tools to support DSPs. Most of these tools have integrated development environment (IDE) support with GUI. These tools will be supported with cycle accurate simulators, optimized compilers, joint test action group (JTAG) target board debugging, and library building tools. JTAG helps to debug the programs, executed on processors. When DSPs are used for extended VoIP and networking operations, the extended tools with Linux OS and company/processor-specific proprietary real-time OS are used.

19.4.4 Operating System Aspects for Network Processors

On network processors, several OS and tool chains are supported. Most third-party tools support several processors. Tools vary with the real-time OS used. Most code running on a network processor use ALU operations and data transfers for packet processing. Hence, C language code is used and the compilers take care of required optimization. In some cases, some sections of the application code may not be executed quickly enough when coded in C. These sections will typically be optimized with in-line assembly or hand-coded assembly that resides as part of the generic C-code.

19.4.5 Operating System Aspects for Network Processor with DSP Extensions

The network processor will typically have a large external memory, usually synchronous dynamic random access memory (SDRAM), synchronous random access memory (SRAM), or double data rate (DDR) SDRAM. The available memory is much larger than algorithmic voice processing code requirements. However, running algorithmic functions from the external memory makes them inefficient. When the processing power is more than adequate to the meet requirements, voice software will be implemented as C code, will be compiled with optimization options, and will reside in slower external memory. In the processing constrained system, voice functions will be optimized to fit in a major part of the internal cache memory or other internal scratchpad memories. If code is present in the cache, it will operate at the highest performance allowed by the processor clock frequency. To get the best performance for voice algorithms, voice code is optimized to take lower memory to make the best use of available internal memory. The voice algorithm code is separately optimized in a cycle accurate simulator that allows the real-time performance of the code, including external memory access latency, to be observed. In practice, a library of algorithmic code is built with assembly, in-line assembly, or optimized C-compiled with optimized memory passes for voice algorithms. These libraries are included with the main networking code. In general, network-processor-based total voice processing is one of the most cost-efficient options using a single processor that is capable of voice and of network processing. Network processors use different operating systems such as VXworks, flavors of Linux, Nucleus, RTEMS, and other proprietary operating systems that provide some type of real-time task scheduling.

19.5 VOICE PROCESSING COMPLEXITY

In this section, an overview on processing requirements is given. The major computational intensive modules for VoIP voice chain are codecs (G729AB, G.723.1, and wideband G.722, G.729.1, G.722.2), echo cancellers for narrowband and wideband, dual-tone multifrequency (DTMF), and PLC. The other modules of voice activity detection (VAD)/comfort noise generation (CNG), jitter buffers, tone generation, and various tone detection take less processing. A fax chain with group-3, V.17 requires higher processing. In super group-3, V.34 requires more processing.

For voice processing, the DSPs selected will typically have multiple multiplier and accumulator (MAC) processing units. The number of parallel multipliers determines the speed of processing. The older families of DSPs such as the 218x [URL (ADI-218x)] and TI-54x [URL (TI-54x)] have a single multiplier unit. Later families, such as the TI-55X and Blackfin, have dual multiplier units. The TI-64X [URL (Encore-G729AB)] and Starcore families have complex processing units that work like more than two independent multiplier units.

Many voice processing algorithms can derive significant performance benefits from dual multiplier processor architecture—in some cases as much as 40% to 50%. For example, consider G.729AB [ITU-T-G.729A (1996), ITU-T-G.729B (1996)] processing from reference [URL (Encore-G729AB)]. A single multiplier processor such as the ADSP218x or the TI-54x requires about 12.5 to 13 MCPS (MCPS is million cycles per second) of processing for G.729AB. It takes about 8 to 9 MCPS on dual MAC units such as the TI-55x and Blackfin [URL (ADI-BF536)]. The processor that work like more than two multiplier-equivalent processing such as MSC8101 [URL (Freescale)] consumes 5.0 MCPS [URL (Encore-G729AB)] for the same G.729AB processing.

The ARM ARM-9E processor is a typical RISC engine with DSP extensions. For the G.729AB codec, this takes 35 MCPS [URL (Encore-G729AB)]— more than even the simple single multiplier DSPs. However, the simpler instruction set used by RISC processors allows higher clock-speed implementations, so the time taken to execute the algorithm may be comparable with a simple DSP. Additional support operations such as jitter buffer, Real-Time Transport Protocal (RTP), other packetization, VoIP signaling, and minimal network functions are executed on a packet basis. These modules require typically 5 to 10 MCPS per voice channel on this type of RISC processor. In general, these functions do not derive any benefit from the DSP extensions of the processor.

NarrowBand Voice and Fax Processing. On a single MAC-based processor like TI-54x, the processing for narrowband codecs such as G.729AB and all other modules can be managed with approximately 30 to 40 MCPS for the whole voice chain processing that includes echo cancellation, DTMF, and so on, assuming the host processor is used for packetization and networking. Fax processing complexity is similar to narrowband codec-based voice chain processing. Several voice chain modules are disabled during fax. The processing performance required by narrowband voice chain processing would also be sufficient for fax [URL (Encore-T38)] over IP processing.

WideBand Voice Processing. On a single MAC-based processor such as the TI-54x, wideband codecs including G.729.1 [ITU-T-G.729.1 (2006)] and G.722.2 [ITU-T-G.722.2 (2003)] require approximately 60 to 65 MCPS per channel, assuming host processor is used for processing of packetization and networking. A wideband codec such as G.722 [ITU-T-G.722 (1988)] requires less processing and can typically be accommodated within the same processing power as the narrowband processing chain from processing. In wideband mode, the echo canceller taps will be doubled because of a sampling frequency increase from 8 to 16 kHz, and this consumes more processing than the narrowband operation. Wideband voice processing also occupies more memory. In general, wideband implementations have to use higher speed MHz processors with more memory.

19.5.1 DSP Arithmetic for Voice Processing

Signal processing operations for voice and fax make use of mainly MAC operations. Several basic operations are given in the C-code of ITU-T codecs (e.g., G.729 codec [ITU-T-G.729 (1996)]). The ability to process these basic operations in minimum (preferably single) cycles makes the voice processing more efficient. A summary of some important operations that help voice processing are listed here:

- Multiplications accumulation, with built-in rounding and saturation
- Fractional format support with a sign usually referred to as q15 or 1.15 formats with a sign and 15-bit fractional part. The results of multiplication are left shifted by 1 bit, creating q31 or 1.31 formats.
- Ability to handle maximum negative numbers to make it positive (e.g., multiplication of two 16-bit negative numbers of 0x8000 (16-bit hexadecimal format number for −1) should result in a positive number). Basic multipliers in a general-purpose processor may not handle this type of operation.
- Arithmetic, mainly addition and subtraction, has to cater to rounding and saturation
- Arithmetic shifts with saturation operations
- Exponential and normalization operations with saturation operations
- Efficient division routines support

Several other important features are available such as parallel data access, addressing modes, circular buffering, bit reversal, and absolute operations. Several network processors upgraded in architecture for DSP extensions are taking care of a major part of the above-listed instructions. Many DSP operations that reflect useful instructions can be found in processor data sheets [URL (ADI-BF536), URL (TI-54x)]. Specific to VoIP, it is important to select a processor that supports voice processing with suitable instructions to reduce the amount of processing cycles.

20

VoIP VOICE QUALITY

For several decades, telephone users have been experiencing public switched telephone network (PSTN) based voice communication quality. Subjectively, this experience is used as the main reference for comparing the voice quality from other voice communication systems. In PSTN-based systems, voice has good intelligibility, acceptable speaker identification, naturalness, and only minor disturbing impairments. The PSTN uses G.711 μ-law and A-law as compression. Compressed bits are sent on synchronous times-division multiplexed (TDM) interfaces. In VoIP voice communication, a similar G.711 is also used end to end as one of the compression codecs. In VoIP, instead of TDM bits, groups of compressed bytes are sent as a packet on the Internet Protocol (IP) network. The VoIP quality closely matches with PSTN quality under the correct conditions of end-to-end G.711 IP packet transmission. VoIP will also use many high-compression codecs like G.729AB, G.723.1A, and iLBC, which causes VoIP voice quality to be lower than PSTN quality even in best end-to-end packet delivery networks. The goal here is how to reach and finally to exceed PSTN quality through VoIP.

VoIP will have to take care of additional impairments to achieve the quality comparable with the PSTN quality. The four main voice impairment contributors popularly referred for VoIP telephony according to TIA-810A [TIA/EIA-810A (2000)] are delay, echo, voice compression, and packet loss. In a recent upgrade of TIA 116A [TIA/EIA-116A (2006)], additional influencing factors such as G.711 Packet Loss Concealment (PLC), transcoding, gateway loss plan,

VoIP Voice and Fax Signal Processing, by Sivannarayana Nagireddi
Copyright © 2008 by John Wiley & Sons, Inc.

and terminal coupling loss (TCLw) are added as other contributors to voice quality. From the deployment perspective, the following major contributors to voice quality can also be considered for VoIP customer premises equipment (CPE) products.

- Analog front ends that can meet compliance with the PSTN specifications like TR-57 transmission
- Call establishing, call progress tone timing matching to PSTN switching specifications, and avoiding audible ticks, hollowness, and voice transients at call establishing and termination phases
- Country- and deployment-specific deviations and parameters matching the local PSTN
- Ability to provide lifeline PSTN capabilities or emergency services
- Diagnostics of front end and total system, voice quality monitoring, and incorporating dynamic voice quality feedback
- Supporting right interfaces, signal level combinations for subscriber end-devices like telephones, cordless phone, DECT, and Bluetooth devices
- Front end ability to support wide bandwidth and processors that can support wideband codec modules

In the PSTN, voice quality also degrades from the use of 32-kbps adaptive differential pulse code modulation (ADPCM) to double the transmission channels on the existing digital channels. Regular channels with G.711 use 64 kbps. At some stage of terminal points, digital circuit multiplication equipment (DCME) performs this operation of creating overload channels. The principal application of ADPCM 24 and 16 kbps are also used in overload channels carrying voice in DCME. The 40-kbps channel carries data modem signals in DCME, especially for modems operating at greater than 4800 kbps. VoIP G.711 can deliver better quality than the PSTN channels created through a DCME operation. VoIP also provides several benefits when compared with PSTN-based services as listed in Table 20.1.

20.1 VOICE QUALITY MEASUREMENTS

For comparing the voice quality, and to arrive at the parameters contributing to the voice quality, it is required to know how voice quality is measured and what are the quality goals. The basic test setup for VoIP voice measurements is given in Chapter 13. In this chapter, voice quality as mean opinion scores (MOS), various classifications, voice quality influencing parameters, and improvements are discussed. Voice quality measurements for MOS are classified as subjective and objective.

A functional representation of some popular voice quality measurement techniques is illustrated in Fig. 20.1. In the figure, voice is shown to be from

Table 20.1. PSTN and VoIP Quality Comparisons

Attributes	PSTN	VoIP
Distortions on analog line	Distortions due to several 1000-feet lines from DLC or CO location	No analog transmission distortions with VoIP calls.
Echo cancellation on national calls	Achieved through loss planning and low delays	Carrier grade echo cancellers are used.
Automatic gain control	Not incorporated	Possible to incorporate for better perception of speech levels or listening quality experience.
Voice quality monitoring	Monitoring such as GR-909 are incorporated into the PSTN	RTCP-XR and GR-909 are incorporated in many VoIP deployments.
Bandwidth or bit rate	64 kbps fixed on digital TDM. DCME channels use 16, 24, 32, and 40 kbps that degrades fax Quality	Variable bandwidth, usually requires more on physical interfaces than the PSTN. Fax services can get more bandwidth or redundancy in transmission.
Fax calls	Performance limited by the end transmission line characteristics	Uses short end lines. Hence, fax delivery can be better using VoIP. However, there could be interoperability issues for sending fax.
Voice and data	Mainly for voice calls, some services may reuse voice channels for data	Internet service and VoIP can scale along with data and media service requirements.
Voice call features	Limited features and expensive for several service features	Several features are offered as free.
Voice interfaces	Limited interfaces	Multiple interfaces and services.
Long distance	Long distance is costly	Usually free or much lower rates.
Transcoding	Multiple levels of transcoding for inter-regional calls	End-to-end direct coding can be employed based on the available support.
Wideband support	Voice calls are of narrowband	Wideband end-to-end voice is possible that can exceed PSTN quality.

the sending gateway to the receiving gateway. The receiving gateway is shown with some more expanded blocks for creating a big picture of the E-model, which is used for R-factor estimation, additional quality metrics, and Real-Time Transport Control Protocol-Extended Reports (RTCP-XR) operation. In the E-model, RTP, RTCP, jitter buffer, and total system signal parameters are used. When on calculating the R-factor and other derived parameters,

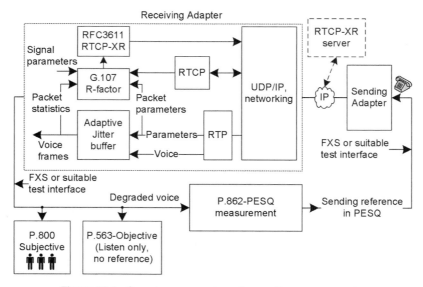

Figure 20.1. Overview on popular voice quality measurements.

RTCP-XR can send packets to the internal applications, destination gateway, and RTCP-XR server. In summary, the nonintrusive R-factor is an objective estimation that resides as part of VoIP implementation, and additional software is required in the gateway for the R-factor estimation. In perceptual evaluation of speech quality (PESQ), instruments like MultiDSLA [URL (DSLAII)] send the reference speech through the VoIP system under test and evaluate the degraded with the reference speech. This measurement is active, and VoIP gateways do not need to know anything about the measurement. In subjective listening, multiple listeners will evaluate the voice quality. In P.563, voice is analyzed entirely on the received degraded signal and the original reference is not required. P.563 is similar to subjective listening, but it is evaluated by the instruments or processors. Each of these techniques arrives at a different scale of voice quality. In a VoIP voice call between A and B, voice measurements are made as half-duplex, which means measurements are made as A to B or B to A, one at a time. Because of the half-duplex listening type of testing, these measurements are referred to as listening quality (LQ) tests. The suffix LQ is appended while presenting the results on half-duplex tests, and objective tests are additionally suffixed with "O" as LQO.

20.1.1 Subjective Measurement Technique

In subjective voice quality evaluation, voice quality MOS is rated by the group of actual male and female listeners. It is the actual listening test for evaluating the MOS. The P.800 and P.830 [ITU-T-P.830 (1996), ITU-T-P.800 (1996)] recommendations are used for assessing the subjective performance of speech codecs. The same tests are extended to the VoIP voice quality. A group of people participates for recording subjective scores. Multiple test phrases are recorded and then test subjects (group of people) listen to them in different conditions. These tests are performed in special rooms with background noises and other environment factors are kept under control for test execution. The test conditions are given in [ITU-T-P.800 (1996)]. The subjective measurement techniques are categorized as absolute category rating (ACR), degradation category rating (DCR), and comparison category rating (CCR).

In ACR, participants listen to recorded speech samples that have been processed through several test connections. A minimum of 16 test subjects (listeners) should participate in the assessment. When listening, users rate the call on a 1 to 5 MOS scale. The average values of the user ratings are considered to generate the overall call quality.

In a DCR test, two speech samples are present. The first speech sample is a reference sample with predefined quality. The sample here refers to speech lasting for several seconds in duration. The other speech sample is a degraded version. Listeners must compare the degraded version with a reference on a degradation scale of 1 to 5. Here, 5 is inaudible degradation and 1 represents worst degradation. The results are summarized as degraded MOS.

In CCR tests, users are asked to listen to two sets of samples, one corresponding to reference and the other to degraded. This test is similar to DCR, except that the order of samples presented to the listeners are changed in different iterations. The order of reference and degraded is not declared to the listener. Listeners are asked to give a comparative rating of a second sample with respect to the first one on a scale of −3 to 3 as per P.800 Annex-D [ITU-T-P.800 (1996)]. In presenting the results, "3" represents much better quality and "−3" represents the worst quality on a relative scale. The quality score is mapped to MOS. The MOS rating allowed is 1 to 5, but a user rating above 4.5 is limited to 4.5.

Subjective tests are involved in procedures, and it is a costly effort. It is limited to less iterations to evaluate any new algorithm or speech codecs. It is difficult to maintain consistency like instrument-based objective tests.

20.1.2 Objective Measurement Techniques

Objective methods are the measurements and calculations. It is expected that results will be consistent across several measurements. Several objective methods exist and are classified as active and passive methods.

- Active monitoring techniques of PESQ [ITU-T-P.862 (2001)]
- Passive monitoring techniques of P.563 and the E-model [ITU-T-P.563 (2004), ITU-T-G.107 (2005)]

Active Monitoring Techniques. Active measurement is called intrusive monitoring or offline monitorings because of involvement of external signals.

In an effort to supplement subjective listening quality, testing with lower cost objective methods are developed. KPN developed the P.861 (this is obsolete now) perceptual speech quality measure (PSQM) for the evaluation of codec performance. British Telecom developed the perceptual analysis measurement system (PAMS) for network measurements. The P.862 PESQ resulted from an ITU competition. The performance of PAMS and a new version of PSQM, PSQM99, were similar so the contributors were invited to combine the algorithms. This resulted in PESQ, which is slightly better than its constituents.

These methods measure distortion introduced by a transmission system and codec by comparing an original reference file sent into the system on a telephone interface with the received impaired signal received on another telephone interface. PSQM was developed for laboratory testing of speech codecs. PAMS and PESQ are designed for network testing. The use of instruments for voice quality is much simpler compared with subjective or passive measurements. Instrument suppliers are also providing the extra-derived parameters to help identifying the sources of degradations through measurements. Refer to some instruments given in Chapter 13 for more details on various features.

While writing this book, PESQ was popularly supported in the instruments. PESQ was approved by the ITU in March 2001 as the P.862 recommendation, replacing P.861 PSQM. The PESQ combined several best merits of PAMS and PSQM. It is accurate in predicting subjective test scores, and it is robust under severe network conditions such as a variable delays, filtering at analog interfaces, and support of both wideband and narrowband. PESQ produces a score that lies on a scale from −0.5 to 4.5. A mapping function from a P.862 PESQ score to an average subjective P.800-LQ MOS score was provided, making it PESQ-LQO [ITU-T-P.862.1 (2003)] for narrowband voice. LQO denotes a listening quality objective. PESQ-LQ lies from 1 to 4.5. A MOS of 4.5 is the maximum quality achieved for a clear undistorted condition. An overview on the PESQ algorithm is given here. It is suggested to refer to the ITU P.862 family of recommendations, software, and some commercial instrument brochures for more details [URL (DSLAII)].

20.1.3   PESQ Measurement

Human auditory perception is the core concept behind PESQ and its predecessors PAMS and PSQM. A perceptual model is used to distinguish correctly between audible and inaudible distortions, and this has proven to be the best way of accurately predicting the audibility and annoyance of complex distortions. In addition to the quantity of distortion, the distribution of audible distortion could make quality predictions much more accurate.

PESQ measures one-way voice quality, which means the half-duplex operation of measurement. It assesses the quality of a distorted speech signal that has been coded and transmitted over the network by comparing it with the original undistorted signal. The original and distorted speech is mapped on to psychophysical representations that match the way humans experience speech.

The quality of the distorted speech is judged based on differences in psycho-physical representations. The PESQ operation makes use of two major classes of logarithmic operations—namely conversion of signals into the psycho-acoustic domain and cognitive modeling. A functional representation of the PESQ algorithm is given in Fig. 20.2. Instrument manufacturers for the PESQ measurement include several extra operations to extract signal analysis parameters and impairments in addition to PESQ measurements.

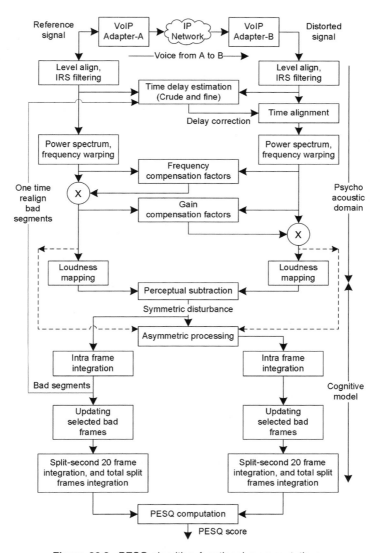

Figure 20.2. PESQ algorithm functional representations.

The processing carried out by the PESQ algorithm includes the stages listed below. Summary steps are given here; several details on the PESQ are given in [ITU-T-P.862 (2001), Rix et al. (2002), Beerends et al. (2002)].

In the first step of processing, both the reference and the degraded signal are scaled to the same constant power level. This scaling is necessary because the reference signal does not have to be at a defined level and the gain of the system under test is unknown before testing. PESQ assumes that the subjective listening level is a constant 79 dBSPL at the ear reference point [ITU-T-P.830 (1996)]. For power normalization, electrical signal levels are normalized to −26 dBov (i.e., −20 dBm as given in the reference [URL (DSLA-usrgd), Malfait et al. (2006)]). A signal-level normalization is applied to both the reference and the degraded signal to bring them to this level.

Perceptual models such as PESQ should take into account the characteristics of the telephone handsets as subjective listening may use telephone handsets. In PESQ, the receive path of the handsets is modeled using an intermediate reference system (IRS) band-pass filter [ITU-T-P.830 (1996)] in the frequency domain. This process takes into account the effects of the electrical and acoustic components of the handset. Both the reference and the degraded signal are IRS filtered.

The system under test may include variable delay. To compare the reference and degraded signals, both signals are time aligned with each other. PESQ aligns overlapping sections of the speech frames. In the first stage, the delay estimation is carried out over the length of files by computing the correlation between the files. The delay obtained in this stage is called crude delay. In the next stage, PESQ applies voice activity detection to the signals to identify required speech segments usually referred to as utterances. The delay estimate between utterances is the fine delay. This process detects delay that is variable over the length of an utterance, as this can be significant in packet-based networks.

The time-aligned reference and degraded signals are transformed into the frequency domain by using a short-term fast Fourier transform (FFT) with a Hanning window over 32-ms frames with 50% overlapping. The powers of original and degraded signals are computed and stored separately. In the next stage of operations, the frequency bands are transformed to bark scale by binning FFT bands. This process warps the frequency scale in Hz to the pitch scale, and the resulting signals are called pitch power densities. In this process, higher bandwidth is used for a high-frequency signal derived through frequency analysis.

The filtering effects in the system under test are equalized by computing a partial compensation factor per each bark bin and by multiplying each frame of the reference signal with this factor. This process equalizes the reference to the degraded signal. The compensation factor is computed as the ratio of degraded signal spectrum to reference signal spectrum. This factor takes into account the filtering at analog components of the network such as telephone handsets. In the second stage of equalization, the frame-by-frame amplitude gain of the system is estimated and used to equalize the degraded signal to the reference signal. In both cases, the equalization is partial and large amounts of filtering or gain variation are not cancelled; therefore, it results in errors being measured. The frequency and gain-equalized pitch power densities are transformed to loudness scale using Zwicker's law [ITU-T-P.862 (2001)]. The resulting time-frequency components are called loudness densities.

The signed difference between the loudness densities for the reference and degraded signals is known as raw disturbance density, which shows any audible differences introduced by the system under test. A masking operation applies a mask factor on the raw disturbance densities that masks the small inaudible distortions in the presence of loud signals. The disturbance density obtained by this process is called absolute or symmetric disturbance density. The symmetric disturbances are integrated over the length of the frame (intraframe). The consecutive frames with a frame disturbance above a threshold are categorized as bad frames. The bad frames may occur because of incorrect time delay estimation or packet drops. On a localized window around bad frames, a new delay estimate is made that is used to recompute the disturbance densities. The minimum of the previous and current disturbances is considered as the final disturbance in that bad frame window.

To model the distortion introduced by the codec used in the network, an asymmetric disturbance density is calculated by multiplying the symmetric disturbance density with an asymmetry factor. The asymmetry factor is the ratio of distorted and the original pitch power densities raised to the power of 1.2. This disturbance density is called an additive or asymmetric disturbance.

Finally, the error parameters are converted to a quality score, which is a linear combination of the average symmetric disturbance value and the average asymmetric disturbance value. From Fig. 20.2, the stages involved from level alignment to the intensity warping on the loudness scale are known as the conversion to the psycho-acoustic domain, and the algorithmic stages from perceptual subtraction to PESQ score computation are known as cognitive modeling.

PESQ gives a score known as the PESQ score in accordance to P.862. The PESQ score is in the range of −0.5 to 4.5. PESQ is correlated to the subjective MOS as 0.94 based on experiments conducted on databases by [Malfait et al. (2006)]. Compared with subjective (actual listeners) scores, PESQ gives better results for poor quality speech and pessimistic results for good quality voice. PESQ-LQ provides better correlation with subjective scores than PESQ on a listening quality scale. PESQ-LQ scores are in range of 1 to 4.5. P862.1 provides a quality mapping between narrowband quality measurements PESQ score and listening quality objective mean opinion score (MOS-LQO). Recommendation P.862.2 provides a quality mapping between wideband quality measurements PESQ score and listening quality objective mean opinion score. More information on these scores can be found in the ITU-T-P.862 series recommendations and in reference [URL (DSLAII)].

PESQ is a half-duplex operation that will not capture accurately on end-to-end delay, echo, loudness loss, sidetone, and listening levels. From the Voice Quality measurement of the VoIP gateway with analog interfaces, the following PESQ-LQO observations are made using DSLA [URL (DSLAII)]. Under the no packet loss condition, the PESQ-LQO score for the G.711 codec is 4.32, G.729A is 3.85, and G.723.1 is 3.75. Another interpretation of these results for packet drop situations and comparison with the E-model are given as part of the R-factor calculations and presented in Table 20.4. In the process of PESQ calculations, several other parameters can be computed. Instrument suppliers provide these parameters as additional features to PESQ measurements [URL (DSLAII)].

20.1.4 Passive Monitoring Technique

In passive monitoring techniques, the reference signal is not present. Two popular methods for passive speech quality monitoring exist. The ITU has standardized a signal based nonintrusive monitoring method, P.563, based on the result of collaboration among three companies, Psytechnics Ltd., Swissqual, and Opticom, which combined the best parameters of three different models. P.563 is a single-ended objective measurement that makes use of a speech production mechanism, and the other speech models make use of listening perception. This algorithm operates on received degraded speech only. It will not need reference speech, and it entirely operates on degraded speech. The measurements through P.563 derive several parameters from received speech classified as noise, artificial speech, and actual speech. An overview on the P.563 single-ended speech-quality assessment operation is given here.

In the absence of a reference signal, the models do not have knowledge of the original signal and assumptions have to be made about the received signal. The P.563 model combines three basic principles for evaluating distortions. The first principle focuses on the human voice production system, modeling the vocal tract as a series of tubes, with abnormal variations of the tubes' sections considered as degradation. The second principle is to reconstruct a clean reference signal from the degraded signal in order to apply a full-reference perceptual model thereafter and to assess distortions unmasked during the reconstruction. The third principle is to identify and to estimate specific distortions encountered in voice channels, such as temporal clipping, robotization, and noise. Listening speech quality is derived from the calculated parameters from the three principles, applying a distortion-dependent weighting.

While writing this chapter, the P.563-based technique was not widely accepted for measurements. P.862 PESQ-based measurements and E-model-based estimations are more popularly accepted. The main advantage of this P.563 technique is its ability to monitor at the degraded end without calling for reference. Thus, it can better monitor long-distance calls outside the laboratory and in deployments, which will be much simpler to conduct than many other measurements. The P.563-based method can also be embedded as part of the receiving gateway similar to E-model and RTCP-XR. P.563 operations can be used on samples that get delivered on the pulse code modulation (PCM) voice interfaces.

More information on the P. 563 technique can be found from P.563 [ITU-T-P.563 (2004)] and [Malfait et al. (2006)]. The MOS score produced by P.563 and other techniques is widely spread and is necessary to average the results of multiple tests to achieve a stable quality metric over multiple results. P.563 is correlated with subjective MOS as 0.85 to 0.9 based on the experiments conducted on a database by [Malfait et al. (2006)], and PESQ is reported as 0.94.

20.2 E-MODEL-BASED VOICE QUALITY ESTIMATION

The E-model is the equipment impairment model described in G.107 [ITU-T-G.107 (2005)]. This model examines signal and packet transmission character-istics to predict voice quality on a linear scale. The objective of the E-model is to determine a transmission quality rating (R) (i.e., the R-factor or R-value that incorporates the "mouth to ear" characteristics of an end-to-end speech path). In the usage, it is also common practice to reference the E-model as the R-model. This model helps in analyzing and identifying the root causes of voice quality degradation. The R-factor is mapped to subjective MOS and to many other voice quality parameters. The typical useful range for the R-factor is 50 to 94 for narrowband telephony. An R-value below 50 is not suitable for continuing call conversation. Voice quality monitoring (VQmon) [URL (Telchemy)] and RTCP-XR packets [Friedman et al. (2003)] in VoIP makes use of E-model parameters. It is widely used for VoIP service quality measurements.

An E-model-based voice quality estimate for multiple voice compression codecs is given in Section 3.7. More details on this topic are available at [ITU-T-G.107 (2005), ITU-T-G.113 (2007), ITU-T-G.108 (1999), ITU-T-G.175 (2000)]. In this section, an overview and some more extensions from the pub-lished literature and recommendations are included. The E-model makes use of several parameters broadly classified under delay, delay variations, echo, noise, and phone characteristics as well as signal transmission, codec, and packet characteristics. In narrowband telephony, the R-factor ranges from 0 to 100, with 100 being the MOS equivalent of 4.5 that is achieved only with direct linear 16-bit samples. Voice compression reduces the R-factor. In an end-to-end digital service such as ISDN with G.711, an R-factor of 93.2 is pos-sible as the highest value. In VoIP service, several impediments contribute to the degradation of the R-factor. In the first version of G.107 recommendation, the R-factor under ideal network conditions was considered to be 94.2. This value was revised to 93.2 in the later versions of G.107 [ITU-T-G.107 (2005)]. From the R-factor, additional parameters can be derived such as MOS, minimum percentage of people able to say Good or Better (GoB), maximum percentage of people that report as Poor or Worse (PoW) quality, and so on. The R-factor is mapped to a subjective voice quality measure MOS using the following equations:

$$MOS = 1 + 0.035R + R(R - 60)(100 - R)(7 \times 10^{-6}) \text{ for } R = 1 \text{ to } 100$$

$$MOS = 4.5 \text{ for } R \geq 100$$

$$MOS = 1 \text{ for } R \leq 0$$

Table 20.2 lists the R-value ranging from 50 to 100 in selected steps. The corresponding MOS, GoB, PoW, and qualitative user satisfaction limits are given. It can be observed that R from 90 to 100 corresponds [ITU-T-G.109 (1999)] to best quality, 80 to 89 is high quality, 70 to 79 is medium, 60 to 69 is low, and 50 to 59 is poor. A rating below 50 indicates unacceptable quality.

In general, deviations will occor between the E-model based MOS and the PESQ-based MOS. The R-factor-based MOS is slightly higher than the PESQ-

Table 20.2. Relation Between R-Value, Corresponding MOS and User Satisfaction for Selected R-Factor Values

R-Value Lower Limit	MOS	Good or Better (GoB % Lower Limit)	Poor or Worse (PoW % Upper Limit)	User Satisfaction
100	4.5	99	0	More than PSTN, comes only with 16-bit linear samples
94	4.42	98	0	Very satisfied (R is 90
92	4.38	98	0	to 100).
90	4.34	97	0	PSTN quality; achieved in VoIP G.711 as best case
88	4.29	96	0	Satisfied
86	4.23	95	1	(R is 80 to 89)
84	4.17	93	1	
82	4.10	92	1	
80	4.02	90	1	
79	3.99	88	2	Some users dissatisfied
78	3.95	87	2	(R is 70 to 79)
76	3.86	84	3	
74	3.78	81	3	
72	3.69	77	5	
70	3.60	73	6	
69	3.55	71	7	Many users dissatisfied
65	3.35	62	11	(R is 60 to 69)
60	3.10	50	17	
55	2.84	38	27	Nearly all users
50	2.58	27	38	dissatisfied (R is 50 to 59)

based MOS for similar impairments. The R-factor-based MOS is represented as listening quality (LQ) and conversational quality (CQ). The MOS score obtained by considering only coding distortions and packet losses is called MOS-LQ. The MOS score obtained by considering delay and loudness impairments in addition to the distortion impairments is called MOS-CQ. These MOS R-factor values are used in RTCP-XR reports [Friedman et al. (2003)].

20.2.1 R-Factor Calculations

The R-factor is a transmission-rating factor for the quality of the voice in VoIP. The R-factor is a scalar prediction that ranges from 0 to 100 for narrowband voice communication. The end equipment used, room noise, losses in the network, delay, packet loss, and compression algorithms used affect the R-value. The value of the R-factor can be computed by the following equation:

$$R = R_0 - I_s - I_d - I_{e\text{-eff}} + A$$

R_0 is the highest value of R that takes into account mainly the signal-to-noise ratio (SNR) value, including noise sources such as circuit noise and room noise and subscriber line noise.

$$R_0 \text{ is a function of } (N_c, SLR, P_s, D_s, RLR, P_r, LSTR)$$

I_s comprises impairments that occur simultaneously with the voice signal. The major factors that contribute to this impairment are loudness ratings of the telephone set, number of quantization distortion units, and side-tone loudness rating.

$$I_s \text{ is a function of } (R_0, SLR, RLR, STMR, TELR, qdu)$$

I_d comprises impairments caused by delay. The factors that contribute to these impairments are the amount of delay present in the network and the values of talker and listener echo loudness ratings.

$$I_d \text{ is a function of } (T, T_r, T_a, RLR, STMR, TELR, WEPL)$$

$I_{e\text{-eff}}$ is the equipment impairment factor, which mainly comprises impairments caused by distortion. The main parameters that contribute to the $I_{e\text{-eff}}$ are the voice compression codec and end-to-end the packet impediments.

$$I_{e\text{-eff}} \text{ is a function of } (I_e, B_{pl}, P_{pl})$$

"A" is the advantage factor, which represents the user tolerance to the degradation of the voice quality. The value of "A" is governed by the end-user communication interface. For wire-bound communication, the value of "A" is zero

Table 20.3. ITU-T Recommendation G.107—Default Values and Permitted Ranges for the Parameters for Use with the E-model R-Factor Calculation [Courtesy: Reproduced with the kind permission from the ITU; International Telecommunication Union, Geneva, www.itu.int.)]

Parameter Symbol	Parameter Name	Default Value	Units	Allowed Range
SLR	Send loudness rating	+8	dB	0 to 18
RLR	Receive loudness rating	+2	dB	−5 to 14
STMR	Side tone masking rating	15	dB	10 to 20
LSTR	Listener side tone rating	18	dB	13 to 23
D_s	D-value of phone at send side	3	—	−3 to 3
D_r	D-value of phone at receive side	3	—	−3 to 3
TELR	Talker echo loudness rating	65	dB	5 to 65
WEPL	Weighted echo path loss	110	dB	5 to 110
T	Mean one-way delay of echo path	0	ms	0 to 500
T_r	Round-trip delay	0	ms	0 to 1000
T_a	Absolute delay in echo-free connection	0	ms	0 to 500
qdu	Number of quantization distortion units	1	—	1 to 14
I_e	Equipment impairment factor	0	—	0 to 40
B_{pl}	Packet loss robustness factor	1	—	1 to 40
P_{pl}	Random packet loss probability	0	percentage	0 to 20
Burst ratio	Burst ratio	1	—	1 to 2
Nc	Circuit noise referred to 0-dBr point	−70	dBm0p	−80 to −40
Nfor	Noise floor at receive side	−64	dBmp	−64
P_s	Room noise at send side	35	dB(A)	35 to 85
P_r	Room noise at receive side	35	dB(A)	35 to 85
A	Advantage factor	0	—	0 to 20

and it is "5" for mobility. The value of "A" only nullifies the contributions from $I_{e\text{-eff}}$, I_d, and I_s, but it cannot increase the R_0 value.

The detailed steps of R-factor calculations and parameters are given in G.107. A summary on parameters used and their range of values with defaults [ITU-T-G.107 (2005), Britt (2007)] are given in Table 20.3. Several details on decibel units used in Table 20.3 are given in references [ITU-T-G.100.1 (2001), TIA/EIA-116 (2001)].

$R_0 - I_s$ value is 93.2 as the default value specified for room noise conditions and signal levels. This value is valid in good implementations of hardware and front ends meeting TR-57 or local PSTN transmission characteristics.

Delay Impairments—I_d. The delay impairment factor deals with impairments that are caused by loss of interactivity between the communicating ends. This loss is from long values of absolute delay and perceivable echo. The I_d is calculated from delay, talker, and listener echo. The talker echo loudness rating

(TELR) is the loudness loss of the speaker's voice reaching the ear as a delayed voice. As the delay increases, the loudness loss should also increase as per G.131 [ITU-T-G.131 (2003)]. Assuming a good echo cancellation (i.e., TELR > 65 dB), the delay impairment factor can be modeled as a function of one-way absolute delay "T_a" in milliseconds. T_a is the cumulative sum of all one-way delay components involved in a voice call. Increase in absolute delay causes users to feel that interactivity in communication is lost. In relation to G.108 [ITU-T-G.108 (1999)], assuming that all other impairments are in control, the high value of one-way absolute delay causes an increase in I_d value. A one-way delay of more than 150 ms causes more degradation to voice quality [ITU-T-G.108 (1999)]. Based on the graph given in G.108, the formulation [Sun and Ifeachor (2003)] for I_d based on one-way delay impairments is expressed as

$$I_d = 0.024(T_a), \text{ for } T_a \leq 177.3\,\text{ms}$$

$$I_d = 0.024(T_a) + 0.11(T_a\text{-}177.3), \text{ for } T_a > 177.3\,\text{ms}$$

The delay contribution dominates after 177.3 ms when the TELR value is ≥65 dB. In G.131 recommendation [ITU-T-G.131 (2003)] figures for a TELR of 65 dB, the delay dominance is shown as starting from 150 ms through R-factor degradation. This value of 150 ms is very close to 177.3 ms.

Equipment Impairment Factor—$I_{e\text{-}eff}$. The equipment impairment factor $I_{e\text{-}eff}$ is computed by considering the voice compression codec used and the end-to-end packet losses. According to ITU-T-G.107 [ITU-T-G.107 (2005)], the effective impairment factor for narrowband codecs is calculated from the following equation for random packet losses:

$$I_{e\text{-}eff} = I_e + (95 - I_e)\frac{P_{pl}}{P_{pl} + B_{pl}} \tag{20.1}$$

In the above equation, I_e is the default equipment impairment value of the codec used [ITU-T-G.113 (2003)] under zero packet loss conditions. The parameter P_{pl} is the random packet loss rate. The random packet loss rate implies that the probability of loss of a packet is independent of the probability of loss of any other packet. The B_{pl} is the packet loss robustness factor assigned to each codec. This value reflects the ability to recover the packet loss using a method of packet loss concealment.

Equipment impairment values for different packet loss conditions were formerly given as codec-dependent tabulated values [ITU-T-G.113 (2003)]. The calculations using parameters I_e, P_{pl}, and B_{pl} along with Eq. (20.1) also provide similar results. With Eq. (20.1), the impairments can be calculated for the required conditions. Under random packet loss, the values of I_e and B_{pl} for different narrowband codecs used in VoIP are given below [ITU-T-G.113 (2001), ITU-T-G.113 (2007)].

- Codec G.711 of 10-ms packetization, $I_e = 0$ for zero percentage packet drop with B_{pl} value of 25.1.
- Codec G.711 of 10-ms packetization, $I_e = 0$ for without PLC and B_{pl} value of 4.3.
- Codec G.729AB of 20-ms packetization, $I_e = 11$ for zero percentage packet drop with B_{pl} value of 19.
- Codec G.723.1A of 30-ms packetization, $I_e = 15$ for zero percentage packet drop with B_{pl} value of 16.1.

Table 20.4 is given for $I_{e\text{-eff}}$, R, and listening MOS for two popular codecs G.711 and G.729AB with voice activity detection (VAD). The values are calculated based on Eq. (20.1). From the table, it can be observed that voice compression creates major degradation. Packet losses degrade voice quality even more compared with other impairments like delay, echo, and signal transmission.

In practical deployments, packet drops are not just random and can follow bursty losses. Bursty losses are dependent on previous packets statistics, and random loss is an independent loss. A bursty loss is a high packet loss in a short time interval. During burst, a high proportion of packets are either lost or discarded because of late arrival. A bursty drop creates more degradation compared with random loss. A bursty equipment impairment factor is given as [ITU-T-G.107 (2005)].

Table 20.4. Random Packet Loss percentage and corresponding MOS for G.711 (10 ms packetization) and G.729AB (20 ms packetization) codecs (PESQ-LQ readings in this table are mean values over several measurements taken using DSLAII on gateways with an FXS interface)

Packet Drop %	G.711 with ITU-Based PLC $I_e = 0$, $B_{pl} = 25.1$			G.711 Measurements on VoIP Gateway	G.729AB, $I_e = 11$, $B_{pl} = 19$			G.729AB Measurements on VoIP Gateway
	$I_{e\text{-eff}}$	R	MOS-LQ	PESQ-LQ Mean	$I_{e\text{-eff}}$	R	MOS-LQ	PESQ-LQ Mean
0	0	93.2	4.41	4.32	11.0	82.2	4.10	3.77
1	3.6	89.6	4.33	4.08	15.2	78.0	3.95	3.62
2	7.0	86.2	4.24	3.83	19.0	74.2	3.79	3.29
3	10.1	83.1	4.14	3.65	22.5	70.7	3.63	3.16
4	13.1	80.1	4.03	3.43	25.6	67.6	3.48	2.96
5	15.8	77.4	3.92	3.49	28.5	64.7	3.34	2.95
6	18.3	74.9	3.82	3.17	31.2	62.0	3.20	2.58
7	20.7	72.5	3.71	3.22	33.6	59.6	3.08	2.60
8	23.0	70.2	3.61	2.96	35.9	57.3	2.96	2.56
9	25.1	68.1	3.51	2.94	38.0	55.2	2.85	2.38
10	27.1	66.1	3.41	2.76	40.0	53.2	2.74	2.17

$$I_{e\text{-}eff} = I_e + (95 - I_e) \cfrac{P_{pl}}{\cfrac{P_{pl}}{BurstR} + B_{pl}} \qquad (20.2)$$

The BurstR parameter is used to capture the burst conditions of the packet loss distribution. The values of BurstR are defined as [ITU-T delayed D.20 (2001), Alexander (2006)].

$$\text{BurstR} = \frac{\text{Average number of consequitively lost packets}}{\text{Average number of consequtively lost packets for random loss}}$$

The BurstR is calculated based on consecutive packet losses relative to random loss [ITU-T delayed D.20 (2001)]. For random packet loss, BurstR = 1, and for bursty packet loss, BurstR is >1, Burst R < 1 corresponds to scatter losses present in the network. BurstR is helpful in measuring short-term burstiness of packet loss.

20.2.2 Bursty Packet Losses

In the G.113 Amendment [ITU-T-G.113 (2007)] under a bursty loss condition, a set of provisional planning values were specified for the G.711 codec with 20-ms packetization, and this assumes that PLC employed is repetition/silence. It is also assumed that packet loss percentage Ppl is ≤ 2%. The B_{pl} value proposed is reduced to "4.8" from the random case of "25.1," and BurstR is set to "1." This assumption is for a specific sample of burst packet loss and may not reflect the impairment caused by burst packet loss in general. As per the G.113 Amendment-2 [ITU-T-G.113 (2007)], it is recommended that under bursty packet loss, the BurstR approach of the E-model Eq. (20.2) [ITU-T-G.107 (2005)] should be employed only for codecs with an efficient codec-state-based PLC with a packet loss robustness factor $B_{pl} \geq 16$. The BurstR approach is based on a two-state Markovian packet loss model, and it captures only short-term consecutive loss dependencies [ITU-T-G.107 (2005)]. The BurstR approach and provisional planning values for $I_{e\text{-}eff}$ given in G.113 [ITU-T-G.113 (2007)] are valid under some constraints. The computation of the equipment impairment factor from the BurstR approach is not yet concluded fully in recommendations and standards. Based on the published literature, overview on packet loss models are presented below.

Packet Loss Models. The mechanisms that lead to packet loss are generally transient in nature. Hence, the packet loss distribution can be shown by packet loss model rather than by a simple set of packet loss count. Packet losses are mainly modeled as random and burst. The random loss in the network can be modeled by the Bernoulli model. Using this loss model, the probability "P_e" of a packet loss is estimated by counting the number of lost packets and by dividing it by the total number of packets transmitted.

Two-State Models. During bursty conditions, the BurstR approach of the I_e calculation [ITU-T-G.107 (2005)] is based on a two-state Markovian packet loss model. In this model, a transition between a "good" state-0 and a "bad" state-1 is used for state transition probabilities [ITU-T-G.1020 (2006)]. The probability "p" denotes a packet dropped given that the previous packet was received. The probability "q" is the probability that a packet will be received given that the previous packet was dropped. "$1 - p$" is the probability that a packet is received given that previous packet is also received. "$1 - q$" is the probability that a packet is lost given that a previous packet is also lost. The parameter BurstR can be derived from a two-state Markov model as [Alexander (2006)].

$$\text{BurstR} = \frac{1}{p + q}$$

The other popular two-state models are Gilbert and Gilbert–Elliott. In a two-state Markov model, state "1" indicates a loss state. In the Gilbert model, the loss state will have an independent loss probability associated with it. In the Gilbert–Elliot model, state "0" will also have an independent loss probability associated with it [Alexander (2006)]. These two-state packet loss models using transient probabilities are used in analysis of packet loss burstiness in terms of consecutive loss. Hence, these packet loss models can capture short-term dependencies between packet losses. These models miss the effects of longer periods of high loss density. Markov-4 model helps with detailed analysis of the packets.

Markov Four-State Model. A Markov four-state model can be used to capture the very short duration consecutive loss events and the longer events of lower loss density. This model divides packets as part of bursts and gaps, which typically represent phases of higher and lower packet loss. The Markov-4 model is also treated as a combination of two, two-state Markov models. As shown in Fig. 20.3(b), these states are identified with gap and burst boundaries. Each two-state Markov model has total four independent transition probabilities. Two more transition probabilities represent a transition between bursts and gaps. The combination of all these states and probabilities provides better analysis with the Markov-4 model.

The states in gap and burst are assumed based on the number of consecutively received packets between two packet loss events. The characterization of burst and gap is decided based on the consecutive received packets of size Gmin. The value of Gmin is 16 [ITU-T-G.1020 (2006)]) for voice applications. A sequence begins and ends with a loss during which it the number of consecutive received packets is less than Gmin; then the two lost packets and sequence are assumed to be part of the burst. The periods between the bursts, where numbers of consecutively received packets are more than Gmin are regarded as gap. The Markov four-state model assumes that a packet can be

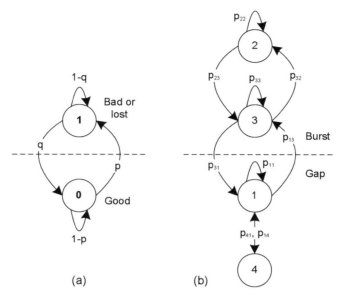

Figure 20.3. Packet loss models. (a) Two-state model. (b) Four-state Markov model.

in one of the four states. In the process of state change, state transitions are created, as follows:

State-1—packet received successfully in a gap
State-2—packet received within a burst
State-3—packet lost within a burst
State-4—isolated packet loss within a gap

State-1 and state-4 are part of the gap. Gap is the indication of good quality and is desired. State-2 and state-3 are part of the burst. Burst is of more losses and bad quality. At each packet loss event, the number of received (good) packets count from the last packet loss event is noted. This value is used to categorize whether the sequence of received packets and the two lost packets are part of gap or part of burst. A set of counters is used to track the number of packet counts present in each state as well as the key transition counts. These counters are updated at each packet loss event during the course of a call. After the call is completed, the remaining transition counts are derived and then normalized to provide the desired independent probabilities. The probabilities p_{11}, p_{14}, p_{41}, p_{13}, p_{22}, p_{23}, p_{33}, p_{32}, and p_{31} marked in Fig. 20.3(b) represent probabilities from one state to another. Complete formulation is available in [ETSI TS 101 329-5 (2000)].

This statistical analysis approach in terms of individual probabilities for a state transition provides the following packet-related statistics [Clark (2001), ETSI TS 101 329-5 (2000)]:

Gap length is the average length in ms of the gaps detected for the stream.

Gap density is the average packet loss percentage in a gap.

Burst length is the average length (ms) of the bursts detected.

Burst density is the average packet loss percentage in a burst.

A Markov-4 model is a combination of two, two-state models of independent probabilities that correspond to gap and burst. The effective equipment impairment factors under the burst conditions as well as the gap conditions are obtained by using loss densities obtained in each state respectively. The equipment impairment factor I_{eg} for the gap condition is obtained by substituting gap loss density in the place of packet loss percentage (P_{pl}) in Eq. (20.1). Similarly for the burst condition, the impairment factor I_{eb} is calculated using the burst loss density in place of P_{pl} in Eq. (20.1). The obtained values of I_{eg} and I_{eb} represent the effective impairments under the gap and burst conditions, respectively. As bursts represent periods of high packet loss, the value of I_{eb} will be high and the R-factor in burst is low.

Note on VAD Packets. For determining a lost or discarded packet near the start or end of an Real-Time Transport Protocol (RTP) session, it is assumed that the RTP session is preceded and followed by at least Gmin received packets. Calculation of burst parameters in silence zones is not correct. A packet loss in silence will not degrade voice quality. It is assumed that VAD is being used and, hence, packet loss reports have to be generated based on speech packets. In practice, several implementations may not use VAD/comfort noise generation (CNG) because of the availability of more bandwidth. In this situation, the statistics based on all the available packets may not represent correct voice quality. Any other information from local silence detections may help for proper estimation of packet loss parameters.

Time-Varying Packet Loss Effects. The packet loss in the network is time varying in nature. As the packet loss behavior varies, the packet sequence switches between gaps (good quality) and burst (bad quality) during the call. Hence, the call quality varies between good and bad. If voice quality changes, the listener will be able to report on it after some time. Assume that the quality-level change from burst (I_{eb}) to gap (I_{eg}) is indicated by I_1 and that the quality-level change from gap (I_{eg}) to burst (I_{eb}) is indicated by I_2. It is assumed that the equipment impairment factor varies between burst period "b" and gap period "g." The values of I_1 and I_2 are expressed as [ETSI TS 101 329-5 (2000)].

$$I_1 = I_{eb} - (I_{eb} - I_2)e^{-b/t_1}$$
$$I_2 = I_{eg} + (I_1 - I_{eg})e^{-g/t_2}$$

where I_{eb} is effective impairment value under burst condition, I_{eg} is the effective impairment value under gap condition, "b" is the burst duration and "g" is the gap duration. The values of t_1 and t_2 were arrived from time-varying assessments, which were subjectively done by a group of users on a 3-minute call in which packet loss was varied from 0% to 25% [Clark (2001)]. In [Clark (2001)], $t_1 = 5$ seconds and $t_2 = 15$ seconds were used. In the recent publication of [Alexander (2006)], the values used are $t_1 = 9$ and $t_2 = 22$ seconds. At this stage, t_1 and t_2 values are perception based and are not regularized through recommendations.

The transient nature of the equipment impairment factor during the call [Carvalho et al. (2005)] can be explained as follows. From an exit value I_2, the I_e impairment factor exponentially increases to I_{eb}, with the time constant of t_1 in the transition between gap periods to burst periods. From an exit value I_1, the I_e impairment factor exponentially decays to I_{eg}, with a time constant of t_2. This transition occurs between a burst period and a gap period. Combining above two equations of I_1 and I_2 gives an expression of I_2 in terms of I_{eg} and I_{eb} [ETSI -TS 101 329-5 (2000)].

$$I_2 = \left(I_{eg}(1 - e^{-g/t_2}) + I_{eb}(1 - e^{-b/t_1}) e^{-g/t_2} \right) / \left(1 - e^{b/t_1 - g/t_2} \right)$$

The average value of the equipment impairment factor is computed by integrating I_1 and I_2 over burst and gap durations. The average impairment factor $I_{e(avg)}$ is expressed as

$$I_{e(avg)} = \frac{1}{(b+g)} \left[bI_{eb} + gI_{eg} - t_1(I_{eb} - I_2)(1 - e^{-b/t_1}) + t_2(I_1 - I_{eg})(1 - e^{-g/t_2}) \right]$$

According to [Carvalho et al. (2005)], the values of I_2 and I_1 are computed at the end of each burst and the value of $I_{e(avg)}$ for that period of the call is calculated. In this way, during the course of a call at the end of each pair of gap and burst periods, the values of $I_{e(avg)}$ for the corresponding periods are calculated. The final $I_{e(avg)}$ will be the weighted average of each instantaneous $I_{e(avg)}$. The weights are the duration in the call of each $I_{e(avg)}$ value. This value can be used as an effective ($I_{e\text{-eff}}$) equipment impairment factor in the R-factor equation to calculate the effects of packet loss and the codec on the transmission quality. Many of these packet loss estimation approaches are base on simulations, perceptions, and the published literature. While writing this chapter, these conclusions were not arrived at in the form of a standard for the time-varying equipment impairments and quality.

Recency Effect. The time location of impairment with in a call can affect the listener's perception of quality. This effect is known as the recency effect. Recency reflects the way that a listener will remember voice quality or the way a listener forgets reduced quality. As part of the statistical analysis of a voice packets stream, the most recent position of the significant packet loss/drop influences the recency effect. Significant drop is defined as 8 packets or more packets lost in 16 consecutive packets. The position of significant loss is used to calculate the recency degradation factor. The tendency of the listener is to remember the most recent events. These events can be modeled with a function that typically decays the recollection over a 30-second interval. Let "y" represent the time delay since the last significant burst, and then the equipment impairment factor considering the recency effect can be written as [ETSI TS 101 329-5 (2000)].

$$I_{e\text{-}eff}(end\ of\ call) = I_{e(avg)} + [0.7(I_1 - I_{e(avg)})]e^{-y/30}$$

The factor "y" can be obtained from the Markov-4 model [ETSI TS 101 329-5 (2000)]. The bursts in the beginning of the voice call are more tolerated.

Toward the end of the call, the same bursts are less tolerated in qualifying the quality. It is important to ensure fewer and short bursts with call progressing time. Burst is always undesirable. Gap is tolerated, and packet loss concealment operations can take care of isolated packet drops of gap.

20.2.3 Improving Voice Quality Based on the E-Model

The following major guidelines have to be followed to improve voice quality based on E-model calculations:

- Making front-end and circuit noise of an end-to-end system to mimic the PSTN system. As an example, the systems can be made to adhere to the TR-57 reference in North America.
- Reducing end-to-end delays to 50 ms and limiting it to less than 100 ms in several deployment situations.
- Echo rejection matching or exceeding the delay-based requirements and perceptions.
- Following country-specific deviations, loss planning, and overall loudness rating.
- Using G.711 with small packetization mainly to help with end-to-end delays.
- Using codecs with smaller "I_e" if G.711 is not possible to use.
- Avoiding the loss of packets irrespective of codec selection.
- Employing packet loss concealment to help retrieve the voice.

20.3 VoIP VOICE QUALITY CONSIDERATIONS

A VoIP solution supports full-chain voice and fax modules as explained in the previous chapters. Many of these algorithms and modules are based on several standards. In VoIP solutions, the main voice quality goals would be to match or exceed PSTN quality. To achieve this goal, several proprietary solutions are incorporated into voice processing. Major influencing items can be analyzed through the voice quality measuring models like the E-model, PESQ, and perceptions based on subjective listening. In VoIP, some critical areas considered for voice quality contribution are given below [Alexander (2006), Bellamy (1991), ITU-T-G.1020 (2006), ITU-T-G.177 (1999), TIA/EIA-810A (2000), TIA/EIA-116A (2006)]. Some of these aspects are also covered in the previous chapters of this book while presenting an individual operation like codec, echo canceller, VAD/CNG, PLC, jitter buffer, and so on. A summary of the major voice quality influencing operations and improvements is given as follows:

- End-to-end delay reduction
- Packet flow impediments in processing
- Adaptive jitter buffer with utilization of silence zones
- Packet loss concealment
- Echo cancellation
- Voice compression codecs
  - Narrowband coding
  - Wideband coding
  - Transcoding tandem operation of codecs
  - Codecs and congestion
- Country-specific deviations
- Signal transmission characteristics
- Transmission loss planning
- Subscriber line interface circuit (*SLIC*)–CODEC interfaces and configurations
- Dual-tone multifrequency (DTMF) rejection as annoyance
- Quality-of-service (QoS) considerations
- GR-909 telephone interface diagnostics
- Voice quality monitoring and RTCP-XR
- Miscellaneous points of voice quality

20.3.1 End-to-End Delay Reduction

As per G.114 [ITU-T-G.114 (2003a)] recommendation, end-to-end delay between talkers to the listener has to be less than 150 ms. End-to-end delay is also known by several other names such as one-way delay, mean one-way delay, and half of round-trip delay. The TIA/EIA-116A document [TIA/EIA-116A (2006)] suggests containing delays to less than 100 ms. The literature [URL (Cisco-delay)] has provided some more detailed breakup on these delays. For inter-regional (international) calls, a delay of 300 ms is considered with an upper limit of 400 ms. Many adaptive jitter buffer algorithms use 400 ms as the default upper boundary. In the previous sections of the E-model, it is indicated that 177.3 ms will be talk turn time. Keeping delays below 177.3 ms prevents frequent double-talk situations.

In qualitative terms, an increase in delays will have several bad quality effects and will not allow the voice conversion with the proper level of emotion. The delayed responses are treated with an element of doubt and as unwilling decisions. More delay locks the talker, and other listeners have to interrupt for initiating the conversation.

Several parameters contribute to an increase in end-to-end delays, such as the algorithms, software implementation, processor architectures, processing power, interprocessor and interfaces communication, codecs and packetization, number of channels processing, other concurrent applications, end-to-end network conditions, physical interfaces, QoS mechanisms-fragmentation of large packets, jitter buffer designs, sampling clock precisions, as well as time-stamp resolution and accuracy. PSTN voice calls operate with minimum delays. VoIP calls typically take 60 to 80 ms more than PSTN calls in good implementations, packetization of 10 to 20 ms, and good network conditions. In VoIP, the effort would be to achieve as low a delay as possible to mimic a PSTN-based call. In practice, delay reduction achieved during the process of increasing competition, which provides better voice quality in comparison with the existing deployments. A limit on the amount of reduction will be reached based on the physical distance issue.

From codec selection, waveform-based or sample-based speech codecs such as G.711, G.726, and G.722 allow for lower delays to be achieved. In VoIP, packets based on small compressed frames will not use Internet bandwidth efficiently. A small packet size of 5, 10, or 20 ms is desired to maintain lower delays and balance on bandwidth utilization. The codec selection and several trade-offs are given in Chapters 3 and 11.

Incremental delay reductions are difficult to achieve. VoIP systems should cater to lower delays by design and while provisioning the service. Internet service providers and VoIP service providers have to ensure proper low delay and minimal IP impediments to the delivery of voice packets. While writing this chapter, intraregional VoIP calls achieve end-to-end delays in the order of 50 to 80 ms.

Echo will be the other main voice-degrading item coupled closely with delay. A increase in delay always creates more echo complaints. The design of the echo canceller gets more stringent with more round-trip delays, which depends on the TELR arrived at in the end-to-end voice call. Good phones and allowed loss settings will accept delays up to 25 ms [ITU-T-G.131 (2003)] without calling for echo cancellers. To arrive at proper numbers and arithmetic on this echo-delay decision boundary, refer to the mean one-way delay-TELR graph given in the G.131 recommendation and in Chapter 6 of this book. In a high-level summary, it is preferred to maintain a one-way delay of 50 to 100 ms.

Voice quality will be under the degradation zone of quality with delays exceeding 150 to 200 ms.

End-to-End Delay Calculation Example with G.729-20-ms Packetization.

End-to-end delays are represented in Fig. 20.4 considering the G.729 codec with 20-ms packetization as well as some of the main delay contributors. Voice entering the telephone SLIC–CODEC interface goes through a 1- to 2-ms delay. The G.729 codec processes on 10-ms frames (marked as F0, F1 ... F9); hence, samples of a 10-ms duration have to be collected before processing in the encoder. It is also called PCM frame delay. The G.729 encoder has a 5-ms look-ahead delay, which causes an equivalent delay of 15 ms to start processing for G.729. Each frame in the encoder takes a few ms for processing, which is represented as the "encoder-processing delay," and the compressed frames are marked as P0, P1 ... P8. In 20-ms RTP packetization, P0 + P1, P2 + P3 compressed payloads are used in creating the final RTP-based VoIP packet. About a 5-ms delay may occur, between the end of the compressed payload and the beginning of sending the packets on the IP network interface marked as "packetization + interface queuing delays." The 20-ms packets composed of two 10-ms payloads are sent on the physical network. The packets are transmitted to the destination via the network. For the local region, this transmission is expected to go through a delay of about 15 ms. In this situation, packet impediments are assumed to be creating lower than 20-ms jitter or no jitter. Jitter buffers typically keep 20- to 50-ms minimum packets even under no network impediment conditions. In Fig. 20.4, 20 ms is considered a minimum jitter buffer delay. At the decoder, jitter buffer output will go through the

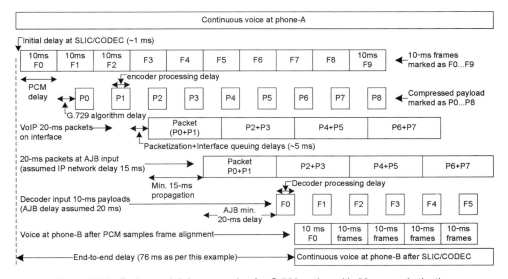

Figure 20.4. End-to-end delay examples for G.729 codec with 20-ms packetization.

decoder and will be played on the telephone interface, and this delay is accounted as 5 ms. The overall delay is marked as 76 ms in this example. For 10-ms based IP packets, this delay will be 66 ms, and 40-ms packets can take 96 ms as per this example diagram. The delays may vary based on the processor architecture and delay estimates at various stages of processing.

Conference Call Delays with an Example.   Voice conferencing can degrade voice quality because of transcoding and increased end-to-end delays. As illustrated in Fig. 20.5(a), in a conference call among users A, B, and C, user C is hosting the conference. Voice samples started from A or B should be routed through the conference combining bridge; in Fig. 20.5(a), the conference mixing is shown closer to user C. Hence, for a conversation between A and B, voice samples from A first go to the conference combining bridge present near C and then they reach B, which includes A-C and C-B delay. In many situations, the delay in conferencing is perceived as two times the long delay. In Fig. 20.5(b), the conference combining is occurring between two users A and B. In this mode, low delays are possible among all users. Overall, the location of the conference bridge and the user's locations are important. It is recommended to manage the conference bridge close to the maximum available users to reduce delay. The configuration indicated in Fig. 20.5(b) reduces delay, consumes less bandwidth on the IP backbone, and runs between inter-regions. The diagrams are shown for three users. These diagrams can be extended for more users. The users A and B are shown connected to the same gateway, but they can be connected on different VoIP gateways or VoIP user terminals that are located in the same region.

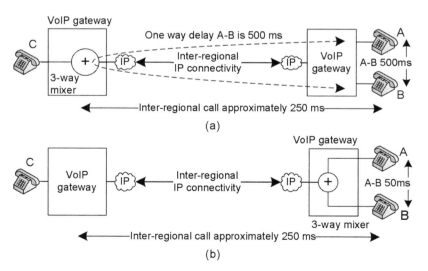

Figure 20.5. Three-way conference example. (a) Mixing at farthest point. (b) Mixing at a point close to maximum number of users.

20.3.2 Packet Flow Impediments in the VoIP System

Jitter buffers will put forth their best effort in playing the packets arrived at their input. The VoIP system (gateway) has to ensure proper packet flow between voice payloads and VoIP system physical interfaces. The physical interfaces are Ethernet, wireless local area network (WLAN), universal serial bus (USB), digital subscriber line (DSL), and so on. In general, users are getting VoIP systems with more processing power and IP network bandwidth. In practice, the applications and interfaces that demand processing and bandwidth are also growing with time. In many implementations, voice and fax chain modules will have dedicated resources. The processing of voice and fax packets is combined with several other applications on the host network processor. On host processing, some applications may block processing to several voice frame durations. This blocking will occur as random events developing IP impediments of packet bursts or packets drop. With several user applications and interfaces working together, it is essential to control voice packet delays inside the VoIP system. In simple terms, the packets from a physical interface to the adaptive jitter buffer (AJB) input must be guaranteed a minimum fixed delay of a few milliseconds in the receive path. The same analogy is applicable to the send path. The payloads created by voice processing should reach physical interfaces without any packet impediments of jitter and drop. A fixed delay in both directions will always occur. Fixed delays are less harmful than the impediments.

20.3.3 AJB with Utilization of Silence Zones

On the network, voice packets encounter variable transit delays caused by variable queue lengths in routers or congestion in traffic. AJB removes the jitter in the arrival times of the packets and delivers packets synchronized with the voice processing algorithms and the PCM interface clock. Jitter buffer algorithms keep the buffering delay as short as possible while optimally using all available packets delivered from the network. Adaptive jitter buffers are also made to cater to various codecs, packet sizes, as well as voice, fax, and fax pass-through modes. Many jitter buffer designs may be lacking the knowledge of speech and silence zones. To adjust the packets, jitter buffers either drop a packet or create a silence on an as-needed basis. This type of packet adjustment may occur in the middle of speech causing the degradation of voice quality.

By implementing detection of silence zones, matching the way PESQ and MOS measuring algorithms identify utterances can help in preserving or improving voice quality. The PESQ algorithm identifies utterances and any adjustments of speech in the middle of an utterance that appear as lower PESQ-MOS. Silence adjustments after end-of-valid utterances will not degrade the PESQ measure. By imposing these conditions, AJB implementation can preserve voice quality and cause PESQ-MOS to be the same even during AJB adjustments, and this type of operation also preserves voice quality in clock drift conditions.

20.3.4 Packet Loss Concealment

End-to-end packet impediments in the VoIP system degrade the perceived voice quality. After ensuring the possible best effort in packet delivery, and jitter buffer operations, the amount of improvement to packet impediments depends on the robustness of the speech codec's PLC algorithm. PLC is given in Chapter 5. Most low-bit-rate codecs have built-in packet loss concealment. The improvements will be fixed with most low-bit-rate codecs. For waveform-based codecs like G.711 and G.726, several PLC options are available. For low packet drop up to 5%, most higher end techniques perform to the same level of improvement. With higher packet drop deployments, a user will go for proprietary decoder-based schemes. Transmitter–receiver-based techniques can perform better for higher drops, but once again, the problem may get exaggerated because of the higher demand of bandwidth with transmitter–receiverd-based techniques. ITU-G.711 PLC given in Chapter 5 is found to be meeting the same performance as linear prediction (LP) and hybrid PLC techniques up to a 5% drop. In recent deployments, several service providers have been targeting a lower than 1% drop. This level is in agreement with the TIA/EIA-116A (2006) recommendations. Thus, the ITU-T G.711 PLC algorithms are performing to the required level without calling for proprietary implementations. For voice quality considerations, it is recommended to incorporate a good PLC scheme, but the effort should be in ensuring low end-to-end delay with no or minimal packet impediments.

20.3.5 Echo Cancellation

The details on Echo cancellation were discussed in Chapter 6. In this section, line echo cancellation in relation to the voice quality is discussed. Line echo is always present with analog telephones. In local PSTN-based systems, echo also is present. It is not perceived in the PSTN because of low end-to-end delays, losses in lines, and programmed padding losses at digital loop carriers (DLCs). In VoIP systems, end-to-end losses are also programmed, but end-to-end delays will be more than the PSTN and will vary with time. Increased delay calls for mandatory incorporation of an echo canceller in VoIP systems. Adaptive echo cancellers remove echo in two stages, as follows:

1. Linear part (also called ERLE, typical echo residue removal of 30 to 35 dB)
2. Nonlinear part to create extra required loss (typical removal of 12 to 24 dB)

When end-to-end delays are comparable with 50 ms, the first stage will be sufficient to remove the linear part of the echo. This arrangement of linear part cancellation will create better voice quality. The nonlinear stage helps in catering to some more rejection for higher end-to-end delays. G.131 provides the requirements on TELR as a function of the mean one-way transmission as given in Chapter 6. It shows that echo becomes more perceivable as delay increases. As an example, a one-way delay of 50 ms demands 40-dB total TELR rejections. A phone contributes to 10 dB loss, and echo canceller, send loudness rating (SLR), receive loudness rating (RLR), and padding losses in signal path have to take care of the remaining 30 dB. Most echo cancellers cancel up to 30 dB with linear part cancellation meeting the requirements for the linear part of the cancellation.

An echo canceller also takes care of many other conditions of comfortable background creation during echo residue shaping, double-talk protection, modem/fax tone detection, convergence monitoring, and generation of required parameters for voice quality monitoring. Many control plane operations along with nonlinear echo cancellation influence echo perceptions. Hence, reducing delay is one of the main goals to help with echo cancellation.

20.3.6 Voice Compression codecs

In the PSTN, voice goes through a cascade of G.711 (μ/A-law) codecs. On international calls, the number of cascade operations may increase. In voice communication between two VoIP gateways, a G.711 voice will go end-to-end causing the number of quantization units to be lower.

Narrowband Coding. VoIP service providers are migrating back to G.711-based solutions because of the availability of more Internet bandwidth. This migration allows VoIP voice quality to approach to PSTN quality. For better interoperability, and to cater to wider deployments, several compression codecs like G.729AB, G.723.1A, G.726, and the GIPS family of codecs will also be supported in VoIP systems.

Wideband Coding. Wideband voice is one of the main approaches in providing voice quality, which is better than the existing narrowband PTSN. PSTN and narrowband VoIP services use 300 to 3400 Hz. Wideband compression uses frequencies from 50 to 7000 Hz. VoIP provides a better experience of voice quality by going for wideband voice. By expanding the limit of bandwidth, significant improvement in intelligibility and quality is achieved. Often fricative sounds such as "s" and "f," which are very hard to distinguish in telephony band situations, sound very natural in wideband speech. Low-delay, wideband speech creates a natural conversation experience to the users.

While writing this chapter, many VoIP systems in the market were interoperating with the G.722 wideband codec. Depending on the deployment requirements, many other G.722 and G.729 wideband codecs may be supported. VoIP signaling and RTP supports the wideband codec family. Wideband-capable telephones are limited in availability at present. It is assumed that wideband acoustic interfaces will be available because of the increasing demand for improved VoIP voice beyond toll quality.

20.3.7 Transcoding and Conference Operation with Codecs

On international (inter-regional) PSTN calls, the number of G.711 cascades could reach four to seven [ITU-T-G.173 (1993)]. In voice communication between two VoIP systems, G.711 voice will move directly between two end systems causing the number of quantization units to be lower. While making calls from the PSTN to mobile phones, it may go through multiple cascaded stages of a global system for mobile communications (GSM) codecs and G.711. In the case of VoIP, voice packets can be sent directly using cell-phones-compatible codecs such as the adaptive multirate (AMR) codecs. This process can reduce the number of quantization stages even for international calls and to non-VoIP systems. It will help in improving overall voice quality. During a conference call, combining occurs on linear samples. Hence, additional transcoding on combined output is mandatory at the conference bridge. In situations in which some users operate in low-compression like G.729AB and other users in G.711, the voice quality degrades because of multiple transcoding operations. It is recommended to use conference mixing at a central location, using low-compression codecs like G.711. The location of a conference bridge and the delays involved are illustrated in Fig. 20.5(a) and 20.5(b). In summary, a VoIP call made as end-to-end with minimal codecs involved in the call will give better quality.

The effect of transcoding or tandeming of speech codecs in a VoIP end-to-end system has an impact on speech quality. The equivalent equipment impairment factor $I_{e\text{-eff}}$ of the entire system changes according to the type of codecs used in tandeming. The I_e factor increases on an additive scale that is equal to the cumulative sum of the equipment impairment factor values of individual codecs under no packet loss conditions [ITU-T-G.113 (2003)]. For example, tandeming of G.726 ($I_e = 7$) with G.729 ($I_e = 10$) results in effective $I_{e\text{-eff}}$ of 17; similarly, tandeming of G.729 ($I_e = 10$) with G.723 ($I_e = 15$) results in I_e of 25. More details on effective $I_{e\text{-eff}}$ values for codecs under tandeming conditions can be found in P.833 [ITU-T-P.833 (2001)].

The effective impairment factor values in tandeming mode for wideband codecs also change on the additive scale. For the wideband mode, the equipment impairment factor $I_{e,wb}$ of the codecs should be considered. The value of R_0 should be taken as 129 [ITU-T-G.107 (2006)]. The tandeming of two G.722 ($I_{e,wb} = 13$ from Chapter 3) codecs results in an effective $I_{e,wb}$ of 26. Hence, when are in tandem operation in the VoIP system, the effective impairment factor increases to cause reduction in the overall R-factor and voice quality.

Cordless Phones and Advantage Factor. A cordless phone has an analog interface with the base station. The base station communicates with the cordless handset using ADPCM, which is also known as G.726 compression. As explained in Chapter 3, G.726 supports multiple rates. In cordless-handset-to-base-station communication, the G.726 codec at 32 kbps is used. As per the E-model, mobile phones will give an advantage factor of 5. The cordless phone G.726 has an equipment impairment factor I_e of "7" for 32 kbps [ITU-T-G.113 (2007)]. As a first-level approximation, the R-factor reduces by 2 (5 −7 = −2) instead of adding "5" as an advantage factor. The aspects of cordless phones and I_e degradations are not seriously considered in the literature. In general, cordless phones do reduce voice quality, and this reduction in quality is approximately compensated with advantage factor "A" in voice quality estimation. It will also work like a tandem operation of the G.726 codec with other codecs used in the VoIP system. In practice, cordless phone G.726 coding can limit voice quality.

20.3.8 Codecs and Congestion

While arriving at trade-offs on packet loss and quality, waveform-based lower compression codecs like G.711 can give better quality than low-bit-rate codecs like G.729AB. In the situations of higher packet loss, forward error correction (FEC)/redundancy techniques are more helpful than decoder-based PLC. Redundancy demands higher bandwidth. In bandwidth-limited systems, redundancy drops packets even more. In such situations, it is worth renegotiating for low-bit-rate codecs instead of continuing in G.711. Codec G.729AB with FEC/redundancy may take lower bandwidth than G.711. Bandwidth requirements for different codecs are given in Chapter 11.

20.3.9 Country-Specific Deviations

The VoIP adapter is like a PSTN central office (CO) to the phone. The VoIP adapter has to emulate the characteristics of the PSTN CO. The front-end SLIC of the VoIP gateway is programmed for required voltages, line conditions ringer equivalence number (REN) drive, impedances, gain/losses, and diagnostic features. The right combination of phones, central offices, and transmission lines can provide better quality. Country-specific deviations and catering to the requirements of multiple countries are given in Chapter 17.

20.3.10 Signal Transmission Characteristics

In VoIP adapters, a foreign exchange subscriber (FXS) interface is used for connecting the telephone. The interface has to meet certain switching and transmission characteristics for better voice quality. Various specifications of transmission requirements are listed in the TR-57 document for North America. Similar requirements will be available in local PSTN standards. Even though the TR-57 standard does not talk about VoIP, meeting its specifications for the FXS interface is expected to result in better speech quality under better network conditions matching closely with the PSTN. TR-57 measurements are for narrowband voice and are classified into two categories, namely signaling or switching characteristics and transmission characteristics.

In a VoIP system, delays encountered in establishing a telephone call are more than that in a PSTN-based system. These delays vary widely based on the distance, network conditions, and the supported VoIP infrastructure. Some signaling events may take much longer than PTSN-based call signaling. Many signaling and call establishing timings in VoIP systems are more delayed than TR-57 switching requirements. These signaling delays may not result as direct degradations on voice quality, but improving on these can create better natural interactions matching closely with the PSTN service.

The documents IEEE STD-743 (1995) and TIA/EIA-470C (2003) provide additional specifications and tests to take care of wideband characterization. It is expected that while perfecting for wideband voice, several new specifications will be added for end-to-end wideband voice quality.

20.3.11   Transmission Loss Planning

The goal of establishing a loss plan for voice communication systems comes from the desire to have the received speech loudness at a comfortable listening level. The received loudness will depend on the speech level of the talker, transmit and receive efficiencies of the voice terminals, as well as loss in the system and intervening network. It is generally accepted that a connection with an overall loudness rating of 10 dB will provide a high degree of satisfaction for most of users. Some more details on the losses and loudness rating are given in Chapter 6.

20.3.12   SLIC–CODEC Interface Configurations

SLIC–CODEC devices communicate samples to the processor on a PCM interface. This interface operates at 8-kHz frame synchronization and accommodates slots based on the requirements. Lower channel systems use a 24- or 32-channel PCM interface. The PCM interface will use A-law, μ-law, or 16-bit linear samples for communication between SLIC–CODEC devices and the processor. Using 16-bit samples minimizes the quantization distortion and helps in getting a better linear part of echo cancellation. In wideband, linear samples support is used as a requirement to get better voice quality.

20.3.13   DTMF Rejection as Annoyance

Various operations of DTMF, tone generation, and tone detection are discussed in Chapter 7. PSTN users are clearly aware that dialed digits are audible. In VoIP because of in-band and out of band combinations, users are annoyed with dialed digits on an established voice call. DTMF rejection and issues are discussed in Chapter 7. The residues and DTMF ticks create undesired disturbances. It is also essential to minimize false digits. A false digit can create wrong operations in a call and disturb in-band voice.

20.3.14 QoS Considerations

Quality of service is fundamental to VoIP network management, branding the voice quality performance at end-devices and ensuring the expected service quality. These considerations are identified with layer-3 and layer-2 techniques. The bottom-line action for upstream QoS is to deliver every voice/fax packet without dropping and with a minimum delay and jitter of a few milliseconds (with a worst case of 5 to 10 ms). The techniques for this QoS will vary with the total system, interfaces, and the provisioned applications. Some issues in the systems will include media packets staying in the packets queue for a long time and encountering varying queue delays (jitter).

Downstream is another most troubling aspect of VoIP systems. From the Internet service provider, the supported downstream bandwidth is usually much higher than the upstream. Hence, many designers ignore upstream voice packet issues. The demand of data bandwidth on the downstream direction is also much higher from users. Several download operations, data, video, and other media will occur simultaneously in downstream. Several members in residence may share the same available bandwidth. In practice, the high-throughput downstream bandwidth is also drained to the congestion point. During congestion, data will be recovered from a Transmission Control Protocol (TCP) based operation, but User Datagram Protocol (UDP) based voice and fax will encounter drops. In downstream, packet impediments will mainly occur at the VoIP or Internet interfaces like DSL and Ethernet.

Several CO devices of Internet service providers (ISPs) may have upstream QoS. Upstream QoS at the ISP CO will work similarly to the downstream QoS of the end-user system. In the absence of any QoS mechanisms from the service provider, it is essential to control the downstream bandwidth requirements from the end-user router and VoIP system. Many statistics from RTCP and RTCP-XR parameters can indicate a packet drop. The reported statistics can be used to control low-priority applications that demands downstream data. In summary, downstream QoS is not a direct approach like upstream QoS. It is essential to incorporate some proprietary techniques that ensure no packet impediments occur from the growing bandwidth requirements of the end user. With increasing awareness of using parameters from RTCP-XR packet details, several simple techniques will be adapted to help in controlling the media packet flow for reducing packet impediments.

20.3.15 GR-909 Telephone Interface Diagnostics

For several decades, PSTN service providers perfected many self-diagnostic and terminal interface monitoring tests through electronic central offices and DLCs. These tests require a telephone service provider to check remotely the first-level health of the telephone interface without going to the customer premises. The important diagnostics specific to telephone interface are as follows:

- Line voltage, DC voltages, and battery voltages
- Receiver off-hook/on-hook
- Ring and ringer equivalence
- Resistive faults
- Loop-back tests

Many of these features are given in GR-909 and SLIC–CODEC manufacturer datasheets. Some proprietary monitoring features have also been added to suit customer requirements. In VoIP, with the incorporation of the PSTN type of diagnostics, it is possible to diagnose the problems remotely through network interfaces and to provide possible updates or solutions. Many service providers are incorporating their own management interface to conduct GR-909 tests on VoIP solutions. Incorporating GR-909 tests may not improve voice quality, but it helps in monitoring the interfaces for quality. In the next sections, voice quality monitoring and RTCP-XR packets are discussed to help in monitoring the voice quality based on signal characteristics and packet flow. Several VoIP software and hardware solutions provide these operations [URL (PIQUA)].

20.3.16   Miscellaneous Aspects of Voice Quality

Some other important items that contribute to the voice quality are given as summary points below:

- Several phone call impediments are given in the manual calls of VoIP voice testing in Chapter 13.
- Ensuring clock precision and making systems perform under wide environmental conditions is essential. Laboratory tests on crystals and clock oscillators may show very less PPM. It is required to ensure better control on PPM under practically possible worst-case conditions.
- PSTN telephone interfaces cater up to required REN and a maximum up to 5 REN. PSTN systems also drive long distance from the DLC. VoIP system interfaces are selected for lower REN to save on the cost of building the customer premises systems, which can degrade voice quality with several parallel phones. This degradation has to be considered while building the total system.
- Several switching and call feature tones have to match closely with local PSTN systems. This matching will improve call interactions and continue to create a similar experience as with PSTN calls.
- Voice clippings have to be contained to less than 0.2% to 0.5% of the active speech [ITU-T-G.116 (1999)]. These clippings can happen with VAD/CNG and in transition between speech and silence.

- Systems that can support for future upgrades of wideband voice and other applications.

20.4 VoIP VOICE QUALITY SUMMARY

VoIP can closely match PSTN quality when taking care of the several improvements listed in the previous sections. A short critical items list is given below:

- Use of G.711 compression
- Low end-to-end delay with no or negligible packet drops
- Improved echo cancellers
- Front-end interfaces, and end-to-end signal transmission meeting, local country requirements

VoIP will achieve an extra benefit because of

- Short loops at the user telephone.
- Global portable numbers—able to keep the same number globally and able to interface directly with digital phones, Bluetooth, digital enhanced cordless telecommunications (DECT), and wireless fidelity (WiFi) phones.
- Continuous voice quality monitoring.
- VoIP G.711 calls can deliver much higher quality than G.726-based overloaded PSTN channels with DCME.

In addition to the above benefits of VoIP, VoIP quality can be made to exceed PSTN quality because of

- Use of wideband codecs to create more natural and comfortable conversation
- Automatic level control to deliver required signal levels
- Elimination of the background disturbances in voice algorithms
- The benefit of short loop lengths at the gateway
- Ability to minimize number of transcoding stages while communicating with different communication links

20.5 VOICE QUALITY MONITORING AND RTCP-XR

On front-end operations, telephone interfaces are monitored using GR-909 diagnostic methods. VQmon is mainly used for monitoring the voice quality from signal and packet transmissions. VQmon, which was popularized by Telchemy, is voice quality monitoring based on the E-model algorithm discussed in previous sections. VQmon and the RTCP-XR framework provide a set of metrics for VoIP performance monitoring and diagnosis. They are used in addition to GR-909 and other proprietary diagnostic methods. It supports both real-time monitoring and postanalysis.

VQmon mainly derives parameters from packet transmission characteristics that include packet impediments on the network and end-to-end delays jitter buffer dynamics as well as signal transmission characteristics that include signal level, noise level, gain, echo rejections, R-factor, and MOS derived from the R-model. The VQmon parameter's role in the RTCP-XR [Friedman et al. (2003)] packet is shown in Fig. 20.6. These parameters are typically updated once every 256 packets. Use of these parameters and applying feedback for improving the voice quality helps the VoIP system to deliver the highest quality under severe conditions.

RTCP-XR sends about 20 parameters categorized as packet loss, discarded metrics, delay metrics, signal-related metrics, call quality or transmission quality metrics, configuration metrics, and jitter buffer parameters. Complete details on these parameters are in RFC3611 [Friedman et al. (2003)]. A short description of parameters is provided here in Fig. 20.6.

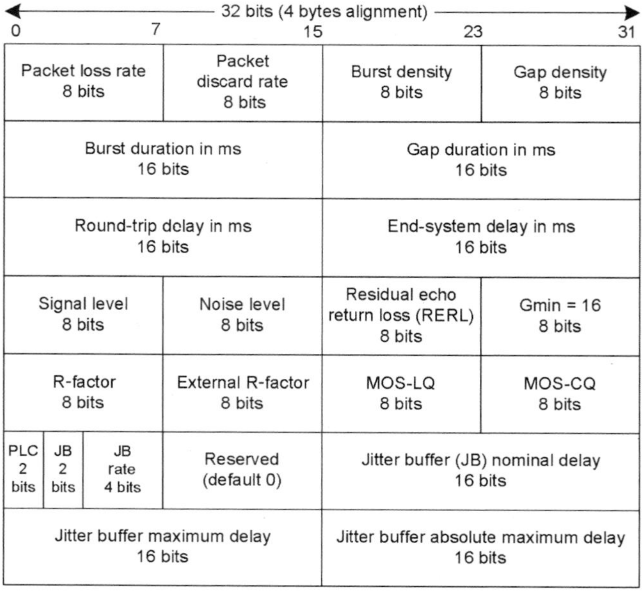

0	7	15	23	31
Packet loss rate 8 bits	Packet discard rate 8 bits	Burst density 8 bits	Gap density 8 bits
Burst duration in ms 16 bits		Gap duration in ms 16 bits	
Round-trip delay in ms 16 bits		End-system delay in ms 16 bits	
Signal level 8 bits	Noise level 8 bits	Residual echo return loss (RERL) 8 bits	Gmin = 16 8 bits
R-factor 8 bits	External R-factor 8 bits	MOS-LQ 8 bits	MOS-CQ 8 bits
PLC 2 bits | JB 2 bits | JB rate 4 bits	Reserved (default 0)	Jitter buffer (JB) nominal delay 16 bits	
Jitter buffer maximum delay 16 bits		Jitter buffer absolute maximum delay 16 bits	

32 bits (4 bytes alignment)

Figure 20.6. RTCP-XR packet with 22 parameters in 28-byte payload.

Packet Loss and Discard Metrics

Loss rate: The fraction of RTP data packets from the source lost since the beginning of reception.

Discard rate: The fraction of RTP data packets from the source that have been discarded since the beginning of reception, because of late or early arrival, under-run, or overflow at the receiving jitter buffer.

Burst Metrics

Burst density: The fraction of RTP data packets within burst periods since the beginning of reception that were either lost or discarded.

Gap density: The fraction of RTP data packets within inter-burst gaps from the beginning of reception that were either lost or discarded.

Burst duration: The mean duration expressed in milliseconds of the burst periods that have occurred since the beginning of reception.

Gap duration: The mean duration expressed in milliseconds of the gap periods that have occurred since the beginning of reception.

Delay Metrics

Round-trip delay: The most recently calculated round-trip time between RTP interfaces, which is expressed in milliseconds. If RTCP is used, then the reported delay value is the time of receipt of the most recent RTCP packet from the source SSRC, minus the last sender report (LSR) time reported in its SR, minus the delay since last SR (DLSR) reported in its SR. A nonzero LSR value is required to calculate round-trip delay.

End-system delay: Sum of the total sample accumulation and encoding delay associated with the sending direction, decoding, and jitter buffer playout buffer delay associated with the receiving direction. It is expressed in milliseconds.

Signal Metrics

Signal level: This level is measured only for packets containing speech energy. It is represented as a signed integer in a two's complement format. A value of 127 indicates that this parameter is unavailable.

$$\text{Signal level} = 10 \, \text{Log10} \, (\text{RMS talk-spurt power in mW}).$$

Noise level: The noise level is defined for the silent period. A value of 127 indicates that this parameter is unavailable. Noise level = 10 Log10 (RMS silence power in mW).

Residual echo return loss: The residual echo return loss value may be measured directly by the VoIP end system's echo canceller or may be estimated by adding the echo return loss (ERL) and echo return loss enhancement (ERLE) provided by the echo canceller. A value of 127 represents that it is undefined.

Call Quality Metrics

R-Factor: The R-factor is a voice quality metric that considers only equipment impairments. A value of 127 indicates that this parameter is unavailable.

Ext R-factor: This metric is defined as including the effects of delay and other parameters. A value of 127 indicates that this parameter is unavailable.

MOS-LQ: This mean opinion score is defined by not including the effects of delay. It is expressed in the range 10 to 50, which corresponds to 10 times scaled MOS. A value of 127 indicates that this parameter is unavailable.

MOS-CQ: This mean opinion score is defined by including the effects of delay. It is expressed in the range of 10 to 50, which corresponds to 10 times MOS. A value of 127 indicates that this parameter is unavailable.

Jitter Buffer Parameters

Nominal delay: Current nominal jitter buffer delay in milliseconds.

Maximum delay: Current maximum jitter buffer delay in milliseconds (in this case, the adaptive jitter buffer is used).

Absolute maximum delay: This delay is the absolute maximum in milliseconds that the adaptive jitter buffer can reach under worst-case conditions. Its value must be set to JB maximum for fixed jitter buffer implementations.

Configuration Parameters

Gmin: The recommended value of 16, which corresponds to a burst/gap decision threshold count of number of consecutive received packets.

Receiver configuration byte: This byte consists of the following information:

Packet loss concealment (PLC). This parameter represents the type of PLC algorithm used. It is represented by 2 bits; Standard (11) / enhanced (10) / disabled (01) / unspecified (00).

Jitter buffer adaptive (JBA). This parameter represents the jitter buffer type represented by 2 bits; 11—Adaptive, 10—Nonadaptive, 01—reserved, 00—Unknown.

Jitter buffer rate (JB rate). This parameter represents the implementation-specific adjustment rate of a jitter buffer in adaptive mode.

RTCP-XR messages containing key call-quality-related metrics are exchanged periodically between IP phones and gateways. This exchange lets a probe or analyzer monitor these metrics midstream to support problem resolution, or be retrieved from a gateway using a simple network management protocol (SNMP). RTCP-XR in a deployment is also marked in Fig. 20.1.

20.6 SUMMARY AND DISCUSSIONS

VoIP users are accustomed to the high quality offered by traditional PSTN service. In many situations, VoIP does not offer consistent voice quality. Voice quality in VoIP networks widely varies as a result of several factors. This chapter is presented on voice quality evaluation and how to avoid service degradations in a VoIP network and how to deliver voice quality that is comparable with, or even exceeds, PSTN levels. In this chapter, voice quality influencing parameters and their improvements are presented. VoIP systems have to cater to multiple countries and multiple customer deployments. QoS mechanisms and monitoring are also required to be incorporated. The goal in narrowband (300 to 3400 Hz) voice quality is to meet PSTN quality. In the wideband case, the perceptions have to exceed PSTN reference quality. Several options for measuring voice quality are available. The passive E-model can be defined as a function of room noise, end equipment impairments, echo, delay, codec, and packet loss. PESQ is a function of coding distortions, errors, noise, filtering, and variable delay. P.563 is a function of noise, filtering, and variable delay as well as distortions caused by channel errors and speech coders. The first level of quality is to ensure by design. Measurements and monitoring will also help in giving feedback. In addition to these several perception-based tests, user experiences in different conditions are also important.

21

VoIP VOICE FAQS

1. What are the important voice band frequencies?
   Narrowband speech is important between 300 and 3400 Hz. The bandwidth is 3100 Hz. It is also referred to as 3k (or 3K) flat band, which means 3000-Hz bandwidth. Wideband is gaining popularity through VoIP and multimedia terminals. Wideband voice frequencies are from 100 to 7000 Hz. Sometimes wideband is also referred to in the range 50 to 7000 Hz.

2. What are dBm and other related dB units?
   The ratio of two quantities in logarithmic scale is represented in deciBel (dB) units. In power representation, when the reference power is 1 milli-Watt, the dB units are denoted as dBm, where "m" represents 1 milliWatt of power. Speech power levels are expressed in dBm units. Noise power is low and expressed using picoWatt units represented as dBrn; 0 dBm = 90 dBrn (rn is referenced for noise power of 1 picoWatt).

3. What are the speech signal levels?
   The undistorted sine wave amplitude in μ-law is 3.17 dBm, which gives quantized sample values in the range of ±8159. The full-scale square wave gives 6.17 dBm. In A-law, sine wave power is 3.14 dBm, and samples are in the range of ±4096. Typical speech levels are at −16 dBm, and ideal channel noise is lower than −68 dBm (22 dBrn). When speech levels are approaching −40 dBm, speech is not perceivable clearly.

VoIP Voice and Fax Signal Processing, by Sivannarayana Nagireddi
Copyright © 2008 by John Wiley & Sons, Inc.

4. What are 3K flat, dBrn, dBrnC, dBp, dBmp and pWp?

   Signals are measured with milli-Watt (mW) reference and noise is measured with pico-Watt reference (pW). The noise measurement without a filter is called 3K flat (approximately 300 to 3400 Hz) relates to 0 dBm (1 mW) = 90 dBrn (10^9 pW). The suffix "rn" in dBrn denotes 1 pW reference noise (rn) power. The noise measured after passing through C-message weightage (band pass filter) is represented as dBrnC (or dBrnc) where 0 dBm = 88 dBrnC. The noise measured after passing through psophometric weightage (filter gives more importance to low frequency compared with the C-message) is represented as dBp, and 0 dBm = 87.5 dBp. The psophometric weightage noise is also represented as dBmp with mW reference (dBp is with pW reference), 0 dBm = −2.5 dBmp. This dBmp representation indicates psophometric weightage band-pass filter reduces the 3K flat noise power by 2.5 dB in the measurement. Another common representation in pico-Watt units is pWp related as 1 pW = 0.562 pWp. The factor $10 \log(0.562)$ is 2.5 dB, which is consistent with dBp and dBmp representation. Some literature is using dBrnp for psophometric noise–avoid using dBrnp representation as the usage is not consistent. The C-message is applicable to North America, and the psophometric is applicable to Europe.

   Summary: 1 mW = 0 dBm = −2.5 dBmp = 90 dBrn = 88 dBrnC = 87.5 dBp

5. What are the ideal channel noise levels?

   In simple interpretation, ideal channel noise is the noise that occurs when phones are in mute. The levels are maintained to lower than 20 dBrnC (lower power than −68 dBm). Practically, they are maintained to 10 to 15 dBrnC.

6. What is G.711 and its relation to the PSTN?

   G.711 is the pulse code modulation (PCM) narrowband voice compression. In a simplified way, it is also mentioned as linear to logarithmic compression. It is the simplest compression for a higher quality of narrowband speech. Narrowband speech is sampled at 8 kHz. Each sample of 12 or 13 bits and sign bit (presented in 16-bit format to G.711 compression) is compressed to 8 bits, producing a total basic bit rate of 64 kbps. Two compression methods, A- and μ-law, are used in all public switched telephone network (PSTN) digital transmission. A-law is used in Europe and Asia. μ-law is popular in North America and Japan. Logarithmic mapping changes slightly between A-law and μ-law.

7. How does a basic PSTN call work?

   The communication among digital loop carrier (DLC), central offices (COs), and multiple COs is a basic PSTN (G.711). A call goes from the telephone to the analog lines ⇔ DLC/CO ⇔ CO switches ⇔ CO/DLC ⇔ the analog lines to the telephone. All digital links are of synchronized time division multiplexing (TDM), and no sample packets are involved.

8. How are basic VoIP-to-VoIP and VoIP-to-PSTN calls made?

   VoIP boxes work as PSTN DLC and CO. A VoIP-to-VoIP call between two phones goes from the telephone ⇔ the analog lines ⇔ VoIP box ⇔ (IP packets) ⇔ VoIP Box ⇔ the analog lines to the telephone. All digital communication is IP packets here.

   A VoIP-to-PSTN call between two phones goes from the telephone ⇔ the analog lines ⇔ VoIP box ⇔ (IP packets) ⇔ VoIP–PSTN gateway box (synchronized TDM) ⇔ PSTN CO and DLC ⇔ the analog lines to the telephone. It is a combination of analog signal, Internet protocol (IP) packets, and synchronized TDM.

9. What are the main voice modules in VoIP?

   - Narrowband compression codecs of G.711, G.729AB, G.723.1A, G.726, Internet low bit rate codec (iLBC), etc.
   - Wideband codecs of G.722, G.729.1, G.722.2, etc.
   - Line echo canceller (G.168).
   - Acoustic echo cancellers (earlier G.167, now the P.340 series recommendations).
   - Dual-tone multifrequency (DTMF) detection and generation (Q.24) and local country-specific standards.
   - Voice activity detection (VAD)/comfort noise generation (CNG) (ITU-T-G.711 Appendix-II) or 1-byte VAD/CNG in G.711.
   - G.711 packet loss concealment (PLC) (ITU-T-G.711 Appendix-I).
   - Automatic gain control (AGC) with P.56, G.136, and G.169.
   - On-hook and off-hook caller ID, frequency shift keying (FSK) generation and detection (ETSI and Telcordia and varies with country).
   - On-hook and off-hook DTMF based caller ID generation and detection (specifications vary with country).
   - Tone generation (E.180, E.185, and local country-based PSTN standards).
   - Call progress tone (specifications vary with country), fax, and modem tone detection.

10. What are the modules in VoIP packetization?

    Real-time transport protocol (RTP)/RTP Control Protocol (RTCP)–(RFC3550/3551), adaptive jitter buffer (AJB) or fixed jitter buffer (FJB), and RTCP-extended reports (RTCP-XR). Usually, the User Datagram Protocol (UDP) and Internet Protocol (IP) are treated as part of the network processing.

11. What are the VoIP signaling modules?

    Session Initiation Protocol (SIP) (RFC3261), H.323 (H.225/Q.931, RAS, and H.245), Megaco (H.248), media gateway control protocol (MGCP–RFC 3435), and their variants. H.323 and MGCP were used in early adaption of VoIP deployments. While preparing this book, SIP and extensions were considered in new deployments.

12. Is there any difference in the voice chains for H.323, MGCP, and SIP?
Not significant. Voice modules on the processor will remain the same, and they need to perform similar functionality. There will be deviations in call establishing on the VoIP signaling and supported VoIP infrastructure.

13. What are the supporting main VoIP and networking modules?
Networking stack is used to create UDP/IP, transmission control protocol (TCP)/IP packets, IP Quality of Service (IPQoS), simple traversal of UDP through NATs (STUN)/transport layer security (TLS) for SIP application layer gateways (ALGs), security functions, drivers for telephone interfaces, voice quality monitoring, GR-909 port diagnostics and remote monitoring, and so on.

14. What is simple traversal of UDP through NATs (STUN)?
The STUN protocol is given in RFC3489. The STUN server allows clients to find out their public address, the type of network address translators (NATs) they are behind, and the Internet side port associated by the NAT with a particular local port. This information is used to set up a VoIP call to a VoIP service provider hosted outside of the local network.

15. What is an ALG?
ALG is the application-level gateway that accepts a connection on one network interface and establishes the cognate connection on another network interface. ALG does not route the packets; they look more deeply into packet headers than the packet firewalls.

16. What is the relevance of G.711 to PSTN and VoIP?
Most PSTN systems use digital switching. In this situation, analog signals of speech are converted to G.711 at the closest DLC or CO. Between multiple DLCs and COs, the same G.711 digital transmission is used. The same G.711 is also used as one of the compression schemes in VoIP creating a possibility for matching VoIP voice quality to PSTN.

17. When G.711 is the best quality, why consider other compression codecs like G.729A?
VoIP adds a lot of headers to basic G.711 payload. On VoIP, the same PSTN 64 kbps of G.711 will use about 126.4 kbps on an Ethernet interface. In bandwidth-limited Internet service, it is required to consider higher compression codecs like G.729A.

18. What are waveform codecs?
Waveform-based codecs do not depend on vocal tract models. A waveform codec can operate sample by sample; hence, they can also be called sample-based codecs. They can work for voice, tones, or any waveform in the voice band. Waveform compression is suitable for sending tones and fax/modem modulations. PCM G.711 and adaptive differential PCM (ADPCM) G.726 are waveform codecs.

19. What is in-band signaling?

    In-band means sending basic call progress and signaling tones in the voice band (300 to 3400 Hz). The basic tones dial, busy, ring back, and so on are in the band 330 to 680 Hz. DTMF (dial digit) tones are from 697 to 1633 Hz. Fax and modem tones are in the band 1100 to 2100 Hz. These tones fall in the voice frequency band.

20. What is out-of band signaling?

    In-place of in-band tones and digits, digit packets, and tone parameters are sent on the IP network. On the IP network, these are message packets, not in-band speech packets. At a destination, based on the event type and parameters received on the IP packet, the suitable in-band tones are regenerated in the proper time slot.

21. Why use out-of-band signaling?

    Waveform codecs allow in-band tones without significant distortion. Some in-band tones create a disturbance to the destination user and can disturb all the participants in the (conference) call. In-band tones passing through high-compression codecs (G.729AB /G.723.1A) disturb the tone characteristics, and hence, the destination may identify in-band signaling with errors. In actual practice, based on the end-to-end capabilities and compression, out-of-band mode is enabled without end-user intervention.

22. Why are the G.729 family of codecs popularly listed in VoIP systems?

    The G.729 family of codecs have been widely adapted worldwide after G.711. The main reasons are wider choice of compatibility/interoperability, higher compression while maintaining acceptable quality, lower complexity, and operate on smaller frames. The G.729 family has wide-band speech codecs like G.729.1, which is backward compatibility to narrowband.

23. Why do we see a big list of codecs in VoIP support—why not one high-compression codec like G.729AB?

    To interoperate with multiple VoIP systems and deployments, many codecs are supported. Each codec has its own distinct advantages. G.726 is scalable in compression from 16 to 40 kbps. G.726 at 16 kbps is not as good as G.729A in quality. G.723.1 provides 6.3 and 5.3 kbps (better compression than 8 kbps in G.729A). It is of a 30-ms frame, and quality is lower than G.729A. G.728 works on very small frames, quality is better, but it takes a lot of processing. G.728 was entered late in to the list. iLBC works better under more packet drop of >5%. In most deployments, the service provider guarantees that end-to-end packet drops are below 1%. iLBC entered late, and G.729AB is already adapted in several deployments.

24. Why are there so many variants in the G.729 family?

    G.729 gained popularity for 8 kbps. G.729A was derived for lower complexity, and it inter-ops with G.729 in normal and VAD enabled mode.

G.729A has very little mean opinion scare (MOS) degradation. G.729AB has a built-in suffix "B" as VAD. Internet bandwidth is growing. G.729E is derived to extract more quality by increasing more bits. It is used at 11.8 kbps. In the wideband codec family, G.729.1 is used. It creates part of the payload that is compatible with G.729 and G.729A. This compatibility helps both wideband and narrowband to inter-op better without calling for duplicate codecs and processing. Most codecs have a VAD and a PLC suffix.

25. Can the G.711 and G.726 codecs support built-in VAD/CNG and PLC? G.711 VAD (ITU-T-G.711 Appendix–II) and ITU-T-G.711 Appendix–I PLC are external, and they can be used with G.711 and G.726 codecs in VoIP.

26. What is G.729E? G.729E is a narrowband codec that works better for music and background. G.729E gives improvement of MOS in narrowband mode. It is not backward compatible with other G.729 codecs. G.729E operates at 11.8 kbps instead of at 8 kbps. It is considered for voice along with video applications, but it is not widely used in VoIP terminals.

27. What are G.723.1, G.723.1A, G.729A, and G.729AB? G.723.1 is the main codec that operates at 6.3 and 5.3 kbps, and suffix A in G.723.1 A indicates the presence of VAD/CNG. In the case of G.729AB, suffix A indicates low complexity and B is for VAD/CNG. Both G.729AB and G.723.1A have built-in VAD/CNG and PLC.

28. What are the options for the speech codec? Based on the deployment, codec selection is decided on multiple factors. Some of the parameters are listed below:
    - Deployment inter-op support with other systems
    - Quality expectations and acceptance in market
    - Number of channels and available bandwidth as well as processing for voice service
    - Network characteristics and the usual network impediments
    - VoIP-to-PSTN gateway support
    - Patents and royalties
    - Future expansion plans for wideband codecs
    - Market, competition, and differentiated quality requirements
    - Amount of voice quality degradation for various accepted network conditions

29. What is the MOS of the voice call and codecs? MOS is the mean opinion score. It is expressed on a scale of 1 to 5. A MOS of 5 is the best possible quality, and 1 is poor quality. The PSTN will send voice on DLC from local distribution points. It will use G.711 (PCM–64 kbps). The MOS for G.711 is 4.3 to 4.4. Under good network

conditions, VoIP can also deliver PSTN quality. The MOS of toll quality voice is 4. MOS is measured on a subjective scale. MOS can above estimated using measurements. A Perceptual Evaluation of Speech Quality (PESQ)-listening MOS is 3.8 for G.729A and 4.32 for G.711.

30. Why is there a difference in reported MOS for the same codec by different instruments?

   All codec implementations must meet the requirements of ITU recommendations and other standards.

   Instruments provide different measurements like Perceptual Speech Quality measure (PSQM), PSQM+, perceptual analysis measurement system (PAMS), PESQ, PESQ-LQ, and rating (R)-factor MOS for speech quality. Suffix LQ is the lineming Quality. Some of these objective measurements when mapped to the subjective do not have the same correlation. The measurements are found to be varying with the input speech characteristics—like male/female, language, duration of speech, distribution of utterances, and silence zones. As an option, MOS is also supplemented with the details of speech files used for additional information.

31. What is algorithmic delay for different codecs?

   Algorithmic delay is related to the speech codec used. Total algorithmic delay is look-ahead delay plus frame size. For G.729A, frame delay is 10 ms and look-ahead delay is 5 ms. In G.723.1, frame delay is 30 ms and look-ahead delay is 7.5 ms. Look ahead is the requirement of future samples to operate on the current frame. To continue with compression, part of the previous and present samples create look-ahead delay.

32. What is the packet loss concealment (PLC) scheme?

   PLC creates synthetic voice during a missing packet (packet drops). PLC takes care of voice recovery for 20 ms and to some extent up to 50 ms. PLC is an integrated operation in several low-compression codecs such as G.729AB and G.723.1A. Waveform-based codecs such as G.711 and G.726 use external PLC. The popular PLC for G.711 and G.726 is ITU-T-G.711 Appendix–I. PLC effectiveness varies with the codec used and the PLC algorithm. PLC introduces additional delay to create a smooth recovery at packet loss transition events.

33. What will happen in the absence of PLC?

   Without PLC, MOS will be low in packet drop situations. PLC creates synthetic voice created from the previous history of the speech. In the absence of PLC, usually silence or previous packets are repeated that will be more annoying compared with planned PLC schemes that go with the selected codecs.

34. What should be the packet drop goal to get reasonably good quality?

   In recent standards, the guidelines ensure that end-to-end packet drop is less than 1%. The quality improvement is dependent on the codec, packet drop percentage, and nature of packet drop such as random and burst.

35. What is VAD/CNG?

    VAD is voice activity detection. On detecting nonvoice (silence), VAD sends a silence insertion description (SID) packet. The VAD option reduces the VoIP voice packets during silence periods. Speech will have about 41% silence periods. In normal speech conversation, one person talks at a time that saves some more Internet bandwidth in half-duplex speech. CNG is comfort noise generation. CNG is created locally as a continuity of the VAD packet to create the feeling that a channel is active and to produce a comfortable perception.

36. What is the gain with VAD/CNG?

    The main advantage of VAD/CNG is the saving of Internet bandwidth. We can send fewer packets for similar performance. The savings could be as much as 41%.

37. What are the disadvantages of using VAD/CNG?

    A MOS score may come down by 0.1 to 0.2 under certain conditions. Voice clipping may distort the starting of speech. In general, when bandwidth savings is not a major requirement, it is recommended to avoid using VAD/CNG.

38. How do you distinguish between VAD/CNG and PLC?

    VAD/CNG is active during an actual silence period. It is dependent on speech and silence zones. VAD/CNG is both a transmitter- and a decoder-based scheme. PLC is active during end-to-end packet drop/loss. PLC can work as either an encoder/decoder scheme or a decoder-only scheme.

39. How many types of echo cancellers are used in the voice processing?

    Mainly two types of echo cancellers—line/electrical/network echo canceller and acoustic echo canceller. Line echo cancellers are for electrically created echo by two- to four-wire hybrids. Acoustic echo cancellers are for speakerphones and mechanical body acoustic coupling. Both echo cancellers will work from a processor.

40. What is talker echo and listener echo in a line echo canceller?

    Echo is the reflection of a signal. When A and B are talking:

    • Talker echo: Person is listening to his or her own echo—meaning person A's speech comes back to A with delay and distortion.

    • Listener echo: Person B's "hello" speech comes up to A for the first time, and the same "hello," which is reduced in strength, comes back to A a second time after completing one more round trip. The presence of listener echo is an indication of a very bad system.

    • Talker echo is more common and easy to remove. In the VoIP adapter, the echo canceller (EC) is used to remove talker echo.

41. Who is helping whom with the echo canceller?

    With near-end echo cancellers, an echo canceller at A helps B, and an echo canceller at B helps A. In customer premises equipment (CPE), "A" keeps the echo canceller and it removes the talker echo of "B," meaning it helps B. When A keeps the echo canceller, and B does not have an echo canceller, A will get his voice back. But B will be comfortable.

42. What are the main modules of an echo canceller?

An adaptive filter based on the least mean square (LMS) or recursive least square (RLS) family of algorithms is the main operation. Several extensions are available for the adaptive filter algorithms. The control plane includes double-talk detection, nonlinear processing (NLP) to remove residue and to create a comfortable background, fax and modem detectors, quality monitoring, and configuration modules.

43. What is the meaning of echo path or tail?

In a conversation between A and B, speech from A will go up to B. At B, echo is created. This echo is like A's voice going through a band-pass filter. This filter impulse response is called an echo path. An echo path is created through subscriber line interface circuit (SLIC)–CODEC (ADC and DAC) hardware, telephone lines, telephone hybrid characteristics, and impedance mismatch.

44. Can we use an echo canceller available in SLIC–CODEC hardware chips?

No. Echo cancellers listed in SLIC–CODEC hardware chips are for forming a good hybrid to match the impedance and to improve the equivalent echo rejection loss (ERL) of the hybrid. This echo canceller is of fixed coefficients and spans less than a millisecond. Different telephones are used with VoIP CPE. We need an additional echo canceller in the VoIP processing. Echo cancellers for the actual echo path will span from 16 to 32 ms, not just 1 ms.

45. If G.729A works with 10-ms and G.723.1A with 30-ms frames, how do we use an 8- to 16-ms echo tail length?

An echo canceller will not see any frame or codec algorithmic delays. In the design, if some frame delays appear, a bulk delay (codec selection-dependent fixed delay) is used in the adaptive filter path. This delay allows for optimal utilization of the tail-span.

46. When do we need long echo span cancellers?

We need long echo cancellers in speakerphones to cater to reverberations of the room/hall and the surroundings. In IP phones and wireless fidelity (WiFi) handsets, the speakerphone operation is also supported. In VoIP adapters, a 16-ms echo span will be sufficient. VoIP services extended on the PSTN will need a long echo span of 48 to 128 ms.

47. What is the relation of full-length, tail-free and long tail-length echo cancellers?

- It varies by line and acoustic echo canceller.
- Line echo canceller: In the marketing terminology, echo cancellers above 32 ms (usually 64 to 128 ms) go by the name long tail length.

- In an acoustic echo canceller, a 64-ms span is short. We need 128 to 256 ms to remove echo in speakerphone mode.
- A full-length echo canceller of 128 ms can remove echo anywhere from 0 to 128 ms. A tail-free echo canceller of 128 ms will select important small windows of dominant echo regions and remove echo. Tail free is of low computation compared with full length.
- Some line echo cancellers use tail free, but most acoustic echo cancellers use full length.

48. What are AGC and ALC?

AGC is the automatic gain control used in several electronic systems that interface with signals. ALC is the automatic level control. AGC and ALC serve the same function, but the ALC name is more common in VoIP. In VoIP systems, level variation (if required) is made slow at the rate of 10 dB per second, and gain variation is limited to 15 dB as per the G.169 recommendation.

49. Why is AGC not popularly used in the VoIP voice applications?

In VoIP, phones are connected at a distance of a few feet, and the remaining part consists of digital IP packets. No signal loss occurs with IP packets. The main PSTN systems do not use AGC, and a user may change gain on an end telephone. VoIP systems are targeted to provide PSTN reference quality.

50. Some products specify ALC as part of voice quality enhancements, what is the reason?

When a talker is delivering voice at low volume, a listener has to put forth additional listening effort that results in lower quality. Low-volume speech may not convey the proper emotions during interactive speech. To improve voice quality, it is useful to apply ALC on a speech level falling below set limits. ALC also helps during three-way conferences.

51. Do we need AGC with a normal phone in speakerphone mode?

AGC is required along with speakerphone to maintain the required sound levels and for stable operation of speakerphone mode. A speakerphone manufacturer will keep AGC inside the phone as part of an acoustic echo canceller. Hence, AGC is not a requirement in VoIP systems. Many handset phones and speakerphones support volume control that works as manual gain control.

52. Some configurations allow gain changes. How come we cannot use AGC?

Different countries have different gain/loss in the send and receive path. Matching gain/loss will maintain a proper loudness rating on phones to enable better voice quality in speaking and listening. Gain/loss configurations are provided on the VoIP system to change the settings as per country/region requirements. AGC may be used on top of the fixed gain/loss in three-way conference calls.

53. In VoIP systems, why is the AGC/ALC block tied around echo canceller modules?

    ALC should not appear in an echo path. It has to be before reference R_{in} or after S_{out}. The gain corrections are applied immediately after echo cancellers. Echo cancellers will have several estimators on signal/power levels. The blocks can be used directly by the ALC. The module ALC as part of the echo canceller can reuse several computations and can ensure the stability of the echo canceller. Hence, ALC will be kept as part of the echo canceller or just adjacent to the echo canceller but not in the echo path.

54. What is DTMF?

    DTMF is dual-tone multifrequency in the frequency range of 697 Hz to 1633 Hz. It is generated when a telephone dial pad is pressed after lifting the phone handset. It is governed by ITU-T Q.24, Q.23, and local country-specific standards.

55. In the DTMF standard, we see 16 digits, but most phones have 12 keys?

    Most telephones use 12 keys of 0, 1, 2, 3, 4, 5, 6, 7, 8, 9, *, and #. Digits A, B, C, and D are used indirectly as call feature buttons on telephones with names. Some of these digits are also used during call wait ID.

56. Some phones will have a pulse/tone dial switch. How do we relate these switches with DTMF?

    Earlier rotary phones were made with pulse dialing. Pulse is an interruption of telephone line battery voltage through a mechanical switch (usually called make and break) while a wheel is recovering. Digit-1 interrupts the telephone TIP-RING line once. Digit-9 interrupts it nine times. Digit-0 is at the end of digit-9 in a rotary phone, and it interrupts the telephone line ten times. Pulse dialing mode supports 10 digits of 1, 2, 3, ... 9, and 0. In keypad-based phones, no mechanical timing of digit-0, occurs and no rotary system interrupts ten times for digit-0. A phone manufacturer regenerates pulses similar to a rotary phone using electronic circuitry. Most of the current operations with PSTN and VoIP use DTMF mode as the default. It is suggested to use a phone in DTMF mode. DTMF helps in fast dialing with fewer errors.

57. While using pulse dialing, sometimes we cannot give responses to interactive voice response (IVR). Why?

    In DTMF-based dialing, every digit takes approximately 100 to 150 ms, including ON–OFF periods of dual tone. In pulse dialing, each pulse takes 100 ms (including make and break). Digit-0 takes 1 s. The inter-digit delays are increased in pulse dialing. Hence, some pulse dial modes will cause IVR systems to fail.

58. Phones connected to a PBX have more than 12 or 16 buttons. The count is more than the DTMF's 16 digits. How are they handled? Can we use an existing PBX phone with VoIP adapters or PSTN?

    Many phones connected to PBX systems consist of four or more wire phones and use more than 16 digits. In some PBX systems, a separate pair of wires is used for dialing and signaling events, which causes it to work for more features than actual DTMF digits. Some PBX phones are called digital phones. In digital phones, send and receive wires are separate. In a two-wire telephone, send and receive are combined and a hybrid inside the telephone separates the send and receive signals. Existing digital PBX phones can not be used with VoIP adapter or PSTN.

59. Pulse dialing has two modes: 10 pps and 20 pps. What is the reason?
Most pulse dialing systems around the world use 10 pulses per second (pps), meaning 10 pulses are dialed in 1 s. Rotary phones use 10 pps. To dial 1 (one pulse), it takes one tenth of a second. To dial 0, it takes a full second. In redial pulse dialing mode, phone systems may use 20 pps. To dial 0, it takes 0.5 seconds in 20-pps mode. This mode is not applicable to the physical rotary type of phones. It is applicable to push-button redial mode. Some fax machines may use high-speed pulse dialing (20 pps).

60. What is the make and break in pulse dialing?
In the PSTN, when a phone handset is lifted from the phone, it makes contact with the exchanges. This contact is called off-hook in recent terminology. In pulse dialing during make, a phone is connected with the central office allowing for current sensing at the central office. Break is like on-hook, and a line is disconnected from the CO for some duration. In a 10-pps system, the typical values of make is 33 ms, and break is 67 ms. In a 20-pps system, make is 16 to 19 ms and break is 31 to 34 ms and may slightly vary with country.

61. What are the most popular algorithms for DTMF tone analysis?
Goertzel is the most popular time-domain algorithm. FFT is a frequency-domain schemes. Teager and Kaiser (TK) is another time-domain algorithm. In general, extended techniques and logic exist on top of these algorithms to provide required DTMF detection performance.

62. Will the same DTMF algorithm work for all countries?
Some countries have incorporated deviations to the DTMF parameters list, such as with power levels, twist, and drift. A configuration based on country selection can be used with the same algorithm. In general, DTMF processing state machines are made robust to handle multiple countries.

63. What are the tests conducted on DTMF modules?
Telcordia, MITEL test tapes, developer/implementer-generated test vectors, several speech, and music files are used. These tests identify both right and wrong operations. Some of these tests send speech signals that closely match with DTMF. The goal is to achieve minimal false detections and to guarantee all the required detections.

64. What is the purpose of DTMF suppression in VoIP? We do not talk about it in regard to the PSTN.

The purpose is to stop in-band DTMF tones when a digit is generated from a phone. The digit information is sent as IP packets in VoIP, and in-band DTMF is suppressed. DTMF suppression is not applicable to the PSTN. In the PSTN all the tones, DTMF, and speech go as an analog signal and TDM bits using G.711.

65. What is MF?

MF is multifrequency that makes use of row–column two-tone combinations. In MF, five-row tones and five-column tones are used. MF is used to communicate between digital telephone exchanges. In VoIP, PSTN interfacing gateways may also use MF. Telephones use DTMF or pulse dialing.

66. What is a tone generation?

Some gateways send tones as events and information in digital codes in IP packets. These code words cannot be played in its form. On telephones, continuous tones are required to be generated. Based on the code and other parameters specific to a country, a two-frequency tone combination is used to synthesize in the voice band. It requires information of two frequencies, power, ON-period, and OFF-period (silence period) as inputs. This scheme is also called call progress tone generation (CPTG).

67. What occurs in caller-ID generation?

Caller-ID information is exchanged between two VOIP gateways in out-of-band signaling in a VoIP call setup. When caller-ID comes as a code (out-of-band signaling in VoIP), the frequency-shift keying (FSK) tone is generated at the right time slot in relation to the ring cadence (ON–OFF periods).

68. What is a VoIP dial tone?

To distinguish from the PSTN dial tone, several VoIP services provide a VoIP dial tone. This tone has higher frequencies compared with the basic PSTN dial tone, and it is created as a combination of several tones to produce melody in listening.

69. What is DTMF caller-ID? What are the countries that support DTMF-based caller ID?

In DTMF caller ID, caller ID information is sent as DTMF tones to the phone. Usually a phone generates DTMF tones during dialing. In DTMF caller ID, digits are sent to the phone to display as caller ID. DTMF-based on-hook caller ID is used in Brazil, Belgium, Denmark, Holland, Finland, Iceland, India, the Netherlands, Saudi Arabia, Sweden, and Uruguay.

70. What is call progress tone detection (CPTD)?

This detection is required when one of the interfaces to a VoIP box is a foreign exchange office (FXO) and is connected to the PSTN. The PSTN gives call progress tones. These tones will not be understood by VoIP signaling. Hence, it is required to convert the in-band tones to the decisive events. A call progress tone detector produces detection bits when it encounters dial, ring-back, busy, reorder, call wait tones, fax, and modem tones. On FXO, caller ID detection is also used. With a foreign exchange subscribes (FXS) interface, DTMF digits as well as fax and modem tone detections are sufficient.

71. What is the wideband voice? Can VoIP systems work for wideband voice?

    Narrowband voice is from 300 to 3400 Hz. Human speech has higher frequencies. Wideband voice is twice the bandwidth of narrowband from 50 to 7000 Hz. Wideband voice creates better perception and natural conversation. Wideband requires some voice modules to be wideband capable, and telephone interfaces have to support wideband acoustics.

72. What are the misconceptions with wideband voice? Clarify these.

    Wideband voice does not require more bandwidth on the network. The maximum network bandwidth is the same as G.711 or lower than G.711. A wideband G.722 high-quality codec requires less processing than G.729AB. Some codecs like G.722.2 and G.729.1 requires more processing, but that can be easily taken care of with the present processors.

73. What is jitter?

    Jitter is the time delay variation or change of delay. In the PSTN, delay is fixed. In VoIP, the packets go through multiple network nodes and packets can appear with different impediments. Jitter is one form of IP impediments.

74. What are the types of jitter buffers? Where are they implemented?

    Jitter buffers are either fixed jitter buffers (FJBs) or adaptive jitter buffers (AJBs). They will reside in the receive path from the IP network and work in close association with Real-Time Transport Protocol (RTP) and RTP control protocol (RTCP) modules. Several implementations keep this operation close to networking. FJB is used with low jitter networks and in fax and modem pass-through mode. AJB is used with voice calls with several impediments. AJB adapts to varying traffic conditions.

75. Can we use the same jitter buffer controls for all channels?

    No. Each VoIP call may go to a different destination with different network conditions. Hence, a separate jitter buffer and different estimation is required in each channel (voice port).

76. What are the expected AJB delays?

    These delays depend on the network conditions. They can work with one packet buffer (in the case of G.711/G.729A with one 10-ms delay). On the maximum side, they can grow up to maximum delay variations or jitter buffer maximum sizes. Typical maximum sizes are of 400 ms.

77. What is the influence of jitter buffer on an echo canceller?

    An echo canceller looks at the near-end pulse code modulation (PCM) interface, SLIC–CODEC, transmission lines, and telephone mismatches. This delay is steady or slowly varying with reference to the decoder path for the established connection. Hence, the presence of jitter buffer or the variation in buffer lengths will not pose a problem. More jitter causes end-to-end delay to increase and demands more care in echo cancellation.

78. What are the lower and upper jitter levels? What is the maximum delay?
The lower limits are usually 20 ms, and some systems cater to 50 to 70 ms as a lower limit. Adaptive jitter buffers grow up to 400 ms in the adaptive process.

79. In some box configurations, we see a 50-ms fixed delay. What does that mean?
Two popular interpretations for this delay exist. The initial delay is 50 ms, and the adaptive Jitter buffer adapts with this initialization. It could be that a fixed jitter buffer threshold of 50 ms can work for most situations.

80. What are the functions of packetization?
VoIP packets have to be transported end-to-end. Voice compression creates raw payload. This payload is used with RTP (12 bytes), User Datagram Protocol (UDP) (8 bytes), and IP (20 bytes) packet headers. Depending on the network interface (like Ethernet or DSL), interface-specific headers are also used.

81. Voice consists of real-time traffic. What ensures that this traffic exists?
RTP mainly creates headers that allow real-time information and sequence numbers. Quality-of-service (QoS) mechanisms and a networking stack also ensures voice priority and go with the least possible delay path.

82. What is the function of RTCP and RTCP-XR?
RTCP is for control information. RTCP shares several statistics of voice call and quality. RTCP-XR is an extension of RTCP that sends about 20 parameters of voice configurations and voice quality monitoring. RTP, AJB, RTP, RTCP-XR, and information from QoS closely work together deliver the best possible packets delivery.

83. What is the network bandwidth for a VoIP voice call?
Bandwith varies by codec selection, selected bit rate in the case of a multirate codec, packetization interval, and selected interface such as Ethernet, DSL, and cable. G.711 on the PSTN requires 64 kbps. The same G.711 with 10-ms packetization requires 126.4 kbps on Ethernet. G.729A of 8-kbps compression at the codec level requires 70.4 kbps on Ethernet with 10-ms packetization. The bandwidth increase is caused by additional headers required to transport voice on the IP network.

84. What are the trade-offs with codec, packetization, and bandwidth?
Higher packetization reduces packet header overhead per second. Hence, it reduces the overall bandwidth requirements. A drop of packet with higher packetization is more harmful, and PLC works better for short

erasures. Higher packetization also increases end-to-end delay that creates degradation of quality.

Higher compression reduces bandwidth and reduces quality in most situations. In network congestion or network bandwidth limitation, it is suggested to go for higher compression—assuming reducing bandwidth requirements are helping to reduce congestion and packet drop. It is better to switch to a G.729A call instead of continuing G.711 with bandwidth issues and packet drops.

85. VAD/CNG saves on bandwidth. Why is this saving not quantified exactly?

    VAD/CNG is dependent on conversation silence zones. It varies according to several factors. Hence, VAD/CNG savings is given in subjective terms. It is also possible to monitor and provide statistics on bandwidth savings from the live call.

86. What is the accepted clock parts per million (PPM) in VoIP systems? What are the preferred options?

    A clock of 50 PPM is acceptable in VoIP calls and is given in several recommendations and RFCs. A clock precision of 32 PPM (Stratum 4) is preferred. Targeting for 4.6 PPM will produce improved quality on long calls and in fax pass-through calls.

87. What is the accepted clock jitter? What are the preferences?

    Maintaining a front-end analog-to-digital conversion (ADC)/digital-to-analog conversion (DAC) (PCM interface) clock or clock that maintains voice sampling and packetization to be kept as lower than 10 ns, preferably to less than 3 ns.

88. What are the clock sources for VoIP systems?

    Simple VoIP adapters use 32- to 50-PPM crystal or clock oscillators. VoIP-to-PSTN gateways achieve stratum-3 at 4.6 PPM. Some gateways achieve a clock from the network time protocol (NTP) and a fallback stand-alone option with a voltage control crystal oscillator (VCXO). Gateways with interfaces like DSL/VDSL may use a network timing reference (NTR) 8-kHz clock and achieve the required high-frequency clock. IEEE-1588-based clock generation is also now under consideration.

89. Why do VoIP solutions have to consider multiple countries?

    Multiple service providers use the products made for VoIP in multiple regions. The front-end telephone interfaces are also catering to programmable matching impedances. The efforts of multiple countries will need to be considered as software upgrades or configurations that can reuse the same hardware.

90. What are the major deviations for multiple countries?

    The VoIP system is like the PSTN central office. The major deviations include call progress tones, ring patterns, caller ID, other call features,

DTMF signal parameters, pulse dialing parameters, telephone ringer equivalent number (REN), telephone deviations (ERL, loudness rating, impedance), special tones specific to VoIP, emergency services, and voice quality expectations. Additional deviations include diagnostics, monitoring, ring and battery voltages, and front-end A/μ-law compression.

91. What are the issues in sending voice and data together? What is the main function to help voice?
In most user applications, the tendency is to use more network bandwidth than is available. As a result, congestion and packet drop occur. When data and voice are sent together, both voice and data packets can be dropped. Data packets are recovered by TCP retransmission, and voice packets may not be recovered because of real-time requirements. PLC helps to improve voice to some extent.

QoS mechanisms help to create the required bandwidth and priority for voice. Monitoring will also occur on voice packet flow to create feedback. When a voice packet is in a queue with a large-size data packet, it can create more delay or delay variation. It is a degradation of quality. Data packets are fragmented to a small size to minimize delay variations.

92. Why is upstream QoS more common in implementations?
Upstream QoS is in the control of the system. Upstream QoS can aggregate voice, data, and forward path traffic from multiple interfaces. Downstream is not really controlled at CPE. The service providers can incorporate QoS that can help with downstream. Some proprietary techniques are reported that implement QoS mechanisms in downstream. In principle, the downstream demand from low-priority traffic can be reduced to maintain the required packet quality for high-priority packets.

93. What are VoIP softphones?
A softphone is a network-interfaced computer that works like a stand-alone VoIP phone. Computer acoustics or speaker/microphone interfaces on a computer are used for user acoustics. Skype and Google-talk on a computer are softphones.

94. What are the popular VoIP architectures from implementation?
VoIP on computers as softphones

VoIP on PC add-on cards

VoIP with digital signal processor (DSP) for voice chain and network processors for networking

VoIP with network processor with extended DSP instructions

VoIP with DSP with extended network interfaces

95. Why is PSTN quality more deterministic and consistent?
The PSTN operates with a few 1000-ft analog signals and the remaining part is synchronous digital transmission. It is a rare event to drop a bit

or sample of voice in the PSTN. The main degrading elements of the PSTN are the analog interfaces from the DLC/CO to the user telephone and the user-end multiple telephone combinations and mismatches.

96. What are the major quality degradation contributors in VoIP?

Voice quality in VoIP is affected by end-to-end packet impediments resulting in more delay and packet drop, capabilities of packet loss concealment techniques, echo annoyance with increase in delay, voice compression, number of transcoding stages, loudness rating and other losses, delay in call establishing interactions, and VoIP as a total hardware software solution.

97. What are the main advantages of VoIP compared with the PSTN?

- No analog loop distortions because of the few feet of wires.
- Voice quality with the G.711 codec can approach PSTN quality under good implementation and the IP network.
- VoIP quality can exceed the PSTN with wideband voice. Wideband voice is not supported in the PSTN. It creates conversation that is more natural.
- Remote monitoring of total voice quality and scalable channels and quality; more bandwidth allows more quality. Scales better along with data and other media services.
- Creates global roaming using an IP network—able to go with the same number globally, able to interface directly with digital phones, Bluetooth, digital enhanced cordless telecommunications (DECT), and WiFi phones.
- Automatic level control to deliver required signal levels.
- End-to-end direct coding (no transcending) can minimize quantization losses.

98. How is voice quality MOS measured?

MOS is the mean opinion score measured as a single number in the range of 1 to 5. MOS is a subjective measure evaluated average Quality reported by a group of listeners. MOS can also be modified from objective measurement with PESQ, the E-model, and P.563.

99. What is the subjective measure. Explain ACR, CCR, and DCR?

In subjective voice quality evaluation, the voice quality MOS is rated by the group of actual male and female listeners. It is the actual listening test for evaluating the MOS. The ITU-T-P.800 and 830 recommendations support this test. Subjective tests are conducted in three ways. ACR, DCR, and CCR.

In absolute category rating (ACR), participants listen to recorded speech samples that have been processed through several test connections. A minimum of 16 test subjects (users) should participate in the assessment. On listening to these samples, users rate the call on a 1-to-5 MOS scale. The MOS 5 is the best and 1 is poor quality. The average values of the user ratings are taken to generate the overall call quality.

In degradation category rating (DCR), two speech samples are present. The first speech sample is a reference sample with a predefined quality. The sample here refers to speech lasting for several seconds in duration. The other speech sample is a degraded version. Listeners must compare the degraded version with reference on a degradation scale of 1 (worst) to 5 (best).

In comparison category rating (CCR), users are asked to listen to two sets of samples—one corresponding to reference and the other to degraded. This test is similar to DCR, except that the order of samples presented to the listeners are changed in each test iteration. Listeners are asked to give a comparative rating of the second sample with respect to the first one on a scale of −3 (worst quality) to 3 (better quality).

100. What is the R-factor?

The R-factor is a rating factor based on the ITU-T-G.107 E-model. It is measured on a linear scale. The R-factor is mapped to MOS. For narrowband voice, the R-factor varies from 1 to 100. For wideband voice, the R-factor is up to 129. A typical range of the R-factor for narrowband to achieve 4.0 to 4.4 MOS is from 80 to 93.2.

101. What is PESQ?

A perceptual (the way listening occurs) model is used to distinguish between audible and inaudible distortions. It accesses the quality of a distorted speech signal by comparing it with the original undistorted signal. The PESQ operation makes use of two major classes of logarithmic operations—namely conversion of signals into psychoacoustic domain and cognitive modeling.

102. What is ITU-T-P.563?

ITU–T–P.563 makes use of a speech production mechanism and model. Most other models use listening perception. This algorithm operates on received speech only. The measurements through P.563 derive several parameters from received speech classified as noise, artificial speech, and actual speech.

103. Why are speech quality measurements not able to account for echo and delay?

Speech quality measurements are of a listening quality, and they are of a half-duplex operation. In R-factor calculations, delay and echo effects are considered.

104. What are the one-way delay goals?

ITU-T-G.114 and other recommendations have set delay goals. In summary, local VoIP calls operate with less than 100 ms. Intraregional of less than 5000 km target for less than 150 ms, but a delay of 100 ms is possible in a good deployment. Intraregional of less than 10,000 km is about

225 ms. Inter-regional of 27,500 km is just over 300 ms. It is recommended to contain delays to less than 400 ms as a worst case. Delays will exceed 400 ms for double-hop satellite communication for hard-to-reach regions. For delays beyond 500 ms, nearly all users will be dissatisfied.

105. What are major sources of delay in VoIP in an end–end voice call?

In a VoIP, the major delay contributions in end–end delay are algorithmic delay, processing delay, packetization delay, network delay, including queuing delay, and buffering delay from hardware interfaces and in jitter buffer. Processing delay is related to processing power and frame duration for processing.

22

BASIC FAX AND FAX OVER IP FAQS

In this chapter, several questions related to basic fax operation, fax calls on the PSTN, fax over IP (FoIP), and various modes of operation are listed. The questions are organized in the order of fax, fax over PSTN, fax over IP, and related connected points on bandwidth, messages, timing, and various modes of operation.

1. What is fax?

   Fax is an acronym for a "facsimile" (facsimile means "a copy") or "tele-facsimile" system. A fax machine scans and sends written or graphical material to the destination fax machine for producing hard copy. A fax makes use of a standard telephone line (or VoIP operation) to transmit electrical signals to a remote fax machine.

2. What are the main functional components of a fax machine?

   A standard facsimile terminal device consists of a paper input device (scanner); a paper output device (printer); accessories like a telephone keypad, telephone line control, display, headset, controller, and fax modem; and a limited amount of processing power.

3. How does a fax machine send pages?

   The fax machine electrically scans and breaks up the document page into picture elements or pixels. These pixels are compressed, coded for error protection, passed through digital modulation, and delivered as a voice

VoIP Voice and Fax Signal Processing, by Sivannarayana Nagireddi
Copyright © 2008 by John Wiley & Sons, Inc.

band analog signal. At the destination, an analog signal is sampled, demodulated, decoded, and given to the printer. A fax makes use of a voice band analog signal. The intermediate transmission is not known to the fax machine. It can be PSTN or VoIP.

4. What are the main signal processing operations inside the fax machine?
In the send direction, picture elements are compressed and coded with different schemes to protect from errors and bits are modulated with digital modulations. Several messages and acknowledgments based on the fax operation are sent as low-bit-rate frequency-shift keying (FSK) modulations. Tones, and caller ID are also used for call progress. The front-end hardware inside the fax machines delivers an analog signal. In the receive direction, a front end gives digital samples. These samples are demodulated for low-speed (messages and indications) and high-speed data. The demodulated bits are used for error correction and decompression before giving them to printer.

5. What are the different fax standards?
Fax devices are mainly classified as Group-1, Group-2, Group-3 (G3), and Group-4 (G4) depending on the image formatting and handling methods inside the fax machine. Group-1 and Group-2 are for analog fax and not relevant at present. G3 is a digital fax on analog lines, and G4 is a digital fax over digital telephones lines.

6. What is the role of the T.30 recommendation, and how does it relate with coding?
T.30 is the main controller of the fax call. The characteristics and the operation of G3 fax devices adhere to the requirements of the T.30 recommendation for a fax control signaling handshake. It is applicable to both the PSTN and fax over IP. Both fax machines will be aware of T.30 signaling. The T.30 fax control signaling messages are coded using a V.21 modem at 300 bps.

7. What are the different coding standards used in G3 and G4 fax machines?
A coding operation is similar to compression and error concealing of fax page pixel bits. A group-3 fax machine follows both ITU-T-T.4 and ITU-T T.6 recommendations for image compression and coding. It uses Modified Huffman (MH), modified read (MR) and modified MR (MMR) coding schemes. A group-3 color fax machine uses Joint bi-level expert group (JBIG) and Joint Photographic experts group (JPEG) color-coding schemes. Group-4 fax follows the ITU-T T.6 standard for image compression and coding. It uses the MMR, Trellis, JPEG, and JBIG coding schemes.

8. What are the main differences between G3 and SG3?
The G3 fax machine supports the maximum data rates of V.17 14.4 kbps. The super-G3 (SG3) fax machines support maximum data rates of V.34 up to 33.6 kbps. All SG3 fax machines support error correction mode (ECM) mode by default. The G3 fax machine answers with an answer (ANS) tone,

whereas the SG3 fax machine answers with an ANS with amplitude modulation and phone reversal (/ANSam) tone during the call setup phase.

9. How many maximum lines can a standard A4 size paper contain?
The G3 fax machine scanner gives 1145 lines of information per page in the vertical direction (while using the lowest vertical resolution of 3.85 lines per mm) and 1728 bits of information per line in the horizontal direction on a standard A4 size piece of paper. It produces (1728)(1145) bits of approximately 2 million bits without any compression. Several compression methods are available. In compression, about 10 times compression is achieved that varies with page and compression scheme.

10. What are the basic principles of fax compression?
Considering black-and-white pixels, a fax page will have several consecutive pixels of white and black. Instead of sending every pixel as one bit, all the consecutive bits (grouped) are sent as one code word. This property is also exploited across multiple lines for two-dimensional compression. In the case of gray-level images, bit planes are derived and compressed to get better compression. Compression methods such as JBIG and JPEG are used with multilevel and multicolor fax compression.

11. What are the typical compression ratios for MH, MR, and MMR, JBIG and JPEG schemes in fax?
Fax compression widely varies based on the image data in a page. Typically, Modified Huffman 5:1; Modified read 7.5:1; and Modified modified read 10:1, JBIG and JPEG from 10:1 to 20:1. JBIG best suits black and white gray scale images and JPEG give more Compression for color images.

12. What is end of line (EOL)?
Each coded line ends with end of line (EOL) in MH and MR coding schemes. The EOL is a unique code word of 11 zeros followed by bit "1" [format: 000000000001]. The EOL plus one tag bit is used as a synchronization code in MR coding. The end of the document page is indicated by sending the six consecutive EOLs in an MH coding scheme and the six consecutive EOLs plus tag bit in the MR coding scheme.

13. What is minimum scan line time (MSLT)?
MSLT defines how much time the receiving machine requires to print a single scan line. The minimum scan line time (MSLT) is set by the receiving fax machine and transmitted to the sending machine during the initial messages handshaking.

14. What is the minimum transmission time per total coded scan line?
The minimum transmission time of a total coded scan line is defined as the total time taken for transmission of the total coded scan line. The minimum transmission time of a total coded scan line is about 20 ms in G3 fax machines. The transmission time varies with image data and resolution.

15. Why are "fill" bits used in a coded line?
A pause may be placed in the message flow by inserting fill bits. A fill can be placed between a line of data and an EOL but never within a line of data. A fill is inserted to ensure a minimum transmission time of data. Format: A variable-length string of zeros.

16. What are the different fax modems used in a G3 facsimile device?
A modem used in a "fax modem" is modulation and demodulation. A fax modem is not a dial-up modem in fax processing. The G3 facsimile device uses V.27ter, V.29, and V.17 modems, and the SG3 fax uses a V.34 modem for image transmission. A V.21 modem is a default fax modem in G3 and SG3 facsimile devices, and it is used for fax handshaking and negotiation.

17. What is V.21?
V.21 is an ITU standard describing a low-speed data transmission rate of 300 bps with FSK modulation for exchange of T.30 fax control information and messages and V.8 signaling messages. Fax image data are not sent on V.21. The messages in V.21 are more robust and can work even if V.21 FSK signal is compressed like voice.

18. What is V.21 (L) and V.21 (H)?
Modems that use full-duplex communication (for example V.34 modem) use two-channel transmission. In V.21, channel-1 is called the V.21 (L) low-band mode that uses 980 and 1180 Hz for mark and space frequencies, and channel-2 is called the V.21 (H) high-band mode that uses 1650 and 1850 Hz. In full-duplex mode, the originating V.21 modem transmits in low band and receives in high band, and the receiving V.21 modem transmits in high band and receives in low band. Group-3 fax machines use channel-2 in both directions.

19. What are V.27ter, V.29, and V.17?
V.27ter is an ITU recommendation for high-speed image data transmission that is used for medium transfer rates of 2400 and 4800 bps. It makes use of phase-shift keying (PSK). V.29 is an ITU recommendation used for medium transfer rates of 7200 and 9600 bps that makes use of Quadrature amplitude modulation (QAM). V.17 is an ITU standard that is used for high transfer rates of high-speed fax page data at 7200, 9600, 12,000, and 14,400 bps. It makes use of Trellis coding and QAM.

20. What are the different rates in an SG3 V.34 fax?
The advanced supergroup-3 fax supports G3 modems (V.27ter, V.29, and V.17) and the V.34 data modem for fax transmission. The V.34 modem operates in both half-duplex and full-duplex mode to support 14 primary data channel rates from 2400 to 33,600 bps in steps of 2400 bps and two control channel rates of 1200 and 2400 bps. A V.34-based fax session starts with 2100 Hz amplitude modulated and phase reversal answering tone. Binary procedural data are exchanged at 1200 or 2400 bps using V.34 in full-duplex mode.

21. What is the role of the V.8 recommendation, and how does it relate with coding?

    The V.8 protocol defines procedures for starting sessions of data transmission and their signaling format. V.8 signaling determines the best mode of operation before the initiation of a modem handshake. V.8 signaling messages are coded using the V.21 modem at a 300-bps rate.

22. What is G3C fax?

    The G3 operation over digital networks like integrated services digital network (ISDN) is described in Annex-C of the T.30 recommendation. This option is known as Group-3C. G3C is mainly designed to be used on ISDN with procedures and signals based on T.30, T.4/T.6, and ECM mode.

23. What is G4 fax?

    Group-4 is mainly designed for an ISDN fax at 64 kbps based on ITU-T-T.6. G4 fax machines are digital and can interface directly with an ISDN line. Group-4 has three sets of classes (class-1, class-2, and class-3) based on pixel (pel) transmission densities. G4 can communicate with G3 with an optional dual-mode G4 class-1 fax that could use a standard G3C or G3 fax modem.

24. What is error correction mode (ECM)?

    ECM is used with fax image data. ECM is an optional transmission mode built into fax machines or fax modems. ECM automatically detects and corrects errors by retransmission of the error data. It helps fax transmission errors caused by telephone line noise or any impairment. Fax machines have the capability to enable or disable this function.

25. Why is ECM mode the most common with SG3 fax machines?

    Generally ECM support is optional for G3-supported fax with MH and MR coding schemes. The use of ECM is mandatory for all facsimile messages using the V.34 half-duplex and full-duplex modulation system because of a higher bit rate. ECM is always used in G3 fax with MMR line encoding because an MMR coding scheme does not use EOL codes for synchronization.

26. Why is ECM used with the MMR coding scheme?

    MMR is two-dimensional compression. The MMR coding does not use EOL codes between scan lines to limit the effect of line errors and for synchronization. Any scan line data corruption would also corrupt the interpretation of any data that followed it. The use of ECM with MMR reduces or eliminates the errors. ECM is also used with JPEG and JBIG image data in G3 color fax machines.

27. What are the block and frame sizes used in error correction mode (ECM)?

    A page of coded data is divided into several small blocks. Each block contains multiple frames up to a maximum of 256 frames. The frame size can be either 256 or 64 octets (bytes) of coded fax data. These values of frame size do not include facsimile control field (FCF) and frame number octet. Therefore, the total length of the high-level data link control (HDLC) information field, including both the FCF and the frame number octet, is 258 or 66 octets.

28. What is training in a fax call?
    Fax machines while establishing capabilities will go through training. Capability and agreed rate do not guarantee the success of a fax at that rate. Training ensures that actual pages can be sent at the agreed rate. If training produces more errors, fax machines renegotiate and retrain at a lower rate before proceeding with actual fax transmission.

29. How does the receiving fax machine respond to line errors in a non-ECM case?
    ECM is retransmission of error frames before switching to the next block. When detecting line errors in non-ECM mode, the receiving fax machine has multiple options to handle errors, as follows:
    • Respond to page reception with the ReTrain Positive (RTP) command. This response causes the transmitting fax to go through the training check (TCF) process before transmitting the next page.
    • Respond to the page reception with the ReTrain Negative (RTN) command. This response causes the transmitting fax to go through the TCF with a lower modulation scheme.
    • Continue with few errors or disconnect immediately when more error lines are received.

30. Describe different phases in a fax call.
    The different phases in a fax call are A, B, C1, C2, D, and E. Phase A is for call setup, including dialing and transmission of fax tones CNG and CED. Phase B is the premessage handshake procedure. In Phase B, calling and answering fax machines identify themselves and exchange capabilities data for transmission parameters selection. In Phase C1, the calling unit sends a test pattern to determine the maximum data rate. The answering unit either accepts the data rate or requests a lower data rate. Phase C2 is the actual fax page (image) transfer at an accepted data rate. Phase D is the post-image handshake procedure (e.g., end-of-message confirmation and multipage document procedures). Phase E is for call release or for switching the call to another mode of operation.

31. What are the ANS family signals or fax and modem tones?
    ANS is $2100 \pm 15\,Hz$, /ANS is $2100 \pm 25\,Hz$ with 450 ± 25-ms phase reversal modulated, and ANSam is 2100 ± 1-Hz tone with amplitude modulation at $15\,Hz$. The /ANSam is $2100 \pm 1\,Hz$, amplitude modulated with $15\,Hz$, and phase reversal modulated with 450 ± 25 ms. Low-speed fax machines (G3) send an ANS tone, whereas high-speed SG3 fax machines send the /ANSam tone during the call setup phase. The tones /ANS and ANSam are used by the dial-up data modem.

32. What are the different answering tone anomalies?

    The T.30 standard directs that the answering fax device send an answer tone of 2100 Hz for approximately 2.6 to 4 seconds before sending the first handshake message. Some fax machines send a 1650-Hz or 1850-Hz tone instead of a 2100-Hz tone, and some may omit the answer tone altogether and just begin with the first handshake message.

33. What is the calling indication (CI)? When will this tone be sent by a fax machine?

    The calling indication is an alternative to call tone (CT) and carries information to permit the selection of call functions (e.g., facsimile or data). V.34-based modems send this tone during the call setup phase. A signal transmitted from the calling modem indicates the general communication function. CI is transmitted with an ON/OFF cadence. The ON periods consist of a repetitive sequence of bits at 300 bps, modulating on a V.21 low-band channel. The CI sequence consists of ten ones followed by synchronization bits and call control bits.

34. What are the different V.8 signals used in V34 faxes?

    The different V.8 signals used in V.34 faxes in the beginning of a fax call phase are listed as follows:

    /ANSam tone: The SG3 fax machine answers with an /ANSam tone during a call setup phase.

    Call menu signal (CM): A signal transmitted from the calling super G3FE primarily to indicate modulation modes available in the calling terminal when V.34 is enabled. CM consists of a repetitive sequence of bits at 300 bps, modulating the V.21 (L) low-band channel.

    CM terminator (CJ): A signal that acknowledges the detection of a JM signal and indicates the end of a CM signal. CJ consists of three consecutive octets of all zeros with start and stop bits, modulating V.21 (L) at 300 bps.

    Joint menu signal (JM): A signal transmitted from the answering super group-3 facsimile equipment (G3FE) primarily to indicate modulation modes available jointly in the calling-and-answering super G3FEs. JM consists of a repetitive sequence of bits at 300 bps, modulating the V.21 (H) the high-band channel defined in V.21.

35. What are nonstandard facilities and capabilities in a fax call?

    The Group-3 facsimile standard permits a facsimile transmitter to request that the facsimile receiver switch to a nonstandard mode of operation if the receiver is equipped with the appropriate proprietary capability. The nonstandard facilities (NSFs)/nonstandard capabilities (NSCs) signal is used to invoke nonstandard features between two facsimile machines made by the same manufacturer. A digital command called the nonstandard facilities setup (NSS) is given in response to NSF/NSC. The features specified by NSF/NSC are not recognized by facsimile machines made by other manufacturers. If a handshake is completed in the NSF/NSC mode, the format for facsimile transmission could be changed for a proprietary mode of operation.

36. What are the various main timing requirements in a fax call?
The T.30 standard specifies various timing requirements:
    - If a response to the message is not received within a specified time usually of within 3 seconds, then the messages are repeated up to three times or until an acknowledgment is received. Some unacknowledged messages like digital identification signal (DIS) are repeated every 3 seconds until the timeout of 35 ± 5 seconds.
    - The fax machine responds with a message confirmation (MCF) command to the message end of message (EOM) and enters into the beginning of phase B. The timeout to enter into phase B after receiving the MCF command is 6 ± 1 second, which is defined as T2 in the T.30 standard.
    - The timeout that a terminal will attempt to alert the local operator in response to a procedural interrupt is 10 ± 5 seconds. Failing to achieve operator intervention, the terminal will discontinue this attempt and shall issue other commands or responses.
    - The receiving fax terminal can respond to a postmessage sequence with receiver not ready (RNR), and the calling terminal then queries the receiving fax terminal with receiver ready (RR). This sequence can be repeated up to 60 ± 5 seconds and then disconnects the call after timeout.

37. What is the duration of the silence period between signals using different modulation modes?
The ITU-T T.30 standard specifies a silence period of 75 ± 20 ms between signals using different modulation modes [e.g., the delay between digital command signal (DCS) and the V.27/V.29/V.17 training sequence]. The silence period between end of TCF and response to training [confirmation to receive (CFR) or failure to train (FTT)] is of 3 seconds.

38. How many times are the commands or messages repeated in a fax call?
Unacknowledged command messages are automatically repeated after listening for a response for 3 seconds. Commands are sent up to three times or until acknowledgment whichever comes first.

39. What is the necessity of using fax over IP?
Voice transmission is converted to an IP network in many deployments. Maintaining separate PSTN telephone lines for fax is an extra cost. The fax over IP application enables standard fax machines to work with packet networks. VoIP adapters and gateways manage fax packet transmission. Fax signals are distorted through normal low-bit-rate compression of G.729A and G.723.1. Hence, separate packets through T.38 and an associated data pump of V.27ter, V.29, V.17, and V.34 manage the fax signal to create fax payload.

40. What is an IAF device?

An Internet-aware facsimile (IAF) is a facsimile machine that can access the Internet directly. It can work as a combined operation of fax machine and VoIP gateway. For voice, the IP phone serves a similar function.

41. What is special about an IAF device?

Usually IAF supports T.38 real-time and T.37 store-and-forward fax. Hence, it is more capable than the fax machine and fax over IP adapter. T.30 timer values may be extended by two or three times when both fax terminals are identified as IAF devices in negotiation. IAF also works as a fax server and hosts several fax functions. The bit number "123" in the DIS/DCS message indicates an IAF device in negotiation.

42. What is the store-and-forward fax over IP, and when it is used?

Store-and-forward fax is similar to e-mails, but at both end terminations, fax machines and computers are used to interface for fax pages. In the store-and-forward mode, the caller sends the fax messages stored on the one server to another server, and finally, it reaches the destination VoIP fax interface either as an e-mail message or as a fax to a standard fax machine.

43. What is T.37?

The ITU-T T.37 recommendation defines a standard method for store-and-forward delivery of fax through an IP network. Fax messages are sent as multipurpose Internet mail extensions (MIME) encoded e-mail messages using a simple mail transfer protocol (SMTP). T.37 can handle storage of fax-like e-mails.

44. What is real-time fax over IP?

Real-time fax over IP works like a regular fax call. In real-time fax over IP, fax machines synchronize and send data over the IP link between the two connections. The two popular modes of sending real-time fax over IP link are T.38 fax relay and fax pass-through, which is similar to a G.711 VoIP voice call.

45. What is fax pass-through?

Fax pass-through sends the fax similar to a VoIP voice using G.711 μ-law or A-law as compression. A-law and μ-law are also used in the PSTN between digital loop carriers and telephone central offices. In VoIP, A-law and μ-law are supported in all the systems. Fax pass-through mode makes use of the G.711 or G.726 at 32/40 kbps to send fax. The performance matches closely to fax over PSTN under the best conditions. This method is also known as in-band fax transmission over VoIP.

46. What are the different voice modules controlled in a fax pass-through call?

Fax pass-through mode works similar to a VoIP voice call with the G.711 codec. On detecting fax tones (CED or /ANSam) by the gateway, a codec will be switched to G.711 (PCMU or PCMA), and some voice modules of voice activity detection (VAD), packet loss concealment (PLC), dual-tone multifrequency (DTMF) rejection, and echo cancellation are disabled for the duration of the fax session.

47. What are the benefits of fax pass-through?
When end-to-end packets delivery is good, fax pass-through creates less end-to-end delay and interoperability improves compared with T.38.

48. What are the main issues in fax pass-through?
Fax pass-through sends A-law or μ-law compressed samples on the IP network. IP calls can have many impediments such as delay, jitter, reorder, loss of packets, and fragmentation of packets. For fax machines, continuity of signal is lost with IP impediments. Fax may not go through with end-to-end packet impediments. Fax pass-through requires much higher bandwidth than T.38-based fax.

49. What is the relevance of RFC2198 for fax pass-through?
RFC2198 defines the format to pack redundant real-time transport protocol (RTP) payload. To compensate for packet loss in the IP network, redundant payload will recover the lost packets. RFC2198 will have one or multiple extra previous payloads that will demand higher bandwidth on the IP network. In some deployments, forward error correction (FEC) as per the RFC2733 scheme is also used to minimize the influence of packet impediments.

50. What is T.38?
T.38 is an ITU standard that defines the procedures necessary to transfer facsimile data and signaling packets in real time over the Internet or on any other IP network. T.38 interfaces demodulation bits from V.17, V.27ter, V.29, and V.34 modules into IP packets and at the receiver gives back packets to bits for modulation.

51. What are the advantages of T.38 in VoIP?
T.38 eliminates the need for high-bit-rate codecs for transmission of fax over IP networks. It is more attractive because the bandwidth used is better compared with a in-band G.711-based fax in VoIP. Fax is half-duplex; hence, IP packets will use one side of the bandwidth. It works well even in poor IP network conditions when employing redundancy and FEC. It is also possible to control the properties of the fax session. T.38 can be sent with three times the redundancy compared with G.711.

52. What is the bandwidth gain of T.38 over G.711?
Considering a fax at the 9600-bps rate, G.711 on the Ethernet interface requires 126.4 kbps in each direction. T.38 requires 24.8 kbps in one direction (send to receive direction) because the fax is half-duplex.

53. What are the disadvantages and concerns of T.38?
Many gateways may not support T.38. Interoperation is the critical issue with a T.38 fax. Many deviations exist across fax machines. Fax machine deviations, delays with T.38, and IP network impediments create several timing issues for T.38 and may force other modes of G.711 pass-through

to be used. These concerns can be eliminated by using redundancy and catering to tolerant design in T.38.

54. How will T.30 work with T.38?

T.30 is an ITU standard for fax signaling. It is a PSTN standard that is applicable to both PSTN and fax over IP. Both fax machines will be aware of T.30 signaling. T.38 is not known to fax machines, and it is internal to the VoIP adapters and the IP network. As part of a VoIP-based fax using T.38, once the call is connected and fax negotiation starts, each gateway takes part in the T.30 signaling with the local fax machines, but negotiation is end-to-end. T.30 signals are encoded into packets and relayed over the IP network using T.38 and other VoIP signaling.

55. What is a fax data pump?

This keyword is used by the engineers for representing total modulation and demodulation modules of a fax. The modules of V.21, V.27ter, V.29, V.17, and V.34 are referred to as a data pump. They take fax samples and demodulate to gives bits. They take bits and modulate to provide fax signals.

56. What are different T.38 version numbers?

The T.38 version numbers are 0, 1, 2, and 3. Internet facsimile protocol (IFP) over transmission control protocol (TCP) support is added in version 1. Version 3 supports a V.34 fax. The ASN.1 notation is modified in version 2. The modified Abstract syntax notation.1 (ASN.1) notation in version 2 and previous notations in version 0 or 1 cannot interoperate with each other. T.38 recommends use of version 2 for indicating the ASN.1 syntax. If no version number is provided in negotiation, the default version 0 is assumed.

57. What is the redundancy applicable to T.38?

Redundancy is used in T.38 in order to recover the lost packets from IP network impediments. User datagram protocol (UDP) transport layer (UDPTL)-based FoIP makes use of the redundancy procedure given in T.38. IFP over the RTP-based fax makes use of a procedure given in RFC2198.

58. What should be the usual redundancy factors?

Redundancy is applicable to both low-speed V.21 HDLC IFP packets and high-speed T.4/T.6 image IFP packets. High-speed redundancy of two to three is very common in deployments. A redundancy factor of four to six is very common for low-speed IFP packets. Compared with pass-through modes, T.38 will consume very less bit rate even for a higher redundancy factor of three.

59. What is the applicability of FEC in T.38?

FEC is forward error correction. The main operation in FEC is an exclusive OR (XOR) operation of present and some ordered previous packets. Normal packets and XOR modified packets are sent on the network. Lost packets can be recovered through packet loss detection logic and XOR operation. Some products use FEC in place of redundancy. FEC is optional in the implementations. Redundancy is used as a commonly supported operation.

60. How does pass-through compare with T.38?

    Pass-through mode is simple for VoIP fax transmission. The inter-op issues will be minimal and work closely with PSTN fax quality. It is not preferred because it demands more bandwidth than T.38. Under packet impediments, fax pass-through fails easily and T.38 performs better than pass-through mode.

61. What is the relevance of FaxLab for T.38?

    FaxLab is a tool that emulates about 166 facsimile device profiles. It can be used to perform compatibility testing among several facsimile devices. It can also perform extensive facsimile protocol analysis and identify protocol errors. It can be used to automate multiple fax calls. FaxLab will use only one analog front end on a TIP-RING interface. Hence, exact fax machine front-end characteristics and impedances are not emulated.

62. What are the fax quality measuring recommendations?

    The ITU-T-E.453 and E.458 recommendations are used for classifying fax image quality, transmission-induced scan line errors, and call success. In general, the ITU-T-E.450 series addresses fax quality. Instruments like FaxLab support these recommendations.

63. Is T.38 applicable with H.323 and SIP?

    T.38 is used with several VoIP signaling protocols. Certain deviations exist with the negotiation and renegotiation phases based on the VoIP protocol used, but the major part of T.38 operations remains the same across all signaling.

64. Why is there no RTP header for T.38 packets?

    When the T.38 protocol was standardized, RTP was an emerging protocol. Initial solutions with fax over IP were perfected with UDPTL based and TCP. Recent revisions of T.38 support RTP. Several boxes in the deployment and market support only UDPTL-based implementation.

65. What is ISDN fax, and how does it work?

    ISDN is four-wire digital telephony. The minimum configuration supports two channels of true 64-kbps voice or fax and one data channel. Voice and fax use 64-kbps channels, and signaling is managed in a data channel. A G4 fax is transmitted on an ISDN fax.

66. What are the different standards used for ISDN fax?

    A list of a few ITU recommendations for ISDN voice and fax are listed here:

    • Standards for G4 fax are ITU-T T.6, T.62, T.62bis, and T.85.

    • Fax test charts: T.20 and T.21.

    • F.161 International Group-4 facsimile service.

    • T.90 characters and protocols for the ISDN terminal.

- F.185, Internet facsimile: Guidelines for the support of the communication of facsimile documents.
- Q.850, usage of cause and location in the Digital Subscriber Signaling System No. 1 and the Signaling System No. 7 ISDN User Part.
- Q.921 operation of LAPD, ISDN User–Network Interface, Data Link Layer Specification.
- X.25 defines only the interface between a subscriber (DTE) and an X.25 network (DCE).
- Q.930: ISDN user–network interface layer 3—general aspects.
- Q.931: ISDN user–network interface layer 3 specification.

67. Why is the UDPTL protocol more popularly used than TCP and RTP in T.38?
UDPTL was the first approved approach for fax over IP. Many implementations in present deployments incorporated UDPTL protocol for fax over IP. T38 over UDPTL takes less bandwidth compared with TCP and RTP.

68. What are the main issues with fax interoperation?
The main issues with fax over IP start with increased end-to-end delay and packet impediments. Several problems in T.38 then multiply as a result of these issues.

69. How is the packet loss issue taken care of in T.38?
For higher packet losses, redundancy and FEC are used in T.38. IFP over UDPTL makes use of redundancy and FEC techniques as per ITU-T-T.38 for error correction. IFP over RTP uses redundancy as per RFC2198 and FEC as per RFC2733 techniques to compensate for a lost packet. When a fax machine is working in ECM-enabled mode, about a 2% packet drop is taken care of by ECM. TCP takes care of packet loss through retransmission.

70. What are the different data rate management methods in T.38?
Two methods of handling TCF are available for determining the high-speed data rate. Method-1 is the local generation of TCF, in which TCF is locally generated by the receiving gateway. This type of locally generating the training data is used in TCP implementations and is optionally used in UDP implementations. Method-2 is transferred TCF used with UDP (UDPTL or RTP) and is not recommended for use with TCP.

71. What is real-time fax with spoofing?
Spoofing techniques are used to extend the delay tolerance of fax machines. Spoofing and jitter compensation allow the fax machines to tolerate network delay without losing communication. These techniques add to the T.30 protocol used by fax machines to communicate, keeping them online beyond their normal T.30 timeout intervals. Fax spoofing inserts pseudo-packets into the data stream to make up for IP network delay that can cause fax device timeouts.

72. Does every fax system manufacturer comply with ITU-T standards, and are there any deviations?

    ITU-T recommendations for fax specify the sequence of operations for fax call setup, handshake messages, and procedures for testing and timing. Most fax systems deviate from the ITU-T recommendations in one way or another. Some of these differences are also created by country-specific deviations. The common anomalies and deviations are listed below:

    - Frame sequence deviations
    - Preamble and flag sequence variations
    - Answer tone anomalies
    - Timing deviations
    - Improper EOM usage
    - Unusual data rate fallback sequences
    - Common training pattern detection algorithms and incorrect duration of TCF
    - Image transmission deviations
    - Echo protect tone usage
    - Image padding and short lines
    - Retrain positive/retrain negative handshake message usage and inability to retransmit messages after receiving RTN
    - Long duration lines
    - Nonstandard disconnect sequences
    - Disconnect (DCN) usage

73. Do all T.38 implementations follow the same packetization format (raw payload excluding the T.38 protocol header)?

    Variations exist in different T.38 implementations. Some implementations send one T.30 signal frame in one packet, and others disassemble it and send it in multiple packets. Therefore, a T.38 implementation should handle both situations and assemble the multiple packets when necessary. This principle applies to image packets as well. Some implementations place an entire HDLC frame (between flags) into a single packet; others may ignore the frame boundaries when inserting the data into packets.

BIBLIOGRAPHY

REFERENCES BY STANDARDS

BT SIN 227 2004. CDS™ Calling Line Identification Service Description. UK: British Telecommunications. Available at: www1.btwebworld.com/sinet/227v3p4.pdf.

DOCSIS 1.1 2005. Data-Over-Cable Service Interface Specifications DOCSIS 1.1 Radio Frequency Interface Specification CM-SP-RFIv1.1-C01-050907. Kentucky: Cable Television Laboratories Inc.

ETSI EN 301 703 V7.0.2 1999-12. Adaptive Multi-Rate (AMR); Speech Processing Functions; General Description. (GSM 06.71 version 7.0.2, Release 1998). France: ETSI.

ETSI EN 301 706 V7.1.1 1999. Comfort Noise Aspects for Adaptive Multi-Rate (AMR) Speech Traffic Channels. (GSM 06.92 version 7.1.1, Release 1998). France: ETSI.

ETSI EN 301 708 V7.1.1 1999. Voice Activity Detector (VAD) for Adaptive Multi-Rate (AMR) Speech Traffic Channels. (GSM 06.94 version 7.1.1, Release 1998). France: ETSI.

ETSI ETS 300 648-01 1997. Public Switched Telephone Network (PSTN); Calling Line Identification Presentation (CLIP) Supplementary Service. France: ETSI.

ETSI ETS 300 659-1 V1.3.1 2001. Access and Terminals (AT); Analogue Access to the Public Switched Telephone Network (PSTN); Subscriber Line Protocol over the Local Loop for Display (and related) Services–Part 1: On-Hook Data transmission. France: ETSI.

ETSI ETS 300 659-2 V1.3.1 2001. Access and Terminals (AT); Analogue Access to the Public Switched Telephone Network (PSTN); Subscriber Line Protocol over the Local Loop for Display (and related) Services–Part 2: Off-Hook Data Transmission. France: ETSI.

ETSI ETS 300 778-1 1997. Public Switched Telephone Network (PSTN): Protocol Over the Local Loop for Display and Related Services: Terminal Equipment Requirements—Part 1: Off-Line Data Transmission. France: ETSI.

ETSI ETS 101 329-5 2000. Quality of Service (QoS) Measurement Methodologies. France: ETSI.

FR-763-01 2006. DTMF Digital Simulation Immunity. New Jersey: Telcordia Technologies.

FT ITS-1 2007. Analogue Subscriber Interface Characteristics ITS1. Eighth edition. France: French Telecom.

FT ITS-3 2005. Interface Technical Specifications—Ringing, Tones and Dialing on Analogue Lines. Fourth edition. France: French Telecom.

FT STI-4 2004. User-Network Interface Characteristics for FSK Data Transmission (V.23). France: French Telecom. Available at: http://www.francetelecom.com/fr/groupe/initiatives/savoirplus/documentation/spec_techniques/att00022723/STI04-ed6_EN.pdf.

GR-30-CORE 1998. LSSGR: Voice Band Data Transmission Interface (FSD 05-01-0100), Section 6.6, Issue 2. New Jersey: Telcordia Technologies.

GR-31-CORE 2000. LSSGR: CLASS^{SM} Feature: Calling Number Delivery (FSD 01-02-1051), Issue 1. New Jersey: Telcordia Technologies.

GR-57-CORE 2001. Functional Criteria for Digital Loop Carrier Systems, Issue 1. New Jersey: Telcordia Technologies. (Note: GR-57-CORE 2001 replaces TR-NWT-000057 1993 version).

GR-506-CORE 1996. LSSGR: Signaling for Analog Interfaces-A Module of LSSGR, FR-64. New Jersey: Telcordia Technologies.

IEEE STD-743 1995. IEEE Standard Equipment Requirements and Measurement Techniques for Analog Transmission Parameters for Telecommunications. New York: IEEE Publications.

ITU-Handbook 1992. ITU Handbook on Telephonometry. Geneva: ITU Publication.

ITU-T Delayed contribution D.20 2001. The Burst Ratio: A Measure of Bursty Packet Loss Effects. Geneva: ITU Publication.

ITU-T Rec. E.164 2005. The International Public Telecommunication Numbering Plan. Geneva: ITU Publication.

ITU-T Rec. E.180/Q.35 1998. Technical Characteristics of Tones for the Telephone Service. Geneva: ITU Publication.

ITU-T Rec. E.182 1998. Operation, Numbering, Routing and Mobile Services—International Operation—Tones in National Signaling Systems. Geneva: ITU Publication.

ITU-T Rec. E.453 1994. Facsimile Image Quality as Corrupted by Transmission-Induced Scan Line Errors. Geneva: ITU Publication.

ITU-T Rec. E.458 1996. Figure of Merit for Facsimile Transmission Performance. Geneva: ITU Publication.

ITU-T Rec. F.185 1998. Internet Facsimile: Guidelines for the Support of the Communication of Facsimile Documents. Geneva: ITU Publication.

ITU-T Rec. G.100.1 2001. The Use of the Decibel and of Relative Levels in Speech Band Telecommunications. Geneva: ITU Publication.

ITU-T Rec. G.107 2005. The E-model, a Computational Model for use in Transmission Planning, Geneva: ITU Publication.

ITU-T Rec. G.107 2006. Amendement-I. New Appendix II-Provisional Impairment Factor Framework for Wideband Speech Transmission. Geneva: ITU Publication.

ITU-T Rec. G.108 1999. Application of the E-Model: A Planning Guide. Geneva: ITU Publication.

ITU-T Rec. G.109 1999. Definition of Categories of Speech Transmission Quality. Geneva: ITU Publication.

ITU-T Rec. G.111 1993. Loudness Ratings (LRs) in an International Connection. Geneva: ITU Publication.

ITU-T Rec. G.113 2002. Appendix I: Provisional Planning Values for the Equipment Impairment Factor Ie and Packet-Loss Robust. Factor Bpl. Geneva: ITU Publication.

ITU-T Rec. G.113 2003. Transmission Impairments due to Speech Processing. Geneva: ITU Publication.

ITU-T Rec. G.113 2006. Amendment 1: New Appendix IV-Provisional Planning Values for the Wideband Equipment Impairment Factor Ie,wb. Geneva: ITU Publication.

ITU-T Rec. G.113 2007. Amendment 2: Revised Appendix I—Provisional Planning Values for the Equipment Impairment Factor Ie and Packet-Loss Robustness Factor Bpl. Geneva: ITU Publication.

ITU-T Rec. G.114 2003a. One-Way Transmission Time. Geneva: ITU Publication.

ITU-T Rec. G.114 2003b. Appendix II: Guidance on One-Way Delay for Voice over IP. Geneva: ITU Publication.

ITU-T Rec. G.116 1999. Transmission Performance Objectives Applicable to End-to-End. Geneva: ITU Publication.

ITU-T Rec. G.121 1993. Loudness Ratings (LRs) of National Systems. Geneva: ITU Publication.

ITU-T Rec. G.122 1993. Influence of National Systems on Stability and Talker Echo in International Connections. Geneva: ITU Publication.

ITU-T Rec. G.126 1993. Listener Echo in Telephone Networks. Geneva: ITU Publication.

ITU-T Rec. G.131 2003. Talker Echo and its Control. Geneva: ITU Publication.

ITU-T Rec. G.161 2004. Interaction Aspects of Signal Processing Network Equipment. Geneva: ITU Publication.

ITU-T Rec. G.164 1988. Echo Suppressors. Geneva: ITU Publication.

ITU-T Rec. G.168 2004. Digital Network Echo Canceller. Geneva: ITU Publication.

ITU-T Rec. G.169 1999. Automatic Level Control Devices. Geneva: ITU Publication.

ITU-T Rec. G.173 1993. Transmission Planning Aspects of the Speech Service in Digital Public and Land Mobile Networks. Geneva: ITU Publication.

ITU-T Rec. G.175 2000. Transmission Planning for Private/Public Network Interconnection of Voice Traffic. Geneva: ITU Publication.

ITU-T Rec. G.177 1999. Transmission Planning for Voice Band Services over Hybrid Internet/PSTN Connections. Geneva: ITU Publication.

ITU-T Rec. G.191 2005. Software Tools for Speech and Audio Coding Standardization. Geneva: ITU Publication.

ITU-T Rec. G.191 STL-2005. ITU-T Software Tool Library 2005 User's Manual. Geneva: ITU Publication.

ITU-T Rec. G.711 1988. Pulse Code Modulation (PCM) of Voice Frequencies. Geneva: ITU Publication.

ITU-T Rec. G.711 Appendix I 1999. A High Quality Low-Complexity Algorithm for Packet Loss Concealment for G.711. Geneva: ITU Publication.

ITU-T Rec. G.711 2000. Appendix-II: A Comfort Noise Payload Definition for ITU-T G.711 use in Packet-Based Multimedia Communication Systems. Geneva: ITU Publication.

ITU-T Rec. G.722 1988. 7 kHz Audio-Coding within 64 kbits/s. Geneva: ITU Publication.

ITU-T-Rec. G.722 Appendix IV 2007. A Low-Complexity Algorithm for Packet Loss Concealment for G.722. Geneva: ITU Publication.

ITU-T Rec. G.722.2 2003. Wideband Coding of Speech at around 16 kbits/s using Adaptive Multi-rate Wideband (AMR-WB). Geneva: ITU Publication.

ITU-T Rec. G.723.1 2006. Dual Rate Speech Coder for Multimedia Communications Transmitting at 5.3 and 6.3 kbits/s. Geneva: ITU Publication.

ITU-T Rec. G.726 1990. 40, 32, 24, 16 kbit/s Adaptive Differential Pulse Code Modulation (ADPCM). Geneva: ITU Publication.

ITU-T Rec. G.728 1992. Coding of Speech at 16 kbits/s using Low-Delay Code Excited Linear Predictions. Geneva: ITU Publication.

ITU-T Rec. G.729 1996. Coding of Speech at 8 kbit/s using Conjugate-Structure Algebraic-Code-Excited Linear-Prediction (CS-ACELP). Geneva: ITU Publication.

ITU-T Rec. G.729A 1996. Coding of Speech at 8 kbit/s using Conjugate-Structure Algebraic-Code-Excited Linear-Prediction (CS-ACELP), Annex A: Reduced Complexity 8 kbit/s CS-ACELP Speech Codec. Geneva: ITU Publication.

ITU-T Rec. G.729B 1996. Annex B, C Source Code and Test Vectors for Implementation Verification of the Algorithm of the G.729 Silence Compression Scheme. Geneva: ITU Publication.

ITU-T Rec. G.729E 1996. Annex E, Coding of Speech at 8 kbits/s using Conjugate-Structure Algebraic-Code-Excited Linear-Prediction (CS-ACELP); Annex E 11.8 kbits/s CS-ACELP Speech Coding Algorithm. Geneva: ITU Publication.

ITU-T Rec. G.729.1 2006. G.729 based Embedded Variable Bit-Rate Coder: An 8–32 kbits/s Scalable Wideband Coder Bit Stream Interoperable with G.729. Geneva: ITU Publication.

ITU-T Rec. G.992.1 1999. Asymmetric Digital Subscriber Line (ADSL) Transceivers. Geneva: ITU Publication.

ITU-T Rec. G.992.3 2005. Asymmetric Digital Subscriber Line Transceivers 2 (ADSL2). Geneva: ITU Publication.

ITU-T Rec. G.992.5 2005. Asymmetric Digital Subscriber Line (ADSL) Transceivers— Extended Bandwidth ADSL2 (ADSL2plus). Geneva: ITU Publication.

ITU-T Rec. G.993.1 2004. Very High Speed Digital Subscriber Line Transceivers. Geneva: ITU Publication.

ITU-T Rec. G.993.2 2006. Very High Speed Digital Subscriber Line Transceivers 2 (VDSL2). Geneva: ITU Publication.

ITU-T Rec. G.1020 2006. Performance Parameter Definitions for Quality of Speech and other Voice Band Applications Utilizing IP Networks. Geneva: ITU Publication.

ITU-T Rec. H.225.0 2005 Call Signaling Protocols and Media Stream Packetization for Packet-Based Multimedia Communication Systems. Geneva: ITU Publication.

ITU-T Rec. H.245 2005. Control Protocol for Multimedia Communication. Geneva: ITU Publication.

ITU-T Rec. H.248 2005. Gateway Control Protocol: Version 3. Geneva: ITU Publication.

ITU-T Rec. H.323 2006. Packet-Based Multimedia Communications Systems. Geneva: ITU Publication.

ITU-T Rec. P.79 1999. Calculation of Loudness Rating for Telephone Sets. Geneva: ITU Publication.

ITU-T Rec. P.310 2003. Transmission Characteristics for Telephone Band (300–3400 Hz) Digital Telephones. Geneva: ITU Publication.

ITU-T Rec. P.311 2005. Transmission Characteristics for Wideband (150–7000 Hz) digital hound ret telephones. Geneva: ITU Publication.

ITU-T Rec. P.340 2000. Transmission Characteristics and Speech Quality Parameters of Hands-Free Terminals. Geneva: ITU Publication.

ITU-T Rec. P.341 2005. Transmission Characteristics for Wideband (150–7000 Hz) Digital Hands-Free Telephony Terminals. Geneva: ITU Publication.

ITU-T Rec. P.563 2004. Single-Ended Method for Objective Speech Quality Assessment in Narrow-Band Telephony Applications. Geneva: ITU Publication.

ITU-T Rec. P.800 1996. Methods for Subjective Determination of Transmission Quality. Geneva: ITU Publication.

ITU-T Rec. P.830 1996. Subjective Performance Assessment of Telephone-Band and Wideband Digital Codecs. Geneva: ITU Publication.

ITU-T Rec. P.833 2001. Methodology for Derivation of Equipment Impairment Factors from Subjective Listening-Only Tests. Geneva: ITU Publication.

ITU-T Rec. P.862 2001. Perceptual Evaluation of Speech Quality (PESQ) an Objective Method for End-to-End Speech Quality Assessment for End-to-End Speech Quality Assessment of Narrow-Band Telephone Networks and Speech Codecs. Geneva: ITU Publication.

ITU-T Rec. P.862.1 2003. Mapping Function for Transforming P.862 Raw Results Scores to MOS-LQO. Geneva: ITU Publication.

ITU-T Rec. Q.23 1988. Technical Features of Push-Button Telephone Sets. Geneva: ITU Publication.

ITU-T Rec. Q.24 1988. Multi-Frequency Push Button Signal Reception. Geneva: ITU Publication.

ITU-T Rec. T.4 2003. Standardization of Group 3 Facsimile Terminals for Document Transmission. Geneva: ITU Publication.

ITU-T Rec. T.6 1988. Facsimile Coding Schemes and Coding Control Functions for Group 4 Facsimile Apparatus. Geneva: ITU Publication.

ITU-T Rec. T.30 1996. Procedure for Document Facsimile Transmission in the General Switched Telephone Network. Geneva: ITU Publication.

ITU-T Rec. T.30 2005. Procedure for Document Facsimile Transmission in the General Switched Telephone Network. Geneva: ITU Publication.

ITU-T Rec. T.37 1998. Procedure for the Transfer of Facsimile Data via Store-and-Forward on the Internet. Geneva: ITU Publication.

ITU-T Rec. T.38 2005. Procedures for Real-Time Group 3 Facsimile Communication over IP Networks. Geneva: ITU Publication.

ITU-T Rec. T.42 1996. Continuous-Tone Colour Representation Method for Facsimile. Geneva: ITU Publication.

ITU-T Rec. T.62 1993. Control Procedures for Teletex and Group 4 Facsimile Services. Geneva: ITU Publication.

ITU-T Rec. T.70 1993. Network-Independent Basic Transport Service for the Telematic Services. Geneva: ITU Publication.

ITU-T Rec. T.81 1993. Information Technology—Digital Compression and Coding of Continuous-Tone Still Images—Requirements and Guidelines. Geneva: ITU Publication.

ITU-T Rec. T.82 1993. Coded Representation of Picture and Audio Information—Progressive Bi-Level Image Compression. Geneva: ITU Publication.

ITU-T Rec. T.503 1991. A Document Application Profile for the Interchange of Group 4 Facsimile Documents. Geneva: ITU Publication.

ITU-T Rec. V.2 1988. Data Communication over the Telephone Network. Geneva: ITU Publication.

ITU-T Rec. V.8 2000. Procedure for Starting Sessions of Data Transmission over the Public Switched Telephone Network. Geneva: ITU Publication.

ITU-T Rec. V.17 1991. A 2-Wire Modem for Facsimile Applications with Rates up to 14400 bit/s. Geneva: ITU Publication.

ITU-T Rec. V.21 1988. 300 bits per Second Duplex Modem Standardized for Use in the General Switched Telephone Network. Geneva: ITU Publication.

ITU-T Rec. V.23 1988. 600/1200 Baud Modem Standardized for Use in the General Switched Telephone Network. Geneva: ITU Publication.

ITU-T Rec. V.25 1996. Automatic Answering Equipment and General Procedure for Automatic Calling Equipment on the General Switched Telephone Network including Procedures for Disabling of Echo Control Devices for both Manually and Automatically Established Calls. Geneva: ITU Publication.

ITU-T Rec. V.26 1988. 2400 bits per second Modem Standardized for use on 4-wire Leased Telephone-Type Circuits. Geneva: ITU Publication.

ITU-T Rec. V.27ter 1988. 4800/2400 bits per Second Modem Standardized for use in the General Switched Telephone Network. Geneva: ITU Publication.

ITU-T Rec. V.29 1988. 9600 bits per Second Modem Standardized for Use on Point-to-Point 4-Wire Leased Telephone-Type Circuits. Geneva: ITU Publications.

ITU-T Rec. V.34 1998. A Modem Operating at Data Signaling Rates of up to 33,600 bits/s for Use on the General Switched Telephone Network and on Leased Point to Point 2-Wire Telephone Type Circuits. Geneva: ITU Publication.

ITU-T Rec. V.42 2002 Error-correcting Procedures for DCEs using Asynchronous-to-Synchronous Conversion. Geneva: ITU Publication.

ITU-T Rec. V.150.1 2003. Modem-over-IP Networks: Procedures for the End-to-End Connections of V-Series DCEs. Geneva: ITU Publication.

ITU-T Rec. Y.1541 2006. Network Performance Objective for IP-Based Services. Geneva: ITU Publication.

T1.521a 2000. Supplement to T1.521-1999, Packet Loss Concealment for Use with ITU-T Recommendation G.711. Washington: Alliance for Telecommunications Industry solutions (ATIS).

TIA/EIA-116 2001. Telecommunications-IP Telephony Equipment-Voice Quality Recommendations for IP Telephony. Virginia: Telecommunications Industry Association (TIA).

TIA/EIA-116A 2006. Telecommunications-IP Telephony Equipment-Voice Quality Recommendations for IP Telephony. Virginia: Telecommunications Industry Association.

TIA/EIA-470C 2003. Telephone Terminal Equipment—Overview of Performance Standards for Analog Telephones (Revision of TIA-470-B). Virginia: Telecommunications Industry Association.

TIA/EIA-470.110-C 2004. Telephone Terminal Equipment—Handset Acoustic Performance Requirements. Virginia: Telecommunications Industry Association.

TIA/EIA-810A 2000. Transmission Requirements for Narrowband Voice over IP and Voice over PCM Digital Wire Line Telephones. Virginia: Telecommunications Industry Association.

TIA/EIA-912 2002. Telecommunications Telephony Equipment Voice Gateway Transmission Requirements. Virginia: Telecommunications Industry Association.

TIA/EIA-920 2002. Transmission requirements for Wideband Digital Wire Line Telephones. Virginia: Telecommunications Industry Association.

TIA/EIA-464C 2001. Telecommunications Multiline Terminal Systems Requirements for PBX Switching Equipment. Virginia: Telecommunications Industry Association.

TR-NWT-000057 1993. Functional Criteria for Digital Loop Carrier Systems, Chapter 6. New Jersey: Bell Communications Research. (Note: Refer GR-57-Core 2001 that replaces TR-NWT-000057).

REFERENCES BY URLS

URL (ADI-218x). ADSP-218x Family Selection Table. Available at: http://www.analog.com/processors/adsp/overview/IST.html. Accessed 2007 Sept 25.

URL (ADI-BF536). ADSP-BF536 Blackfin Processor with Embedded Network Connectivity. Available at: http://www.analog.com/UploadedFiles/Data_Sheets/ADSP-BF534_BF536_BF537.pdf. Accessed 2007 Dec 27.

URL (ADI-VoIP). Design Your Own VoIP Solution with a Blackfin Processor-Add Enhancements Later. Available at: http://www.analog.com/library/analogDialogue/archives/40-04/blackfin_voip.pdf. Accessed 2007 Sept 25.

URL (ADI Vol-2). Digital Signal Processing Applications using the ADSP-2100 Family, Volume 2. http://www.analog.com/UploadedFiles/Associated_Docs/60899921adsp2100vol2.zip. Accessed 2008 May 20.

URL (Advent-5120). Telephone Line Testers. Available at: http://www.adventinstruments.com/products/p-5120.htm#Applications. Accessed 2007 July 22.

URL (Advent-CID1). FSK Based Caller ID Simulator Software. Available at: http://adventinstruments.com/Wdf/swup_bul/b991018b.pdf. Accessed 2007 Oct 23.

URL (Advent-CID2) Overview of Caller ID Standard and Market in Taiwan. Available at: http://www.adventinstruments.com/resources/giart/giSTD3a.htm. Accessed 2007 Dec 30.

URL (Ameritec). Voice over Packet Test Applications. Available at: http://www.ameritec.com/downloads/whitepapers/VoP%20Test%20Applications%20-%20GMOS,%20G-PSQM,%20and%20G-PESQ.pdf. Accessed 2007 Oct 13.

URL (Astrek). Astrek Announcements. Available at: http://www.asterisk.org/. Accessed 2007 Sept 25.

URL (ATIS). Facsimile (FAX). Available at: http://www.atis.org/tg2k/_facsimile.html. Accessed 2007 Oct 12.

URL (Audiocodes-fax). Fax Relay over Packet Networks-White Paper. Available at: http://www.audiocodes.com/objects/FaxRelay.pdf. Accessed 2007 Oct 12.

URL (Brekeke). A SIP Registrar and SIP Proxy Server. Available at: http://www.brekeke.com/. Accessed 2008 Dec 27.

URL (Broadcom). BCM1111 Product Brief. Available at: http://www.broadcom.com/collateral/pb/1111-PB01-R.pdf. Accessed 2007 Sept 25.

URL (Canata). Peter J Davidson. V.34 Fax: Superior Performance and Cost Savings. Available at: http://www.cantata.com/whitepapers/pdf/v34_whitepaper.pdf. Accessed 2007 Oct 12.

URL (CED). Tackling Fax over IP. Available at: http://www.cedmagazine.com/tackling-fax-over-ip.aspx. Accessed 2007 Nov 9.

URL (CID-FAQ). Caller ID FAQ. Available at: http://www.ainslie.org.uk/callerid/cli_faq.htm. Accessed 2007 Oct 14.

URL (Cisco-cable) Understanding Upstream Modulation Profiles. Available at: http://www.cisco.com/warp/public/cc/so/neso/ns289/mdpth_ov.pdf. Accessed 2007 Dec 30.

URL (Cisco-CID). Caller ID. Available at: http://www.cisco.com/univercd/cc/td/doc/product/software/ios121/121newft/121t/121t3/clid_t4.pdf. Accessed 2007 Oct 14.

URL (Cisco-coding). Waveform Coding Techniques. Available at: http://www.cisco.com/warp/public/788/signalling/waveform_coding.html. Accessed 2007 July 16.

URL (Cisco-delay). Understanding Delay in Packet Voice Networks. Available at: http://www.cisco.com/warp/public/788/voip/delay-details.pdf Accessed 2007 Oct 20.

URL (Cisco-EC). Echo Analysis for Voice over IP. Available at: http://www.cisco.com/univercd/cc/td/doc/cisintwk/intsolns/voipsol/ea_isd.pdf. Accessed 2007 Oct 26.

URL (Cisco-fax1). Fax Services Overview. Available at: http://www.cisco.com/univercd/cc/td/doc/product/software/ios122/122newft/122t/122t11/faxapp/fxappdoc.pdf. Accessed 2007 Aug 29.

URL (Cisco-fax2). Fax Services. Available at: http://www.cisco.com/univercd/cc/td/doc/cisintwk/intsolns/voipsol/fax_isd.pdf. Accessed 2007 Aug 29.

URL (Cisco-fax3). Cisco IOS Fax Services over IP Configuration Guide. Available at: http://www.cisco.com/application/pdf/en/us/guest/products/ps4068/c1091/ccmigration_09186a00801446ef.pdf. Accessed 2007 Dec 16.

URL (Cisco-fax4). Fax and Modem Services over IP Overview. Available at: http://www.cisco.com/univercd/cc/td/doc/product/software/ios123/123cgcr/vvfax_c/faxmodem/fxappdoc.pdf.

URL (Cisco-G168). NextPort Dual-Filter G.168 Echo Canceller White Paper. Available at: http://www.cisco.com/univercd/cc/td/doc/product/access/acs_serv/as5400/sw_conf/g168nxt.pdf. Accessed 2007 Oct 26.

URL (Cisco-impedance). Analog Voice Port Best Match Impedance Setting Choice. Available at: http://www.cisco.com/warp/public/788/voip/impedance_choice.pdf. Accessed 2007 Sept 15.

URL (Cisco-modem). Modem Support for VoIP. Available at: http://www.cisco.com/en/US/docs/ios/12_3/vvf_c/cisco_ios_voice_configuration_library_glossary/vclmodem.html. Accessed 2008 March 27.

URL (Cisco-QoS). Using QoS to Optimize Voice Quality in VoIP Networks-White Paper. Available at: http://www.cisco.com/warp/public/779/smbiz/community/qos_voip.html. Accessed 2007 Oct 12.

URL (Commetrex). Commetrex Announces T.38 Interoperability Lab. Available at: http://www.commetrex.com/press_releases/01212002T38Lab.html. Accessed 2007 Sept 24.

URL (dialup). Dial-up Modem Standards. Available at: http://www.webopedia.com/quick_ref/dialup_modem_standards.asp. Accessed 2007 Aug 29.

URL (Ditech) Echo Basis Tutorial. Available at: http://www.ditechnetworks.com/learningCenter/echoBasics.html. Accessed 2007 Dec 16.

URL (DSLAII). Digital Speech Level Analyzer II. Available at: http://www.malden.co.uk/dsla.htm. Accessed 2007 Oct 13.

URL (DSLA-usrgd). Digital Speech Level Analyser User Guide, Revision 4.4. Available at: http://www.malden.co.uk/downloads/dslausrgd.pdf. Accessed 2007 Oct 23.

URL (EECIS-NTP). Network Timing Protocol General Overview. Available at: http://www.eecis.udel.edu/~mills/database/brief/overview/overview.pdf. Accessed 2007 Oct 14.

URL (Elect-matching). Impedance Matching: A Primer. Available at: http://www1.electusdistribution.com.au/images_uploaded/impmatch.pdf. Accessed 2007 Sept 15.

URL (ETSI-news). ETSI News. Available at: http://www.etsi.org/WebSite/NewsandEvents/newsandevents.aspx. Accessed 2007 Sept 24.

URL (Encore-G729AB). ITU-T-G.729AB. Available at: http://www.ncoretech.com/products/ip/speech/g729ab.html. Accessed 2007 Sept 25.

URL (Encore-T38). Fax over IP (T.38). Available at: http://www.ncoretech.com/products/ip/fax/t38.html. Accessed 2007 Sept 25.

URL (Fax-theory) Facsimile Theory. Available at: http://www.geocities.com/SiliconValley/Vista/7055/fax.htm. Accessed 2007 Dec 18.

URL (Freescale). MCS8101 Networked Digital Signal Processor Rev.18 08/2005. Available at: http://www.freescale.com/files/dsp/doc/data_sheet/MSC8101.pdf?pspll=1. Accessed 2007 Sept 25.

URL (Fujitsu-IP). IP Phone Chip. Available at: http://www.fujitsu.com/downloads/MICRO/fma/pdf/ip_phone.pdf. Accessed 2007 July 12.

URL (Gips-G711). GIPS Enhanced G.711 Improved Sound Quality with Standard Codec Compatibility. Available at: http://www.gipscorp.com/files/english/datasheets/EG711.pdf. Accessed 2007 Sept 9.

URL (GL). Echo Canceller Testing. Available at: http://www.gl.com/echocan.html?source=google#voiceandvqt. Accessed 2007 Oct 26.

URL (HDLC). HDLC Protocol. Available at: http://www.interfacebus.com/HDLC_Protocol_Description.html. Accessed 2007 Oct 12.

URL (IDT). Overview of IDT Wide Area Network PLLs. Available at: http://www.electronicsweekly.com/Articles/2006/08/14/39474/Overview+of+IDT+wide+area+network+PLLs.htm. Accessed 2007 Oct 14.

URL (IEC-ADSL). Extending Asymmetric Digital Subscriber Line (ADSL) Services to Remote Digital Loop Carrier (DLC) Locations. Available at: http://www.iec.org/online/tutorials/adsl_dlc/topic02.html?Next.x=33&Next.y=19. Accessed 2007 Oct 22.

URL (IEC-DLC). Digital Loop Carrier (DLC). Available at: http://www.iec.org/online/tutorials/dlc/. Accessed 2007 Oct 22.

URL (IEEE-802.1Q). Virtual Bridged Local Area Networks. Available at: http://standards.ieee.org/getieee802/download/802.1Q-2003.pdf. Accessed 2007 Oct 14.

URL (iLBC). GIPS iLBC. Available at: http://www.gipscorp.com/files/english/white_papers/iLBC.WP.pdf. Accessed 2007 July 16.

URL (Ikanos-Fusiv). Fusiv Vx150-IKF6833-1 Gateway Processor for Broadband Gateway Systems. Available at: http://www.ikanos.com/lib/assets/pdfs/solutions/Fusiv_Vx150.pdf. Accessed 2007 Sept 25.

URL (iPCM). GIPS iPCM-wb. Available at: http://www.gipscorp.com/files/english/datasheets/iPCM-wb.pdf. Accessed 2007 Nov 24.

URL (iSAC). GIPS iSAC. Available at: http://www.gipscorp.com/files/english/datasheets/iSAC.pdf. Accessed 2007 July 16.

URL (ISDN). ISDN Definitions. Available at: http://www.ralphb.net/ISDN/defs.html. Accessed 2007 July 12.

URL (Julian). Julian Penketh Providing Touch-Tone services in Voice over Packet Networks. Available at: http://www.ee.ucl.ac.uk/lcs/papers2000/lcs044.pdf. Accessed 2007 Nov 17.

URL (Legerity-Ring). Am79R70/79/100/101Ringing SLIC Device Technical Overview. Available at: http://www.legerity.com/getfile.php?bpd_51_AN_RSLIC-Technical-Overview_C1_ID080158.pdf. Accessed 2007 Nov 13.

URL (Legerity-WB). Available at: http://www.legerity.com/products.php?sid=&bpid=36#documentation. Accessed 2007 Oct 19.

URL (Linux-docs). Appendix C: baud vs bps, A Simple Example. Available at: http://www.linuxdocs.org/HOWTOs/Modem-HOWTO-22.html. Accessed 2007 Oct 12.

URL (Maine). Facsimile Theory. Available at: http://home.maine.rr.com/randylinscott/Fax.htm. Accessed 2007 Oct 12.

URL (Microtronix-470C). Microtronix ANSI/TIA-470-C Test Systems. Available at: http://microtronix.ca/470c_countrykit.html. Accessed 2007 Sept 15.

URL (Microtronix-501). Model 501 Telephone Test System. Available at: http://www.microtronix.ca/wireline.html. Accessed 2007 Sept 15.

URL (Microtronix-country). Country Kits. Available at: http://www.microtronix.ca/countrykits.html Accessed 2007 Sept 15.

URL (Midcom). Singing Return Loss. Available at: http://www.midcom-inc.com/Tech/singing_rl.asp. Accessed 2007 Oct 22.

URL (Mindspeed). Comcerto 500 Series VoIP Processors. Available at: http://www.mindspeed.com/web/download/download.jsp?docId=28041. Accessed 2007 Sept 25.

URL (MIPS32). MIPS Technologies, 32-bit cores. Available at: http://www.mips.com/products/cores/32-bit-cores/. Accessed 2007 Nov 17.

URL (Multitech). Fax overview Developer's Reference. Available at: www.multitech.com/DOCUMENTS/Collateral/manuals/S000265B.pdf. Accessed 2007 Oct 15.

URL (National-Eth). Ethernet Packet Format. Available at: http://www.national.com/AU/design/courses/260/index.htm?start_file=pac06/01pac06.htm. Accessed 2007 Oct 14.

URL (Netiq). A Handbook for Successful VoIP Deployment: Network Testing, QoS, and More. Available at: http://download.netiq.com/CMS/NetIQ_Handbook_for_Successful_VoIp_Deployment.pdf. Accessed 2007 Sept 24.

URL (Newport-BW). VoIP Bandwidth Calculations, Newport Networks. Available at: http://www.newport-networks.com/cust-docs/52-VoIP-Bandwidth.pdf. Accessed 2007 Oct 14.

URL (NMS-Access). Open Access Family of Boards and Software. Available at: http://www.nmscommunications.com/DevPlatforms/OpenAccess/default.htm. Accessed 2007 Sept 24.

URL (NTT-E). Telephone Service Interface. Fifth edition. Japan: NTT. Available at: http://www.ntt-east.co.jp/gisanshi/analog/edit5e.pdf. Accessed 2007 July 12.

URL (Opticom-P563). P.563 Single Ended Speech Quality Measure. Available at: http://opticom.de/download/02SpecSheet_P563_06-03-03.pdf. Accessed 2007 Nov 14.

URL (Opticom-PESQ). PESQ–Opticom. Available at: http://www.opticom.de/technology/pesq.html. Accessed 2007 July 16.

URL (Orion-EC). Voice Quality Enhancement (VQE) & Echo Canceller. Available at: http://www.oriontelecom.com/brochures/ds3_ec.pdf. Accessed 2007 Oct 26.

URL (OSU-Eth). Ethernet Frame. Available at: http://8help.osu.edu/wks/sysadm_course/html/sysadm-326.html. Accessed 2007 Oct 14.

URL (PCM). Understanding PCM Coding—Application note-AN574.1. Available at: http://www.powerdesigners.com/InfoWeb/design_center/Appnotes_Archive/an574.pdf. Accessed 2007 July 16.

URL (pic). Caller ID Basics. Available at: www.picbasic.co.uk/support/tml_callerid_cnt.pdf. Accessed 2007 Oct 14.

URL (PIQUA). PIQUA System Quality Management for IP-Based Services. Available at: http://focus.ti.com/pdfs/bcg/PIQUA_product_bulletin.pdf. Accessed 2007 Oct 20.

URL (POTS). Analog Telephony—POTS Basics. Available at: http://www.cisco.com/warp/public/788/signalling/tipandring.gif. Accessed 2007 July 12.

URL (Protocols-ATM). ATM. Available at: http://www.protocols.com/pbook/atm.htm. Accessed 2007 Oct 12.

URL (Qualitylogic). Fax/Telephony Test Tools. Available at: http://www.qualitylogic. com/genoa_test_tools/fax/faxlab.html. Accessed 2007 Aug 29.

URL (Quartz). Plug-in NTP server for OSA 5581C GPS GPS-SR or OSA 5240. Available at: http://www.oscilloquartz.com/file/pdf/TCU_03_screen.pdf. Accessed 2007 Oct 14.

URL (Qualitylogic). Fax/Telephony Test Tools. Available at: http://www.qualitylogic. com/genoa_test_tools/fax/faxlab.html. Accessed 2007 Dec 27.

URL (Radvision). Radvision SIP server. Available at: http://www.radvision.com/Products/Developer/SIPServer/. Accessed 2007 Dec 27.

URL (RFC791). RFC0791, Internet Protocol. September 1981. Available at: http://www.ietf.org/rfc/rfc0791.txt?number=791. Accessed 2007 Oct 12.

URL (RFC822). RFC822, Standard for the Format of ARPA Internet Text Messages. 1982. Available at: http://www.ietf.org/rfc/rfc0822.txt?number=822. Accessed 2007 Oct 14.

URL (RFC791). RFC0791, Internet Protocol. 1981. Available at: http://www.ietf.org/rfc/rfc0791.txt?number=791. Accessed 2007 Oct 14.

URL (Sage935). Sage Instruments 935AT Advanced Technology Communication Test Set. Available at: http://www.sageinst.com/935AT.html. Accessed 2007 Oct 13.

URL (Sage-Balance). Measuring the Longitudinal Balance of 2-Wire Equipment. Available at: http://www.sageinst.com/downloads/925an/longball.pdf. Accessed 2007 Oct 22.

URL (Sage-fax) Information on Sage's Fax Emulator 2004. Available at: http://www.sageinst.com/downloads/Faxtrx.pdf. Accessed 2007 Oct 12.

URL (Sagem-RG) Broadband devices–Sagem Communications. Available at: http://www.sagem-communications.com/index.php?id=52&L=0. Accessed 2008 Mar 27.

URL (Sage-options). Sage Instruments 935AT Test Options Guide. Available at: http://www.sageinst.com/downloads/935at/935optionsguide1.pdf. Accessed 2007 Oct 22.

URL (Sage-PAR). Voice Frequency Measurements. Available at: http://www.sageinst. com/downloads/925an/vfmeas.pdf. Accessed 2007 Nov 23.

URL (Sangoma). Sangoma Digital Telephony Cards. Available at: http://www.sangoma. com/main/products/digitaltelephony. Accessed 2007 Sept 25.

URL (Seemix). Digital Telephone Hybrid. Available at: http://seemix.no/downloads/dhy02digital.pdf. Accessed 2007 Sept 15.

URL (Si3015). Si3215 PROSLIC Programmable CMOS SLIC/CODEC with Ringing/Battery Volatge Generation. Available at: http://www.silabs.com/public/documents/tpub_doc/dsheet/Wireline/ProSLIC/en/Si3215_Data_Sheet.pdf. Accessed 2007 Sept 15.

URL (Silab-DAA). Silicon Laboratories, Global Voice DAA. Available at: http://www.silabs.com/public/documents/tpub_doc/dsheet/Wireline/Silicon_DAA/en/si3050.pdf. Accessed 2007 Sept 15.

URL (SJ Labs). SJ Phone a VoIP Soft Phone. Available at: http://www.sjlabs.com/sjp.html. Accessed 2007 Dec 27.

URL (Skype).Skype VoIP Soft Phone. Available at: http://www.skype.com/download/skype/windows/. Accessed 2007 Dec 27.

URL (SLIC-WB). ProSLIC Programmable Wideband SLIC/Codec with Ringing/Battery Voltage Generation. Available at: http://www.silabs.com/public/documents/tpub_doc/dsheet/Wireline/ProSLIC/en/si3216.pdf. Accessed 2007 Sept 15.

URL (Spirent-IP). IP Wave Network Impairments Emulator. Available at: http://www.spirent.com/documents/385.pdf. Accessed 2007 Oct 13.

URL (Spirent-Protocol). Spirent Protocol Tester Next-Generation VoIP Signaling & IMS Protocol Testing. Available at: http://www.spirentcom.com/analysis/technology.cfm?media=7&ws=325&ss=209. Accessed 2007 Oct 13.

URL (SPRA073). DSP Solutions for Telephony and Data/Facsimile Modems: Application Book. Available at: http://focus.ti.com/lit/an/spra073/spra073.pdf. Accessed 2007 Oct 12.

URL (SPRA080). Implementation of an FSK Modem Using the TMS320C17, Application report SPRA080. Available at: http://focus.ti.com/lit/an/spra080/spra080.pdf. Accessed 2008 May 21.

URL (SPRA129). Digital Voice Echo Canceller with a TMS32020 Application Report SPRA729. Available at: http://focus.ti.com/lit/an/spra129/spra129.pdf. Accessed 2007 Oct 26.

URL (SPRA188). Implementing a Line Echo Canceller using the Block Update and NLMS Algorithms on the TMS32054x DSP, Application report SPRA188. Available at: http://focus.ti.com/lit/an/spra188/spra188.pdf. Accessed 2007 Oct 26.

URL (SPRA576). SPRA576 Application Report, Implementation of Echo Control for ITU G.165/DECT on TMS320c62xx processors, Application report SPRA576. Available at: http://focus.ti.com/lit/an/spra576/spra576.pdf. Accessed 2007 Oct 26.

URL (Surf-com). V.150 Modem over IP White Paper. Surf Communication Solutions. Available at: http://www.surf-com.com/pdfs/V.150_MoIP_WP.pdf. Accessed 2007 Aug 29.

URL (Symmetricom). Primary Reference Sources (PRSs). Available at: http://ngn.symmetricom.com/products/primary_reference_sources/primary_reference_sources.asp. Accessed 2007 Oct 14.

URL (T1/E1). Tutorial on T1/E1 Alarming, Dropping, and Inserting. Available at: http://www.commsdesign.com/design_corner/showArticle.jhtml?articleID=16501900. Accessed 2007 July 12.

URL (Telchemy). VQmon/EP Call Quality Monitoring for IP Phone and IP/TDM Gateways. Available at: http://www.telchemy.com/vqmonep.html. Accessed 2007 Oct 15.

URL (TI-54x). TMS320VC5421 Fixed-Point Digital Signal Processor. Available at: http://focus.ti.com/lit/ds/symlink/tms320vc5421.pdf. Accessed 2007 Sept 25.

URL (TIA-496B). TIA/EIA-496-B. Interface Between Data Circuits-Terminal Equipment (DCE) and the Public Switched Telephone Network (PSTN). Available at: http://ftp.tiaonline.org/tr-30/tr303/Public/9809Riverside/39809095.doc. Accessed 2007 July 12.

URL (TIAonline-2003). DC Resistance Requirements–Correspondence. Available at: http://ftp.tiaonline.org/tr-41/tr4136inactive/Public/2003-05-LakeBuenaVista/

TR41.3.6-03-05-009(word)-T1E1.3Liaison-DCResistance,JBress,AST.pdf. Accessed 2007 Sept 16.

URL (TIAonline-2006). Research of Ringer Loads in Other Standards. Available at: http://ftp.tiaonline.org/tr-41/tr41.1.1/Public/2006-02-PalmSprings/TR41.1.1-06-02-006-RingerLoadResearch,DMcKinnon,AST.pdf. Accessed 2007 Sept 16.

URL (TI-impedance). Active Output Impedance for ADSL Line Drivers. Available at: http://focus.ti.com/lit/an/slyt108/slyt108.pdf. Accessed 2007 Sept 15.

URL (TI-PCM). A-law and Mu-law Companding Implementations using TMS320C54x. Available at: http://focus.ti.com/lit/an/spra163a/spra163a.pdf. Accessed 2007 July 16.

URL (Tldp-modem). Modem-Howto. Available at: http://tldp.org/HOWTO/Modem-HOWTO.html#toc1.7. Accessed 2007 Aug 29.

URL (Tolly). VoIP Capable Infrastructure (Quality of Service). Available at: http://www.tolly.com/TVProgDetail.aspx?ProgID=10501. Accessed 2007 Sept 24.

URL (TTA). TTA News. Available at: http://www.tta.or.kr/English/new/main/index.htm. Accessed 2007 Sept 24.

URL (Verisilicon). Verisilicon Products & Solutions. Available at: http://www.verisilicon.com/en/products.asp?id=26: Accessed 2007 Sept 25.

URL (VoIP-Supply). VoIP Supply. Available at: http://www.voipsupply.com/home.php. Accessed 2007 July 12.

URL (WinSLAC). WinSLAC Software User's Guide. Available at: http://www.nalanda.nitc.ac.in/industry/AppNotes/AMD/22646.pdf. Accessed 2007 Sept 15.

URL (Wireshark). Network Toolkit. Available at: http://www.wireshark.org/. Accessed 2007 Sept 24.

REFERENCES BY AUTHOR NAMES

ADI Vol-1 (1990). Digital Signal Processing Applications using the ADSP-2100 Family, Volume 1. Englewood Cliffs NJ: Prentice Hall Publications. Available at: http://www.analog.com/UploadedFiles/Associated_Docs/2127342adsp2100vol1.zip. Accessed 2007 Dec 26.

Alexander R (2006). Speech Quality of VOIP Assessment and Prediction. New York: Wiley.

Almes G, Kalidindi S, Zekauskas, M (1999). RFC2679, A One-Way-Delay Metric for IPPM. Available at http://www.ietf.org/rfc/rfc2679.txt?number=2679. Accessed 2007 Oct 14.

Al-Naimi K (2003). A Robust Noise and Echo Canceller. Eurospeech, pp.1437–1440.

Andreasen F, Foster B (2003). RFC3435, Media Gateway Control Protocol (MGCP), Version 1.0. http://www.ietf.org/rfc/rfc3435.txt?number=3435. Accessed 2007 Oct 14.

Beerends JG, Hekstra AP, Rix AW, Hollier MP (2002). Perceptual Evaluation of Speech Quality (PESQ), the New ITU Standard for End-to-End Speech Quality Assessment. Part II—Psychoacoustic Model. Available at: http://www.mp3-tech.org/programmer/docs/2001-P03b.pdf. Accessed 2007 Oct 23.

Bellamy JC (1991). Digital Telephony. Second edition. New York: John Wiley Sons.

Benesty J, Gansler T, Morgan DR, Sondhi MM, Gay SL (2001). Advances in Network and Acoustic Echo Cancellation. Berlin: Springer.

Berg K (2000). Fax over IP. Thesis Report, Rapid City, SD: South Dakota School of Mines and Technology. Available at: http://www.phonet.cz/archiv/dok_cizi/ BergThesis.pdf. Accessed 2008 May 21.

Bingham ACJ (1988). The Theory and Practices of Modem Design. New York: Wiley.

Black UD (1998). ATM Volume I: Foundation for Broadband Networks. Englewood Cliffs, NJ: Prentice Hall.

Black UD (1999). QOS in Wide Area Networks. Englewood Cliffs, NJ: Prentice-Hall.

Blake S, Black D, Carlson M, Davies E, Wang Z, Weiss W (1998). RFC2475, An Architecture for Differentiated Services. Available at: http://www.ietf.org/rfc/rfc2475. txt?number=2475. Accessed 2007 Oct 12.

Borys A (2001). Nonlinear Aspects of Telecommunications, Discrete Volterra Series and Nonlinear Echo Cancellation. Boca Raton, FL: CRC Press.

Bourget F (2003). Generating Noise in VoIP. Available at: http://www.commsdesign. com/showArticle.jhtml?articleID=16500759. Accessed 2007 Oct 26.

Brannon B (2004). Understanding the Effects of Clock Jitter and Phase Noise on Sampled Systems, EDN. Available at: http://www.edn.com/contents/images/484488. pdf. Accessed 2007 Oct 14.

Britt R (2007). G.1070V2007, E-Model Parameters and Calculations–Revised from First version of G.1070 based on G.107V 2006. Available at: http://www.e-model. org/E-ModelV2006+MM-ModelV2007.xls. Accessed (Sept 7 2007).

Carvalho L, Mota E, Aguiar R, Lima AF, de Souza JN (2005). An E-Model Implementation for Speech Quality Evaluation in VoIP Systems. Proceedings of the 10th IEEE Symposium on Computers and Communications (ISCC'05), pp. 933–938.

Casner S, Jacobson V (1999). RFC2508, Compressing IP/UDP/RTP Headers for Low-Speed Serial Links. Available at: http://www.ietf.org/rfc/rfc2508.txt?number=2508. Accessed 2007 Oct 14.

Clark AD (2001). Modeling the Effects of Burst Packet Loss and Recency on Subjective Voice Quality. Internet Telephony Workshop.

Demichelis C, Chimento P (2002). RFC3393, IP Packet Delay Variation Metric for IP Performance Metrics (IPPM). Available at: http://www.ietf.org/rfc/rfc3393. txt?number=3393. Accessed 2007 Oct 14.

Donohue D, Mallory D, Salhoff K (2006). Cisco Voice Gateways and Gatekeepers Sanlose, CA: Cisco Press.

Dyba RA, He PP, Pessoa LFC (2004). Network Echo Cancellers and Freescale Solutions Using the StarCore™ SC140 Core. Freescale semiconductor application note, AN2598.

Elsabrouty M, Bouchard M, Aboulnasr T (2004). Receiver-Based Packet Loss Concealment for Pulse Code Modulation (PCM G.711) Coder. Signal Process. Vol. 84, pp. 663–667.

Emiya V, Pessoa LFC, Vallot D, Melles D (2004). Generic Tone Detection using Teager-Kaiser Energy Operators using Starcore SC140 Core. Freescale Semiconductors Application note AN-2384. Available at: http://www.freescale.com/files/dsp/doc/ app_note/AN2384.pdf. Accessed 2007 Oct 14.

Ferguson P, Huston G (1998). Quality of Service: Delivering QoS on the Internet and in Corporate Networks. New York: Wiley.

Fiebrink R (2004). An Exploration of the Teager Operator. Thesis Report, Montreal, Puebec, Canada: McGill University. Available at: http://www.music.mcgill.ca/~rebecca/605/final_paper.pdf. Accessed 2007 July 21.

Foster B, Andreasen F (2003). RFC3660, Basic Media Gateway Control Protocol (MGCP) Packages. Available at: http://www.ietf.org/rfc/rfc3660.txt?number=3660. Accessed 2007 Nov 16.

Freeman RL (1996). Telecommunication System Engineering. Third edition. New York: John Wiley.

Friedman T, Caceres R, Clark AD (2003). RFC3611, RTP Control Protocol Extended Reports (RTCP-XR). Available at: http://www.ietf.org/rfc/rfc3611.txt?number=3611. Accessed 2007 Oct 14.

Funkai K, Nakamura K (2005). An Improvement of ITU-T G.711 Packet Loss Concealment. Proceedings of 9th IASTED International Conference.

Gay SL, Benesty J (2000). Acoustic Signal Processing for Telecommunications. Norwell, MA: Kluwer Academic Publisher.

Goldberg R, Riek L (2000). A Practical Handbook of Speech Coders. Boca Raton, FL: CRC Press.

Goodman D, Jaffe O, Lockhart G, Wong W (1986). Waveform Substitution Techniques for Recovering Missing Speech Segments in Packet Voice Communications. IEEE Trans. Acoustics Speech Signal Process. Vol. 34, No. 6, pp. 1440–1448.

Goralski W (1998). ADSL and DSL Technologies. New York: McGraw-Hill.

Gross G, Kaycee M, Lin A, Malis A, Stephens J (1998). RFC2364, PPP Over AAL5. Available at: http://www.ietf.org/rfc/rfc2364.txt?number=2364. Accessed 2007 Oct 12.

Gunduzhan E, Momtahan K (2001). A Linear Prediction Based Packet Loss Concealment Algorithm for PCM Coded Speech. IEEE Trans. Speech Audio Processing. Vol. 9, No. 8, pp. 778–785.

Gustafsson F (2001). Adaptive Filtering and Change Detection. New York: Wiley.

Gustafsson S, Martin R, Jax P, Vary P (2002). A Psychoacoustic Approach to Combined Acoustic Echo Cancellation and Noise Reduction. IEEE Trans. Speech Audio Process. Vol. 10, No. 5, pp. 245–256.

Handley M, Jacobson V (1998). RFC2327, SDP: Session Description Protocol. Available at: http://www.ietf.org/rfc/rfc2327.txt?number=2327. Accessed 2007 Oct 12.

Haykin S (1996). Adaptive Filter Theory. Englewood-Cliffs, NJ: Prentice Hall.

Haykin S, Widrow B (2003). Least-Mean-Square Adaptive Filters. New York: Wiley Inter-science.

Heinanen J (1993). RFC1483, Multi protocol Encapsulation over ATM Adaptation Layer 5. Available at: http://www.ietf.org/rfc/rfc1483.txt?number=1483. Accessed 2007 Oct 14.

Hersent O, Petit JP, Gurle D (2005). Beyond VoIP Protocols: Understanding Voice Technology and Networking Techniques for IP Telephony. New York: Wiley.

Jacobson V, Karels MJ (1988). Congestion Avoidance and Control. Available at: http://ee.lbl.gov/papers/congavoid.pdf. Accessed 2007 Oct 4.

Kaiser JF (1990). On a Simple Algorithm to Calculate the Energy of a Signal. Proceedings of IEEE ICASSP-90, Vol. 1, pp. 381–384.

Kondoz AM (1999). Digital Speech. Chichester, UK: John Wiley and Sons.

LaBarba LH (2000). The Force Behind Fragmentation. Available at: http://telephonyonline.com/mag/telecom_force_behind_fragmentation/. Accessed 2007 Oct 12.

Laubach M, Halpern J (1998). RFC2225, Classical IP and ARP over ATM, April 1998—IPoA Specific RFC. Available at: http://www.ietf.org/rfc/rfc2225.txt?number=2225. Accessed 2007 Oct 14.

Madisetti VK, Williams DB (1998). The Digital Signal Processing Handbook. Piscataway, NJ: IEEE Press.

Mahfuz E (2001). Packet Loss Concealment for Voice Transmission over IP Networks. Master of Engineering Thesis. Montreal, Canada: McGill University.

Malfait L, Berger J, Kastner M (2006). P.563-The ITU-T Standard for Single-Ended Speech Quality Assessment. IEEE Trans. on Audio Speech and Lang. Process., Vol. 14, No. 6, pp.1924–1934.

Mamakos L, Lidl K, Evarts J, Carrel D, Simone D, Wheeler R (1999). RFC2516, A Method for Transmitting PPP over Ethernet (PPPoE). Available at: http://www.ietf.org/rfc/rfc2516.txt?number=2516. Accessed 2007 Oct 12.

Mathews VJ, Sicuranza GL (2000). Polynomial Signal Processing. New York: Wiley.

McConnell KR, Bodson D, Urban S (1999). Fax: Facsimile Technology and Systems. Norwood, MA: Artech House.

McNeill KM, Liu M, Rodriguez JJ (2006). An Adaptive Jitter Buffer Play-out Scheme to Improve VoIP Quality in Wireless Networks. Military Communications Conference, MILCOM.

McPherson D, Dykes B (2001). RFC3069, VLAN Aggregation for Efficient IP Address Allocation. Available at: http://www.ietf.org/rfc/rfc3069.txt?number=3069. Accessed 2007 Oct 14.

Mills DL (1992). RFC1305, Network Time Protocol (Version 3) Specification, Implementation and Analysis. Available at: http://www.ietf.org/rfc/rfc1305.txt?number=1305. Accessed 2007 Oct 14.

Mills DL (1996). RFC2030, Simple Network Time Protocol (SNTP) Version 4 for IPv4, IPv6 and OSI. Available at: http://www.ietf.org/rfc/rfc2030.txt?number=2030. Accessed 2007 Oct 14.

Mimura K, Yokoyama K, Satoh T, Watanabe K, Kanaide C (2005). RFC4161, Guidelines for Optional Services for Internet Fax Gateways. Available at: http://www.ietf.org/rfc/rfc4161.txt?number=4161. Accessed 2007 Oct 12.

Mitel-tapes (1980). MITEL Semiconductor DTMF Receiver Test Cassette, CM 7291.

Moon SB, Kurose J, Towsley D (1998). Packet Audio Playout Delay Adjustment: Performance Bounds and Algorithms. Multimedia Systems, Vol. 6, pp. 17–28.

Nagireddi S (2006). Accessing Voice Quality in VoIP Telephones. Available at: http://www.electronicproducts.com/ShowPage.asp?FileName=fa-ikanos-voip.apr2007.html. Accessed 2007 Nov 9.

Nichols K, Blake S, Baker F, Black D (1998). RFC2474, Definition of the Differentiated Services Field (DS Field) in the IPv4 and IPv6 Headers. Available at: http://www.ietf.org/rfc/rfc2474.txt?number=2474. Accessed 2007 Oct 12.

Nikhil J (2000). Signal Compression-Coding of Speech, Audio, Text, Image and Video. Singapore: World Scientific.

Paxon V, Almes G, Mahdavi J, Mathis M (1998). RFC2330, Framework for IP Performance Metrics. Available at: http://www.ietf.org/rfc/rfc2330.txt?number=2330. Accessed 2007 Oct 14.

Paxon V, Allman M (2000). RFC2988, Computing TCP's Retransmission Timer. Available at: http://www.ietf.org/rfc/rfc2988.txt?number=2988. Accessed 2007 Oct 12.

Perkins C, Kouvelas I, Hodson O, Hardman V, Handley M, Bolot J C, Vega-Garcia A, Fosse-Parisis S (1997). RFC2198, RTP Payload for Redundant Audio Data. Available at: http://www.ietf.org/rfc/rfc2198.txt?number=2198. Accessed 2007 Oct 12.

Perkins C, Hodson O (1998). RFC2354, Options for Repair of Streaming Media. Available at: http://www.ietf.org/rfc/rfc2354.txt?number=2354. Accessed 2007 Sept 7.

Perkins C, Hodson O, Hardman V (1998). A Survey of Packet Loss Recovery Techniques of Streaming Audio. IEEE Network. Vol.12, No.5. pp. 40–48.

Pessoa LFC, Su WW, Aziz AU, Gan KC (2004). Tone Event Detection in Packet Telephony Using the StarCore SC140 Core. Available at: http://www.freescale.com/files/dsp/doc/app_note/AN2775.pdf. Accessed 2007 Oct 14.

Pinto J, Christensen KJ (1999). An Algorithm for Playout of Packet Voice Based on Adaptive Adjustment of Talk Spurt Silence Periods. Proceedings of 24th Annual IEEE International Conference on Local Computer Networks, pp 224–231.

Postel J (1980). RFC0768, User Datagram Protocol. Available at: http://www.ietf.org/rfc/rfc0768.txt?number=768. Accessed 2007 Oct 12.

Postel J, ed. (1981). RFC0793, Transmission Control Protocol Specification. Available at: http://www.ietf.org/rfc/rfc0793.txt?number=793. Accessed 2007 Oct 14.

Ramjee R, Kurose J, Towsley D (1994). Adaptive Playout for Packetized Audio Applications in Wide Area Networks. Proceedings of IEEE INFOCOM. Available at: http://www.cs.columbia.edu/~hgs/papers/Ramj94_Adaptive.pdf. Accessed 2007 Oct 4.

Rix AW, Hollier M P, Hekstra AP, Beerends JG (2002). Perceptual Evaluation of Speech Quality (PESQ), the New ITU Standard for End-to-End Speech Quality Assessment. Part I—Time Alignment. Available at http://www.mp3-tech.org/programmer/docs/2001-P03a.pdf. Accessed 2007 Oct 23.

Rose MT, Cass DE (1987). RFC1006, ISO Transport Services on Top of TCP: Version 3. Available at: http://www.faqs.org/rfcs/rfc1006.html. Accessed 2007 Oct 12.

Rosenberg J, Schulzrinne H (1999). RFC2733, An RTP Payload Format for Generic Forward Error Correction. Available at: http://www.ietf.org/rfc/rfc2733.txt?number=2733. Accessed 2007 Oct 12.

Rosenberg J, Schulzrinne H, Camarillo G, Johnston A, Peterson J, Sparks R, Handley M, Schooler E (2002). RFC3261, SIP: Session Initiation Protocol. Available at: http://www.ietf.org/rfc/rfc3261.txt?number=3261. Accessed 2007 Oct 12.

Schmer G (2000). DTMF Tone Generation and Detection: An Implementation Using TMS32054x-Application Report SPRA096A. Available at: http://focus.ti.com/lit/an/spra096a/spra096a.pdf. Accessed 2007 July 19.

Schulzrinne H, Petrack S (2000). RFC2833, RTP Payload for DTMF Digits, Telephony Tones and telephony Signals. Available at: http://www.ietf.org/rfc/rfc2833.txt?number=2833. Accessed 2007 July 19.

Schulzrinne H, Casner S, Frederick R, Jacobson V (2003). RFC3550, RTP A Transport Protocol for Real-Time Applications. Available at: http://www.ietf.org/rfc/rfc3550.txt?number=3550. Accessed 2007 Oct 12.

Schulzrinne H, Casner S (2003). RFC3551, RTP Profile for Audio and Video Conferences with Minimal Control. Available at: http://www.ietf.org/rfc/rfc3551.txt?number=3551. Accessed 2007 Oct 14.

Sollaud A (2006). RFC4749, RTP Payload Format for the G.729.1 Audio Codec. Available at: http://www.ietf.org/rfc/rfc4749.txt. Accessed 2007 Nov 9.

Sparks R (2003). RFC3515, The Session Initiation Protocol (SIP) Refer Method. Available at: http://www.ietf.org/rfc/rfc3515.txt?number=3515. Accessed 2007 Oct 14.

Sun LF, Ifeachor EC (2003). New Methods for Voice Quality Evaluation for IP Networks. ITC 18. Berlin, Germany, pp. 1201–1210.

Tseng KK, Lai YC, Lin YD (2004). Perceptual Codec and Interaction Aware Playout Algorithms and Quality Measurement for VoIP Systems. IEEE Trans. On Consumer Electronics, Vol. 50, No. 1, pp. 297–305.

Vidyanathan PP (1992). Multi Rate Systems and Filter Banks. Englewood Cliffs, NJ: Prentice Hall.

Wasem OJ, Goodman DJ, Dvorak CA, Page HG (1988). The Effect of Waveform Substitution on the Quality of PCM Packet Communications. IEEE Trans. Acoustics, Speech, Signal. Process Vol. 36, No. 3, pp. 342–348.

Zopf R (2002). RFC3389, Real-time Transport Protocol (RTP) Payload for Comfort Noise (CN). Available at: http://www.ietf.org/rfc/rfc3389.txt?number=3389. Accessed 2007 Oct 14.

INDEX